Nicolai N. Vorob'ev

Foundations of Game Theory

Noncooperative Games

Translated from the Russian
by Ralph P. Boas

Springer Basel AG

Author:

Professor Nicolai N. Vorob'ev
St. Petersburg Institute for
Economics and Mathematics
Russian Academy of Sciences
Serpuchovskaya Street 38
198147 St. Petersburg, Russia

Originally published as «Osnovy teorii igr. Beskoalitsionnye igry»

A CIP catalogue record for this book is available from the
Library of Congress, Washington D.C., USA

Deutsche Bibliothek Cataloging-in-Publication Data
Vorob'ev, Nikolaj N.:
Foundations of game theory: noncooperative games / Nicolai
N. Vorob'ev. Transl. from the Russian by Ralph P. Boas. –
Basel ; Boston ; Berlin : Birkhäuser, 1994
Einheitssacht.: Osnovy teorii igr. Beskoalitsionnye igry (engl.)
ISBN 978-3-7643-2378-3 ISBN 978-3-0348-8514-0 (eBook)
DOI 10.1007/978-3-0348-8514-0

Originally published by Birkhäuser Verlag in 1994
Typesetting, layout and TEX-hacks: *mathScreen online*, CH-4123 Allschwil, Switzerland
Printed on acid-free paper produced from chlorine-free pulp

ISBN 978-3-7643-2378-3

9 8 7 6 5 4 3 2 1

Table of Contents

Preface to the English Edition

The English edition differs only slightly from the Russian original. The main structural difference is that all the material on the theory of finite noncooperative games has been collected in Chapter 2, with renumbering of the material of the remaining chapters. New sections have been added in this chapter: devoted to general questions of equilibrium theory in nondegenerate games, subsections 3.9–3.17, by N.N. Vorob'ev, Jr.; and § 4, by A.G. Chernyakov; and § 5, by N.N. Vorob'ev, Jr., on the computational complexity of the process of finding equilibrium points in finite games. It should also be mentioned that subsections 3.12–3.14 in Chapter 1 were written by E.B. Yanovskaya especially for the Russian edition.

The author regrets that the present edition does not reflect the important game-theoretical achievements presented in the splendid monographs by E. van Damme (on the refinement of equilibrium principles for finite games), as well as those by J.C. Harsanyi and R. Selten, and by W. Güth and B. Kalkofen (on equilibrium selection). When the Russian edition was being written, these directions in game theory had not yet attained their final form, which appeared only in quite recent monographs; the present author has had to resist the temptation of attempting to produce an elementary exposition of the new theories for the English edition; readers of this edition will find only brief mention of the new material. In the light of these and other recent researches in game theory the present book acquires more and more features of an introductory (though hardly an elementary) text-book.

N.N. Vorob'ev

I have been faced with the task of writing in Russian on a topic for which no turns of speech or expressions have yet been established by usage.

V.Ya. Bunyakovskiĭ,
Foundations of the Mathematical
Theory of Probability, 1846.

Thus, you should not expect an exhaustive history and theory of the bead game. Even authors more learned and skillful than I would not be able to provide this today. It remains a problem for the future ....

H. Hesse, The Bead Game, 1943

Foreword

We may say that the mathematical theory of games originated in 1929, with the publication of J. von Neumann's paper, "On the theory of games of strategy". It found its original form as a new mathematical discipline with the appearance in 1944 of the fundamental monograph "The Theory of Games and Economic Behavior", by J. von Neumann and O. Morgenstern. By now there are more than ten thousand papers and over a hundred monographs devoted to game theory. In spite of so many publications, there are not, at present, any accounts of the foundations of game theory beyond elementary questions. The main aim of the present book is to fill this gap.

Game theory is a new branch of mathematics, and information about its problems, achievements, and technical difficulties is still not widely disseminated. Consequently the author has tried to present it so that a reader should not need any previous knowledge either of the facts of game theory or even of formulations of any game-theoretic problems. Moreover, from the point of view of contemporary mathematics, the author has confined himself to the use of comparatively elementary methods, so that it would seem that any professional mathematician, independently of his specialties and types of activity, will have little difficulty in understanding both the results and the methods.The author did not intend to appeal to readers who are only concerned with pure mathematics as such, but rather to address those who specialize in applications of mathematics in such fields as economics, psychology, social psychology, or law. However, he assumes that preparation to the extent, say, of that of students in an advanced program in economic cybernetics will allow them to understand the contents of the book except, perhaps, in a few places. The same can be said about specialists who apply mathematics in defence activities, etc.

It is possible that some psychological difficulties may arise for mathematical readers, since the mathematical methods that are applied in game theory have, for the most part, been developed for other reasons than the solution of game-theoretical problems. Hence it not infrequently turns out that special and rather complicated mathematical methods have been used for the solution of individual, concrete, and relatively simple problems. Introduced ad hoc, these are ineptly transferred to other game-theoretic problems (even to those that, it would seem, should be solved in a similar way); and it often seems to be difficult to localize, in any concretely applicable way, the kernel of a general game-theoretic method.

Something that is not altogether commonplace is the logical structure of much game-theoretic reasoning, which is distinguished by its unusual "relativization": here the reader (that is, someone who, in mathematical texts, is ordinarily labeled "we") has to alternately adopt first the point of view of one of the interested sides, then the other, and then finally look at the question with the eyes of a disinterested investigator. The author has tried to arrange the exposition so that it will help the reader overcome these difficulties.

The book consists of an Introduction, and four chapters devoted respectively to noncooperative games, finite noncooperative games, two-person zero-sum games, and matrix games. The material in these sections of the book is organized so that they can be read in any combination and any order, depending on the reader's objectives, interests, and mathematical training. For someone who is reading the book in the direct order, Chapters 2, 3, and 4 are devoted to special cases of the propositions presented earlier; for someone who reads them in the opposite order, Chapters 4, 3, and 2 contain more general propositions and constructions. For its part, the Introduction attempts to give the reader a course in the fundamental concepts and mathematical problems of game theory as a whole, but does not provide complete preparation for reading the rest of the book. However, one does not have to be familiar with the Introduction in order to understand the rest of the book.

For the purposes of the rather representative coverage of the theory of noncooperative games provided in each of the expositions, general questions are introduced only by the study of selected individual features, after which in turn a few are subjected to a more detailed analysis, and so on. It seems to the author that in this way he has succeeded in combining a general description of the most significant directions in the theory of noncooperative games with more detailed accounts of special topics, together with complete analyses of some individual problems.

Therefore, to paraphrase Chekhov's famous phrase, in the course of the book we have collected a whole arsenal of guns that fire (or may fire) only beyond its range. However, the treelike hierarchical ("telescopic") structure of the book, resulting from the fragmentation and incompleteness of the exposition, could perhaps only have been overcome by greatly increasing its size.

The details of the content of game theory make the metaphysical point of view especially important, and force us to consider them in more detail than is usual in expositions of other branches of mathematics. It also appears necessary to concentrate not only on what is understood, but also on what is *not* understood by game theory in its present state, because the most fantastic ideas about it are current. The Introduction contains a rather detailed explanation of the fundamental methodological questions of game theory, including the motives for describing the book precisely as "foundations" of game theory with the subtitle "Noncooperative games". The author intends to give an account in further monographs of other branches of game theory which could also be considered fundamental for this branch of mathematics, but which seem logically subordinate as compared with the theory of noncooperative games. These branches of game theory are briefly mentioned in the appropriate sections of the Introduction.

The historical and bibliographical notes are confined to special sections located at the ends of the Introduction and of the corresponding chapters. The bibliography at the end of the book makes no pretence to completeness, but does try to indicate the sources from which material has been drawn, as well as some game-theoretic questions that do not appear in the text, but were used somewhere in its development.

The chapters are divided into sections, and the sections into subsections, with binary numbering that starts over in each chapter. Binary numbering is also used in each chapter, for formulas. However, propositions are not individually numbered, but are cited by the corresponding subsections. The ends of proofs of theorems or lemmas, and the ends of definitions, are indicated by the symbol □.

The author is grateful to the Division of Mathematical Methods of the Institute of Socio-Economic Problems of the Academy of Sciences of the USSR, where this book was written, and to A.A. Korbut for the effort that he expended in improving it. The author will be grateful for any and all critical remarks.

Vyritsa, December 1982 N.N. Vorob'ev

Introduction

1. Mathematization of informal ideas. Every science originates concepts about the real world that are rich in content and as adequate as possible. It is natural to refer to these concepts as *informal models* of the corresponding phenomena. In contrast to other sciences, mathematics originates its own, formalized (in essence, symbolic), concepts about the phenomena of the real world; these are known as *mathematical models*.

The historical development of any particular epistemological process, that is, a process of passing from ignorance to knowledge, begins with informal concepts about some objective phenomenon, and then — on the basis of the already available informal concepts — proceeds to corresponding mathematical concepts. This process of passing from an informal concept to a mathematical one, from an informal model to a mathematical one, is what is usually described as the *mathematization* of informal concepts or models. The systematic process of mathematization of the concepts of a science is called the *mathematization* of that science.

2. In essence, the fundamental (let us say, the "strategic") problem of mathematics as a whole is the mathematization of the totality of the informal concepts about the real world. A gradual solution of this problem is ensured by establishing an enormous number of mathematical (as we say, "purely mathematical") facts at various levels of complexity and generality, by establishing mathematical theories that synthesize the facts, providing a general view of them and an important source of additional facts. To establish such facts and construct such theories is the specific content of mathematics. Each new class of phenomena of the real world, for which informal concepts begin to be mathematicized, predetermines the introduction into mathematics of new facts and of theories that unify them.

For an intuitively clear example, we may consider the mathematization of displacements of physical objects in physical space, a subject whose intensive study was undertaken in the sixteenth century. It was in the course of this study that there developed, starting from the coordinate method, the concept of a function and the calculi (differential and integral), which led to the creation of mathematical analysis in approximately the form in which we now know it. Although the rise of every new field of science (including new branches of mathematics) has, as a rule, a number of different sources, it is nevertheless clear that interest in algebraic equations and their solutions in radicals, in the late eighteenth and early nineteenth centuries, was undoubtedly fostered by the problem of determining the eigenvalues of the systems of equations that arise in problems of celestial mechanics. Thus, even Galois theory, together with many later developments in algebra, is embedded in the mathematization of the concepts of mechanical motion. (It is pertinent to recall here the connection between geometry and group theory, established in F. Klein's Erlanger Programm.) An account of this development could evidently be extended to include functional analysis, topology, and many other modern branches of mathematics.

Thus, on the basis of the mathematization of the idea of the displacement of physical objects in physical space, there developed a significant and essential part of traditional mathematics. Just this, together with the accumulation of the corresponding mathematical facts, and the development of a unification of these mathematical theories, have led to the creation of the physico-mathematical sciences in their present form, and appear as the most important prerequisites for the scientific and technical revolution as we know it today.

At the same time, one of the distinctive features of the development of mathematics is that it formulates more general concepts and develops more powerful methods than are necessary for solving the original problems. Consequently mathematics is ready, at the next appearance of new scientific requirements in their turn, for the solution of completely new problems, and ready to study completely new structures (although, naturally, between the readiness and the skill there may be a very wide zone of scientific obstacles). In particular, the experience that has been acquired by mathematics during the mathematization of the informal concepts mentioned above gives us the possibility of mathematicizing the concepts required by essentially new phenomena.

3. Systematization of approaches. Whenever a branch of mathematics comes into contact with its nonmathematical environment, the methodological and metaphysical questions begin to play an especially important role. This happens, in particular, when one comes to consider the mathematization of informal concepts that involve real processes; or, conversely, informal interpretations of formal mathematical structures; or, finally, various approaches, possibilities, and difficulties in the practical use of mathematical results.

Game theory deals with the modelling of socio-economic processes, and is oriented toward socio-economic applications. Consequently, the methodological and metaphysical aspect is even more essential for game-theory than for other branches of mathematics, and demands more careful consideration.

At this point it becomes pertinent to explain the mechanism, which we may call the methodological instrumentation, that it will be convenient to use in what follows. We emphasize that this explanation will be merely schematic, and will display only the basic connections among the objects and concepts that we consider.

For all the informal variety of objects and processes whose models are considered in the branches of mathematics that were mentioned in **2**, the characteristic features are the objectivity of physical existence, the spontaneity of appearance and development, the dependence on objective physical regularities; and, what is most important for us to realize here, the lack of dependence on any subjective activities or on *any interests*. Deliberate intervention in their existence and development can consist, to use mathematical language, only of the imposition of "initial conditions" subject to invariable general regularities.

The apparent purposefulness of such objects and processes is sometimes suggested, but is completely illusory: it represents the notorious teleological point of view, and can be considered only as a possible interpretation of objective cause-and-effect relationships. Consequently the cognition of such processes is directed solely to the revelation and analysis of cause-and-effect relationships connected with them, and is inevitably only descriptive. We may say that this cognition is limited to its *descriptive* aspect. We shall sometimes denote this aspect by the symbol D. Its results always turn out to be the establishment of some feature of an object, and the description of the object both as such, and as connected with some other objects. The descriptive aspect of cognition does not include either any evaluation of what is cognized, or any action by the cognizing individual based on the information that has been acquired. The descriptive aspect of cognition does not ask questions about what is perceived (such as how "good" it may be), or about what action to take with the object of changing it in one direction or another. Consequently, in the descriptive aspect of cognition, the sole criterion of truth is, to a considerable extent, the noncontradictory character of deductions.

4. The descriptive aspect of cognition has played, and continues to play, an important role in the development of the sciences, especially the natural sciences. At the same time, the real world does contain phenomena that cannot be overlooked, cognition of which cannot and should not be restricted to the descriptive aspect. Among these are phenomena connected with the social and other applied sciences.

Above all, as Karl Marx put it in the eleventh thesis on Feuerbach, "Philosophers had interpreted the world in different ways; but the point is to change it."*) Consequently the process of cognition must include, and actually include explicitly, an individual's energetic, practical, and creative activity in changing the world, however local this change may be. To a greater or lesser extent, this principle applies to all cognizable objects, independently of their nature.

The aspect of cognition that considers an individual's conscious change, and in particular, construction and creation of objects that did not exist at the beginning of the cognitive process, is naturally said to be *constructive*. We shall denote it by the symbol C. The constructive aspect of cognition involves answering the question, "What does it do?" (for the cognizing individual, in connection with the object of cognition). It is just this aspect of cognition that is exhibited in the practical stage. Here the constructions that have been created, whether technical, symbolic, or theoretical, serve as the concrete patterns in which practice, in the widest sense of the word, appears as the test of validity. Notice that we may also associate with the constructive aspect of cognition the organization of the subject

*) K. Marx and F. Engels, Theses on Feuerbach, Collected Works, vol. 3, p. 1. In German, "Die Philosophen haben die Welt verschieden interpretiert, es kommt darauf an, sie zu verändern."

matter of scientific theories, as well as of classes of objects that can be topics in them.

We are not inclined to reckon among instances of the constructive aspect all the changes that are imposed on an object by our studying it, as in the study of phenomena of the microscopic world. At the present level of scientific development, this problem has become more physical than epistemological.

5. In addition, there are still other distinctive features of the cognition of socio-economic phenomena. These activities involve people, as well as organized groups of people, with various interests. From the point of view of furthering these interests, under the constraints of the outcomes of socio-economic processes, there arise sensible and appropriate questions as to whether the outcomes are desirable or undesirable. However, participants in the majority of socio-economic processes generally have discordant interests, and this discordance can have an extremely wide range, from rigid antagonism, through clearly expressed a priori preferences, to compromise and even unanimity.

We must add, to the basis of the discordant interests of the participants in socio-economic phenomena, a description of conflicts. In understanding these there inevitably arise fundamental questions that go beyond the descriptive aspects that we have discussed above. In particular, we shall be primarily interested in the *normative* aspect of cognition. For this aspect, we are concerned with reasonableness, appropriateness, and preferences; we use for this reasonableness the term "optimality", and denote this aspect by the symbol N. Here it is only natural to speak of the optimal actions of participants in conflicts if these actions lead to optimal outcomes. In a sense, the concept of optimality is fundamental for the whole of game theory, and, in particular, for this entire book.

By analogy with the preceding discussion, we may say that the normative aspect of cognition is connected with the question, "What does optimality mean?" Here we must consider how this question differs from the similar, and notorious, question, "What does 'good' mean?" From a grammatical point of view, we may consider "optimal" to be the superlative of "good", "very good", "could not be better". However, there is an informal sense in which the optimal is objective (although only relative), whereas the "good" is subjective (and absolute in this subjectivity). The "good" may be capable of improvement and therefore not "optimal"; the "optimal" can be even "bad", but incapable of being improved. We must keep our eyes on both aspects of optimality: its objectivity and its relativity.

Of course, aside from all these aspects of cognition, there are also aspects that we shall disregard. Here we mention only the prognostic aspect of cognition, associated with the question, "What will occur?", only because we shall later reject it (**13**).

We emphasize that the aspects of cognition that have been discussed here, as well as other aspects, are involved in the cognition of any kind of object.

Naturally the relative role of each aspect depends on the specific object that is being investigated.

6. We now make the following observation. In each of the aspects of cognition (of any object) that we have considered, we can separate out subsequent, more specific, secondary aspects; we may call these second-order aspects. Evidently this process of formulating subaspects can be iterated, to provide aspects of any desired order. Some of these aspects have already been formulated and named. For example, the secondary constructive aspect of the first descriptive aspect (naturally denoted by D-C) is essentially the creation of an information base for a problem that is being investigated; the secondary aspect D-N consists of questions about the reasonableness and the qualitative nature (that is, the adequacy) of a description; and the third-order aspect D-N-D is the methodology of adequate descriptions (D-N-C refers to the construction of adequate descriptions). With a view to eventual systematic study of the normative aspect of understanding conflicts, we notice that N-D studies optimality as such; N-N is concerned with a reasonable (appropriate) definition of optimality and optimal behaviour; N-C, with problems of constructing concepts of optimality; N-C-C, with reasonable definitions of these concepts and behaviors; N-D-D, with methodological descriptions of what is optimal; N-D-C, with the question of what actually defines (or describes) optimality; etc.

7. Mathematical models of making optimal decisions under conditions of conflict, as the subject of game theory. Each of the aspects and subaspects of cognition that we have mentioned can perhaps be summarized in terms of two words: informal (verbal) and formal (mathematical).

Informal versions of optimality under conditions of conflict (i.e., when there are competing interests) have arisen and have long been studied. They are present in the most varied scientific disciplines and in such practical activities as economics and the protection of the environment; military affairs and disarmament problems; the methods and problems of the advance of science and technology; politics and law; biology and medicine; sociology and social psychology; and many other disciplines. Moreover, in many fields within these disciplines, the notions of a conflict and of optimality in conflict situations are leading ideas, in the sense that to neglect them makes the whole field pointless. It is enough to refer to activities for which conflict is of the very essence: for example, the military activity of troops; political disputes; law-suits; competition in a market economy; making an urgent medical decision when the diagnosis is uncertain; or, finally, the simultaneous comparison of objects with respect to several criteria that are not directly comparable (comparisons of this kind appear, for example, in determining the ranking of a team in a competition that involves complex athletic events, or analogous activities).

Instead of summarizing the historical and philosophical arguments that support the necessity, or at least the expediency, of mathematicizing important specific ideas about the real world (for example, normative studies of conflicts), we shall merely refer to the generally accepted fruitfulness of the mathematization (discussed above) of ideas about the displacement of physical objects in physical space, and also to the successful penetration of mathematical methods into military affairs, economics, and sociology. Furthermore, we shall start from the proposition that it is important to mathematicize ideas about social and economic phenomena, of course respecting all the informal ideas that are relevant to them.

For mathematical models to have scientific validity, they must be adequate, that is, they must reflect the essential aspects of reality.

The presence of conflict, in one form or another, in one or another of its internal aspects, and in striving for optimality, are characteristic of all socio-economic phenomena. In essence, precisely these socio-economic phenomena differ from phenomena of any other kind that arise in human society; for example, from purely technical or technico-economic phenomena.

Therefore, in mathematical models of socio-economic phenomena and processes, the models must, for the sake of adequacy, reflect the inherent conflicts and the related ideas of optimality. This means, in turn, that in each model it is necessary to have a mathematical submodel of optimality in the presence of conflict.

The mathematical problems connected with these models involve establishing a correspondence between the formal characteristics of a *conflict* (and, of course, of optimality) and the formal characteristics of *optimal behavior* of the participants in the conflict.

8. We now introduce an important simplifying restriction. For normative cognition of phenomena that involve conflicts, it is essential to recall the secondary constructive aspect N-C, described above, which comprises the study of the purposeful activities of the participants in the conflicts. As we said in **6**, this aspect is characterized by participants' asking questions like "how to act" or "what to do" if their actions are to qualify as optimal (for example, what action is to be planned in order to produce the desired effect). In spite of its theoretical interest and practical importance, we refrain from analyzing this aspect in full generality, and reduce the question of the performance of optimal *actions* to the question of the determination and nature of optimal *decisions*. In the methodological sense, this strongly recalls the abstract notion of potential realizability that is so common in mathematics, where no essential distinction is made between existence in principle and actual existence. (As we know, without this abstraction we could not operate with sufficiently large positive integers.) In practical terms, this abstraction means that a question on the technique of optimal performance of optimal actions has been transferred to experts in special disciplines. To some extent, this transfer suggests the division of labor between the designer, as the originator of a project, and the development by a technician of a method for realizing it. The inverse influence, of technology on construction, corresponds (in our situation) to the possibility of having a plan include the details of its realization. Another, perhaps more demonstrative, example is the comparison of the work of a specialist in exterior ballistics with the activity of a sharpshooter.

What we have just said leads to the problem of creating and elaborating *a theory of mathematical models of optimal decision making in conflict situations.* Game theory is just such a theory.

It follows from the preceding definition that game theory, which has to do with optimal decision making, is connected with the normative (but by no means the descriptive) aspect of the cognition of conflicts. Moreover, it is not concerned with the description of any particular games in the everyday sense of the word, although such games are also conflicts or imitations of conflicts. In just the same way, game theory touches only to a limited extent on the constructive aspect of the cognition of actual conflicts (and of games that imitate them), and does not claim to provide a guide to winning, although it may indeed promote winning.

9. Fundamental concepts of game theory. Since game theory is accepted as a part of mathematics, a description of its content ought to be formulated in a sufficiently rigorous mathematical way. This means that all the concepts and terms involved in its definition ought to be unambiguously explained and interpreted. If we are going to avoid any discussion of either mathematical models or models in general, we must face the fact that the description of game theory at the end of the preceding section contains three terms that demand exact definitions:

1) decision making,
2) a conflict,
3) the optimality of a decision.

In what follows, we shall successively discuss the precise semantics of each of these terms. We shall consider that the explanation of each term has been carried to a sufficiently rigorous and generally intelligible level if it is presented in terms of a set-theoretic construction, of the nature of a "structure" in the sense of Bourbaki.*)

10. Decision making. For an individual K to make a decision can be interpreted simply as the choice of an element (decision) x_K from the set x_K of all ("admissible") decisions.

However, at first glance this utterly simple construction is pregnant with all the set-theoretical paradoxes that are connected with the *axiom of choice* (Zermelo's axiom). This has to do, first of all, with the fact that in the very formulation of the problem that we have posed, that of developing game theory as a branch of mathematics, we must attend to the making of decisions in conflict situations, that

*) Elements of Mathematics. General Topology. Fundamental structures, Addison-Wesley, Reading, Mass., 1966.

is, first of all to multilateral decision making, when there is a whole set $\mathfrak{K}$ of decision makers K and we have to consider the collection of decisions $x_K \in \mathfrak{x}_K$ that each $K \in \mathfrak{K}$ makes (or selects). The fact that, in actual socio-economic phenomena, the set of decision makers (that is, the set $\mathfrak{K}$) is finite, is of little assistance: in mathematical models it is customary to replace finite sets that have a large number of elements by infinite sets (or even continua), since infinity is the most convenient model of "largeness".

In addition, decision making can have a functional nature for an individual when a "parametric decision" is to be made, where each value of the parameter describes the specific condition under which a partial (or conditional) decision is made. Let P be the set of values of the parameter, and let the set of admissible (partial, conditional) decisions by the decision maker K, under the specific conditions $p \in P$, be $\mathfrak{x}_K^{(p)}$. Then every (we may say, a priori) decision for K is a collection $\{x_K^{(p)}\}_{p \in P}$, where $x_K^{(p)} \in \mathfrak{x}_K^{(p)}$. In particular, the characteristic values p can be the times at which some partial decision is made.

Consequently, an individual's decision making can be thought of literally (as the selection of an element from a set) only when we impose proper restrictions; otherwise it demands some sort of supplementary construction.

The informal types of strategies $x_k \in \mathfrak{x}_K$ can be of various kinds; however, we do not try to enumerate all the possibilities but concentrate on one that touches on the use of random (or stochastic) elements in game theory.

The elements of a set $\mathfrak{x}_K$ can be *randomized* acts of K, that is, probability measures distributed over some set of actual acts of K (which we call, for obvious reasons, K's *pure* strategies). On the other hand, the consequences of a pure, unrandomized, strategy can turn out to be random. These two ways of introducing a random element into an individual's actions must be distinguished like, say, the outcomes of two different events: (1) firing or not firing a shot at a target with probability $1/2$ in either case, given that the target will certainly be hit if a shot is fired; (2) actually firing a shot at the target, with probability $1/2$ of hitting it.

Let us consider a game in the everyday sense of the word, that is, as a model of a process of competitive nature that imitates a conflict. If the game incorporates a random component, but only as an event of the second kind, and an individual's choice is restricted to one of only two alternatives, namely, participating or not participating in the game, we say that the game is a *game of chance*. Evidently the question of optimal decisions simply does not arise in games of chance. Consequently the discussion of mathematical models of games of chance is, strictly speaking, not relevant to game theory (although such games do lead to many theoretical problems in probability, some of which are both difficult and important).

To underline the difference between games of chance and the games whose mathematical models are discussed in this book, the latter are often referred to as *games of strategy*. We note that the appearance of an event of the second kind in a game of strategy does not necessarily transform the game into a game of chance. Such games may be referred to as *stochastic.**)

At the same time, it turns out that in games of strategy (whether or not they are stochastic), it is often expedient for the players to use randomized strategies (that is, to introduce random events of the first kind into a game). This is the feature that has led to such a close connection between game theory and the theory of probability that the theory of games of strategy has long since come to be thought of as a branch of the theory of probability.

We note, in conclusion, that the question of a possible joint randomization of the actions of several individuals who form overlapping coalitions is quite nontrivial; it leads to a special kind of problems whose discussion would lead outside the domain of this book.

11. Conflicts (definition of a game). We naturally think of a conflict as a phenomenon in connection with which there are sensible questions about who participates, and in what way; what the outcomes will be; and who is interested in the outcomes, and in what way. Consequently, even in a descriptive mathematical model of a conflict, we must pay attention to the following components.

In the first place, we must establish the roles that are played by the parties, thought of as collections of people, that make the decision. It is natural to call these parties *coalitions of activity*. Here the term "coalition" is used to emphasize the possible complexity and degree of structure of the group of individuals who make the decisions. A whole association may appear in this role; furthermore, associations consisting of several coalitions of activity may, in general, overlap. Notice that a conflict may involve only a single coalition of activity; moreover, from a game-theoretical point of view, this case is far from unimportant. We shall denote the set of coalitions of activity by $\mathfrak{K}_a$.

In the second place, it is necessary to take account of the possibilities of conflict among the parties, that is, to indicate just which decisions can be made by each coalition of activity $K \in \mathfrak{K}_a$. These decisions are said to be *coalitional strategies* of C. We denote the set of these strategies by x_K. There may be connections of some kind between the strategies of different coalitions. The outcome of a conflict is determined in correspondence with the assumption made above of an abstract potential realization by all coalitions of their strategies, taking account of all stipulations about their connections (if any), and is called a *situation*. Consequently the set of situations can be thought of as defined by a subset x of the Cartesian product $\prod_{K \in \mathfrak{K}_a} x_K$.

In the third place, it is necessary to specify the parties that have some interests in common. It is natural to call these interest-coalitions. Like coalitions of activity,

*) Roulette is a typical example of a game of chance in the sense just defined. Chess is a typical game of strategy. The game of dominos, in common with most card games, requires the players not only to know the rules, but also to play with some skill. Therefore, from the game-theoretic point of view, such games are not pure games of chance.

interest-coalitions generally form associations. The members of these associations are unified by their common interests. Different interest-coalitions may overlap; moreover, one such association may be included in another. As actual informal interpretations we may mention the problems of combinations of regional and local interests, or of personal and group interests, as well as interests of society as a whole. The set of all interest-coalitions in a conflict will be denoted by $\mathfrak{K}_i$.

In the fourth place, it is necessary to describe the interests (that is, the objectives) of the parties interested in the conflict (interest-coalitions). This means that for each interest-coalition in $\mathfrak{K}_i$ there should be given a *binary preference relation* R_K on the set x of all situations (it is convenient to take this relation to be reflexive and to think of it as not being strict.) Generally speaking, the relation R_K may be supposed to be rather arbitrary; and, in particular, not every situation has to appear in some pair of these relations. Nevertheless, it is rather natural to suppose that R_K is transitive. As usual, the nonstrict preference relation R_K generates an *indifference relation* $I_K = R_K \cap R_K^{-1}$, as well as a relation of *strict preference* $P_K = R_K \setminus I_K$ $(= R_K \setminus R_K^{-1})$.

On the basis of what we have said, we may take

$$\Gamma = \langle \mathfrak{K}_a, \{x_K\}_{K \in \mathfrak{K}_a}, x, \mathfrak{K}_i, \{R_K\}_{K \in \mathfrak{K}_i} \rangle \tag{0.1}$$

as a formal description of a conflict; here $\mathfrak{K}_a$, $\{x_K\}_{K \in \mathfrak{K}_a}$ and $\mathfrak{K}_i$ are arbitrary sets,

$$x \subset \prod_{K \in \mathfrak{K}_a} x_k \quad \text{and} \quad R_K \subset x \times x, \; K \in \mathfrak{K}_i.$$

We call a system of this kind a *game*. Game theory is concerned with the study of games in precisely this sense of the word. In essentials, the description of a game by a system (0.1) is a formal presentation of what are usually called the rules of the game.

We notice that a game, so defined, is actually a structure in Bourbaki's sense (see the footnote in **9**).

Sometimes one supplements the preceding definition of a game by the concept of the *outcome* of a game, by adjoining to (0.1) a set $\mathcal{A}$ whose elements are called outcomes of the game, and defining a single-valued mapping $\alpha : x \to \mathcal{A}$. If this is done, preferences are defined on the set of outcomes. It is clear that this approach, whether or not it seems useful or convenient, is formally identical with (0.1). Later on, we shall simply identify "outcome" with "situation".

The preceding definition of a conflict was purely descriptive. A truly constructive definition (that is, a definition in the D-C sense of **6**) ought also to contain precise identifications of all the components of (0.1). Its realization as part of a model of a conflict is (as we noticed in **6**) an information base for the model. The development of this secondary aspect is essential for applications of game theory, but will not be discussed in this volume.

12. The concept of optimality. We still have to give a mathematical definition of the optimality of solutions corresponding to the N-D aspect of conflicts. This, however, turns out to be much more difficult than the mathematization that we have outlined for the concepts of "decision" and "conflicts". The main reason is that, for the time being, we do not possess sufficiently well-developed or well-defined informal concepts about optimality. Hence we have to be satisfied with ideas that are very general and, therefore, inevitably rather poor.

We begin with the almost trivial reflection that any informal notion of optimality under conflict, a notion that is applicable to an arbitrary class of conflicts, consists of a rule for choosing, for each conflict in that class, a set of outcomes (or situations) that we have declared to be optimal.

This means, in particular, that optimality can be ascribed only to decisions that constitute optimal outcomes. Thus, if the set of optimal outcomes (or situations) can be described in terms of outcomes (or situations) only, we may speak of optimal joint decisions of a coalition of activity; if the set can be described in terms of coalitional strategies, we speak of optimal strategies; and so on.

In mathematicized form, the representation of optimality that we have described takes the form of a mapping φ that assigns, to every game Γ of the form (0.1), belonging to some fixed class $\mathfrak{G}$ of games, a subset x of its outcomes:

$$\varphi\Gamma \subset x, \ \Gamma \in \mathfrak{G}. \tag{0.2}$$

Then we say that φ is an *optimality principle* for $\mathfrak{G}$, and that the set $\varphi\Gamma$ of outcomes is a *realization* of this principle for Γ or a *solution* of Γ under the optimality principle φ.

We emphasize that there ordinarily are two aspects inherent in the realization of an optimality principle. First, every solution $\varphi\Gamma$ can involve more than one outcome (or situation) of the game; we then speak of a *multivalued solution*. Second, a game $\Gamma \in G$ can have more than one solution $\varphi\Gamma$; we then speak of nonuniqueness of the solution.

It is clear, however, that for a relation φ to reflect an idea of optimality for a class $\mathfrak{G}$ of games, it is necessary for φ to have some additional formal properties that are mathematical expressions of additional (informal) ideas about optimality. Thus, after having taken a short step further, we find ourselves back at the original formulation of the problem. At the same time, as we have said previously, we may nevertheless speak about such important, informal, and intuitively natural ideas about optimality as, for example, profitability, stability, and fairness, and try to express them formally in terms of properties of the mapping φ in (0.2). If we accept that these formal properties actually reflect natural informal features of optimality, their formulations provide the meaning of the axioms of optimality. Consequently the problem of a formal definition of optimality becomes the problem of constructing an axiom system, so that the N-D aspect becomes an N-C aspect. The choice of intuitively reasonable properties of optimality, and the problem of characterizing optimality as such, and as needing to be presented in a set of axioms, belongs to the third aspect N-C-N.

For example, a rather natural idea would be to think of a class $\mathfrak{G}$ of games as a category, and of an optimality principle φ as a functor from $\mathfrak{G}$ to the category of sets (of outcomes of games that belong to $\mathfrak{G}$), and to formulate an optimality axiom in categorical terms.

However, the very idea of an axiomatized representation of a concept leads to nonuniqueness of the concept, because of possible variations in the axiom system. In the present case there are at least two sources of such variability.

In the first place, none of the concepts of profitability, stability, and fairness, that we mentioned above, is an a priori concept, or has a single meaning, so that any of them could be axiomatized in various logically distinguishable ways.

In the second place, if we try to formulate an axiom of optimality, we encounter surprisingly soon (in contrast to the axioms of geometry, for example) a saturation, a contradiction in the resulting axiom system, displayed in the nonrealizability of the optimality principles that are defined by these axioms (that is, in the possibility that the image $\varphi\Gamma$ is empty). Consequently it is, as a rule, impossible to embody, in a single principle of optimality, both one an some other informal features of optimality. What could be accomplished is at most to provide a set of alternative optimality principles, one having one feature; another, another; and so on.

It follows that it is simply impossible to have a single, absolute, universal optimality principle for any rather extensive class of games. Each optimality principle φ is applicable to a restricted class of games and reflects a limited set of features of the informal idea of optimality.

Among such features we should notice the principle of organizing a collective preference (one kind of optimality) on the basis of individual preferences; and, in the opposite direction, that of individual optimality (which is usually included in the concept of fairness) on the basis of collective optimality.

13. Let us make another comment on principles of optimality. In addition to the aspects of cognition that were recalled above, we may also speak of its *prognostic* aspect, whose connection with conflicts is the prediction of their outcomes, or, what amounts to the same thing, the prediction of the strategies that the coalitions of activity will actually use. Game theory, at least in its present state, is *not concerned* with predictions of such a kind.

In particular, both now and later we shall not touch questions connected with the prediction by a coalition of any kind of strategy that another coalition might choose on the basis of considerations (or principles) of optimality.

Of course, what we have said does not exclude predictive applications of game theory (for example, forming predictions), but this is already a completely different problem.

14. Mathematical problems of game theory. The fundamental features of the problems of game theory follow from its description as a mathematical theory of models of optimal decision making that are connected with conflicts. Theoretically, problems of game theory consist, for each class $\mathfrak{G}$ of games defined by (0.1) and selected under certain considerations (the reasonable construction of classes of games belongs to aspect C-N; one such purpose may be, for example, to "complete" a class of games, that is, to describe it in purely structural terms); to elaborate some optimality principle φ of (0.2) and to establish its functional properties (that is, to elucidate the relations between games Γ and sets $\varphi\Gamma$ of their optimal outcomes).

The spectra of these relations can be very diverse. As the logically weakest of these we may consider the problem of realizability of some optimality principle φ for a class $\mathfrak{G}$ of games, that is, the question of the nonemptiness of the set $\varphi\Gamma$ of outcomes for all games $\Gamma \in \mathfrak{G}$ or, more generally, the description of the subclass $\mathfrak{G}$ of the games $\Gamma \in \mathfrak{G}$, for which $\varphi\Gamma \neq \varphi$. As the strongest relation, we might consider the problem of a complete description of the set of realizations of $\varphi\Gamma$ for games in $\mathfrak{G}$, that is, to give a more or less constructive description (aspect N-D-C) of sets $\varphi\Gamma$ for each $\Gamma \in \mathfrak{G}$.

It is natural also to discuss the further development of this circle of problems. Let a class $\mathfrak{G}$ of games be a structure (again, in the sense of Bourbaki), that is, let there be given on it various sorts of operations, relations, or morphisms of one game on another, etc. For definiteness, we shall think of a binary relation $*$ ("composition") on games. The following question arises: is it possible to represent, for games Γ_1 and $\Gamma_2 \in \mathfrak{G}$, the set $\varphi(\Gamma_1 * \Gamma_2)$ of outcomes of the game $\Gamma_1 * \Gamma_2$ in terms of the sets of outcomes $\varphi\Gamma_1$ of Γ_1 and $\varphi\Gamma_2$ of Γ_2; and, if such a representation turns out to be possible, how does $\varphi(\Gamma_1 * \Gamma_2)$ depend on $\varphi\Gamma_1$ and $\varphi\Gamma_2$? In this sense we may speak of calculi of games, topological spaces of games, measure spaces of games, categories of games, etc. In particular, it is natural to consider classes of games as categories, to construct their functors on other categories of games, or on categories that are not game-theoretic (for example, on the category of groups, in a way similar to what was done, at one time, in the construction of homology and homotopy groups for various classes of topological spaces).

In this book, however, we shall not specifically discuss this wide circle of questions, considering it to be outside the limits of the foundations of game theory; our main interest will, as a rule, be in individual games. Our occasional discussions of "subgames" of games (and of the simplest instances of the construction of more complicated games out of simpler ones) does not mean that we are trying to construct a systematic calculus of games; and the occasional comment, that some property of a game is inherent to "almost all" games, does not mean that we are going to consider topological or measurable spaces of games.

Finally, we note that, in contrast to many other branches of mathematics, the question of the uniqueness of solutions of games, under various principles of optimality, plays only a rather modest role. For many classes of games, what is significant (in the sense mentioned above of being typical) is not the uniqueness of a solution, but the existence of only a finite number of solutions.

15. A very typical class of game-theoretical problems arises from attempts to overcome the nonrealizability of individual principles of optimality for various games. The general method for solving such problems is no different from the general method of overcoming unsolvability in other branches of mathematics: it consists of the introduction of "generalized" objects. However, the nontriviality of this scheme requires a more detailed discussion.

Let us consider a class $\mathfrak{G}$ of games that includes games that are not solvable under some principle $\varphi_{\mathfrak{G}}$ of optimality. Let us suppose, however, that we know another class $\mathfrak{H}$ of games that are solvable in terms of their own optimality principle $\varphi_{\mathfrak{H}}$.

We now construct a mapping $f : \mathfrak{G} \to \mathfrak{H}$, after which situations $x \in x_{\mathfrak{H}}$ can be interpretated, for each game $f\Gamma$ ($\Gamma \in \mathfrak{G}$), as situations in Γ. It is natural to denote this translation by f^*. Then the set $\varphi_{\mathfrak{H}} f\Gamma$ of situations, that are understood to be interpreted as $f^* \varphi_{\mathfrak{H}} f\Gamma$, can be considered as a "generalized" solution $\varphi^*_{\mathfrak{G}}\Gamma$ of Γ. We can describe this by the following diagram:

$$\begin{array}{ccc} \Gamma & \longrightarrow & \varphi^*_{\mathfrak{G}}\Gamma = f^* \varphi_{\mathfrak{H}} f\Gamma \\ \Big\downarrow & & \Big\uparrow \\ f\Gamma & \longrightarrow & \varphi_{\mathfrak{H}} f\Gamma \end{array}$$

For us really to be able to call $\varphi^*_{\mathfrak{G}}\Gamma$ a *generalized* solution of Γ, it is evidently necessary for the extended game to satisfy $\varphi_{\mathfrak{G}}\Gamma \subset \varphi^*_{\mathfrak{G}}\Gamma$.

Usually the mapping f is constructed in such a way that, in $f\Gamma \in \mathfrak{H}$, the set of strategies of each coalition contains the set of its strategies in the original game $\Gamma \in \mathfrak{G}$. Consequently the transition f from Γ to $f\Gamma$ is naturally called an *extension*. From the category-theoretical point of view, we can perceive a functorial aspect of such an f.

If $\mathfrak{G}$ is sufficiently complete in the sense outlined in **14**, we may suppose that $\mathfrak{H} \subset \mathfrak{G}$; then $\varphi_{\mathfrak{H}}$ is a natural restriction of $\varphi_{\mathfrak{G}}$ and we can now omit any mention of $\varphi_{\mathfrak{H}}$.

Up to now, in practice, only extensions of functional type were to be considered; now functions of the set $x_K^{\mathcal{A}}$ are to be considered as generalized strategies of K, where $\mathcal{A}$ can be a set of elements of rather varied kinds.

Thus, if $\mathcal{A}$ is a set Ω of elementary events, the generalized strategies consist of measurable functions on Ω with values in x_K, that is, random variables on x_K. These are said to be *mixed strategies*, and the corresponding extensions of the game are *mixed extensions*. Ordinarily we identify mixed strategies of K with measures on x_K. It is mixed strategies that we had in mind in **10** when we spoke of randomization of strategies.

If $\mathscr{A}$ is the set x_L of strategies of some other coalition L, the generalized strategies will be functions: the responses of K to the strategies of L. These are known as *metastrategies* of K, and the corresponding extension of the game is a *meta-extension*.

If $\mathscr{A}$ is the set of positive integers, the generalized strategies are sequences of strategies of K. They are called *superstrategies*, and the resulting games are *supergames*.

In **4**, Chapter 3, we shall also discuss finitely additive extensions of games. The question of the functional nature of the finitely additive strategies involved in them is still not entirely clear.

16. Noncooperative games. The concept of a game, as defined in (0.1), is extremely general. Since it has a very broad logical extent, it necessarily has a very poor content, no matter from what point of view we examine it. In particular, there is little to be said about principles of optimality associated with the class of all such games.

From the purest point of view, not complicated by either formal or informal ideas, the essence of game theory as a theory of mathematical models of optimal decision-making in conflicts is embodied precisely in the theory of *noncooperative games*, in which the sets of coalitions of activity, and of interest-coalitions, coincide; in this case both are said to be *players*, and the set of players is usually denoted by I. Thus, in noncooperative games it is assumed that

$$\mathfrak{K}_a = \mathfrak{K}_i = I. \tag{0.3}$$

Correspondingly, the set of strategies of each player $i \in I$ will be denoted by x_i; and the player's preferences, by R_i.

The following restriction, which we impose on the games under consideration, will be the absence of "forbidden situations", that is,

$$x = \prod_{i \in I} x_i. \tag{0.4}$$

If we take account of (0.3) and (0.4), the description of a game in (0.1) can be rewritten in the form

$$\Gamma = \langle I, \{x_i\}_{i \in I}, \{R_i\}_{i \in I} \rangle. \tag{0.5}$$

We notice that a noncooperative game Γ, in which the set I consists of a single player i, is actually a relation R_i on the set $x_i = x$. Consequently the theory of relations can be considered as a very special case of game theory. However, game theory develops its subject matter in a different direction from the theory of relations.

The "noncooperativity" of games of the form (0.5) means that nothing is said in their rules (compare **11**) about any particular coalitions of players (subsets of I), assigning them any strategical or combinatorial properties. In addition to this, in the study of noncooperative games it will be convenient to deal, not only with individual players, but with sets of players, which we call *coalitions*. We emphasize that, within the framework of the theory of noncooperative games, coalitions are completely devoid of any (as one is accustomed to say) "system properties". In other words, the strategic possibilities for a coalition K are completely determined by the strategic possibilities of the players $i \in K$ (in formal language, the set x_K of cooperative strategies of a coalition K can be thought of as the Cartesian product $\prod_{i \in K} x_i$ of the sets of individual strategies of the players in the coalition); also, the preference relations of each coalition K (holding or not holding for certain pairs of situations) are completely determined by the preferences (for these pairs of situations) of the players in K. Some versions of the descriptions of coalitional preferences in terms of individual preferences will be discussed in the next section.

17. Optimality principles in noncooperative games. For noncooperative games of the form (0.5), there is a natural description of a rather extensive class of optimality principles.

The essential nature of optimization of a situation x in a noncooperative game Γ of the form (0.5) consists of fixing a set $\mathfrak{K}$ of coalitions; and, for each coalition $K \in \mathfrak{K}$, describing, first, the class of situations which K could bring about as alternatives to x (that is, the situations into which K can transform x by the use of its own possibilities); and second, a preference relation with which K may compare a situation in this class with the situation x. The designation of such a class of situations is said to be an *effectiveness condition* for the coalition K, and the description of a relation on it is a *preferability condition* for K.

Corresponding to what we have said for noncooperative games (0.5), we can define, in a natural way, a rather wide class of optimality principles.

Let $K \subset I$ be any coalition, and x a situation in Γ. A situation $y = \{y_i\}_{i \in I}$ is said to be *effectively attainable* for K from the situation $x = \{x_i\}_{i \in I}$ if players in K can effect a transformation of x into y by changing their own strategies, that is, if $y_i \neq x_i$ only for $i \in K$. Evidently this definition of effective attainability is an equivalence relation, and we may speak of effectiveness classes, for K, of situations in Γ. An effectiveness class of Γ for K, containing a situation x, will be denoted by $\mathscr{E}_\Gamma(x, K)$.

We now turn to a discussion of preferability relations. It is based on the following simple concept: a coalitional preference is naturally expressed by alternative versions (or disjunctions) of combinations (or conjunctions) of strong or weak individual preferences of its participants. Formally speaking, we can choose, for every coalition $K \subset I$, a set S_K of indices and a set of pairs

$$\mathfrak{L}(K) = \{\langle L_P^S, L_R^S \rangle : s \in S_K\}$$

of subsets of K. We introduce a preference $R_{K,\mathfrak{L}(K)}$ for K by setting, for two situations x and y of Γ,

$$xR_{K,\mathfrak{L}(K)}y \Leftrightarrow \bigvee_{S\in S_K}\left(\left(\bigwedge_{i\in L_P^S} xP_iy\right)\wedge\left(\bigwedge_{i\in L_R^S} xR_iy\right)\right). \tag{0.6}$$

As is easily verified, the coalitional preference, so defined, is in general neither transitive nor complete (that is, "linear"). Moreover, any postulated completeness of any coalitional preference, expressed by individuals, is accompanied by paradoxical consequences. We shall not linger over this circle of questions, which constitutes the subject of the special *theory of group decisions*, one of the directions in which game theory can develop.

Since xP_iy implies xR_iy, in the definition of $R_{K,\mathfrak{L}(K)}$ we may restrict ourselves to the case when $L_P^S \cap L_R^S = \emptyset$. Evidently, we may, without loss of generality, impose additional restrictions on the family $\mathfrak{L}(K)$ of pairs of sets.

A situation x in a game Γ is said to be $\mathfrak{L}(K)$-*optimal* if there is no situation $y \in \mathscr{E}_\Gamma(x, K)$ for which $yR_{K,\mathfrak{L}(K)}x$.

Now let $\mathfrak{K}$ be any set of coalitions in Γ and let $\mathfrak{L}(\mathfrak{K}) = \{\mathfrak{L}(K) : K \in \mathfrak{K}\}$ be the corresponding set of pairs of their subcoalitions. A situation x in Γ is said to be $\mathfrak{L}(\mathfrak{K})$-*optimal* if it is $\mathfrak{L}(K)$-optimal for every $K \in \mathfrak{K}$. The set of all $\mathfrak{L}(K)$-optimal situations in Γ will be denoted by $\mathscr{C}_{\mathfrak{L}(K)}(\Gamma)$.

A mapping $\varphi_{\mathfrak{L}(\mathfrak{K})}$ which establishes a correspondence between each noncooperative game Γ and the set $\mathscr{C}_{\mathfrak{L}(\mathfrak{K})}(\Gamma)$ can, in the light of what we said in **12**, be thought of as an optimality principle. Evidently this principle corresponds to an intuitive idea of stability and deserves the name of an optimality principle to a greater degree than the abstract (and in essence rather arbitrary) mapping φ in **12**.

All such optimality principles will be called $\mathfrak{K}$-*optimality principles*.

At the present time, a rather small number of $\mathfrak{K}$-optimality principles are in use. Let us notice the following ones.

If, in $\mathfrak{L}(K)$, we have $S = K$, $L_P^i = \{i\}$, and $L_R^i = K \setminus \{i\}$ $(i \in K)$, the property of $\mathfrak{L}(\mathfrak{K})$-optimality is called *Pareto K-optimality*. If, in addition, $\mathfrak{K}$ consists of a single coalition K, that is, if $I = \{K\}$, the property of $\mathfrak{L}(I)$-optimality is known as Pareto K-*stability*. The optimality principle that consists of K-stability for every $K \subset I$ is known as the principle of *strong stability*.

If, for some $\mathfrak{K} \subset 2^I$, and for every $K \in \mathfrak{K}$, we have $S = \{s\}$, $L_P^S = K$, and $L_R^S = \emptyset$, the property of $\mathfrak{L}(\mathfrak{K})$-optimality is called the K-*equilibrium-property*. If, in particular, $K = 2^I$, a $\mathfrak{K}$-equilibrium is called a *strong equilibrium*.

Informally, Pareto K-optimality corresponds to the case when, in a coalition K, preference for a situation x over the situation y means that *every player* in K prefers x to y, and, for *at least one player*, the preference is *strict*. Correspondingly, we may say that the situation x is Pareto K-optimal if the coalition K cannot change to a situation that would be preferred to x by all players in K, and the preference is strict for at least one player.

Analogously, a K-equilibrium corresponds to the case when a preference for a situation x over y means for K that x is *strictly* preferred to y by *each player* in K. Consequently, in a K-equilibrium situation it is not possible to change to a situation that would be strictly preferred by all players in K.

Evidently, for every individual player i the concepts of Pareto i-optimality and i-equilibrium coincide. If $\mathfrak{K} = I$ (that is, $\mathfrak{K}$ consists of the one-element coalitions, namely players), and if, for every $K = \{i\}$, $\mathfrak{L}(K) = \mathfrak{L}(i)$ implies $S = \{i\}$ (evidently, in this case, $|S| > 1$ is meaningless), $L_P^i = i$ and $L_R^i = \emptyset$, the property of $\mathfrak{L}(K)$-optimality is known as *equilibrium in the sense of Nash*.

This book is almost entirely devoted to studying equilibrium in the sense of Nash for noncooperative games.

18. Noncooperative games with payoffs. In essence, it is precisely the class of noncooperative games described by (0.5), without forbidden situations, that will be considered as logically fundamental. However, the "basic" class of games in contemporary game theory is actually a somewhat special class of games; the games described in (0.5) will be called, for short, "general" noncooperative games.

Let us consider additional conditions which these general noncooperative games may satisfy.

First, we may suppose that a preference relation is given in the form of payoff functions:

$$H_i : x \to \mathbf{R}, \tag{0.7}$$

and we then set, for a player $i \in I$ and situations $x, y \in x$,

$$xR_i y \text{ if and only if } H_i(y) \leqq H_i(x). \tag{0.8}$$

The problem of the representation of preferences by payoff functions is the subject of a whole theory, which is known as *utility theory* and has generated an extensive literature; here we shall not spend any time on this circle of problems.

Noncooperative games of the form (0.5), in which the relations R_i are replaced by functions H_i that satisfy (0.8), are known as *noncooperative games with payoffs*. We may define them as systems

$$\Gamma = \langle I, \{x_i\}_{i\in I}, \{H_i\}_{i\in I}\rangle, \tag{0.9}$$

where H_i are the functions in (0.7).

If the set I consists of the single player i, it follows that, in the noncooperative game Γ, the relation (0.9) reduces to a function,

$$H_i : x = x_i \to \mathbf{R}.$$

For a noncooperative game Γ as in (0.9), a "play of a game in Γ" has the following simple and natural interpretation: each player $i \in I$ chooses a strategy

$x_i \in x_i$ and, in the resulting situation, receives the payoff $H_i(x)$. We notice at once that the calculation of the value of a player's payoff function can be made in the most varied units: in cash, if it is a question of a purely economic result; in units of time, if it is a question of loss of time; in probabilities, if it is a question of the likelihood of a patient's surviving a medical procedure; in loss of forces or materiel, if the payoff is the infliction of damage on a military opponent; etc. Moreover, there are also interesting games in which the payoffs to different players are expressed in qualitatively different, incomparable, units. Consequently we shall measure the payoffs to the players in units of some *abstract utility* or, more precisely: in units of increase (or decrease) of a player's utility as the result of playing the game. We do not consider the question of measuring utility; this question belongs to the D-C aspect.

Finally, it is usual to impose two more restrictions on noncooperative games: finiteness of the set I of players and boundedness of the payoff functions H_i $(i \in I)$. These conditions are rather technical; they allow us to study games that satisfy them by using relatively elementary mathematical tools. On the other hand, the transition from games with payoffs of the form (0.9) to more general games with preferences of the form (0.5) is more fundamental from the game-theoretic point of view, since it affects the structure of optimality principles.

It is precisely games of the form (0.9), with a finite set of players and bounded payoff functions, that are usually called *noncooperative games*. They constitute the main topic of this book, and we shall presently devote Chapter 1 to them. Giving up any of the special features of this class of games that distinguish it from the general noncooperative games of (0.5) leads to generalized noncooperative games. Thus one may consider noncooperative games "with forbidden situations", "with an infinite set of players", " with preferences", or "with unbounded payoff functions".

19. Mixed extensions of noncooperative games. We continue the discussion initiated in **15**. Many noncooperative games (with payoffs) have no (Nash) equilibrium situations; consequently we encounter the question of appropriate extensions of such games in some class of solvable games. It turns out that if, in a noncooperative game, the sets of players' strategies form compact convex subsets of certain topological linear spaces, and the payoff functions are concave and continuous, it follows that the game has equilibrium situations (equilibrium points). Consequently it seems natural to choose, as the mapping f of **15**, the embedding of spaces of strategies in appropriate linear spaces, to form their convex hulls, and then to extend the payoff functions by linearity to the new set of situations. Operations of this kind, performed on games, are known as *mixed extensions* of the games.

Mixed extensions of games lend themselves to informal interpretations. For example, elements of the convex hulls of sets of player's strategies, known as *mixed strategies*, can be thought of as random variables with values in these sets of strategies, collectively independent for all the players (of course, here it is necessary that the outcomes of the sets of strategies have a measurable structure), and the payoffs can be thought of as their expectations. Thus the formulation of the definition of a mixed extension of a noncooperative game is equivalent to the definition given at the end of **1.15**. We may notice that players' mixed strategies are just the randomized actions discussed in **10**.

20. A particular class of noncooperative games: two-person zero-sum games. Some important types of games arise as classes of noncooperative games as defined in (0.9).

In the first place, in the noncooperative games defined in (0.9) we may impose *restrictions* of one kind or another on their *components* of games. Such procedures of forming special concepts and investigating their individual theories are quite common in mathematics. For example, we might specialize sets x_i of strategies by describing their cardinality or their structure. In particular, if the sets x_i are finite, the game Γ is said to be *finite*. A discussion of the theory of finite noncooperative games will be given in Chapter 2.

From the game-theoretic point of view, a case in principle is that in which the set of players consists of two elements: $I = \{1,2\}$, and in every situation $x \in x$ we have $H_1(x) = -H_2(x)$ (that is, in every situation, one player's gain is equal to the other player's loss). Games of this kind are called *two-person zero-sum* games (or *antagonistic* games). In these games, it is evidently sufficient to assign the payoff for just one player; we call this the *payoff function* of the game and denote it by H. These games are discussed in Chapter 3.

The Nash equilibrium principle for general noncooperative games (see **17**) becomes the *maximin principle* for two-person zero-sum games: the players try to maximize their minimum payoffs (i.e., the payoffs that they can expect to receive under all circumstances). The maximin principle is sometimes also called the principle of the *greatest guaranteed result*. If the maximin principle is in effect in a two-person game, the payoff function H has the same value in all situations that are optimal in the sense of this principle; we call this the *value* of the game, and denote it by $v(\Gamma)$ or v_Γ.

The theory of two-person zero-sum games is conceptually simpler than other branches of the theory of noncooperative games. It has a special place in the general theory of noncooperative games, one that recalls, in some aspects, the theory of Abelian groups in general group theory. The mathematical machinery used in the study of two-person zero-sum games is closely related to that used in conventional branches of mathematics. Finally, the informal interpretations of two-person zero-sum games are more intuitive, and the informational content of the game-theoretic models is more comprehensible, in the two-person zero-sum case. Because of all this, the theory of two-person zero-sum games has now been developed in significantly more detail than the general theory. Moreover, it appears that among the unconversants the idea is widespread that the theory of two-person zero-sum games (or, in the best case, of "non-strict" two-person zero-sum games) is all that there is to game theory.

Not only does the theory of two-person zero-sum games have independent interest, both theoretical and applied, but also, since it has been developed further than other branches of the general theory of noncooperative games, it can serve, and in fact does serve, as a source of game-theoretic problems, for the creation of game-theoretic intuition, and even for the adaptation of already existing mathematical methods to the purposes of game theory. Chapter 3 of this book is devoted to problems in the theory of two-person zero-sum games.

Finite two-person zero-sum games are called *matrix* games because we can evidently display the values of the payoff functions as matrices. Matrix games are discussed in Chapter 4. For a similar reason, finite two-person noncooperative games (not necessarily zero-sum) are called *bimatrix* games. In this connection, it seems appropriate to mention other classes of noncooperative games. This will be done at corresponding places in the book.

21. Cooperative theory. In the second place, the noncooperative game in (0.9) can be studied from a *special point of view*.

For example, we might set, for every $K \subset I$,

$$\mathfrak{x}_K = \prod_{i \in K} \mathfrak{x}_i \quad \text{and} \quad H^{(K)}(x) = \sum_{i \in K} H_i(x) \quad \text{for } x \in \mathfrak{x}$$

and consider the two-person zero-sum games $\Gamma(K)$ of each coalition K against the complementary coalition $I \setminus K$, where $\mathfrak{x}_K$ and $\mathfrak{x}_{I \setminus K}$ are the sets of strategies of the coalitions of players, the payoff function is $H^{(K)}$ (evidently, situations in Γ and $\Gamma(K)$ are formally identical), and each of the two players K and $I \setminus K$ acts in accordance with the maximin principle. The value $v(K)$ of each game $\Gamma(K)$ can be thought of as "winning possibilities" of K. Thus there is carried out, so to speak, a "factorization" of a noncooperative game over the family of two-person zero-sum games that it generates.

The theory of noncooperative games, restricted to the class of problems that arise under this approach, is known as *cooperative theory*, and the functions $v : 2^I \to \mathbf{R}$ are said to be *characteristic functions*. Characteristic functions are objects of a game-theoretic nature. In contrast to noncooperative games, here the role of situations is played by *imputations*, that is, vectors whose components are the portions obtained by the players after the distribution of the common payment $v(I)$. A characteristic function, considered together with a given set of imputations, is known as a *cooperative game*.

A cooperative game can also be thought of, not only as a factorization of a noncooperative game of the form (0.9), but also as a special case of general games of the form (0.1). Here $\mathfrak{K}_a$ consists of a single coalition, namely the set I of players; and the set $\mathfrak{x}_I$ of strategies of I is the set of imputations. The imputations occur here as situations as well; the interest-coalitions are the individual players $i \in I$, and, for each player, the preference of an imputation is evaluated as the component of the imputation corresponding to that player.

In accordance with what was said in **12**, optimality principles in cooperative games are obtained by stating, for each cooperative game Γ, a (possibly empty) set $\varphi\Gamma$ of imputations in it. Some of these principles duplicate the principles in **17**. For example, if we set, for any $K \subset I$, $S_K = \{s\}$, $L_P^S = K$, and $L_R^S = \emptyset$, in analogy to what we did in the definition of an equilibrium, we call the realization of such an optimality principle a *core*. There are various other optimality principles in cooperative games.

22. Concretization of strategies (dynamic games). In the third place, in the definition (0.9) of a noncooperative game we may specify not only the components of the game as mathematical objects, but also the *mathematical nature of their elements*, providing them with some kind of structure. In particular, this applies to the elements of the sets x_i, that is, to the players' strategies. Since there arise questions that are connected with the fundamental nature of noncooperative games, we should consider these questions in more detail.

From the formal description of noncooperative games in (0.5) and (0.9), and their important interpretation in **17**, it is clear that the play of a game of this kind consists of the players' choosing their strategies, and then performing single actions independently, that is, with each player remaining ignorant of any information about the other players' strategies. When we want to emphasize this aspect of noncooperative games, we call them *games in normal form.*

Though it may not be obvious that our assumption that the players make single, independent, choices of strategies (that the game is in normal form) is the most general, it includes nevertheless as special cases various possibilities for the amounts of information that the players can use during the game, so that the players can select strategies according to the other players' successive actions, assuming that these are partial solutions that can be modified (as described, for example, in **10**), etc. In reality, the representation of a game as a game in normal form (that is, exactly in the form (0.9)) usually turns out to be quite simple in spite of descriptions of sets of strategies that often appear to be quite complicated. We are also not logically led outside the class of noncooperative games (in normal form) by the consideration of "hierarchical features" of games, where the players are arranged in a certain order, and the strategies available to players at a "lower level" are determined by the strategies of the "higher" players. (What we have said does not apply to games with forbidden situations, in which the players' strategies are related in mtual, so to speak, simultaneous ways, and do not lend themselves to being overcome by the methods we have described.) The following argument may clarify the situation.

Suppose that a game is so arranged that, at various stages in it, players may obtain or lose information and make partial decisions that depend on the information that they have available at each stage. This means that, during the game, a player may find himself in various states of information that are sets of his actual physical states, indistinguishable for him, as soon as he finds himself in any of these states (in particular, information-states of the players may depend on the strategies of other players that are more highly placed in the hierarchy). In such cases the players' strategies are functions, whose domains are their information-states, and whose values are the decisions available for them in these information-states. It is intuitively clear that every plan of a player's behavior reduces to the choice of such a functional strategy.

Here is a simple example. Let a game consist of three steps:

a) player 1 adopts an available strategy x and communicates this to player 2;

b) player 2 adopts one of the available strategies y, on the basis of knowing x;

c) the players receive certain payoffs that depend on x and y.

In this game, the information-states of player 2 consist of the chosen strategy of player 1, and the strategies available to player 2 are the possible functions that assign the value y to the value x. Therefore player 2 actually selects a *strategy* $y(\cdot)$ independently of any specific x, although after x has been selected the particular *decision* $y(x)$ will in general depend on the specific value of x.

Games in which there are special features of the information-structure form a class of solvable games which may be a basis for the extensions that we call meta-extensions (see **15**). The players' metastrategies in these games can be interpreted in a natural way as actions of a functional of a stipulated type, consisting of the choice of a response to the actions of the other players.

23. Concretizations of strategies (games with variable decisions). The optimality principles described in **17** are based on the comparison of certain situations with others that are effectively attainable from them. Here the versions of preference relations that we have described do not distinguish, for any coalition, between initial situations and those that arise from a coalition's changing its strategies. In other words for a coalition (or player) to compare and evaluate situations, it is indifferent whether a situation was originally present or arose from changes made by a coalition (or player) in the initial strategies, or even as a result of deviations from their initial pledges.

It may seem that taking account of the difference between the situations of one kind or another leads to a more general game-theoretic formulation. However, this is not actually the case.

Let us consider the following game. Let each player $i \in I$ be allowed to announce $k_i - 1$ times the adoption of certain decisions $y_i^{(l_i)} \in \mathcal{Y}_i$ ($l_i = 1, \ldots, k-1$), and finally (at the kth step) choose the decision $y_i^{(k_i)}$, also in $\mathcal{Y}_i$. We suppose that the payoff F_i to player i is a function of the whole sequence $y_i^{(1)}, \ldots, y^{(k_i)}$ of that player's preliminary decisions (so to speak, of the player's fluctuations), and, besides, of the final decisions $y_i^{(k_i)}$ of the other players $j \neq i$.

If we now set

$$x_i = y_i^{k_i} \quad \text{for all} \quad i \in I$$

(here the power is to be taken in the sense of a Cartesian product), and

$$H_i(x) = F_i(y_i^{(1)}, \ldots, y_i^{(k_i)}, \{y_j^{(k_j)}\}_{j \neq i}) \text{ for } i \in I \text{ and } x \in x,$$

we obtain a representation of the game in the form (0.9).

This example, as well as the example in **22**, show that the *normal form* (that is, a game in the form (0.9) or at least in (0.5)) is able to represent many kinds of game-theoretic constructions that, at first sight, seem to be more general. Apparently, the universality of this possibility can be expressed as a very general *informal thesis*. Its informality is predetermined by the informality of the very concept of the representation of noncooperative games and of the very idea of "games". Nevertheless, for every precise description of a game, one that admits the transfer of information, the multistage concept of players' decisions, or, finally, a hierarchical structure of the set of players, the thesis can be converted into a mathematical formulation which, as experience shows, can be precisely established in each individual case.

24. Game-theoretic models. A special form of concretization of the concept of a noncooperative game is given by narrower classes of games that we shall call *game-theoretic models*. These classes of games and their components share some unified informal interpretations, discussion of which (in combination with formal analysis) leads to heuristic inference of the form of solutions for these games. The theories of such game-theoretic models become special branches of the general theory of games.

Let us illustrate the preceding remark by two examples.

25. Choice of intensiveness of activities. As a basis for a game-theoretic model to consider, there is a widespread, repeatedly observed, and, strictly speaking, very natural phenomenon, that nevertheless seems somewhat paradoxical at first glance.

Let us suppose that a small number of players undertake to engage in an activity that gives them an advantage that does no harm to the other players; but if a large number of players engage in the same activity, they disadvantage not only all the other players, but also themselves. The question arises, to what extent is it appropriate for the players to engage in this activity? This problem has a natural generalization to the case when each player can choose not only whether or not to engage in the activity, but to what extent.

A formal version of this informal model is as follows.

Let us consider a noncooperative game Γ as in (0.9), where each strategy x_i of player i can be thought of as an amount (intensiveness, duration) of engaging in some activity. Let us suppose that the amount can be specified by a real number. In the simplest case $x_i = \{0, 1\}$ (that is, either the activity is engaged in, or not). In many cases it is enough to take $x_i = [0, 1]$. Let $\varphi_i(x_i)$ denote the expenditure of player $i \in I = \{1, \ldots, n\}$ in using the strategy x_i. We may suppose that φ_i is a monotonic function. As a result of the players' creating the situation

$x = (x_1, \dots, x_n)$, there is created an informal state common to all the players, which we may describe by the value of a function, $f : x \to \mathbf{R}$, nondecreasing in each variable x_i. The value $f(x)$ of this function is estimated for each player by a sort of individual utility h_i, which we naturally assume to be nondecreasing. Moreover, the situation x may affect each player $i \in I$, causing loss to that player of $g_i(x)$ units of utility; the functions g_i should not increase for an x_j for which player j is friendly to player i, and not decrease for an x_j for which player i is hostile. As a result, the payoff function H_i in the game-theoretic model can be represented in the following form:

$$H_i(x) = -\varphi_i(x_i) - g_i(x)h_i(f(x)). \tag{0.10}$$

What is typical in this model is not the particular form of the functions on the right-hand side of (0.10), but their properties of being monotonic or convex, and of their specific combinations in the payoff function.

26. This game-theoretic model, as a game Γ as in (0.9) with the payoff function (0.10), can be given an interesting alternative interpretation.

Let $I = \{1, \dots, n\}$ be a set of users of natural resources, who may pollute some natural object; let x_i be the polluting effect of activity i; and let $\varphi_i(x)$ be the expense of reducing this pollution to some admissible level x_i^0. As before, let $x = (x_1, \dots, x_n)$, denote by $f(x)$ the total amount of pollution, and denote by $h_i(f(x))$ the contribution of i to the general pollution $f(x)$. Then we have a game with the payoff function

$$H_i(x) = -\varphi_i(x_i) - h_i(f(x)).$$

In § 2 of Chapter 3 we shall discuss one of the simplest versions of this model.

Furthermore, let I be a set of producers of some product (for example, of petroleum cartels), and let player i produce the amount x_i (within some prescribed limits), incurring costs $\varphi_i(x_i)$. In situation x, there is a total amount $s = x_1 + \dots + x_n$ of the product in the market; this will be sold at price $f(s)$, depending on the supply for s. In the situation x, players i sell amounts x_i of the product such that the payoff function H_i describing their profits will have the form

$$H_i(x) = -\varphi_i(x_i) + x_i f(s).$$

Finally, let I be a set of armed states, x_i the level of armaments of side i, and $\varphi_i(x_i)$ the level of expenditure of side i on these. Let us denote by $f(x)$ a general measure of the danger from the increase of military potential in the world (say, the general equivalent of megatons of TNT or of some estimate of the possibility or probability of starting a military conflict), and by $g_i(x)$ the losses of side i if it becomes involved in a conflict; it is natural to suppose that g_i decreases with x_i and increases with the other variables. In this game the payoff to player i has the form

$$H_i(x) = -\varphi_i(x_i) + g_i(x)f(x).$$

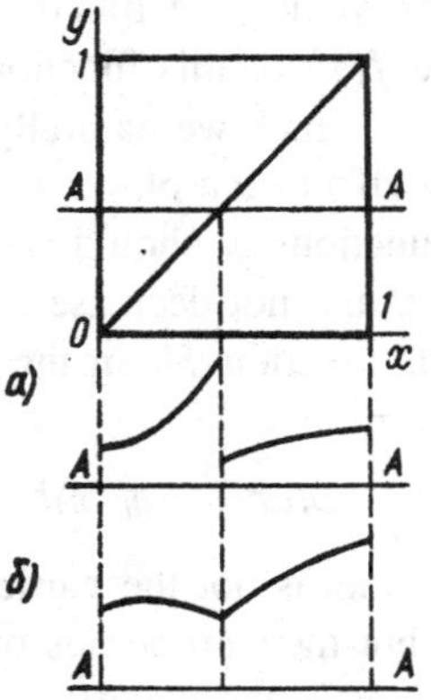

Fig. 0.1

27. Games of timing. Again let Γ be a noncooperative game as in (0.9), $x_i = [0, 1]$ for $i \in I = \{1, \ldots, n\}$. Then the set $x = x_1 \times \ldots \times x_n$ of situations in Γ will be the n-dimensional unit cube.

Let us understand a strategy $x_i \in x_i$ to be the choice by player i of an instant at which to perform some action. We shall suppose that the course of successive actions of the players is defined by a permutation π, that is, the arrangement

$$x_{\pi_1} \geqq x_{\pi_2} \geqq \ldots \geqq x_{\pi_n} \tag{0.11}$$

describes the qualitative picture of the situation. This appears as rather natural, since the performing of an action by one of the players can have a significant influence on the effect of the other players' actions. Consequently we assume that, in the sets x of situations described by (0.11) for different permutations π, the payoff functions H_i are given in $n!$ domains in essentially different ways, and, in particular, there may be discontinuities in passing from one domain to another.

For example, if the players are firms that produce the same kind of product in a strictly competitive economy, the profits of each player depend in an essential way on the sequence in which the firms introduce new and innovative technologies. In this case, on every set described by (0.11) each function $H_i : x_i \times x_{I\setminus i} \to \mathbf{R}$ may be supposed to be continuous in all its variables and increasing in x_i (since making an investment in new technology is advantageous until competition forces a change), and decreasing in the variables x_j $(j \neq i)$ (since an advantage for the competition becomes a disadvantage for "us"). It has a jump if the situation crosses the boundary of a domain (0.11) at which one of the competitors forestalls player i.

Two-person zero-sum games of this kind are the subject of the very highly developed theory of duels (games of timing), of which a fragment forms the content of § 9, Chapter 3. The form of the payoff function in this case is sketched in Figure 0.1,a).

Another version of the theory is suggested by the case when the payoff function H_i is continuous on the whole set x of situations, its first partial derivatives have jumps at the boundaries of sets of the form (0.11), and, on the boundary of each domain, H_i is concave in the variable strategies. In the two-person zero-sum case we have the theory of what is called a butterfly-shaped game (Figure 0.1,b)), which is similar in structure to the theory of duels.

28. Analysis of various classes of games, and also of specific games, including the models discussed above, requires, as a rule, quite complicated and difficult mathematical methods. To a significant extent, the complications arise from the nature of the problems themselves and the poor fit between game-theoretic problems and traditional mathematical methods, which have been developed mainly for the solution of problems in physics and physical technology, that is, for entirely different purposes.

The intensive development of game theory, the accumulation of a large number of game-theoretic facts, and the development of game-theoretic intuition now allow us to hope that in the foreseeable future there will exist a new, mathematical, game-theoretical, "instrumentation" especially oriented toward the systematic solution of problems with socio-economic content.

Historical notes and references for the Foreword and Introduction

Foreword. The origin of game theory goes back to Bachet de Mezirac [1], to a letter to P. Fermat from B. Pascal on 29 July 1654 (see Fermat's Collected Works (Fermat [1])), and R. Montmort [1].

The idea of creating a mathematical theory of conflicts (that is, game theory) was in the air at the beginning of the 20th century, as witness the papers of C.L. Bouton [1], E.H. Moore [1], E. Zermelo [1], E. Borel [1,2,3], and H. Steinhaus [1], and the monograph [1] by E. Lasker (at the time world chess champion), which G. Klaus [2] has called the philosophical precursor of game theory. (We note that the theoretical work of another perennial chess champion, M.M. Botvinnik [1,2], also has some game-theoretic interest, although it is not directly connected with game theory.) This period concludes with J. von Neumann's address of 7 December 1926 to the Göttingen Mathematical Society, which was published in [1]. These results were briefly discussed in von Neumann [2]. For the questions of scientific priority that were raised in this connection by Borel and von Neumann, see M. Fréchet [1,2], and von Neumann [3]. On von Neumann's work on game theory see H.W. Kuhn and A.W. Tucker [2] and O. Morgenstern [4].

The monograph by von Neumann and Morgenstern first appeared in 1944, and was republished with additions (among them, the extremely important axiomatic theory of utility) in 1947 and 1953, and has been reprinted (most recently, in 1980). The Russian edition of 1970 has a new introduction by Morgenstern.

A historical survey of game theory is given in N.N. Vorob'ev [10] (for a more detailed discussion see his [21]). A more complete bibliography of game theory up to the present is given in the indices edited by Vorob'ev [27,28].

Among texts on game theory, we may mention E. Burger [1], McKinsey [2], G. Owen [2], Vorob'ev [20], K. Manteuffel and D. Stumpe [2], E.G. Davydov [1], E. Szép and F. Forgó [1,2], M. Suzuki [1], and A.J. Jones [1]. For deeper presentations, see the books by S. Karlin [3], J.-P. Aubin [1], and J. Rosenmüller [1,3]. For the sake of completeness, we also mention the book by T. Parthasarathy and T.E.S. Raghavan [1] devoted completely to two-person games. The books by A. Wald [2] and D. Blackwell and M.A. Girshick [1] on the connection between game theory and mathematical statistics are still of interest in spite of their age. There are a rather large number of popular books and pamphlets that present game theory in an elementary way. The following examples deserve special mention: A. Rapoport [1,2], as well as the book by M.A. Davis [1]. R.D. Luce and H. Raiffa give an extensive survey of game theory. In this connection, we also mention the survey article by Vorob'ev [2].

The expository principle "from the general to the particular" that is used in the present book seems to have been used previously only in E. Burger's book [1].

Work on game theory has been published in many mathematical, economic, cybernetic, etc., journals. In 1972 a specialized international journal devoted to game theory was founded: the International Journal of Game Theory, published by Physica-Verlag (Heidelberg).

Methodological questions in game theory are discussed in the books by von Neumann and Morgenstern [1] and G. Klaus [4], and also in articles by Vorob'ev [3,5,9,12–14,19,23], Klaus [2,3], and G. Libsher [1]. In this connection, we mention a collection edited by M. Shubik [C14]. It is also appropriate to mention a short methodological note by von Neumann [8].

Introduction. 1–7. There is a vast literature on the mathematization both of concrete and informal ideas and of whole sciences. In connection with this presentation see the monographs by G. Klaus [1,4] and the articles by Vorob'ev [13,23]. The analysis of aspects of cognition was outlined in Vorob'ev [23]. On the role and objectives of game theory in the mathematization of social and economic ideas, see the books by von Neumann and Morgenstern [1], Luce and Raiffa [1] and Morgenstern [2], and articles by Morgenstern [1,3].

8. Game theory, as a mathematical discipline that reduces the problems of carrying out optimal actions to those of finding and establishing the nature of optimal decisions, can be contrasted with the practice of gambling, the object of which is the multistage empirical choice of practicable alternatives. See, in this connection, the book by A.B. Krushevskiĭ [1].

9. An analysis of the fundamental concepts of game theory and its problems was given by Vorob'ev in [11].

10. That the paradoxes of set theory for the case of functional strategies are not "idle worries" can be seen from the results of D. Gale and F.M. Stewart [1] (see also Stotskiĭ [1]). The necessary refinement that overcomes the paradox was discovered by J. Mycielski and H. Steinhaus [1]. General mathematical awareness of this fact is perceivable from Mycielski [1,2] and Mycielski and S. Swierckzowski [1].

The first presentation of the probability-theoretical approach to games of chance was given by C. Huygens [1]; and to more general problems of decision making by P.S. Laplace [1]. A very deep mathematical theory of games of chance is presented in the book by L. Dubins and L. Savage [1]; and instructive historical and methodological considerations of games of chance, in the book by L.E. Maĭstrov [1]. Joint randomization of the actions of coalitions is discussed by Vorob'ev [6].

11. The description in **11** of a general class of games was given by Vorob'ev [11]; E.Ĭ. Vilkas [14] proposed a formally different general concept of games. Moreover, these general alternatives are formally embeddable in each other. V.S. Ilichev [1] considered a peculiar class of "polyantagonistic" games in which there are only two sides with opposite interests, but each of them consists of several agents (players) who act independently.

12. Apparently it was E. Zermelo [1] who first called attention to the nontrivial semantics of optimality. He posed the following problem: "Whether one could calculate with mathematical objectivity, or even give a participant some idea of, the value of a possible position in the game, as well as of the best move in this position: information without which the player would have to eliminate both subjective and psychological guesses and the opinions of 'the perfect player', etc.?" The maximin (or minimax) principle was formulated by von Neumann [1], and in his monograph with Morgenstern [1] appear (though only implicitly) various optimality principles (for cooperative games).

In connection with the rapid appearance of contradictions in axioms of optimality, there is a very instructive example of the "paradox of the dictator" (see K.J. Arrow [1]; for a detailed analysis see R.D. Luce and H. Raiffa [1]). J. Milnor [1] found a contradiction in the set of very natural axioms defining very simple optimality principles. In the same paper he gave an example of an optimality principle that satisfies a given set of completely "reasonable" although not intuitive axioms. Such a version of optimality may be figuratively called "synthetic" (in the sense that it does not occur "naturally", that is in people's intuition).

A very broad concept of optimality that admits a variety of specialized versions was proposed by E.Ĭ. Vilkas [1]. A related discussion of several optimality principles, which constituted a fundamental step in the direction of the construction of a calculus of optimality principles, was given in, for example, the work of E.B. Yanovskaya [20] and [22].

The papers [1–14] of Vilkas are devoted to the formalization of the problem of selecting game-theoretic optimality principles.

13. For the prognostic aspect of cognition and applications of game theory in prognostics, see Vorob'ev [23].

14. The first survey of the problems of game theory was given by H.W. Kuhn and A.W. Tucker [1]. The problems of game theory discussed here add precision to, and develop, the ideas introduced by Vorob'ev [11]. Individual statements that can be considered as steps in the construction of a calculus of games (composition and factorization of games) already appear in von Neumann and Morgenstern [1]. D. Gale and S. Sherman [1] initiated the use of Euclidean spaces of games; measure spaces of games were discussed by K. Goldberg, A.J. Goldman and Neumann [1], and also by M. Dresher [1]; and categories of games, by V.E. Lapitskiĭ [1,2]. Classes of games as indeterminate games were suggested by Vorob'ev [4].

15. The first example of the extension (in fact, of mixed extension) of games was given by Waldegrave (according to information given by R. Montmort [1]). The systematic consideration of mixed extensions was proposed by E. Borel [1], and the solvability of matrix games in the mixed strategies, in the sense of a maximum principle, was established by von Neumann [1].

Metagames were introduced by N. Howard [1,2], N.S. Kukushkin [1,2], and Vorob'ev [24]. Motivation for the introduction of supergames was given in the book by R.D. Luce and H. Raiffa [1] (pp. 136–147).

The idea of the functorial character of extensions of games is implicit in the papers of E.B. Yanovskaya [17,19] and discussed in detail by V.E. Lapitskiĭ [1,2].

16. The first work on noncooperative games in connection with preferences appears to have been given by R. Farquharson [1], and their systematic investigation was undertaken by V.V. Rosen [1–4]. The foundations of a general theory of games with preferences are contained in the papers of Yanovskaya [12,17,19]. A generalization of noncooperative games was considered by E. Marchi [2–4]. A peculiar version of the degeneracy of noncooperative games was investigated by E.B. Yanovskaya [6] and S.A. Orlovskiĭ [2,3]. We also call attention to a paper by V.V. Podinovskiĭ [2]. For an algebraic approach to games with preferences see O.V. Shimelfenig [1].

17. The original formulation of the problem of optimality principles in noncooperative games was given by von Neumann [1]. A detailed investigation of what optimality principles in games should be, as well as specific proposals about their development in individual cases, are given in von Neumann and Morgenstern [1]. Nash [1] suggested the consideration of equilibrium situations, and R. Farquharson [1] generalized this concept to the case when, in (0.6), $\mathfrak{K} = \{K : |K| < n/2\}$, $|S| = 1$, $L_R^S(K) = K$ and $L_P^S(K) = \emptyset$.

A survey of approaches to the concept of optimality in noncooperative games was given by Vilkas [9,11,13–15]. Marchi [5] gave a general discussion of equilibria.

A systematic exposition of the theory of grouped solutions is given in Marchi's book [1]. A profound investigation of Pareto-optimal solutions (actually in a non-game-theoretical context) is given in the book by V.V. Podinovskiĭ and V.D. Nogin [1].

18. The general definition of noncooperative games with payoffs was formulated by von Neumann [1]. On games with forbidden situations, see G. Debreu [1], N.N. Vorob'ev and I.V. Romanovskiĭ [1], Wu Wen-tzün [2], and S.A. Orlovskiĭ [1,4,5]. For games with infinitely many players, the reader may consult the monograph of R.J. Auman and L.S. Shapley [1] and the papers by E.B. Yanovskaya [15] and A.Ya. Kiruta [1]. Games with unbounded payoff functions were discussed by S. Karlin [1] and E.B. Yanovskaya [2]. As a class of noncooperative games intermediate between games with payoffs and games with preferences, one can consider games with lexicographic payoffs. These were discussed by Podinovskiĭ [1]. In this connection we may also mention the papers by P.C. Fishburn [2], Yanovskaya [16], and G.N. Beltadze [1,2]. To the same circle of questions there belongs optimization by sequentially applied criteria, for which see V.V. Podinovskiĭ and V.M. Gavrilov's book [1].

The mathematical theory of utility originated in the supplement to the second edition of von Neumann and Morgenstern [1]. The books by Fishburn [1,3] and many other publications are devoted to it. One can become acquainted with the foundations of utility theory in the books by R.D. Luce and H.I. Raiffa [1] or Fishburn [1,3]. On the connection between utility theory and the theory of decision making, see E.Ĭ. Vilkas [10].

19. The possibility of a fixed stochastic dependence of mixed strategies of players in mixed extensions of noncooperative games was investigated by V.L. Kreps [1]. The exceptional character (and in a certain sense, uniqueness) of a mixed extension of games was discovered by V.E. Lapitskiĭ [2]. For mixed extensions of games in which the player's payoffs in random situations are not defined by expectations but in some other way, see J.G. Kemeny and G.L. Thompson [1].

20, 21. The foundations of the theory of strategic information are in M. Sakaguchi [1]. The consideration of the class of two-person zero-sum games as a subclass of the noncooperative games with payoffs and "factorized" realization of the cooperative point of view is the essential content of von Neumann [1]. Indeed, a large part of the book by von Neumann and Morgenstern [1] is devoted to the cooperative theory. Brief surveys of the results are given in W.F. Lucas [1] and A.I. Sobolev [2], and a detailed exposition in J. Rosenmüller [2]. Here we should also mention the earlier book by Rosenmüller [1].

Chapters 3 and 4 of the present book are completely devoted to two-person zero-sum games, and the main notes on it are given in these chapters.

The parallelism between the general theory of noncooperative games and cooperative theory, which we mentioned at the end of **20**, was discovered by E.B. Yanovskaya [9,20].

22. The foundations of the concept of dynamical (positional) games were introduced by H.W. Kuhn [1] and C. Berge [1]. A fundamental direction in the theory of dynamical games was opened by the work of L.S. Shapley [1]. On hierarchical games, see the book by H. von Stackelberg [1], and also articles by U.B. Germeĭer [3], N.N. Moiseev [1], Germeĭer and Moiseev [1], Germeĭer and I.A. Vatel [1], and V.N. Lagunov [1].

23. Games in which the players are allowed to change their solutions once were considered by A.Ya. Kiruta [1].

24–25. The simplest game-theoretic models of pollution and protection of the environment were described by Vorob'ev [20]. The model of a competitive market goes back to O. Cournot [1], was proposed in a game-theoretical way by E. Burger [1] and then analyzed by F. Szidarovszky [1,3]. The idea of developing a game-theoretic model of disarmament and armament-control apparently belongs to Morgenstern [1]; and N. Howard [1] actually constructed such a model, although it was rather simplified.

The class of games considered by V.A. Gorelik [1] can serve as a model of a class of games that are special cases of game-theoretic models of the same kind.

27. For an exposition of two-person zero-sum games of duel type and of butterfly-shaped games, see Karlin's monograph [3]. It contains a discussion of the history of the problem, and an extensive bibliography. One of the generalizations of the theory of duels is the theory of "truels", games of timing for three (or more) players. On this, see D.M. Kilgour [1].

Chapter 1
Noncooperative Games

§1 Noncooperative games and their components

1.1. Noncooperative game and its components

Definition. A *noncooperative game* is a triple

$$\Gamma = \langle I, \{x_i\}_{i \in I}, \{H_i\}_{i \in I} \rangle, \tag{1.1}$$

where I is a finite set whose elements are called *players* (from now on, unless the contrary is stated, we take $I = \{1, 2, \ldots, n\}$); x_i ($i \in I$) are arbitrary pairwise disjoint sets, whose elements are called the *strategies* of the corresponding players $i \in I$. The ordered sets $\{x_i\}_{i \in I}$ of strategies of the players, i.e., the elements of the Cartesian product

$$x = \prod_{i \in I} x_i \tag{1.2}$$

are called *situations* (or, synonymously, n-tuples) in the game Γ; the bounded functions

$$H_i : x \to \mathbf{R} \tag{1.3}$$

corresponding to the players $i \in I$ are called the *payoff functions* of the players, and their values on the situations are the *payoffs* in these situations. □

Each of two players i and j may carry out "physically" one and the same action, but, from the "strategic" point of view, it is far from indifferent exactly who did it. This is why we assume that the sets of strategies x_i and x_j are disjoint for $i \neq j$, considering the action z as the pair $(z, i) \in x_i$ or $(z, j) \in x_j$. However, in cases where there is no ambiguity we shall not state explicitly whether a strategy belongs to one player or another, and we even speak of "identical" sets of strategies for different players.

In what follows, unless the contrary is explicitly stated, the components (listed in (1.1)) of the noncooperative game Γ will be the set of players in the game, the sets of their strategies, and the payoff functions. If we are considering several games at the same time, we label the games and indicate their components by the same labels, or use the same labels as indices for the components of each game.

In essence, the components of a noncooperative game as in (1.1) form an exhaustive, if somewhat peculiar, set of the "rules" of the game.

Notice that if $n = 1$ the noncooperative game Γ defined by (1.1) reduces to a real function

$$H_1 : x = x_1 \to \mathbf{R}.$$

Consequently many problems of the theory of noncooperative games can be thought of as extentions of corresponding problems for functions.

1.2. In game theory, we of course make no assumptions about the informal nature of the components of a game. At the same time, it is natural to visualize "a play of the game Γ" defined by (1.1) as (let us say, for definiteness) the simultaneous and independent selection by each player $i \in I$ of an individual strategy $x_i \in \mathfrak{x}_i$, followed in the resulting situation $x \in \mathfrak{x}$ by the payoff $H_i(x)$ to each player i.

Thus in a noncooperative game each player functions both as an active source and as an interested party. Consequently it makes sense not to exclude from consideration, a priori, either, on the one hand, the noncooperative games in which some player has only a single strategy, i.e. cannot actually influence the outcome of the game (this is of interest, since such a player can nevertheless have various interests in the realization of one situation or another); or, on the other hand, games in which the payoff to a player is the same in all situations, i.e. that player actually has no interest in the outcome of the game (this case can be of interest because such a player can act so as to affect the payoffs to other players).

We shall actually adopt a somewhat intermediate point of view. We ordinarily suppose that a player in a noncooperative game has at least two strategies, and do not make any assumption about the actual variability of the payoff function.

1.3. Examples of some classes of noncooperative games. Without making any attempt at an exhaustive classification, we present some important classes of noncooperative games. Although this is done here purely for the sake of illustration, we shall subsequently investigate these classes of games systematically.

Definition. A noncooperative game Γ as in (1.1) is said to be *finite* if the sets $\mathfrak{x}_i$ of strategies of all the players x_i are finite.

The finite noncooperative games constitute the simplest, as well as an important and natural, class of noncooperative games. Their study, to which we shall devote Chapters 2 and 4, is an essential part of game theory.

A finite noncooperative game is said to be a *bimatrix game* if its set I consists of two players (i.e., $I = \{1, 2\}$). □

The latter name is explained by the following natural way of describing a game of this class. If we set up two tables, in which the rows are labeled with the strategies of player 1, and the columns are labeled with the strategies of player 2, then the cells of the tables correspond to the situations in the game. If the cells in the first table are filled in with the values of the payoff function of player 1, and those in the second table, with the values of the payoff function of player 2, we will have a pair of real matrices that completely describe the game. These matrices are called the *payoff matrices* of the bimatrix game.

A bimatrix game with $m \times n$ payoff matrices is called an $m \times n$ *bimatrix game*. We denote a bimatrix game with payoff matrices A and B by $\Gamma(A, B)$ or $\Gamma_{A,B}$. The ith row of a matrix M is denoted by $M_{i\cdot}$, and its jth column, by $M_{\cdot j}$.

1.4. Definition. A noncooperative game Γ defined by (1.1) is a *constant-sum game* if, for each situation $x \in \mathfrak{x}$,

$$\sum_{i \in I} H_i(x) = c \tag{1.4}$$

where c is a constant. □

Noncooperative constant-sum games can be considered, for example, as models of economic processes without production or consumption, involving only exchanges of existing commodities, or their redistribution.

A game Γ is said to be a *zero-sum game* if the constant c in (1.4) is zero.

A zero-sum noncooperative game with only two players: $I = \{1, 2\}$ is called a *two-person zero-sum* (or *antagonistic) game.* □

It follows that in a two-person zero-sum game the payoff (i.e., the gain) to one player is equal to minus the payoff (i.e., the loss) to the other: $H_1(x_1, x_2) = -H_2(x_1, x_2)$.

Since, in a two-person zero-sum game, the payoff function of player 2 is completely determined by the payoff function of player 1, when we describe the game we do not have to mention H_2 at all, and we can drop the subscript 1 from H_1, so that we can call this function the *payoff function of the game* (remembering, however, that it describes a *gain* by the first player and a *loss* by the second).

If we know that a game is a two-person zero-sum game, it can be defined as a triple

$$\langle \mathfrak{x}, \mathfrak{y}, H \rangle, \tag{1.5}$$

where $\mathfrak{x}$ and $\mathfrak{y}$ are disjoint sets (of strategies of the players), and $H : \mathfrak{x} \times \mathfrak{y} \to \mathbf{R}$.

Definition. A two-person zero-sum bimatrix game is called a *matrix game.* □

The reason for this terminology is that a bimatrix game is a two-person zero-sum game if and only if $B = -A$, so that it is specified just by the payoff matrix A of player 1. (Its elements can evidently also be thought of as the losses of player 2.) We then call A the *payoff matrix* of the game. We denote a matrix game with payoff matrix A by $\Gamma(A)$ or by Γ_A.

1.5. Coalitions. Any subset of the set of all players in a noncooperative game will be called a *coalition.* This is just a convenient auxiliary concept the definition of noncooperative games. Our introduction and use of it does not mean that we are going to consider cooperative games, which are not discussed in this book, or even any "coalitional" approach to noncooperative games. To put it in formal language, the noncooperative character of the games that we consider is clear from the fact that the grouping of players into coalitions is purely set-theoretical, completely independent of their interests and their actions; and the strategic possibilities and interests of a coalition are simply combinations of those of the players involved in it.

Further, in cases where no ambiguity will arise, we shall drop the distinction between a one-element set $\{\alpha\}$ and its unique element α, and use, for example, expressions like $A \cup \alpha$ (instead of the more accurate $A \cup \{\alpha\}$), and so on. In particular, we shall often not distinguish between a single-player coalition $\{i\}$ and the player i, and correspondingly write i instead of $\{i\}$. At the same time, we must emphasize that it will be necessary to distinguish between the family $\{\{i\}\}_{i \in K}$ of the one-element coalitions that are involved in K and the coalition $K = \{i\}_{i \in K}$ itself.

1.6. If φ_i are objects of any kind that are in one-to-one correspondence with the players $i \in I$, we denote by φ_K the collection of objects corresponding to the players in the coalition $K \subset I$. We then call the collection φ_K a *K-vector*, and call φ_i its *components*. The possible appearance of identical objects φ_i and φ_j with $i \neq j$ will not lead to any confusion.

In particular, when we consider the strategies x_i of the players $i \in K$, we shall speak of the K-vector x_K composed of the components x_i, calling it the *coalitional strategy* of K. In this connection, the strategies of the player i will sometimes be called the *individual strategies* of this player.

We denote by $\boldsymbol{x}_K$ the set of coalitional strategies of the coalition K in the game Γ of (1.1). Evidently

$$\boldsymbol{x}_K = \prod_{i \in K} \boldsymbol{x}_i .$$

In particular, the situation x can be denoted by x_I, and the set of all situations by $\boldsymbol{x}_I$. These notations will be used occasionally.

Notice especially the coalitions of the form $I \setminus i$ (i.e., those consisting of all players other than i). We shall write $x_{I \setminus i} = x^{(i)}$; and a strategy of the coalition $I \setminus i$ will sometimes be denoted by $x^{(i)}$.

If $K \subset L \subset I$ and $x_L \in \boldsymbol{x}_L$, we shall denote by x_K the "projection" of x_L on the set $\boldsymbol{x}_K$ of coalitional strategies of K. In particular, when $L = I$ we may speak of the projection of a situation, and when $K = i$ and $x_L \in \boldsymbol{x}_L$ (or $x \in \boldsymbol{x}$) we shall understand by x_i the *individual strategy* of i which is a component of the strategy x_L (or situation x).

1.7. Another important special instance of a K-vector is that in which the components corresponding to elements $i \in K$ are real numbers. The set of such K-vectors is a subset of Euclidean space which we denote by $\mathbf{R}^K$ (and its origin, by $\mathbf{0}^K$).

In addition, when we consider the payoffs of players $i \in K$ in the situation x, we shall speak of the K-vector $H_K(x) \in \mathbf{R}^K$ that consists of the components of $H_i(x)$. For a set $\boldsymbol{y} \subset \boldsymbol{x}$ of situations we set

$$\{H_{K(x)} : x \in \boldsymbol{y}\} = \mathfrak{W}_{K,\Gamma}(\boldsymbol{y}),$$

and call this subset of $\mathbf{R}^K$ the set of realizable payoff vectors of K on the set y of situations in the game Γ. In particular, we may speak of the set $\mathfrak{W}_{I,\Gamma}(y)$ of realizable I-vector payoffs of the players on y in Γ and of the $\mathfrak{W}(\Gamma)$ of realizable vector payoffs in Γ.

If Γ is a constant-sum game, the set $\mathfrak{W}(\Gamma)$ lies on the hypersurface defined by (1.4), which for a two-person zero-sum game becomes a straight line with the equation $x + y = 0$.

If we make the convention that the coalition K itself is a K-vector, we may, chiefly for brevity of notation, set

$$\sum_{i \in K} H_i(x) = H_K(x)K^T$$

(where ^{T}, as usual, denotes the transpose).

1.8. Structures on sets of strategies. In a formal description of a noncooperative game in the form of a system (1.1), the sets x_i of strategies of the players (and therefore also the sets x_K of coalitional strategies) are not obliged a priori to have either any kind of structure or properties connected with their structure, except purely set-theoretic properties (cardinality, for example). In the same way, a payoff function H_i on the product x is nothing but a mapping of this product into the set of real numbers. It is natural to suppose, as is usually (but not always) done, that the set of values taken by each of these functions is bounded. We shall suppose that this assumption has been made.

In particular, there may be no topology given on the sets x_i, so that the question of the analytic properties of the functions H_i, for example their continuity, simply does not arise.

Nevertheless, it is often both natural and profitable to consider sets of strategies of players in games together with some sort of structure on these sets. They are often assumed to be measurable, as well as topological or linear, spaces. Structures on sets of strategies of players in a noncooperative game can be thought of as supplementing (or making more precise) the rules describing the representation of games in the form (1.1).

1.9. Let us introduce the concept of a topological game.

Definition. A noncooperative game Γ as in (1.1) is *topological* if each set x_i ($i \in I$) of strategies is a topological space (i.e., if a system of neighborhoods $\mathscr{X}_i$ is given in it). □

Such an assignment of a topology on the spaces x_i of strategies of the game Γ, unconnected with the rules of Γ and independent of them (thus, so to speak, a supplementary rule of Γ) is sometimes called the *extrinsic* topology of x_i, in contrast to the semi-intrinsic and intrinsic topologies on x_i, which will be discussed at the beginning of § 6.

For sets of individual strategies of players in a topological game, the topology can be given by the usual rule for the topological product of the sets x_K of coalitional strategies, and in particular for the set x of all situations.

More precisely, let the topology on each set x_i be defined by a system of neighborhoods $\mathcal{X}_i$. Let us form the family of neighborhoods $\mathcal{X}_K$ in x_k generated by all products of the form $\prod_{i\in K} y_i$, where $y_i \in \mathcal{X}_i$, for $i \in K$. For the sake of clarity, we call such a topology a *product* topology. In particular, we may speak of the product topology on the set $x_I = x$ of all situations.

If the game Γ is topological, the names of properties of the spaces of the players' strategies and of the payoff functions in this topology are sometimes used to describe the game itself. Thus one speaks of *compact* games if the spaces of strategies are compact, of *continuous* games if all payoff functions are *continuous* functions of the situations, of (upper- or lower-) semicontinuous games, etc.

In particular, spaces of strategies might be *metric*. If a metric ρ_i is given on each x_i, these metrics induce on the set of situations the metric ρ_I for which

$$\rho_I(x, y) = \max_{i\in I} \rho_i(x_i, y_i).$$

One of the "endogenic" processes for metrizing the space x_i $(i \in I)$ of strategies in a game Γ (as described in (1.1)) by using the payoff functions H_i $(i \in I)$ will be mentioned as a concept at the end of this section, and described in § 6.

A very natural class of noncooperative games consists of the topological games whose spaces of players' strategies are also compact and whose payoff functions are continuous (in the same topology). The study of this class of games, as well as of games with "restricted discontinuous" payoff functions, has been the subject of many papers, the results of which are presented in part in §§ 3 and 5 of this chapter and in Chapter 3. From a formal point of view, the finite noncooperative games (see 1.3) also belong to this class.

1.10. Definition. A noncooperative game Γ as in (1.1) is said to be *measurable* if

1) on each set x_i of strategies there is defined a σ-algebra Ξ_i of measurable sets that makes x_i into a measure space;
2) on the set x of situations there is defined a σ-algebra Ξ generated by the σ-algebras Ξ_i;
3) the payoff functions H_i are Ξ- measurable. □

With this definition, we can define a noncooperative game as a triple

$$\Gamma = \langle I, \{\langle x_i, \Xi_i\rangle\}_{i\in I}, \{H_i\}_{i\in I}\rangle. \tag{1.6}$$

For a triple (1.6) with given H_i to be a measurable game it is necessary for the σ-algebras Ξ_i to be sufficiently "rich". Consequently, in order to construct a measurable game these σ-algebras may have to be "built up" in order to have measurable payoff functions.

On the other hand, if a game Γ is topological, then its topological structure can already generate its measurable structure in a rather natural way. Thus, Ξ_i is often taken to be a *Borel* σ-algebra (i.e., generated by the open sets of $\mathfrak{x}_i$). In particular, if $\mathfrak{x}_i$ is finite (and has the discrete topology), Ξ_i can be taken to be the family of all subsets of $\mathfrak{x}_i$.

Fortunately the Borel σ-algebras usually turn out to be extensive enough for most payoff functions to be measurable.

In cases when the measurable structure of a noncooperative game is determined by the Borel σ-algebras of subsets of the strategy spaces of the players, the measurable noncooperative game (1.6) can simply be written in the form (1.1).

1.11. In the overwhelming majority of cases considered to date, sets $\mathfrak{x}_i$ of players' strategies in noncooperative games, if not finite, can be regarded as subsets of linear (and even Euclidean) spaces. Moreover, these sets are usually assumed to be convex, and the payoff functions are given the corresponding properties of convexity or concavity (see 3.9 and later sections, also 3.3 of Chapter 3 and later sections). Nevertheless, under some special assumptions there is simply no need for the linear structure of sets of strategies, since, aside from the complete transparency of the operations, it has not been necessary up to now to carry over the linear structure from spaces of players' strategies to spaces of coalitional strategies or to the space of all situations (in view of the noncooperative character of the games we consider).

Also, algebraic structures on sets of players' strategies have not been considered up to now (of course, except for examples in which the sets are simply specific mathematical objects).

1.12. Relations for games. Subgames. Let us notice some of the simplest relations that can connect noncooperative games.

Definition. If

$$\Gamma = \langle I, \{\mathfrak{x}_i\}_{i\in I}, \{H_i\}_{i\in I}\rangle, \quad \Gamma' = \langle I, \{\mathfrak{x}'_i\}_{i\in I}, \{H'_i\}_{i\in I}\rangle$$

are two noncooperative games with coinciding sets of players, if also $\mathfrak{x}'_i \subset \mathfrak{x}_i$ for each player i, and if each payoff function H'_i is the *restriction* of H_i to the set $\mathfrak{x}'$ of situations, the game Γ' is said to be a *subgame* of Γ, and Γ is an *extension* of Γ'. Since a subgame Γ' of Γ is completely determined by the set $\mathfrak{x}'$ of its situations, it can be called an $\mathfrak{x}'$*-subgame* of Γ. □

1.13. Strategically and affinely equivalent games.

Definition. Two noncooperative games with the same sets of players and their sets of strategies,

$$\Gamma = \langle I, \{\mathfrak{x}_i\}_{i\in I}, \{H_i\}_{i\in I}\rangle, \quad \Gamma' = \langle I, \{\mathfrak{x}_i\}_{i\in I}, \{H'_i\}_{i\in I}\rangle, \qquad (1.7)$$

are said to be *strategically equivalent* if there are continuous strictly increasing functions $\varphi_i : \mathbf{R} \to \mathbf{R}$ such that for every situation $x \in \boldsymbol{x}$ and player $i \in I$ we have

$$H_i'(x) = \varphi_i(H_i(x)).$$

The noncooperative games Γ and Γ' in (1.7) are said to be *affinely equivalent* if there are positive numbers a_i and real numbers c_i $(i \in I)$ such that for every $x \in \boldsymbol{x}$ and $i \in I$ we have

$$H_i'(x) = a_i H_i(x) + c_i. \qquad \square$$

The informal difference between strategically equivalent games is, generally speaking, the difference between the different scales of measurement which the players use to measure their payoffs. For affine equivalence these scales are uniform. Thus the differences between affinely equivalent games Γ and Γ' are just the differences c_i between the "original capital" of player i on entry into game Γ or game Γ', and the ratio a_i of the units, by which the player's utility is measured for participation in Γ, to the corresponding units for participation in Γ'.

Strategic equivalence between games Γ and Γ' will be denoted by $\Gamma \sim \Gamma'$.

We see directly from the definition that the relation $\sim$ has the following properties:

1°. Reflexivity: $\Gamma \sim \Gamma$

2°. Symmetry: $\Gamma \sim \Gamma'$ implies $\Gamma' \sim \Gamma$.

3°. Transitivity: $\Gamma \sim \Gamma'$ and $\Gamma' \sim \Gamma''$ imply $\Gamma \sim \Gamma''$.

Hence this relation is indeed an equivalence relation, so that its name is justified. It separates the set of all noncooperative games with common sets of players and strategies into pairwise disjoint classes of strategically equivalent games.

We shall sometimes use the term *strategic-equivalence transformation* for the transition from a noncooperative game to a strategically equivalent game.

Similarly, we have reflexivity, symmetry, and transitivity for affine equivalence. This relation separates the family of all noncooperative games with the same sets of players and strategies into classes of affine equivalence (evidently smaller than the classes of strategic equivalence). We may use the term *affine-equivalence transformation* for the transition from a noncooperative game to an affinely equivalent game.

1.14. ε_I-homomorphisms. It is natural to introduce morphisms from one noncooperative game to another. Since we are dealing with a class of noncooperative games with payoffs, we may, looking ahead, speak of approximate morphisms of games. Along the way there naturally arise categories for which the objects are games. *)

---

*) On the fundamental notions of category theory, see, for example, S. Mac Lane, *Categories for the Working Mathematician*, Springer-Verlag, New York-Berlin, 1971.

Considering now a multiple-valued mapping $f : A \to B$ (where the notation implies that $f(a) \neq \emptyset$, for all $a \in A$), we say that it is *invertible* if

$$f(A) = \bigcup_{a \in A} f(a) = B,$$

and its *inverse* is the mapping $f^{-1} : B \to A$, so that for $b \in B$

$$f^{-1}(b) = \{a : b \in f(a)\}$$

(to avoid misunderstanding, we note that $(f^{-1})^{-1}(a)$ may contain, besides a, additional elements of A).

Now let

$$\Gamma = \langle I, \{x_i\}_{i \in I}, \{H_i\}_{i \in I} \rangle, \quad \Gamma' = \langle I', \{x_i\}_{i \in I'}, \{H_i'\}_{i \in I'} \rangle \tag{1.8}$$

be two noncooperative games. Consider the invertible mappings

$$\pi_I : I \to I' \tag{1.9}$$

(in what follows we consider only the case when the mapping is one-to-one, so that $\pi_I i$ is a completely determined player in I') and

$$\pi_i : x_i \to x'_{\pi_I i} \text{ for } i \in I. \tag{1.10}$$

We denote the union of these mappings by π.

We can extend the mapping π to the set x of all the situations, to the set 2^I of the coalitions, and also to the family $\mathfrak{K}$ of these, setting

$$\pi x = \pi(x_1, \ldots, x_n) = (\pi x_1, \ldots, \pi x_n) \quad \text{for} \quad x \in x,$$

$$\pi K = \{\pi i : i \in K\} \quad \text{for} \quad K \subset I,$$

$$\pi \mathfrak{K} = \{\pi K : K \in \mathfrak{K}\} \quad \text{for} \quad \mathfrak{K} \subset 2^I.$$

Definition. If $\varepsilon_I \geqq O^I$, the mapping π defined in (1.9) is called an ε_I*-semihomomorphism* of Γ on Γ' as in (1.8) if, for every $i \in I$,

$$\sup_{x \in x} \sup_{x' \in \pi x} (H_i(x) - H'_{\pi i}(x')) \leqq \varepsilon_i. \tag{1.11}$$

It is called an ε_I*-homomorphism* of Γ on Γ' if, for every $i \in I$,

$$\sup_{x \in x} \sup_{x' \in \pi x} |H_i(x) - H'_{\pi i}(x')| \leqq \varepsilon_i. \tag{1.12}$$

A mapping π is called a *semihomomorphism* (a *homomorphism*) if there is an $\varepsilon_I \geqq O^I$ such that π is an ε_I-semihomomorphism (respectively, ε_I-homorphism). If $\varepsilon_I = \varepsilon^I$ we call an ε_I-homomorphism an ε*-homomorphism.*

If there is an ε_I-homomorphism (or semihomomorphism) mapping π in a one-to-one way on each set x_i of strategies, it is called an ε_I*-isomorphism* (or *semi-isomorphism*).

If $\varepsilon_I = O^I$, i.e.

$$H_i(x) = H'_{\pi i}(x') \text{ for all } i \in I,\ x \in \boldsymbol{x} \text{ and } x' \in \pi x, \tag{1.13}$$

the ε_I-homomorphism (or semihomomorphism) is called an *exact homomorphism* (or *exact semihomomorphism*), and an ε_I-isomorphism is an *exact isomorphism*. However, if there is strict inequality in (1.11) (or 1.12), then an ε_I-semihomomorphism (or ε_I-homomorphism) is called an *ε_I-strict semihomomorphism* (or *ε_I-strict homomorphism*).

A homomorphism of a game on a subgame is called an *endomorphism*, and an isomorphism of a game on itself is an *automorphism*. □

An ε_I-homomorphism plays about the same role for a noncooperative game as an ε-almost-period plays for a real function.*)

Here again it is appropriate to notice that a noncooperative game with a single player is a real-valued function.

Informally, an ε_I-semihomomorphism π of a game Γ on Γ' as in (1.8) means that the assumption by a player i in the game Γ of the role of the player πi in the game Γ' and the choice by a player in the new role of any strategy $\pi'_i \in \pi x_i \subset \boldsymbol{x}'_{\pi i}$ instead of $x_i \in \boldsymbol{x}_i$ makes the payoff to each player change by at most ε_j. Consequently, the concept of an ε_I-semihomomorphism includes intrinsically an idea of being optimal, in the sense of not being capable of improvement, or of being stable.

An ε_I-homomorphism π of Γ on Γ' as in (1.8) means that in the transition described above of "players' change from game Γ to game Γ'", the payoff to the player j changes at most by ε_j.

Interpretations of the variants, described in the preceding definitions, of the general concepts of an ε_I-semihomomorphism and ε_I-homomorphism, can be made in the natural way.

Among the exact automorphisms, we mention the *identity automorphism* π^0 of the game Γ, for which $\pi^0_\Gamma x_i = x_i$ for all $i \in I$ and $x_i \in \boldsymbol{x}_i$.

1.15. Let us concentrate especially on the case of an exact homomorphism. If π is an exact homomorphism of Γ on Γ' as in (1.8), and if for strategies $x_i^{(1)}$ and $x_i^{(2)} \in \boldsymbol{x}_i$ we have $\pi x_i^{(1)} \cap \pi x_i^{(2)} \neq \emptyset$, then for every pair of situations $x \in \boldsymbol{x}$, $x' \in \pi x$; strategies $x^*_{\pi i} \in \pi x_i^{(1)} \cap \pi x_i^{(2)}$; and every player $j \in I$, we must have, because of (1.12),

$$H_j(x \parallel x_i^{(1)}) = H_{\pi j}(x' \parallel x^*_{\pi j}) = H_j(x \parallel x_i^{(2)}).$$

This means that, from the point of view of every player, the strategies $x_i^{(1)}$ and $x_i^{(2)}$ are equivalent. It is natural not to consider them as different strategies of a player, but rather as different instances of the same strategy.

Thus the role of any player i in the game Γ differs from the role of the player πi in an exact homomorphism $\pi\Gamma$ only by a possible choice between formally different but essentially equivalent strategies.

---

*) See, for example, B.M. Levitan, Almost periodic functions, Moscow, 1953, p. 10.

Consequently, from the game-theoretic point of view, a noncooperative game is essentially no different from any of its transforms under an exact homomorphism.

We sometimes denote exact homomorphy between Γ and Γ' by $\Gamma \approx \Gamma'$.

Definition. We say that a game Γ is *asymptotically homomorphic* to the game Γ' if, for every vector $\varepsilon_I > \mathbf{0}^I$, there is an ε_I-homomorphism from Γ to Γ'.

We sometimes indicate asymptotic homomorphy between Γ and Γ' by the notation $\Gamma \underset{A}{\approx} \Gamma'$.

In the light of what has just been said, we can interpret an asymptotic homomorphism as an "arbitrarily small distinction" between the two games.

It is clear that if Γ' is asymptotically homomorphic to Γ, but not homomorphic to it, there must be infinitely many different ε_I-homomorphisms of Γ on Γ'. It follows, in particular, that the concepts of homomorphism and asymptotic homomorphism coincide for finite games.

On the contrary, for infinite games the difference between homomorphism and asymptotic homomorphism can be illustrated by the simplest examples. Thus, in the case of a single player, when the noncooperative game reduces to a real function on the set of strategies of this player (and the case of a nontrivial game can be covered without difficulty by deliberately complicating the construction), we can take

$$x = \{1, 2, \dots\},\ H(x) = 1/x,$$

and

$$x' = \{1, 2, \dots; 0\},\ H'(x) = \begin{cases} 1/x & \text{if } x \neq 0, \\ 0 & \text{if } x = 0. \end{cases}$$

We then take any $\varepsilon > 0$, $x_\varepsilon > 1/\varepsilon$, and set

$$\pi_\varepsilon(x) = \begin{cases} x & \text{if } x \leqq x_\varepsilon, \\ 0 & \text{if } x > x_\varepsilon. \end{cases}$$

Evidently π_ε is an ε-homomorphism.

Suppose now that a mapping $\pi : x \to x'$ is an exact homomorphism from Γ to Γ', and that $x_0 \in \pi^{-1}(0)$. Then

$$H(x_0) - H'(\pi x_0) = 1/x_0 \neq 0,$$

which contradicts the assumption that π is an exact homomorphism.

1.16. We denote the class of all ε_I-homomorphisms from Γ to Γ' as in (1.8) by $\mathrm{Hom}_{\varepsilon_I}(\Gamma, \Gamma')$, and the whole class of homomorphisms from Γ to Γ' by $\mathrm{Hom}(\Gamma, \Gamma')$. Evidently it follows from $\varepsilon_I \leqq \varepsilon'_I$ that

$$\mathrm{Hom}_{\varepsilon_I}(\Gamma, \Gamma') \subset \mathrm{Hom}_{\varepsilon'_I}(\Gamma, \Gamma')$$

and besides that

$$\mathrm{Hom}(\Gamma, \Gamma') = \bigcup_{\varepsilon_I \geqq \mathbf{0}^I} \mathrm{Hom}_{\varepsilon_I}(\Gamma, \Gamma').$$

Let us establish some properties of the classes $\mathrm{Hom}_{\varepsilon_I}(\Gamma, \Gamma')$ of ε_I-homomorphisms.

1°. Transitivity. For every $\varepsilon_I, \varepsilon'_I \geqq \mathbf{0}^I$, there is an ε''_I such that

$$\mathrm{Hom}_{\varepsilon_I}(\Gamma,\Gamma') \times \mathrm{Hom}_{\varepsilon'_{I'}}(\Gamma',\Gamma'') \subset \mathrm{Hom}_{\varepsilon''_I}(\Gamma,\Gamma'').$$

Take $\pi \in \mathrm{Hom}_{\varepsilon_I}(\Gamma,\Gamma')$ and $\pi' \in \mathrm{Hom}_{\varepsilon'_{I'}}(\Gamma',\Gamma'')$, and form the mapping $\pi_I^{-1} : I' \to I$. Then, as is easily verified, we will have

$$\pi'\pi \in \mathrm{Hom}_{(\varepsilon_I + \varepsilon'_{\pi^{-1}I'})}(\Gamma,\Gamma'').$$

From this transitivity of homomorphisms there follows the transitivity of the relation of asymptotic homomorphy of games. In fact, let $\Gamma' \underset{A}{\approx} \Gamma$ and $\Gamma'' \underset{A}{\approx} \Gamma'$, and let $\varepsilon_I \geq \mathbf{0}^I$. Then, taking $\pi \in \mathrm{Hom}_{\varepsilon_I/2}(\Gamma,\Gamma')$ and $\pi' \in \mathrm{Hom}_{\pi\varepsilon_I/2}(\Gamma',\Gamma'')$, we obtain $\pi'\pi \in \mathrm{Hom}_{\varepsilon_I}(\Gamma,\Gamma'')$.

The transitivity of exact homomorphy can be established without any difficulty.

2°. Identity homomorphisms. For every game Γ there is a homomorphism $\pi_I^0 \in \mathrm{Hom}_{\mathbf{0}^I}(\Gamma,\Gamma)$ such that for all Γ and $\pi \in \mathrm{Hom}_{\varepsilon_I}(\Gamma,\Gamma')$ we have $\pi\pi_\Gamma^0 = \pi$ and necessarily $\pi_\Gamma^0\pi' = \pi'$ for every $\pi' \in \mathrm{Hom}_{\varepsilon'_{I'}}(\Gamma',\Gamma)$.

For the proof it is enough to notice that the identity homomorphism of games Γ is such a homomorphism π^0.

It follows from our having established these two properties of homomorphism classes that we are really dealing with a category whose objects are games with a given number of players, and the morphisms for each pair (Γ,Γ') of objects are elements of the set $\mathrm{Hom}(\Gamma,\Gamma')$.

This category has two further properties.

3°. Universality. For all games Γ and Γ' (with the same number of players), $\mathrm{Hom}(\Gamma,\Gamma') \neq \emptyset$.

It is enough to choose an arbitrary mapping π as in (1.9)–(1.10), satisfying (1.12), and to set, for each $i \in I$,

$$\varepsilon_I \geqq \sup_{x\in \boldsymbol{x}} \sup_{x'\in\pi x} |H_i(x) - H'_{\pi i}(x')|.$$

4°. Symmetry. Let Γ and Γ' be two noncooperative games and let $\pi \in \mathrm{Hom}_{\varepsilon_I}(\Gamma,\Gamma')$. Then π is invertible (see 1.14) and $\pi^{-1} \in \mathrm{Hom}_{\varepsilon_{\pi I}}(\Gamma',\Gamma)$.

By hypothesis we have (1.12) for every $i \in I$. Hence, for every situation $x \in \boldsymbol{x}$ we will have

$$\sup_{x'\in\pi x} |H_i(x) - H_{i'}(x') \leqq \varepsilon_i, \text{ where } i' = \pi i. \tag{1.14}$$

But π is invertible. Therefore $x' \in \pi x$ and $i' = \pi i$ are equivalent to $x \in \pi^{-1}x$ and $i = \pi^{-1}i'$, so that (1.14) can be rewritten as

$$\sup_{x\in\pi^{-1}x'} |H_i(x) - H_{i'}(x')| \leqq \varepsilon_i, \text{ where } i = \pi^{-1}i',$$

as was to be proved.

It follows, in particular, that the relations of homomorphy of games, their exact homomorphy, and asymptotic homomorphy are reflexive, symmetric, and transitive. Thus they separate the class of games with the same set of players into classes of mutually homomorphic games (there is only one such class), of exactly homomorphic games and of asymptotically homomorphic games.

In what follows, in all discussions connected with the existence of one kind or another of homomorphisms, of Hom classes, and of homomorphy and asymptotic or exact homomorphy of games, we shall use, in place of the actual descriptions of the Hom classes, merely the properties 1°–4°, which, moreover, are of the nature of axioms. Therefore all further discussions will remain valid for any category $\mathcal{K}$ whose objects are games with a given number of players and whose morphisms satisfy conditions 1°–4°. For this category, we denote the homomorphism classes by $\mathrm{Hom}^{\mathcal{K}}$, the exact homomorphy of Γ and Γ' by $\overset{\mathcal{K}}{\approx}$, and asymptotic homomorphy by $\underset{A}{\overset{\mathcal{K}}{\approx}}$.

1.17. Consider, for two games Γ and Γ' (with the same number of players),

$$\rho_{\mathcal{K}}(\Gamma,\Gamma') = \inf_{\varepsilon_I}\{\bar{\varepsilon}_I : \mathrm{Hom}^{\mathcal{K}}_{\varepsilon_I}(\Gamma,\Gamma') \neq \emptyset\}, \tag{1.15}$$

where $\bar{\varepsilon}_I$ denotes the maximal component*) of the vector ε_I: $\bar{\varepsilon}_i = \max_{i\in I} \varepsilon_i$.

Theorem. *The function $\rho_{\mathcal{K}}$ defined by (1.15) on a class of pairs of games with the same number of players is a pseudometric, i.e. this function*

a) *is defined for every pair of games Γ and Γ';*

b) *is nonnegative: $\rho_{\mathcal{K}}(\Gamma,\Gamma') \geqq 0$;*

c) *does not break up objects: $\rho_{\mathcal{K}}(\Gamma,\Gamma) = 0$;*

d) *is symmetric: $\rho_{\mathcal{K}}(\Gamma,\Gamma') = \rho_{\mathcal{K}}(\Gamma',\Gamma)$;*

e) *satisfies the triangle inequality:*

$$\rho_{\mathcal{K}}(\Gamma,\Gamma'') \leqq \rho_{\mathcal{K}}(\Gamma,\Gamma') + \rho_{\mathcal{K}}(\Gamma',\Gamma'').$$

Proof. a) follows because under the conditions on the category $\mathcal{K}$ (universality property) for every pair of games Γ and Γ' there is a vector $\varepsilon \geq \mathbf{0}^I$ such that there is a nonempty set after the inf sign in (1.15);

b) is evident;

c) (1.15) means that

$$\{\varepsilon_I : \mathrm{Hom}^{\mathcal{K}}_{\varepsilon_I}(\Gamma,\Gamma') \neq \emptyset\} = \{\varepsilon_{I'} : \mathrm{Hom}^{\mathcal{K}}_{\varepsilon_{I'}}(\Gamma',\Gamma) \neq \emptyset\}.$$

---

*) Here we are taking the norm of a vector ε_I to be its maximal component. This is more transparent in the game-theoretic setting than the usual norm $\|\varepsilon_I\|$ (the Euclidean norm).

But from the coincidence of the sets of vectors there follows the equality of the sets of values of their maximal components and thus of the infima of these maximal components, i.e. $\rho_{\mathcal{K}}(\Gamma,\Gamma') = \rho_{\mathcal{K}}(\Gamma',\Gamma)$.

d) Let us consider three games Γ, Γ', and Γ'' (with the same numbers of players), and choose $\pi \in \overset{\mathcal{K}}{\mathrm{Hom}}(\Gamma,\Gamma')$ and $\pi' \in \overset{\mathcal{K}}{\mathrm{Hom}}(\Gamma',\Gamma'')$ arbitrarily. For definiteness, let $\pi \in \overset{K}{\mathrm{Hom}}_{\varepsilon_I}(\Gamma,\Gamma')$ and $\pi' \in \overset{\mathcal{K}}{\mathrm{Hom}}_{\varepsilon'_{I'}}(\Gamma',\Gamma'')$. Then, by what we noticed in 1.15,

$$\pi'\pi \in \overset{\mathcal{K}}{\mathrm{Hom}}_{(\varepsilon_I+\varepsilon'_{\pi^{-1}I'})}(\Gamma,\Gamma'').$$

But

$$\overline{\varepsilon_I + \varepsilon'_{\pi^{-1}I'}} \leqq \overline{\varepsilon_I} + \overline{\varepsilon'}_{\pi^{-1}I'}$$

and therefore, if we take the infimum on the left,

$$\rho_{\mathcal{K}}(\Gamma,\Gamma'') \leq \overline{\varepsilon_I} + \overline{\varepsilon'}_{\pi^{-1}I}.$$

Since the homomorphisms π and π' were chosen arbitrarily, we may take the infimum on the left here; this proves what was required. □

1.18. It follows in an obvious way from the very definition of an asymptotic homomorphism that $\rho_{\mathcal{K}}(\Gamma,\Gamma') = 0$ if and only if $\Gamma \underset{A}{\overset{\mathcal{K}}{\approx}} \Gamma'$.

Let $\Gamma \underset{A}{\overset{\mathcal{K}}{\approx}} \Gamma'$ and $\bar{\Gamma}$ be any games with the same number of players. Then it follows from the triangle axiom that

$$\rho_{\mathcal{K}}(\Gamma,\bar{\Gamma}) \leq \rho_{\mathcal{K}}(\Gamma,\Gamma') + \rho_{\mathcal{K}}(\Gamma',\bar{\Gamma}) = \rho_{\mathcal{K}}(\Gamma',\bar{\Gamma}),$$

which, together with symmetric reasoning, yields

$$\rho_{\mathcal{K}}(\Gamma,\bar{\Gamma}) = \rho_{\mathcal{K}}(\Gamma',\bar{\Gamma}).$$

Therefore the distance $\rho_{\mathcal{K}}$ between games depends not just on the games themselves, but rather on the $\underset{A}{\overset{\mathcal{K}}{\approx}}$-classes that contain them. Thus the function $\rho_{\mathcal{K}}$ can be considered to be defined on the set of pairs of $\underset{A}{\overset{\mathcal{K}}{\approx}}$-classes of games; and from 1.17, combined with what we just said, it follows that $\rho_{\mathcal{K}}$ is a metric on this set of classes of games.

However, according to what was said in 1.15, the $\underset{A}{\overset{\mathcal{K}}{\approx}}$-classes contain practically indistinguishable games. Therefore we may consider $\rho_{\mathcal{K}}$ as describing real differences between games; it can be thought of as a metric in this capacity.

1.19. We conclude the study of this group of questions by introducing the definitions of some concepts connected with automorphisms of noncooperative games.

Definition. In a noncooperative game Γ as in (1.1), let each coalition K correspond to a set t_K of situations. This correspondence is said to be *covariant* if $\pi t_K = t\pi_K$ for every automorphism π of Γ.

A set t of situations in the game Γ is said to be *symmetric* if $\pi t = t$ for every automorphism π of Γ.

In particular, a situation x is said to be symmetric if $\pi x = x$ for every automorphism π. □

The symmetry of every set t_α ($\alpha \in A$) of situations implies the symmetry of their intersection. In fact,

$$\pi \bigcap_{\alpha \in \mathcal{A}} t_\alpha = \bigcap_{\alpha \in \mathcal{A}} \pi t_\alpha = \bigcap_{\alpha \in \mathcal{A}} t_\alpha .$$

As an important example we mention the isomorphy of zero-sum two-person games, which we may call (mirror) symmetry.

Definition. Two zero-sum two-person games $\Gamma = \langle x, y, H \rangle$ and $\Gamma' = \langle x', y', H' \rangle$ are said to be (*mirror*) *symmetric* if $x' = y$, $y' = x$ and $H'(y, x) = -H(x, y)$ for all $x \in x$ and $y \in y$.

A zero-sum two-person game is said to be *symmetric* if it is (mirror) symmetric to itself. □

Informally, the transition from a zero-sum two-person game to a (mirror) symmetric game means an interchange of the roles of the players (for example, interchange of colors in a chess game).

1.20. Factorization of games. One can describe a rather large number of transformations of noncooperative games which can naturally be called factorizations. We shall discuss one of these.

If a coalition $L \subset I$ in a noncooperative game Γ as in (1.1) fixes one of its strategies $x_L^0 \in x_L$, we may consider the noncooperative game

$$\Gamma / x_L^0 = \langle I \setminus L, \{x_i\}_{i \in I \setminus L}, \{H_i / x_L^0\}_{i \in I \setminus L} \rangle, \qquad (1.16)$$

where the function $H_i / x_L^0 : x_{I \setminus L} \to \mathbf{R}$ is such that

$$(H_i / x_L^0)(x_{I \setminus L}) = H_i(x_L^0, x_{I \setminus L})$$

for every $x_{I \setminus L} \in x_{I \setminus L}$.

We will say that (1.16) is the *factorization* of Γ by the coalitional strategy $x_L^0 \in x_L$.

§2 Optimality principles in noncooperative games

2.1. Formulation of the question. We raise the question of the optimality of situations in noncooperative games of the form

$$\Gamma = \langle I, \{x_i\}_{i \in I}, \{H_i\}_{i \in I} \rangle. \tag{2.1}$$

By an *optimality principle* for a class $\mathfrak{G}$ of noncooperative games (either the class of all noncooperative games, or some part of it) we understand a mapping φ which establishes a correspondence between each game $\Gamma \in \mathfrak{G}$ and a subset of the set x of its situations, and expresses mathematically some meaningful feature of the (intuitive) concept of optimality. This feature may be expressed in terms of utility, stability, or fairness.

Here the situations in $\varphi\Gamma$ are said to be *realizations* of the optimality principle φ by the game Γ or *solutions* of Γ in terms of the optimality principle φ. Correspondingly, if $\varphi\Gamma \neq \emptyset$, then φ is said to be *realizable* on Γ, and Γ is *solvable* in the sense of φ.

In view of the ambiguity of the semantics of the concept of optimality, there can be very diversified optimality principles.

Just for amusement, we may note that even to the point of view that "come what may, everything is O.K." there corresponds an optimality principle φ^0, expressed by $\varphi^0\Gamma = x$.

At the same time, in certain cases optimality principles seem to be unmistakable. For example, if there is only one player in $\Gamma(I = \{1\})$, and thus the game, as we noted in 1.1, is formally a real-valued function $H_1 \colon x = x_1 \to \mathbf{R}$, it is difficult to imagine an optimality principle for such a game, other than the tendency of its single player to try for the maximum payoff (since no other procedure that the player might follow does not be represented in this model). Hence we may suppose that in this case $\varphi\Gamma = \arg\max H_1$.

It is natural to suppose that every reasonable optimality principle in a noncooperative game Γ must rely on the principle that we just described, that of optimizing the payoff, and must reduce to it as soon as Γ is reduced to a game with a single player.

Thus, for example, we can think of the principle that a situation x^* is optimal if each player receives the maximal payoff, i.e. if

$$H_i(x^*) = \max_{x \in x} H_i(x). \tag{2.2}$$

To this principle, there corresponds the mapping φ with $\varphi\Gamma = \emptyset$ for the overwhelming majority of noncooperative games; i.e., this principle is practically unrealizable.

At the same time, even (2.2) corresponds to informal concepts of optimality, although it expresses them in an extremely strong and therefore contradictory form. For, (2.2) reflects the idea of the profitability of the situation x^* for each player i, since in this situation each player would receive more than in any other situation. At the same time, (2.2) reflects the idea of the stability of x^*, since no individual player is interested in replacing x^* by any other situation.

Finally, the situation x^* can be treated from the point of view of fairness, since it satisfies, in the framework of the game Γ, the principle of each player's claiming the same (namely, the maximum) measure.

What we have said gives us a reason for considering as optimal for a coalition K the situations that are most preferable for it, that is, those in which the players in K would obtain the best (in some sense) payoffs and would have no desire to change the situation, to the extent that the change would be within the limits of the strategic possibilities.

Besides "exact" optimality, it is desirable also to consider "ε-optimality", in which, although some coalition may observe an instability of the optimal situation, this can nevertheless be avoided by paying an "ε-penalty". This concept of ε-optimality can be useful for a number of reasons. In the first place, it is imperative if, because of the noncompactness of the sets of strategies and situations, or discontinuities of the payoff functions, the exact maximum is inaccessible, but some approximation to it is feasible. Furthermore, there may be difficulties in the way of the determination of exact optimality, although approximations to it can be found, and we may consider these approximations to an exact principle of optimality as realizations of a corresponding approximate principle. In addition, the actual calculation of a realization of an exact optimality principle can be accompanied by errors, although the distorted solution may nevertheless be approximately optimal. Here it is important to emphasize that quantitative estimates of an approximation to an optimum may turn out to be different for different players.

2.2. Fundamental concepts and notation. We begin the formal description of the preceding concepts of optimality by introducing the system of notation which will be used from now on.

Let x be a situation in a game Γ as in (2.1), $K \subset I$ a coalition in Γ, and x'_K a coalitional strategy (see 1.6) of this coalition. Then we denote by $x_K \underset{K}{\|} x'_K$ (or, if no ambiguity will result, $x\|x_K$) the situation obtained from x by replacing the individual strategies of the players in K by their individual strategies that were used in forming the coalitional strategy x'_K.

The players' degrees of preference of situations in the game are evaluated by the payoffs in the situations; and the degrees of preference for each coalition K, by the payoff K-vectors. For componentwise comparison of vectors by magnitude, we use the following notation.

If φ_K and ψ_K are two K-vectors in $\mathbf{R}^K$, in accordance with the usual practice we write

$$\varphi_K < \psi_K \quad \text{if } \varphi_i < \psi_i \quad \text{for all } i \in K,$$

$$\varphi_K \leq \psi_K \quad \text{if } \varphi_i \leqq \psi_i \quad \text{for all } i \in K \text{ and } \varphi_K \neq \psi_K,$$

$$\varphi_K \leqq \psi_K \quad \text{if } \varphi_i \leqq \psi_i \quad \text{for all } i \in K.$$

In developing this system of notation we shall set, for any real K-vector $\varepsilon_K \geqq 0$,

$$\varphi_K \overset{\varepsilon_K}{<} \psi_K \quad \text{if} \quad \varphi_K < \psi_K + \varepsilon_K,$$

$$\varphi_K \overset{\varepsilon_K}{\leq} \psi_K \quad \text{if} \quad \varphi_K \leq \psi_K + \varepsilon_K.$$

In this connection, we shall sometimes use the term "ε-maximum of the set $\{a_\alpha\}_{\alpha \in \mathcal{A}}$" for a number a_ε such that $a_\alpha \leqq a_\varepsilon + \varepsilon$ for every $\alpha \in \mathcal{A}$.

2.3. Preferability and dominance. Definition. Let there be given a noncooperative game Γ as in (2.1), and in it a coalition K and situations x and y. We say that x *K-dominates* y if the following conditions are satisfied:

E) *Effectiveness* for K:

The situation y has the form $x \| y$, where $y_K \in x_K$ (the set of all situations of this form is naturally denoted by $x \| x_K$).

P) *Preferability* for K:

$$H_K(y) \leq H_K(x).$$

We shall also say that a situation x *strictly K-dominates* the situation y if the effectiveness condition is satisfied for K, and

P_s) *Strict preferability*:

$$H_K(y) < H_K(x).$$

Now let ε_K be a nonnegative real K-vector. We say that a situation x ε_K-*K-dominates* the situation y if the effectiveness condition is satisfied for K and in addition the condition

P_ε) ε_K-*preferability*:

$$H_K(y) \overset{\varepsilon_K}{\leq} H_K(x).$$

Finally we say that a situation x *strictly* ε_K-*K-dominates* the situation y if the effectiveness condition is satisfied for K, and in addition the condition

$\text{P}_{\varepsilon s}$) *Strict* ε_K-*preferability*:

$$H_K(y) \overset{\varepsilon_K}{<} H_K(x). \qquad \square$$

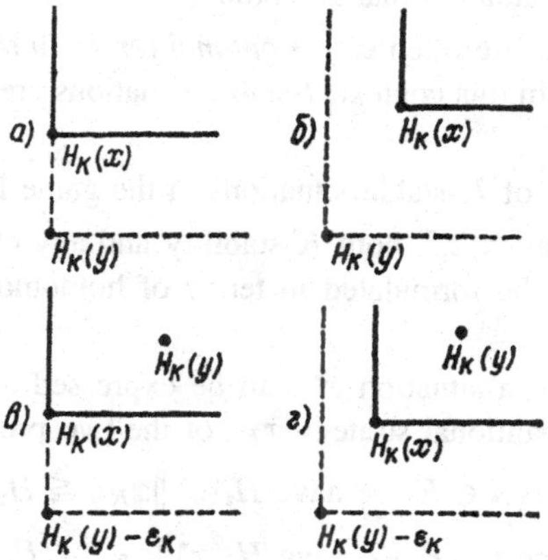

Fig. 1.1

A geometric representation of all these versions of preference conditions is presented in the diagrams in Fig. 1.1.

The situation x is, informally speaking, more preferable for K than situation y if no player in K is interested in the change from x to y and no one is against a change from y to x. If this preference for x as against y is strict, then all the players in K are for a change from y to x, and against a change from x to y. In the case of the ε-versions of the preference conditions, these statements about possible changes are preserved subject to the condition of the payment of an "ε-penalty" for a change from x to y, and the receipt of an "ε-premium" for a change from y to x.

All these versions of domination of situations can be formulated in a natural way in terms of semihomomorphisms and homomorphisms (see 1.14).

For example, let Γ be a game as in (1.8), let $K \subset I$, let x and y be situations in Γ, and let Γ', also as in (1.8), be the subgame of Γ in which

$$x_i' = \begin{cases} (x_i \setminus x_i) \cup y_i & \text{for } i \in K, \\ x_i & \text{for } i \notin K. \end{cases}$$

Consider the mapping π described in (1.9) and (1.10), which sastifies the condition

$$\pi_i x_i = y_i \qquad \text{for } i \in K,$$

and in other respects is the identity. Then a necessary and sufficient condition for the K-domination of y by x in Γ is that π is an exact semihomomorphism.

In what follows, we shall generally use the more traditional terminology based on domination rather than semihomomorphism.

For the sake of brevity, we shall sometimes call an unspecified one of these preference conditions a $^\circ$-preferability; denote an inequality sign indicating a $^\circ$-preference condition by $<^\circ$, and use $^\circ$-domination for a $^\circ$-preferability version of domination of situations.

2.4. Optimality of stability type. The nondomination of situations in any of the versions described above leads to a formal representation of the optimality of situations.

Definition. In a noncooperative game Γ as in (2.1), a situation x^* is said to be *K-stable* if there is no situation that K-dominates it.

K-stable situations are often called *optimal for K in the sense of Pareto* (or *Pareto optimal* for *K*). In this context, I-stable situations are simply called *Pareto optimal.*

We denote the set of K-stable situations in the game Γ by $\mathscr{C}_K(\Gamma)$. □

In the same way as in 2.3, both K-stability and any of its variants that will be mentioned later can be formulated in terms of homomorphisms and semihomorphisms of games.

The K-stability of a situation x^* can be expressed as the alternative (i.e., disjunction), for any coalitional strategy x_K, of the two propositions:

$$\text{for every } i \in K \text{ we have } H_i(x^* \| x_K) \leqq H_i(x^*), \tag{2.3}$$

$$\text{for some } i \in K \text{ we have } H_i(x^* \| x_K) < H_i(x^*). \tag{2.4}$$

It is evident that under (2.4) we can replace (2.3) by:

$$\text{for every } i \in K \text{ we have } H_i(x^* \| x_K) = H_i(x^*).$$

Finally, K-stability can be expressed in the form of an implication:

if there is a player $i \in K$ for whom $H_i(x^* \| x_K) > H_i(x^*)$,
then there is a player for whom $H_i(x^* \| x_K) < H_i(x^*)$.

The last condition for the K-stability of a situation has a very intuitive interpretation: in a K-stable situation, the coalition K can increase the payoff of one of its members only at the expense of another member.

For example, in any constant sum noncooperative game (see 1.4), and in particular in any zero-sum two-person game, all situations are I-stable (Pareto optimal).

In the case when the coalition K consists of a single player i, a situation x^* turns out to be i-stable if, for each of its (individual) strategies,

$$H_i(x^* \| x_i) \leqq H_i(x^*). \tag{2.5}$$

An i-stable situation is sometimes called *i-acceptable*: a situation is said to be acceptable to a player only if changing the player's strategy cannot increase the payoff to that player.

The i-stability of a situation x^* means that

$$(H_K(x^*) + R_+^K) \cap \mathfrak{W}_K(x^* \| x_K) = \{H_K(x^*)\},$$

i.e., that the nonnegative orthant of $\mathbf{R}^K$, translated by the vector $H_K(x^*)$, contains no realizations of the vector payoffs to K other than $H_K(x^*)$. In particular, for $x^* \in \mathscr{C}_K(\Gamma)$ all the points of $H_K(x^*)$ must lie on the boundary (that is, on the northeast boundary) of the set $\mathfrak{W}_K(x^* \| x_K)$. For example, if K consists of two players, and $\mathfrak{W}_K(\Gamma)$ is the shaded area in Figure 1.2, then the set of points $H_K^{(x^*)}$ for $x^* \in \mathscr{C}_K(\Gamma)$ is represented by the heavy lines.

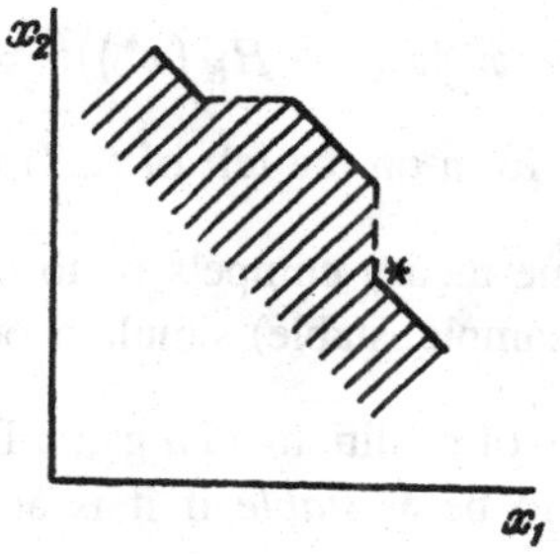

Fig. 1.2

2.5. The following theorem can be useful for an analytic characterization of the K-stability of a situation.

Theorem. *It is sufficient for* $x^* \in \mathscr{C}_K(\Gamma)$ *that we can find a function* $\varphi : \mathbf{R}^K \to \mathbf{R}$, *strictly monotonic in each real argument, such that the function*

$$\varphi(H_K(x^* \| x_K) - H_K(x^*))$$

attains a maximum at the point $\mathbf{0}^K$.

If the set $\mathfrak{W}_K(x^* \| x_K)$ *is convex, the function* φ *can be taken to be linear, and the condition of the existence of a linear function* φ *with the indicated property turns out to be necessary.*

Proof. Let $x^* \notin \mathscr{C}_K(\Gamma)$. Then there is an $\tilde{x}_K$ such that

$$H_K(x^* \| \tilde{x}_K) \geq H_K(x^*). \tag{2.6}$$

Let $K = \{i_1, \dots, i_k\}$. Then

$$\begin{aligned}
&\varphi(H_K(x^* \| \tilde{x}_K) - H_K(x^*))\varphi(\mathbf{0}^K) \\
&= \varphi(H_K(x^* \| \tilde{x}_K) - H_K(x^*)) - \varphi(H_{K\setminus i_k}(x^* \| \tilde{x}_K) - H_{K\setminus i_k}(x^*), \mathbf{0}^{i_k}) + \dots \\
&\dots + \varphi(H_{\{i_1,i_2\}}(x^* \| \tilde{x}_K) - H_{\{i_1,i_2\}}(x^*), \mathbf{0}^{K\setminus\{i_1,i_2\}}) \\
&\quad - \varphi(H_{i_1}(x^* \| \tilde{x}_K) - H_{i_1}(x^*), \mathbf{0}^{K\setminus i_1}) \\
&\quad + \varphi(H_{i_1}(x^* \| \tilde{x}_K) - H_{i_1}(x^*), \mathbf{0}^{K\setminus i_1}) - \varphi(\mathbf{0}^K).
\end{aligned}$$

In view of (2.6) and the assumed strict monotonicity of φ, at least one of the terms in the right is positive, and the others are nonnegative. Therefore

$$\varphi(H_K(x^* \| \tilde{x}_K) - H_K(x^*)) - \varphi(\mathbf{0}^K) > 0,$$

but this contradicts the maximal property of the point $\mathbf{0}^K$.

Now let the set $\mathfrak{W}_K(x^* \| x_K)$ be convex and $x^* \in \mathscr{C}_K(\Gamma)$. Since $H_K(x^*)$ belongs to the boundary of $\mathfrak{W}_K(x^* \| x_K)$, through $H_K(x^*)$ there passes at least one supporting hyperplane

$$c_K(H_K - H_K(x^*))^T = 0 \tag{2.7}$$

of $\mathfrak{W}_K(x^*\|x_K)$. Since the points of $H_K(x^*)$ belong to the northeast part of the boundary of $\mathfrak{W}_K(x^*\|x_K)$, we must have $c_K \geq 0$, and therefore

$$c_K(H_K(x^*\|x_K) - H_K(x^*))^T \leqq 0.$$

Consequently the linear form on the left of (2.7) is the required function. □

2.6. The very essence of game theory compels us to consider situations that are in some sense optimal (for example, stable) simultaneously for several coalitions.

Definition. Let $\mathfrak{K}$ be a family of coalitions in a game Γ as in (2.1) (i.e., $\mathfrak{K} \subset 2^I$). A situation x^* in Γ is said to be *$\mathfrak{K}$-stable* if it is $\mathfrak{K}$-stable for every coalition $K \in \mathfrak{K}$.

The set of all $\mathfrak{K}$-stable situations in Γ is denoted by $\mathscr{C}_{\mathfrak{K}}(\Gamma)$:

$$\mathscr{C}_{\mathfrak{K}}(\Gamma) = \bigcap_{K \in \mathfrak{K}} \mathscr{C}_K(\Gamma).$$

The transition $\mathscr{C}_{\mathfrak{K}}$ from a noncooperative game Γ to the set $\mathscr{C}_{\mathfrak{K}}(\Gamma)$ of its situations is called the *principle of $\mathfrak{K}$-optimality of stability type.* □

A logical analysis of the concept of $\mathfrak{K}$-stability follows from the expressions for K-stability presented in 2.3.

The $\mathfrak{K}$-stability of a situation admits the following natural interpretation. Suppose that the players agree among themselves to choose strategies that form a situation x^*. If x^* turns out to be $\mathfrak{K}$-stable, then no coalition K among those in the family $\mathfrak{K}$ will be interested as a whole in breaking the agreement: no deviation of the players in K from the agreed x^* can lead to an increase in the payoff to any of the players in K unless the payoffs to the other players are decreased. This means that the players in K have no basis for engaging *the whole coalition* in any "conspiracy" with the aim of breaking the agreement on x^*.

Conversely, let us suppose that a situation $\tilde{x}$ has been agreed on, but one that is not $\mathfrak{K}$-stable. In this case there are, by definition, a coalition $K \in \mathfrak{K}$ and a coalitional strategy $x_K \in \mathrm{x}_K$ of its members such that in the situation $\tilde{x}\|x_K$ none of the players in K loses in comparison with $\tilde{x}$, but someone knowingly gains. This means that the agreement, a condition of which was situation $\tilde{x}$, will be broken provided that its observance is not supported by some supplementary clauses.

2.7. Optimality of equilibrium type. We now turn to an optimality principle based on *strict* K-dominance (see 2.3).

Definition. In a noncooperative game as in (2.3), a situation is said to be a *K-equilibrium* situation if it has no strictly K-dominating situations. □

We shall sometimes use the general term *K-optimal* for K-stable and K-equilibrium situations.

The set of all K-equilibrium situations in a game Γ will be denoted by $\mathscr{R}_K(\Gamma)$. It follows directly from the definitions of K-equilibrium and K-stable situations (in 2.4) that $\mathscr{C}_K(\Gamma) \subset \mathscr{R}_K(\Gamma)$. To say that a situation x^* is a K-equilibrium situation means that for every coalitional strategy x_K of the coalition K there is a player $i \in K$ such that

$$H_i(x^*\|x_K) \leqq H_i(x^*). \tag{2.8}$$

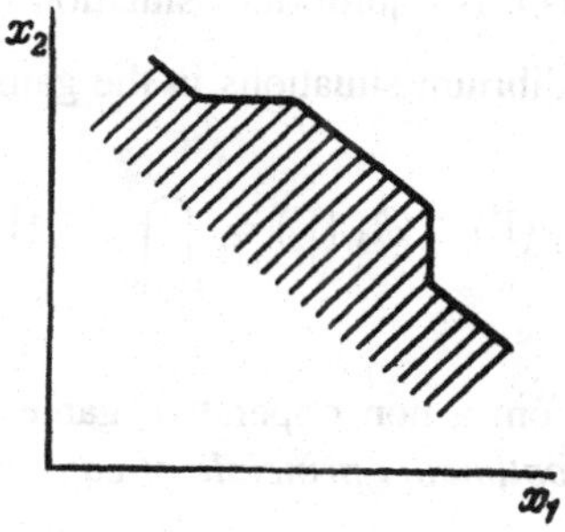

Fig. 1.3

Note that (2.8) follows from either of the alternative statements (2.3) and (2.4). If the coalition K has only the single player i, then (2.8) is equivalent to (2.5), so that i-stability coincides with the i-equilibrium property. Therefore, according to what was said above, we may call this property of a situation *i-optimality.*

The K-equilibrium property of a situation can be interpreted as the impossibility of having K, by its own efforts, increase the payoffs to each member of the coalition K.

Geometrically, the K-equilibrium property of the situation x^* means that

$$(H_K(x^*) + \text{int}\, \mathbf{R}_+^K) \cap \mathfrak{W}_K(x^* \| \mathfrak{x}_K) = \emptyset,$$

i.e., that all points of the nonnegative orthant $\mathbf{R}_+^K$ of the space $\mathbf{R}^K$, translated by the vector $H_K(x^*)$, which are realizations of the vector payoffs to K, lie on the boundary of the translated orthant (see Fig. 1.3, which should be compared with Fig. 1.2).

2.8. In analogy with the theorems of 2.5, we have the following theorem for K-equilibrium situations.

Theorem. *A sufficient condition for $x^* \in \mathfrak{R}_K(\Gamma)$ in a game Γ as in (2.1) is that there is a function $\varphi : \mathbf{R}^K \to \mathbf{R}$ which is strictly monotonic in one of its arguments, such that the difference*

$$\varphi(H_K(x^* \| x_K) - H_K(x^*))$$

attains a maximum at the point $\mathbf{0}^K$.

If the set $\mathfrak{W}_K(x^* \| x_K)$ is convex, then φ can be taken to be linear, and the condition that there exists a linear function with these properties turns out to be necessary.

The proof follows from the proof of the theorem in 2.5, with appropriate modifications.

2.9. Definition. Let $\mathfrak{K} \subset 2^I$. A situation x^* in the game is said to be a $\mathfrak{K}$-*equilibrium situation* if it is a K-equilibrium situation for every coalition $K \in \mathfrak{K}$.

The set of all $\mathfrak{K}$-equilibrium situations in the game Γ is denoted by $\mathcal{R}_{\mathfrak{K}}(\Gamma)$. Evidently

$$\mathcal{C}_{\mathfrak{K}}(\Gamma) \subset \mathcal{R}_{\mathfrak{K}}(\Gamma) = \bigcap_{K \in \mathfrak{K}} \mathcal{R}_K(\Gamma). \tag{2.9}$$

The transition $\mathcal{R}_{\mathfrak{K}}$ from a noncooperative game Γ to the set $\mathcal{R}_{\mathfrak{K}}(\Gamma)$ of its situations is called the *$\mathfrak{K}$-optimality principle of equilibrium type.*

Sometimes, for the sake of brevity, the principles of optimality, either of stability or equilibrium type, will be called simply *$\mathfrak{K}$-principles.* □

An informal interpretation of the $\mathfrak{K}$-equilibrium property of a situation differs only slightly from the interpretation of its $\mathfrak{K}$-stability. Under the conditions of an agreement by the coalitions in $\mathfrak{K}$ to form a $\mathfrak{K}$-equilibrium situation x^*, at least one of the players in each coalition $K \in \mathfrak{K}$ has no interest in breaking this agreement on K: no deviation by the players in K from their self-imposed obligation to x^* under the agreement can lead to a simultaneous increase of the payoffs to all the players in K. This means that, in case the whole coalition K enters into a conspiracy to violate the agreement on x^*, at least one member of the coalition will remain indifferent to the aims of the conspirators.

2.10. The problems of the theory of noncooperative games are mainly concerned with establishing realizations of $\mathfrak{K}$-principles of optimality of stability or equilibrium type for certain classes of noncooperative games, and finding realizations of them. Here the following case is especially important.

Let $\mathfrak{K} = K$, i.e., every coalition in $\mathfrak{K}$ is a player from K (this is not to be confused with the case $\mathfrak{K} = \{K\}$, when the family $\mathfrak{K}$ consists of the single coalition K). Then, in accordance with what we said in 2.7 about our system of notation, we shall have

$$\mathcal{C}_{\{K\}}(\Gamma) = \mathcal{R}_K(\Gamma) = \bigcap_{i \in K} \mathcal{C}_i(\Gamma).$$

In this case the $\mathfrak{K}$-principles of optimality of stability type coincide with the optimality principles of equilibrium type.

Definition. A situation from $\mathcal{C}_I(\Gamma) = \mathcal{C}(\Gamma)$ is said to be an *equilibrium situation in the sense of Nash (a Nash equilibrium situation)* in the game Γ.

The Nash equilibrium of a situation x^* means that, in the game Γ, for every player $i \in I$ and every strategy $x_i \in \mathfrak{x}_i$, it must be true that

$$H_i(x^* \| x_i) \leqq H_i(x^*).$$

A Nash equilibrium situation can be thought of informally as an agreement by the players which it does not pay any player to break.

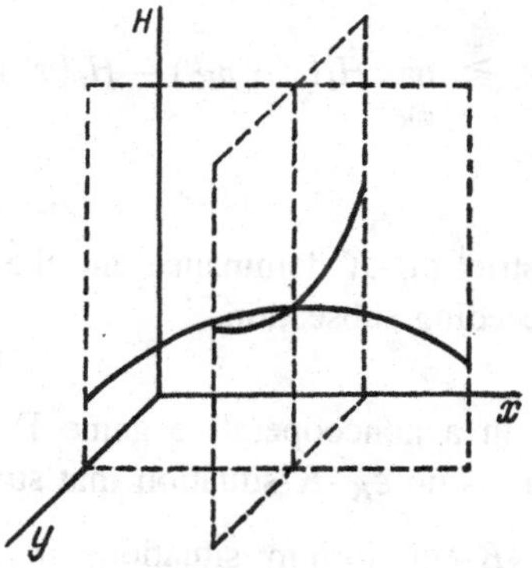

Fig. 1.4

2.11. Equilibrium in two-person zero-sum games. A situation (x^*, y^*) in a two-person zero-sum game $\Gamma = \langle \boldsymbol{x}, \boldsymbol{y}, H \rangle$ is an equilibrium situation if $H(x^*, y^*) \geqq H(x, y^*)$ for all $x \in \boldsymbol{x}$ and $-H(x^*, y^*) \geqq -H(x^*, y)$ for all $y \in \boldsymbol{y}$. These two inequalities can be rewritten in the form

$$H(x, y^*) \leqq H(x^*, y^*) \leqq H(x^*, y)$$

for all $x \in \boldsymbol{x}$ and $y \in \boldsymbol{y}$.

This description of equilibrium situations in two-person zero-sum games provides the basis for calling them *saddle points* of the payoff function: every deviation from (x^*, y^*) in the x-coordinate direction leads only to a decrease in the value of H, and, in the y-direction, only to an increase (Fig. 1.4).

2.12. ε_K-K-optimality. Finally, we shall be concerned with optimality principles based on variants of ε_K-K-domination.

Definition. A situation in a noncooperative game Γ as in (2.1) is said to be ε_K-K-*stable* if no situation ε_K-K-dominates it.

The set of all ε_K-K-stable situations in a game Γ is denoted by $\mathscr{C}_K^{\varepsilon_K}(\Gamma)$. □

Having $x^{\varepsilon_K} \in \mathscr{C}_K^{\varepsilon_K}(\Gamma)$ means that for every coalitional strategy x of the coalition K and for every player $i \in K$ we have

$$H_i(x^{\varepsilon_K} \| x_K) - \varepsilon_i \leqq H_i(x^{\varepsilon_K}), \tag{2.10}$$

or for some $i \in K$,

$$H_i(x^{\varepsilon_K} \| x_K) - \varepsilon_i < H_i(x^{\varepsilon_K}). \tag{2.11}$$

The ε_K-K-stability of a situation can be interpreted as the impossibility, in this situation, of the coalition's being able, by its own efforts, to increase the payoff to *at least one player* $i \in K$ by more than the corresponding ε_i, and the payoffs to each of the other players $j \in K$ by no less than the corresponding ε_j.

In essence, every situation for an arbitrary coalition K is ε_K-K-stable provided that the components ε_K of the K-vector are sufficiently large. That is to say, $x^* \in \mathscr{C}_K^{\varepsilon_K}(\Gamma)$ for $\varepsilon_K \geqq 0$ if each component of the K-vector satisfies the inequality

$$\varepsilon_i \geqq \max_{x_K} H_i(x^* \| x_K) - H_i(x^*).$$

2.13. The discussion of strict ε_K-K-dominance and the ε_K-K-equilibrium based on it is similar to the preceding subsection.

Definition. A situation in a noncooperative game Γ as in (2.1) is said to be ε_K-*K-equilibrium* if there is no ε_K-K-situation that strictly dominates it.

The set of all ε_K-K-equilibrium situations in the game is denoted by $\mathscr{R}_K^{\varepsilon_K}(\Gamma)$. □

The property that $x^{\varepsilon_K} \in \mathscr{R}_K^{\varepsilon_K}(\Gamma)$ means that, for every coalitional strategy, x_K of K there is a player $i \in K$ such that

$$H_i(x^{\varepsilon_K} \| x_K) - \varepsilon_K \leqq H_i(x^{\varepsilon_K}). \tag{2.12}$$

The ε_K-K-equilibrium property of a situation can be interpreted as the impossibility, in this situation, of the coalition's being able, by its own efforts, to increase the payoff to *each of the players* $i \in K$ by more than ε_i.

It is evident that for $0 \leqq \varepsilon_K \leqq \varepsilon'_K$ we must have

$$\begin{array}{ccccc} \mathscr{C}_K^{0^K}(\Gamma) & \subset & \mathscr{C}_K^{\varepsilon_K}(\Gamma) & \subset & \mathscr{C}_K^{\varepsilon'_K}(\Gamma) \\ \cap & & \cap & & \cap \\ \mathscr{R}_K^{0^K}(\Gamma) & \subset & \mathscr{R}_K^{\varepsilon_K}(\Gamma) & \subset & \mathscr{R}_K^{\varepsilon'_K}(\Gamma) \end{array} \tag{2.13}$$

If ε_K has at least one sufficiently large component, then every situation x^* in Γ has the ε_K-equilibrium property. That is to say, if there is, in the situation x^*, a player $i \in K$ for whom

$$\varepsilon_i \geqq \max_{x_K} H_i(x^* \| x_K) - H_i(x^*),$$

then $x^* \in \mathscr{R}_K^{\varepsilon_K}(\Gamma)$.

2.14. $\varepsilon_{\mathfrak{K}}$**-$\mathfrak{K}$-optimality.** Next, for a family $\mathfrak{K} \in 2^I$ of coalitions we denote by $\varepsilon_{\mathfrak{K}}$ the family of K-vectors $\varepsilon_K \geqq \mathbf{0}^K$ for $K \in \mathfrak{K}$. The transition from ε_K-optimality for individual coalitions K to $\varepsilon_{\mathfrak{K}}$-optimality for the family $\mathfrak{K}$ is carried out just like the transition from exact optimality of coalitions to the exact optimality of the family to which they belong. Consequently we restrict ourselves to formulations.

Definition. If $K \in 2^I$ and $\varepsilon_{\mathfrak{K}} \geqq \mathbf{0}^{\mathfrak{K}}$, the situation $x^{\varepsilon_{\mathfrak{K}}}$ is said to be $\varepsilon_{\mathfrak{K}}$-$\mathfrak{K}$-*optimal*, if it is ε_K-K-optimal for every $K \in \mathfrak{K}$. (Here we may think of optimality either in the sense of stability or in the equilibrium sense.)

The set of all $\varepsilon_{\mathfrak{K}}$-$\mathfrak{K}$-stable situations in Γ is denoted by $\mathscr{C}_{\mathfrak{K}}^{\varepsilon_{\mathfrak{K}}}(\Gamma)$, and the set of $\varepsilon_{\mathfrak{K}}$-$\mathfrak{K}$-equilibrium situations, by $\mathscr{R}_{\mathfrak{K}}^{\varepsilon_{\mathfrak{K}}}(\Gamma)$. □

Sets of the form $\mathscr{C}_{\mathfrak{K}}$ and $\mathscr{R}_{\mathfrak{K}}$ can be ordered by inclusion in a pattern similar to (2.13), which is, however, a natural corollary of (2.13).

As we noted in 2.1, the consideration of ε_K-K-optimality for arbitrary K-vectors $\varepsilon_K \geqq \mathbf{0}^K$ is of interest because the "ε-penalties" and "ε-rewards" (see 2.3) appearing in the components of the vector ε_K can be different for different players. Moreover, they can be incompatible with each other, being expressed in different "natural" units of utility. In practice, however, $\varepsilon_{\mathfrak{K}}$-$\mathfrak{K}$-optimal situations have up to the present been encountered in game theory only in contexts like "for every $\varepsilon_{\mathfrak{K}} \geqq \mathbf{0}^{\mathfrak{K}}$ there is an $\varepsilon_{\mathfrak{K}}$-$\mathfrak{K}$-optimal situation (of, let us say, some form)". In these cases, in view of the finiteness of the set of players in the game, and the possibility of determining, in this connection, some $\varepsilon \in (0, \varepsilon_{i,K})$ for every $i \in K \in \mathfrak{K}$, it is enough to consider, not arbitrary systems of K-vectors ε_K, but only those whose components are all the same. Consequently, from now on (except for some discussion of two-person games in 3.8 and § 1 of Chapter 3) we shall speak only of ε-optimal situations, taking ε to be a real number and supposing that all the components of each vector $\varepsilon_K \in \varepsilon_{\mathfrak{K}}$ are equal to ε.

2.15. In the case when $\mathfrak{K} = I$, the concepts of $\varepsilon_{\mathfrak{K}}$-$\mathfrak{K}$-stability and $\varepsilon_{\mathfrak{K}}$-$\mathfrak{K}$-equilibrium coincide, just as for the exact versions of optimality principles. Situations in $\mathscr{C}_{\mathfrak{K}}^{\varepsilon_{\mathfrak{K}}}(\Gamma) = \mathscr{R}_{\mathfrak{K}}^{\varepsilon_{\mathfrak{K}}}(\Gamma)$ are then said to be I-equilibrium situations (in the sense of Nash). Since we shall not consider any other versions of I-equilibrium situations, we shall sometimes not mention Nash when referring to the I-equilibrium property.

If $\mathfrak{K} = I$, the situation x^ε with $\varepsilon = (\varepsilon_1, \dots, \varepsilon_n)$ will be an ε-equilibrium situation if

$$H_i(x^\varepsilon \| x_i) \leqq H_i(x^\varepsilon) + \varepsilon_i, \qquad i = 1, \dots, n,$$

and if we are considering a two-person zero-sum game $\Gamma = \langle \boldsymbol{x}, \boldsymbol{y}, H \rangle$ and $\varepsilon = (\varepsilon_1, \varepsilon_2)$, the ε-equilibrium situation $(x_\varepsilon, y_\varepsilon)$ will be an ε-saddle point of a payoff function H for which

$$H(x, y_\varepsilon) - \varepsilon_1 \leqq H(x_\varepsilon, y_\varepsilon) \leqq H(x_\varepsilon, y) + \varepsilon_2$$

for all $x \in \boldsymbol{x}$ and $y \in \boldsymbol{y}$.

A necessary and sufficient condition for $\varepsilon = (\varepsilon_1, \varepsilon_2)$ to be a saddle point for the situation $(x_\varepsilon, y_\varepsilon)$ is that

$$\varepsilon_1 \geqq \max_{x \in \boldsymbol{x}} H(x, y_\varepsilon) - H(x_\varepsilon, y_\varepsilon),$$

$$\varepsilon_2 \geqq H(x_\varepsilon, y_\varepsilon) - \min_{y \in \boldsymbol{y}} H(x_\varepsilon, y).$$

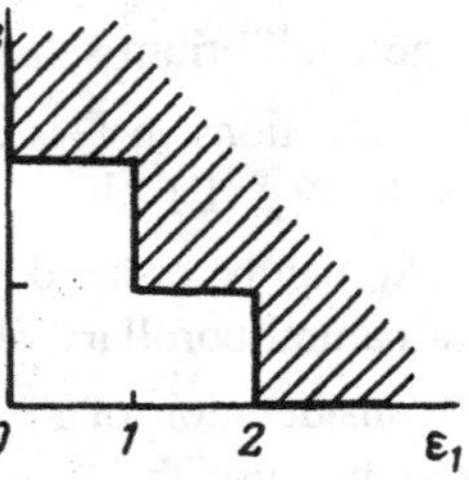

Fig. 1.5

The set of all $\varepsilon = (\varepsilon_1, \varepsilon_2)$ for which $\mathscr{C}^{\varepsilon}(\Gamma) \neq \emptyset$ has the property of "upper exhaustion" (i.e., when $\varepsilon' \geqq \varepsilon$ the condition $\mathscr{C}^{\varepsilon}(\Gamma) \neq \emptyset$ implies $\mathscr{C}^{\varepsilon'}(\Gamma) \neq \emptyset$), but otherwise has rather arbitrary form. For example, for the matrix game Γ_A for which

$$A = \begin{bmatrix} 1 & 0 & 2 \\ 2 & 2 & 0 \end{bmatrix},$$

the set of such vectors ε is sketched in Fig. 1.5.

2.16. Explicit examples of games and their solutions. We present some examples that show that even in the simplest matrix and bimatrix games there arise various possibilities in the realization of optimality principles and differences in the intuitive understanding (or misunderstanding) of these realizations.

Example 1. Content. Each of players 1 and 2 chooses one of the numbers 0 or 1, after which player 1 receives the sum of the chosen numbers from player 2.

Model. This game is modeled by the 2×2 matrix game Γ_A with the payoff matrix

$$\begin{array}{cc} & \begin{array}{cc} 0 & 1 \end{array} \\ A = & \begin{bmatrix} 0 & 1 \\ \boxed{1} & 2 \end{bmatrix} \begin{array}{c} 0 \\ 1 \end{array} \end{array}$$

(here and later, the borders of the matrix merely indicate the informal features of the strategies and do not refer to the elements of the matrix, which are the payoffs).

Analysis. In this game there is just one equilibrium situation (saddle point): (1,0). Since this is a two-person zero-sum game, all of its situations are Pareto-optimal.

Example 2. Content. Having drank that glass, Mr. Pickwick took another, just to see whether there was any orange peel in the punch, because orange peel always disagreed with him.*)

---

*) Charles Dickens, The Posthumous Papers of the Pickwick Club, Oxford University Press, London, 1948, p. 256 (first ed., 1837).

Model. As a model for this conflict we may take the matrix game of Nature (player 1) against Pickwick (player 2), with the following payoff matrix:

		Nature chooses: There is peel	Nature chooses: There is no peel
Pickwick chooses:	to drink punch	−2	1
	not to drink	[0]	0

Analysis. Here we have an equilibrium situation. It consists of the presence of peel in the punch for Nature and abstention for Pickwick.

Comment. In the course of the story, both Pickwick and Nature deviate from the equilibrium situation and arrive at the situation (drink the punch, no peel) that is even more appropriate for a hero. Here we have a game-theoretic model of some features of a comic.

Example 3. "Prisoner's dilemma". Content. Two gangsters (players 1 and 2), suspected of the commission of a serious crime, are isolated from each other in preliminary confinement. In the absence of direct evidence, the success or failure of the charge depends on the gangsters' confession (c) or denial (d). If both gangsters confess (situation (c,c)), they will be found guilty with extenuating circumstances and sentenced to (say) 8 years' imprisonment. If neither one confesses (situation (d,d)), they will be cleared of the principal charge, but nevertheless, in view of their established complicity in less serious offences, they will be sentenced to imprisonment for 1 year. Finally, if only one confesses (situation (d,c) or (c,d)), the one who confesses goes free (this is the law under the conditions of the problem) and the one who does not confess has to serve the maximum term, 10 years.

Model. This model is a 2×2 bimatrix game, where

$$A = \begin{array}{c} \begin{matrix} \mathrm{c} & \mathrm{d} \end{matrix} \\ \begin{bmatrix} -8 & 0 \\ -10 & -1 \end{bmatrix} \end{array} \begin{matrix} \mathrm{c} \\ \mathrm{d} \end{matrix} \qquad B = \begin{array}{c} \begin{matrix} \mathrm{c} & \mathrm{d} \end{matrix} \\ \begin{bmatrix} -8 & -10 \\ 0 & -1 \end{bmatrix} \end{array} \begin{matrix} \mathrm{c} \\ \mathrm{d} \end{matrix}$$

Analysis. This game has, as is easily seen, the equilibrium point (c,c); the other three situations are $\{1,2\}$-optimal (i.e., Pareto optimal). It is clear that the equilibrium situation (c,c), although acceptable for both players, is in this case quite unprofitable for them; the situation (d,c) is unacceptable for player 1 (is not 1-optimal); the situation (c,d) is unacceptable, in the same way, for player 2; and the situation (d,d) is unacceptable for both players, i.e. very unstable. Here we have the problem of a combination of Pareto optimality with equilibrium (i.e., to put it informally, of profitability and stability).

Example 4. The "odd-even" game. Content. Player 1 conceals either an even (e) or an odd (o) number of small objects, and player 2 tries to guess their parity. In case of success (situation (e,e) or (o,o)) player 2 receives 1 unit from player 1; in case of failure (situation (e,o) or (o,e)) player 2 pays player 1 one unit.

Model. The model is the 2×2 matrix game Γ_A, where

$$A = \begin{array}{c} \begin{array}{cc} \mathrm{e} & o \end{array} \\ \begin{bmatrix} -1 & 1 \\ 1 & -1 \end{bmatrix} \end{array} \begin{array}{c} \mathrm{e} \\ o \end{array}$$

Analysis. As for every two-person zero-sum game, here every situation is $\{1,2\}$-optimal (Pareto optimal). In this game there are, as is easily verified, no equilibrium situations.

Example 5. "Battle of the sexes." Content. Husband and wife (players 1 and 2) agree that in the evening they will go to the ballet (b) or the football game (f). Going together to the ballet (situation (b,b)) gives the husband moderate enjoyment (estimated as 1), but more to the wife (estimated as 2). Going together to the game (situation (f,f)) provides enjoyment with the opposite estimates (husband 2, wife 1). Finally, if there is no agreement (situation (b,f) or (f,b)), there is no enjoyment to either (payoffs zero).

Model. The model is the 2×2 bimatrix game $\Gamma(A, B)$ with

$$A = \begin{array}{c} \begin{array}{cc} \mathrm{b} & \mathrm{f} \end{array} \\ \begin{bmatrix} 1 & 0 \\ 0 & 2 \end{bmatrix} \end{array} \begin{array}{c} \mathrm{b}, \\ \mathrm{f} \end{array} \qquad B = \begin{array}{c} \begin{array}{cc} \mathrm{b} & \mathrm{f} \end{array} \\ \begin{bmatrix} 2 & 0 \\ 0 & 1 \end{bmatrix} \end{array} \begin{array}{c} \mathrm{b} \\ \mathrm{f} \end{array}$$

Analysis. Here situations (b,b) and (f,f) are simultaneously equilibrium and Pareto-optimal situations. There remains the open question of an optimal choice of one of these two situations. A possible solution is to suppose that part of the satisfaction obtained by one player can be transferred to the other.

Example 6. "A game on the open square". Content. Players 1 and 2 choose, respectively, numbers x and y on the interval (0,1), after which player 1 receives the payoff $(x + y)$ from player 2. This is evidently a continuous generalization of the game in example 1.

Model. The model is the two-person zero-sum game $\Gamma = \langle x, y, H \rangle$ for which

$$x = y = (0, 1), \quad H(x, y) = x + y.$$

The game gets its name because the set of all situations in it is the Cartesian product $(0, 1) \times (0, 1)$, i.e., the interior of the unit square.

Analysis. Here, evidently, $\mathscr{C}(\Gamma) = \emptyset$, but, by choosing $x_\varepsilon = 1 - \varepsilon$ and $y_\varepsilon = \varepsilon$, the players can come arbitrarily close to a payoff function with unit value. For every $\varepsilon > 0$, the situation $(x_\varepsilon, y_\varepsilon)$ is an ε-equilibrium situation, i.e. one belonging to $\mathscr{C}_{\{1,2\}}(\Gamma)$.

2.17. Properties of optimality principles. Optimality principles for noncooperative games, among them the $\mathfrak{K}$-optimality and $\varepsilon_{\mathfrak{K}}$-$\mathfrak{K}$-optimality of situations, have a number of general properties of interaction with the structures of games, as well as properties of invariance with respect to the basic relations that have been introduced for noncooperative games: transition to a subgame (1.12), strategic (including affine) equivalence (1.13), homomorphisms, semihomomorphisms, and all their many varieties (1.14), as well as factorization (1.20).

These properties are, in the first place, characteristics of an optimality principle as such, and can be taken as axioms for its description (for two-person zero-sum games, this will be discussed in more detail in § 2, Chapter 3). It is a priori clear that if all the properties that we are going to discuss have different optimality principles, the set of their properties cannot form a complete axiom system for any one of these principles. We should also notice that some of these properties are also possessed by other optimality principles, including some that have nothing to do with noncooperative games.

In the second place, these properties turn out to be useful in finding realizations of optimality principles. Numerous examples will be found throughout this book (see in particular the example in 6.14).

Moreover, one can also discuss certain properties of realizations of optimality principles, such as their equilibrium with respect to structures on games.

2.18. We begin with the simplest of the optimality properties, one connected with the transition to a subgame, and known as *independence of irrelevant alternatives*. This consists in saying that removing nonoptimal situations from a game does not affect the optimality of what is left.

We shall denote by $\mathfrak{D}^\circ_{\mathfrak{K}}(\Gamma)$ any one of the sets $\mathscr{C}_{\mathfrak{K}}(\Gamma)$, $\mathscr{R}_{\mathfrak{K}}(\Gamma)$, $\mathscr{C}^{\varepsilon_{\mathfrak{K}}}_{\mathfrak{K}}(\Gamma)$, $\mathscr{R}^{\varepsilon_{\mathfrak{K}}}_{\mathfrak{K}}(\Gamma)$. If it happens that $\mathfrak{K}$ consists of a single coalition K, we shall use the notation $\mathfrak{D}^\circ_{\mathfrak{K}}(\Gamma)$. For the symbol $<^\circ$ see the end of 2.3. Then we have the following theorem.

Theorem. *If* Γ' *is an* x'*-subgame of the game* Γ *as in* (2.1), *then*

$$\mathfrak{D}^\circ\mathfrak{K}(\Gamma)\cap x' \subset \mathfrak{D}^*_{\mathfrak{K}}(\Gamma'). \tag{2.14}$$

Proof. Select $K \in \mathfrak{K}$; if $\mathfrak{D}^\circ_{\mathfrak{K}}(\Gamma')\cap x' = \emptyset$ the conclusion of the theorem is trivial. Let $x^* \in \mathfrak{D}^\circ_{\mathfrak{K}}(\Gamma)\cap x'$. This means that, among the coalitional strategies in x_K there is no strategy x_K such that

$$H_K(x^*_K) <^\circ H_K(x^* \| x_K)$$

(where $<^\circ$ stands for the corresponding symbol $<$, $\leq$, $\overset{\varepsilon_K}{\leq}$, or $\overset{\varepsilon_K}{<}$).

Moreover, there is no such strategy in x', i.e. $x^* \in \mathfrak{D}^\circ_K(\Gamma)$. □

The converse of (2.14) does not necessarily hold. For example, for the matrix game Γ with payoff matrix

$$A = \begin{array}{c} \begin{matrix} y_1 & y_2 & y_3 \end{matrix} \\ \left[\begin{matrix} \left[\begin{matrix} 1 & 1 \\ 1 & 1 \end{matrix}\right] & \begin{matrix} 2 \\ 0 \end{matrix} \\ \begin{matrix} 0 & 2 \end{matrix} & 1 \end{matrix}\right] \end{array} \begin{matrix} x_1 \\ x_2 \\ x_3 \end{matrix}.$$

and its subgame Γ' with the indicated payoff submatrix, we will have

$$\begin{aligned} \mathscr{C}(\Gamma) \cap (x' \times y') &= (x_1, y_1), \\ \mathscr{C}(\Gamma') &= x' \times y'. \end{aligned}$$

2.19. A strategic equivalence transformation, and in particular an affine equivalence transformation (see 1.13), preserves the optimality of situations.

Theorem. *If* Γ *and* Γ' *are noncooperative games as in* (1.7), $\Gamma \sim \Gamma'$, *and* $\mathfrak{K} \subset 2^I$, *then* $\mathfrak{D}^\circ_{\mathfrak{K}}(\Gamma) = \mathfrak{D}^\circ_{\mathfrak{K}}(\Gamma')$.

The proof is evident. □

2.20. The $\mathfrak{K}$-optimality principles of $\mathfrak{K}$-stability, the $\mathfrak{K}$-equilibrium property, and also their ε-versions, are covariant (see 1.19) in the following sense.

Theorem. *For every noncooperative game* Γ *and each of its coalitions* K, *the set* $\mathfrak{D}^\circ(\Gamma)$ *of its situations is covariant, i.e. for every automorphism* π *of* Γ *we have*

$$\pi\mathfrak{D}^\circ_K(\Gamma) = \mathfrak{D}^\circ_{\pi K}(\Gamma).$$

Proof. Let $x^* \in \mathfrak{D}^{\varepsilon_K}_K(\Gamma)$, but let there be an $\tilde{x}_{\pi K}$ for which

$$H_{\pi K}(x' * \|\tilde{x}_{\pi K}) \overset{\varepsilon_K}{>} H_{\pi K}(\pi x^*).$$

It follows from the definition of an automorphism that we must have

$$H_K(x^* \|\pi^{-1}(\tilde{x}_{\pi K})) \overset{\varepsilon_K}{>} H_K(x^*).$$

Since $\pi^{-1}(\tilde{x}_{\pi K})) \in x_K$, this contradicts the ε-K-optimality of x^*. □

Corollary. *For every family* $\mathfrak{K}$ *of coalitions the set of situations is covariant.*

In fact,

$$\pi\mathfrak{D}^0_{\mathfrak{K}}(\Gamma) = \pi \bigcap_{K \in \mathfrak{K}} \mathfrak{D}^0_K(\Gamma) = \bigcap_{K \in \mathfrak{K}} \pi\mathfrak{D}^\circ_K(\Gamma) = \bigcap_{\pi K \in \pi\mathfrak{K}} \mathfrak{D}^\circ_{\pi K}(\Gamma) = \mathfrak{D}^\circ_{\pi K}(\Gamma).$$

It does not, however, follow from the theorem on covariance that there must exist any symmetric situations in the set of $\mathfrak{K}$-stable (and equally not in sets of ε-$\mathfrak{K}$-stability). For example, in the "Battle of the sexes" game (see 2.16) there is a single automorphism π other than the identity; for this situations

$$\pi 1 = 2, \quad \pi 2 = 1,$$

$$\pi \mathrm{f} = \mathrm{b}, \quad \pi \mathrm{b} = \mathrm{f}.$$

The set $\{(\mathrm{f},\mathrm{f}),(\mathrm{b},\mathrm{b})\}$ of equilibrium situations in this game is symmetric, but there is no symmetric equilibrium situation.

This proposition will not contradict the existence, to be presented in 7.4, of a symmetric equilibrium situation in an extension of this game.

2.21. We may also consider "approximate covariance" of $\mathfrak{K}$-optimality principles for ε_I-homomorphisms of games (see 1.14).

Theorem. *If* π *is an* ε_I*-homomorphism from the game* Γ *to the game* Γ' *in* (1.8), $x^* \in x$ *and* $x^{*'} \in \pi x^*$, *then*

$$x^* \in \mathscr{C}_K^{\varepsilon'_K}(\Gamma) \quad \textit{implies} \quad x^{*'} \in \mathscr{C}_{\pi K}^{(\varepsilon'_{\pi K}+2\varepsilon_{\pi K})}(\Gamma'),$$

$$x^* \in \mathfrak{R}_K^{\varepsilon'_K}(\Gamma) \quad \textit{implies} \quad x^{*'} \in \mathfrak{R}_{\pi K}^{(\varepsilon'_{\pi K}+2\varepsilon_{\pi K})}(\Gamma').$$

Proof. Choose $x'_{\pi K} \in x'_{\pi K}$ and $x_K \in \pi^{-1} x'_{\pi K}$ arbitrarily. It follows from the inequality

$$H_i(x^* \| x_K) - \varepsilon'_i \leqq H_i(x^*)$$

by the definition of an ε_I-homomorphism (formula (1.12)) that

$$H'_{\pi i}(x^{*'} \| x_{\pi K}) - (\varepsilon'_{\pi i} + 2\varepsilon_{\pi i}) \leqq H'_{\pi i}(x^{*'})$$

and similarly from

$$H_i(x^* \| x_K) - \varepsilon'_i < H_i(x^*)$$

that

$$H'_{\pi i}(x^{*'} \| x'_{\pi K}) - (\varepsilon'_{\pi i} + 2\varepsilon_{\pi i}) < H'_{\pi i}(x^{*'}).$$

It remains only to recall that K-stability is characterized by inequalities (2.10) and (2.11); and K-equilibrium, by (2.12). □

This theorem does not lend itself to a direct extension in terms of semihomomorphisms. However, an analog is valid; this will be discussed in 2.24.

2.22. The operation of passing from a noncooperative game to the set of its K-equilibrium situations has the closure property.

Theorem. *Let a noncooperative game* Γ *as in* (2.1) *be topological, and let the payoff functions be continuous on the set of situations. Then for every* $K \subset I$ *the operation* $\Gamma \to \mathfrak{R}_K(\Gamma)$ *is closed, in the following sense.*

Consider a sequence of games $\Gamma^{(1)}, \Gamma^{(2)}, \ldots$, *with the same number of players as in* Γ *and a sequence of situations*

$$x^{(k)} \in \mathfrak{R}_{K^{(k)}}^{\varepsilon_K^{(k)}}(\Gamma^{(k)}), \quad k = 1, 2, \ldots, \tag{2.15}$$

and let

a)
$$\lim_{k\to\infty} \rho_{\mathcal{K}}(\Gamma^{(k)}, \Gamma) = 0 \tag{2.16}$$

(this condition means that there are $\varepsilon^{(k)}$-homomorphisms $\pi^{(k)}$ of the games $\Gamma^{(k)}$ on Γ for which $\bar{\varepsilon}^{(k)} \to 0$).

b) *For a sequence $\bar{x}^{(k)} \in \pi^{(k)} x^{(k)}$ $(k = 1, 2, \ldots)$*

$$\lim_{k\to\infty} \bar{x}^{(k)} = x^{(0)} \in x; \tag{2.17}$$

c) *The coalition $K^{(0)}$ appears infinitely often among the coalitions $\pi^{(1)} K^{(1)}$, $\pi^{(2)} K^{(2)}$, ... (without loss of generality, we may suppose that $\pi(k) K^{(k)} = K^{(0)}$ for $k = 1, 2, \ldots$).*

d)
$$\lim_{k\to\infty} \varepsilon^{(k)}_{\pi^{(k)} K^{(k)}} = \varepsilon^{(0)}_{K^{(0)}}. \tag{2.18}$$

Then

$$x^{(0)} \in \mathfrak{R}^{\varepsilon^{(0)}_{K^{(0)}}}_{K^{(0)}}(\Gamma).$$

Proof. By the preceding theorem, it follows from (2.17) and (2.16) that

$$\bar{x}^{(k)} \in \mathfrak{R}^{\varepsilon^*}_{\pi^{(k)} K^{(k)}}(\Gamma),$$

where

$$\varepsilon^* = \varepsilon^{(k)}_{\pi^{(k)} K^{(k)}} + 2\varepsilon^{(k)}_{\pi^{(k)} K^{(k)}}.$$

In addition, in view of (2.16) and (2.18), it must be true that

$$\lim_{k\to\infty} (\varepsilon^{(k)}_{\pi^{(k)} K^{(k)}} + 2\varepsilon^{(k)}_{\pi^{(k)} K^{(k)}}) = \varepsilon^{(0)}_{K^{(0)}},$$

and it remains only to refer to the theorem in 2.18. □

2.23. The analogous proposition for the operator $\mathscr{C}_K$ turns out to be false (essentially, because of the evident failure of preserving strict inequalities in passing to the limit).

The following simple example illustrates this phenomenon.

Example. Consider a variable two-player game $\Gamma^{(\alpha)}$ $(\alpha \geq 0)$

$$\Gamma^{(\alpha)} = \langle\{1, 2\}, (x, y), \{H_1^{(\alpha)}, H_2^{(\alpha)}\}\rangle,$$

in which

$$x = y = [0, 1],$$

$$H_1^{(\alpha)}(x, y) = (1 + ax)(1 - y),$$

$$H_2^{(\alpha)}(x, y) = (1 + ay)(1 - x).$$

The set $\mathfrak{W}_I(\Gamma^{(\alpha)})$ is sketched in Fig. 1.6 for $\alpha > 0$. In accordance with the theorem in 2.5, the set of vector payoffs to the players in the situations in $\mathscr{C}_I(\Gamma^{(\alpha)})$ now forms the northeast boundary of $\mathfrak{W}_I(\Gamma^{(\alpha)})$. A particular vector payoff is attained in a situation of the form $(0, y_0)$, where $y_0 < 1$.

Passing to the limit as $\alpha \to 0$, we obtain the set $\mathfrak{W}_I(\Gamma^{(0)}$, whose northeast boundary is the dashed line in Fig. 1.6. Here the limit of the sequence of points $\bar{H}_I^{(\alpha)}$ is the point $\bar{H}_I^{(0)}$, whose coordinates do not give the vector payoff for any situation in $\mathscr{C}_I(\Gamma^{(0)})$, by the same theorem in 2.5 (and specifically because decreasing y_0 while leaving $x = 0$ leads to an increase in $H_1^{(0)}$ with constant $H_2^{(0)}$).

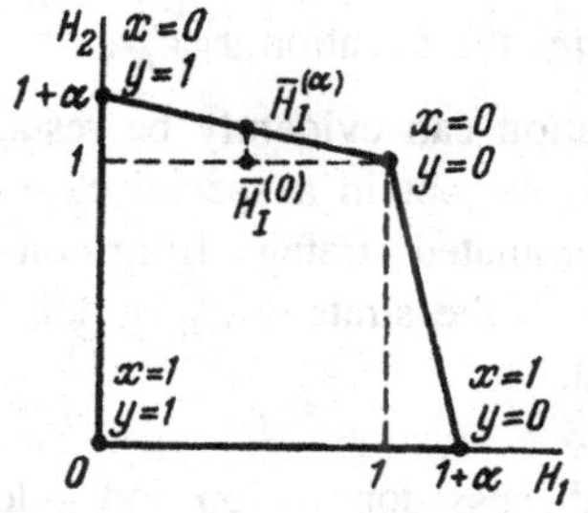

Fig. 1.6

2.24. ε-$\mathfrak{R}$-optimality and semihomomorphisms. Under the conditions of semihomomorphisms, the payoffs to the players can increase unboundedly. Consequently the connection of semihomomorphisms with optimality can appear only in cases when the payoffs to the players in game-theoretic situations are bounded above for some other reason. For example, this happens when the semihomomorphism is a semi-endomorphism. In this case there is a theorem, which is in some sense a converse of the theorem in 2.18, connected with independence of irrelevant alternatives.

Theorem. *If π is an ε_I-semi-endomorphism of the game Γ on a subgame Γ', $x^* \in x, x^{*'} \in \pi x^*$ and $K \subset I$, then*

$$x^{*'} \in \mathscr{C}_{\pi K}^{\varepsilon'_{\pi K}}(\Gamma) \quad \textit{implies} \quad x^* \in \mathscr{C}_K^{(\varepsilon'_K + \varepsilon_K)}(\Gamma);$$

$$x^{*'} \in \mathscr{R}_{\pi K}^{\varepsilon'_{\pi K}}(\Gamma) \quad \textit{implies} \quad x^* \in \mathscr{R}^{(\varepsilon'_K + \varepsilon_K)}(\Gamma).$$

Proof. We suppose for simplicity that the mapping π is the identity on I. Choose arbitrarily $x_K \in x_K$ and $x'_K \in \pi x_K$. From the inequality

$$H_i(x^* \| x'_K) - \varepsilon'_i \leqq H_i(x^{*'}) \tag{2.19}$$

it follows by the definition of an ε_I-semihomomorphism (formula (1.11)) that

$$H_i(x^* \| x_K) - (\varepsilon'_i + \varepsilon_i) \leqq H_i(x^{*'}).$$

A similar discussion applies when there is strict inequality in (2.19).

It remains only to recall inequalities (2.10)–(2.12). □

2.25. Domination of strategies. Finally we introduce the concept of the domination of coalitional strategies in noncooperative games, a concept which must be distinguished from the concept of the domination of situations, discussed in 2.3.

Definition. In a noncooperative game Γ as in (2.1), let $K \subset I$, and $x'_K, x''_K \in \boldsymbol{x}_K$. We say that the strategy x'_K $^\circ$*-dominates* the strategy x''_K if, for every situation x, the situation $x \| x'_K$ $^\circ$*-dominates* the situation $x \| x''_K$.

The concept of domination can evidently be restated in terms of semihomomorphisms. If we do this, we obtain a special case of the theorem of 2.24, which lets us exclude the dominated strategy from consideration. In fact, if the strategy $x'_K \in \boldsymbol{x}_K$ $^\circ$-dominates the strategy $x''_K \in \boldsymbol{x}_K$, a situation of the form $x \| x''_K$ cannot be K-$^\circ$-optimal.

2.26. In particular, we can discuss domination and ε-domination for individual strategies of each player (here strict domination coincides with domination). In this case it is useful to introduce the concept of weak domination.

Definition. In a noncooperative game Γ as in (2.1), let $i \in I$ and $x'_i, x''_i \in \boldsymbol{x}_i$. We say that x'_i *weakly dominates* x''_i if, for every situation,

$$H_i(x \| x''_i) \leqq H_i(x \| x'_i). \qquad \square \tag{2.20}$$

If x'_i weakly dominates x''_i and $x^* \| x''_i \in \mathscr{C}_i(\Gamma)$, then $x^* \| x'_i \in \mathscr{C}_i(\Gamma)$. In fact, by hypothesis we have

$$H_i(x^* \| y_i) \leqq H_i(x^* \| x''_i) \text{ for every } y_i \in \boldsymbol{x}_i.$$

Together with (2.20) this yields

$$H_i(x^* \| y_i) \leqq H_i(x^* \| x'_i) \text{ for every } y_i \in \boldsymbol{x}_i,$$

i.e. $x^* \| x_i \in \mathscr{C}_i(\Gamma)$.

2.27. The following useful proposition is related to the factorization of noncooperative games in the sense of 1.20.

Theorem. *In a noncooperative game as in* (2.1), *let* $K \subset I$, $L \cap K = \emptyset$ *and* $x^\circ_L \in \boldsymbol{x}_L$. *Then a necessary and sufficient condition for the situation* $x^* \underset{L}{\|} x^\circ_L$ *to be* K-$^\circ$*-optimal in the game* Γ *is that in the factorization* Γ / x°_L *of* Γ *the situation* $x^*_{I \setminus L}$ *is* K-$^\circ$*-optimal.*

This proposition can be immediately extended to $\mathfrak{K}$-$^\circ$-optimality of situations in the games Γ and Γ / x°_L for every family $\mathfrak{K}$ of coalitions for which $(\cup \mathfrak{K}) \cap L = \emptyset$.

Hence it follows that when we consider realizations of $\mathfrak{K}$-$^\circ$-optimality principles (whether exact, or "ε-principles") it is enough to restrict our attention to the cases in which $\cup \mathfrak{K}$ is the set I of all players.

For topological games (see 1.9) with continuous payoff functions, the set $\mathscr{R}$ of ε_K-K-equilibrium situations is closed.

Theorem. *If* Γ *in* (2.1) *is a noncooperative topological game with continuous payoff functions;* $\varepsilon_K^{(1)}, \varepsilon_K^{(2)}, \ldots$ *a sequence of nonnegative vectors with a limit* $\varepsilon_K^{(0)}$; *and*

$$x^{(K)} \in \mathfrak{R}_K^{\varepsilon_K^{(k)}}(\Gamma), \quad k = 1, 2, \ldots \tag{2.21}$$

$$\lim_{k\to\infty} x^{(k)} = x^{(0)}, \tag{2.22}$$

we have

$$x^{(0)} \in \mathfrak{R}_K^{\varepsilon_K^{(0)}}(\Gamma). \tag{2.23}$$

Proof. Condition (2.21) means that for every strategy $x_K \in x_K$, and for some player $i \in K$, we will have

$$H_i(x^{(k)} \| x_K) - \varepsilon_i^{(k)} \leqq H_i(x^{(k)}). \tag{2.24}$$

If we take the limit on k, we see that because of the assumed continuity of the payoff functions we must have

$$\lim_{k\to\infty} H_i(x^{(k)} \| x_K) = H_I(x^{(0)} \| x_K)$$

(including the case when $x_K = x_K^{(k)}$).

Let us suppose that

$$H_I(x^{(0)} \| x_K) - \varepsilon_i^{(0)} > H_i(x_i^{(0)}) \quad \text{for all } i \in K. \tag{2.25}$$

Then, in view of (2.22) and the finiteness of the set of players, this means that from some k onward we will have

$$H_i(x^{(k)} \| x_K) - \varepsilon_i^{(k)} > H_i(x_i^{(k)}) \quad \text{for all } i \in K,$$

which however, contradicts (2.21). Therefore (2.22) is impossible. This contradiction shows that (2.23) is valid. □

Corollary. *From the theorem just proved, it follows immediately that sets of the form* $\mathfrak{R}_{\mathfrak{K}}^{\varepsilon_{\mathfrak{K}}}(\Gamma)$ *are closed, among them the set of Nash equilibrium situations.*

Remark. For sets $\mathscr{C}_K^{\varepsilon_K}$ of ε_K-K-stable situations the preceding theorem is, generally speaking, false. Thus, for example, the set $\mathscr{C}_I^0(\Gamma)$ sketched in Fig. 1.2 is not closed: the limit point $*$ of this set does not belong to the set.

§3 Realizability of $\mathfrak{K}$-optimality principles

3.1. K-stability. As we see from the examples in 2.16, $\mathfrak{K}$-optimality principles may turn out to be unrealizable even for noncooperative games of a very simple structure.

It is clear that the realizability (or, on the contrary, nonrealizability) of $\mathfrak{K}$-optimality principles as either models of stability or as models of equilibrium for some class $\mathfrak{G}$ of games depends, on one hand, on the properties of the principle itself (i.e., ultimately on the structure of the family $\mathfrak{K}$ of coalitions); and on the other hand, on the properties of the games that make up the class $\mathfrak{G}$. For all that, it is natural that for a wide class of families $\mathfrak{K}$ there must be more restrictions on the class of games $\mathfrak{G}$.

In the simplest cases, the family $\mathfrak{K}$ consists of the single coalition K, and $\varepsilon_{\mathfrak{K}}$-$\mathfrak{K}$-stability for every $\varepsilon_{\mathfrak{K}} \geqq 0$ becomes ε_K-K-stability.

3.2. Theorem. *Let*

$$\Gamma = \langle I, \{x_i\}_{i \in I}, \{H_i\}_{i \in I} \rangle \tag{3.1}$$

be a noncooperative game, $K \subset I$, and let x be any situation in Γ. Then for every $\varepsilon > 0^K$ there is a coalitional strategy $x_K^\varepsilon \in \boldsymbol{x}_K$ such that the situation $x \| x_K^\varepsilon$ is ε-K-stable in Γ and K-dominates x.

Proof. Let $\mathcal{H}_K^0$ denote the set of vector payoffs in Γ for all situations that K-dominate x^0. The closure $\bar{\mathcal{H}}_K^0$ of this set is compact. Let us enumerate the players in K as $i_1, \ldots, i_k$, and successively maximize the components $\bar{H}_K \in \bar{\mathcal{H}}_K^0$ with respect to the corresponding variables $x_{i_1}, \ldots, x_{i_k}$. Let the final maximum be attained for a vector $H_K^0 \in \bar{\mathcal{H}}_K^0$. Evidently

$$H_K(x^0 \| x_K) \leqq H_K^0 \text{ for all } x_K \in \boldsymbol{x}_K.$$

However, among the situations $x^0 \| x_K$ there are those in which the payoffs to the players in K are arbitrarily close to H_K^0. The strategy $x_K \in \boldsymbol{x}_K$ for which

$$H_K(x^0 \| x_K^\varepsilon) \geqq H_K^0 - \varepsilon,$$

is evidently the one required. □

Corollary. *If the game Γ in* (3.2) *has the property that $H_K^0 \in \mathcal{H}_K^0$ (in particular, if Γ is finite), then for every situation x there is a coalitional strategy x_K^0 of the coalition K such that the situation $x \| x_K^0$ is K-stable in Γ (and K-dominates x).*

3.3. Quasipartition. We are going to extend the preceding theorem to a somewhat wider class of families $\mathfrak{K}$ of coalitions. In this connection, here and later, in conformity with what was said in 2.27, we confine ourselves to the case in which $\cup \mathfrak{K} = I$.

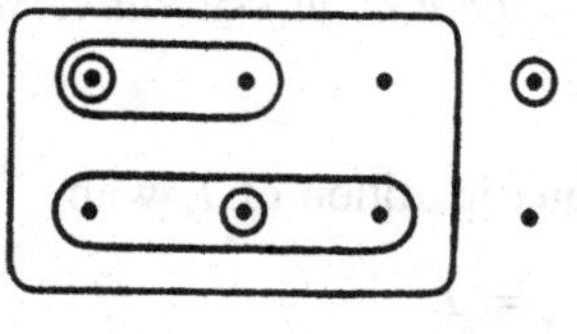
Fig. 1.7

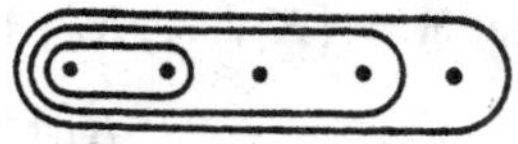
Fig. 1.8

Definition. A family $\mathfrak{K}$ of coalitions is a *quasipartition* of the set I of players if $K, L \in \mathfrak{K}$ and $K \cap L = \emptyset$ imply that either $K \subset L$ or $L \subset K$. □

An example of a quasipartition of the set I is sketched in Figure 1.7. In particular, every partition of a set is its quasipartition.

The theory of $\mathfrak{K}$-optimality principles of noncooperative games is quite meaningful in the case when the family $\mathfrak{K}$ of coalitions is a quasipartition (see § 4).

It is evident that if $\mathfrak{K}$ is a quasipartition of I and $K \subset I$, and also for every $L \notin \mathfrak{K}$ it follows from $K \cap L = \emptyset$ that either $K \subset L$ or $L \subset K$, then $\{L\} \cup \mathfrak{K}$ is also a quasipartition of I.

A coalition K in the quasipartition $\mathfrak{K}$ is said to be *maximal* if there is no coalition in $\mathfrak{K}$ which is different from K and contains K.

A sequence

$$\emptyset = \mathfrak{K}_0 \subset \mathfrak{K}_1 \subset \ldots \subset \mathfrak{K}_r = \mathfrak{K}$$

of quasipartitions is said to be *normal* if every difference $\mathfrak{K}_{k+1} \setminus \mathfrak{K}_k$ $(k \geq 0)$ consists of a single coalition K_k which is maximal in $\mathfrak{K}_{k+1}$. Evidently every quasipartition is the end of at least one normal sequence of quasipartitions. In order to construct it, it is enough to take the number r of coalitions in $\mathfrak{K}$, find in $\mathfrak{K} = \mathfrak{K}_r$ the maximal coalition $K = K_r$, set $\mathfrak{K} \setminus \{K_r\} = \mathfrak{K}_{r-1}$, and iterate this process.

Since the number r depends only on the quasipartition $\mathfrak{K}$ and not on the choice of the normal sequence of quasipartitions, we shall call it the *length* of $\mathfrak{K}$.

The existence, for each quasipartition, of normal sequences makes it convenient to carry out inductions on the length.

3.4. For the present, we shall be interested in quasipartition of a special form.

Definition. A quasidecomposition $\mathfrak{K}$ of I is said to be *ordered* if $K, L \in \mathfrak{K}$ implies $K \cap L \neq \emptyset$.

It is clear that the coalitions in a quasipartition can be arranged in just one way in increasing order (see Fig. 1.8). This means that for all ordered quasipartitions, and only for these, there is a unique normal sequence.

Theorem. *Let Γ be a noncooperative game, and let $\mathfrak{K}$ be an ordered quasipartition of the set of players in Γ. Then for every $\varepsilon > 0$ there is an ε-$\mathfrak{K}$-stable situation in Γ.*

Proof. Let $\mathfrak{K} = \{K_1, \ldots, K_k\}$ be an ordered quasipartition of I, with

$$K_1 \subset K_2 \subset \ldots \subset K_k = I.$$

We find, in accordance with the theorem of 3.2, a vector $H_1^\varepsilon \in \mathfrak{W}(\Gamma)$ which is not ε-K_1-dominated by any other vectors in $\mathfrak{W}(\Gamma)$. We denote the realization of this situation by $x_{(1)}^\varepsilon$. This situation is ε-K_1-optimal. This is the first step in our inductive construction.

Now suppose that we have constructed situations $x_{(t)}^\varepsilon$ which are ε-K_t-optimal for $t = 1, \ldots, l$. By the theorem in 3.2 there is a situation $x_{(l+1)}^\varepsilon$ which K_{l+1}-dominates $x_{(l)}^\varepsilon$ and is ε-K_{l+1}-optimal. Since it K_l-dominates the ε-K_t-optimal situation $x_{(l)}^\varepsilon$, it must itself be ε-K_t-optimal for $t = 1, \ldots, l$. The situation $x_{(l+1)}^\varepsilon$ is the one required, so that the induction is complete. □

In accordance with the corollary of the theorem in 3.2, if K_t^T furnishes the maximum of the sums H_{K_t} on the sets

$$\mathscr{H}_{K_t} = \{H_{K_t}(y) : H_{K_t}(x) \leqq H_{K_t}(x \| y_{K_t})\},$$

we may set $\varepsilon = 0$ in the theorem just proved, and assert the existence in Γ of $\mathfrak{K}$-stable situations for the ordered quasipartition $\mathfrak{K}$.

In particular, this is true for a finite game Γ.

3.5. $\mathfrak{K}$-equilibrium. It is natural that $\mathfrak{K}$-optimality principles of equilibrium type should have realizations for wider classes $\mathfrak{K}$ of families of coalitions than $\mathfrak{K}$-principles of stability type.

Definition. A family $\mathfrak{K}$ of subsets of I is called a *star* if $\cap \mathfrak{K} \neq \emptyset$.

If no two sets that appear in two maximal stars in $\mathfrak{K}$ have elements in common, the family $\mathfrak{K}$ is said to be *star-shaped.*

Evidently every ordered quasipartition is a star.

Theorem. *Let Γ be a noncooperative game as in* (3.1), *and let $\mathfrak{K}$ be a family of coalitions in Γ, forming a star. Then for every $\varepsilon_K > 0^{\mathfrak{K}}$ the game Γ contains an $\varepsilon_{\mathfrak{K}}$-$\mathfrak{K}$-equilibrium situation.*

Proof. According to the theorem in 2.27, we may suppose without loss of generality that $\cup \mathfrak{K} = I$.

Choose an arbitrary $i_0 \in \cap \mathfrak{K}$ and consider a situation x^ε for which

$$H_{i_0}(x^\varepsilon) > \sup_{x \in \boldsymbol{x}} H_{i_0}(x) - \varepsilon.$$

It follows from this inequality that for every $K \in \mathfrak{K}$ it is impossible to find a situation y for which

$$H_K(x^\varepsilon) < H_K(y) - \varepsilon_K,$$

i.e., one such that x^ε is an ε_K-K-equilibrium situation for every $K \in \mathfrak{K}$. □

In the case when Γ has the property that the intersection $\cap\mathfrak{K}$ of the stars contains a player i_0 for whom the function H_{i_0} attains its maximum at x (in particular, this is the case for every finite game), we may take $\varepsilon_{\mathfrak{K}} = 0^{\mathfrak{K}}$ in the statement and proof of the theorem: there is a $\mathfrak{K}$-equilibrium situation in Γ.

3.6. Analysis of unsolvable games. The two preceding theorems cannot be directly improved: if the quasipartition is not ordered, there are games without $\mathfrak{K}$-stable situations, and if $\mathfrak{K}$ is not a star, there are games without $\mathfrak{K}$-equilibrium situations. Typical examples of such games are those described in 2.16. For these games we set $\mathfrak{K} = \{1, 2, \{1, 2\}\}$.

The absence of equilibrium situations from such games as "odd-even" keeps the players from making an agreement which none of the players is interested in violating.

In prisoner's dilemma, the state of affairs is even more complicated: here the players can agree that both will choose the strategy d, and neither one will have any basis for violating this agreement. However, the situation (d,d) is very unfavourable for both players, and they would be inclined to switch to the situation (c,c) which dominates it. But the situation (c,c) is not an equilibrium situation, and therefore an agreement to switch to it is unstable: each player will be inclined to replace the strategy d by c, and if both players do this, the game "skids" back to the unfavourable situation (c,c).

These examples correspond to different reasons for a noncooperative game to lack $\mathfrak{K}$-stable or $\mathfrak{K}$-equilibrium situations for some family $\mathfrak{K}$ of coalitions.

Here the facts are essentially as follows.

Let us suppose that it is expedient for a player (or coalition) to select strategy $x^{(1)}$ against one of the combinations of strategies of the opponent, but a different strategy $x^{(2)}$ against a different combination. If the player (or coalition) will *know* all the combinations of the strategies of the opponent, the choice of a reasonable strategy presents no difficulty. However, here there must exist a priori a rule for selecting strategies, i.e., a function that in the case at hand makes strategy $x^{(1)}$ correspond to the first combination of the opponent's strategies, and $x^{(2)}$ correspond to the second. In the case when the player (or coalition) cannot foresee this combination of the opponent's strategies, it is natural to expect the choice of an intermediate strategy. If the set of the player's (or coalition's) strategies is convex, there is always such an intermediate strategy. If, in addition, the player's (or coalition's) payoff function is concave, the value of the intermediate strategy is rather large (we may say, acceptably large).

3.7. Games with an informed player. In some classes of games and for some families $\mathfrak{K}$ of coalitions, the $\mathfrak{K}$-stable (or $\varepsilon_{\mathfrak{K}}$-$\mathfrak{K}$-stable) situations and, in particular, equilibrium situations, occur in a "universal" (i.e., in a sense, trivial) form. We give two examples.

Let A and B be two sets. By A^B we shall mean some family, depending on the topological or other structures of A and B, of functions from B to A. In particular, if no structure is given on A and B, then A^B denotes the set of all functions from B to A.

Definition. In a two-person noncooperative game

$$\Gamma = \langle \{1,2\}, \{\boldsymbol{x}, \boldsymbol{y}\}, \{H_1, H_2\}\rangle \tag{3.2}$$

player 2 is said to be *informed* if the set $\boldsymbol{y}$ of that player's strategies has the form $\boldsymbol{a}^{\boldsymbol{x}}$, i.e., consists of functions defined on $\boldsymbol{x}$ with values in some set $\boldsymbol{a}$, and if

$$H_i(x, y) = H_i(x, y(x)), \qquad i = 1, 2,$$

for all $x \in \boldsymbol{x}$ and $y \in \boldsymbol{y}$.

A similar definition can be given for player 1. □

To say that a player in a two-person noncooperative game is informed means that this player's strategy is the selection of an action on the basis of knowing the opponent's strategy, and that the payoffs to the players depend, not on the plans of the informed player (i.e., not on this player's strategies) but only on their realization, i.e. on the selected action.

3.8. Theorem. *Every two-person game Γ as in (3.2) with an informed player has an ε-equilibrium situation for every $\varepsilon = (\varepsilon_1, \varepsilon_2) > 0$.*

Proof. Let us suppose for definiteness that player 2 is informed.

We fix an $\varepsilon > 0$ and look for an ε-equilibrium situation in Γ.

Let player 1 have selected one of the available strategies, say x. We shall find an $a_{\varepsilon_2} (= a_{\varepsilon_2}(x)) \in \boldsymbol{a}$ for which

$$H_2(x, a_{\varepsilon_2}) + \varepsilon_2 > \sup_{a \in \boldsymbol{a}} H_2(x, a). \tag{3.3}$$

Evidently the reasonable (let us say, "ε_2-reasonable") behavior for player 2 will consist of the choice for each x of a corresponding $a_{\varepsilon_2}(x)$ that satisfies (3.3), i.e., in the choice of a function y: $\boldsymbol{x} \to \boldsymbol{a}$. Let us denote this function by y_{ε_2}. Here the payoff to player 1 will be equal to $H_1(x, y_{\varepsilon_2}(x))$ and a reasonable ("ε_1-reasonable") choice will be a situation x_{ε_1} with

$$H_1(x_{\varepsilon_1}, y_{\varepsilon_2}(x_{\varepsilon_1})) + \varepsilon_1 > \sup_{x \in \boldsymbol{x}} H_1(x, y_{\varepsilon_2}(x)). \tag{3.4}$$

It is clear that the pair $(x_{\varepsilon_1}, y_{\varepsilon_2})$ provides an ε-equilibrium situation in the game (3.2). In fact,

$$H_1(x_{\varepsilon_1}, y_{\varepsilon_2}) = H_1(x_{\varepsilon_1}, y_{\varepsilon_2}(x_{\varepsilon_1})) > H_1(x, y_{\varepsilon_2}(x)) - \varepsilon_1 = H_1(x, y_{\varepsilon_2}) - \varepsilon_1,$$

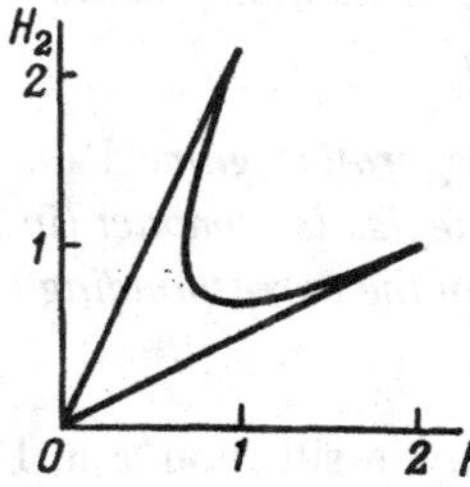

Fig. 1.9

by the definition of x_{ε_1} and in accordance with (3.4). In addition, by (3.3) we must have

$$H_2(x_{\varepsilon_1}, y_{\varepsilon_2}) = H_2(x_{\varepsilon_1}, y_{\varepsilon_2}(x_{\varepsilon_1})) > \sup_{a \in \mathfrak{a}} H_2(x_{\varepsilon_1}, a) - \varepsilon_2 \geqq H_2(x_{\varepsilon_1}, y) - \varepsilon_1,$$

whatever y may be. □

3.9. Concave games. Another possible way for equilibrium situations to arise is provided by a theorem on a class of games described by the following definition.

Definition. Let $\mathfrak{z}$ be a linear space. A function $f : \mathfrak{z} \to \mathbf{R}$ is said to be *concave* if, for all $z', z'' \in \mathfrak{z}$ and $\lambda \in [0, 1]$,

$$\lambda f(z') + (1 - \lambda) f(z'') \leqq f(\lambda z' + (1 - \lambda) z'').$$

It is *strictly concave* if the inequality is strict for $z' \neq z''$ and $\lambda \in (0, 1)$. A function is *convex* if the opposite inequality is satisfied.

A function f: $\mathfrak{z} \to \mathbf{R}$ is *quasiconcave* if the set $\{z : f(z) \geqq \alpha\}$ is convex for every $\alpha \in \mathbf{R}$. It is *quasiconvex* if the set $\{z : f(z) \leqq \alpha\}$ is convex for every $\alpha \in \mathbf{R}$.

A noncooperative game as in (3.1) is said to be *concave* (or *quasiconcave*) if the sets $\mathfrak{x}_i$ of strategies of each player $i \in I$ are convex subsets of linear spaces, and each of the payoff functions H_i is a concave (or quasiconcave) function of the strategies $x_i \in \mathfrak{x}_i$ for arbitrary but fixed strategies of the other players. □

In order to avoid misunderstanding, we note at the same time that for a concave game Γ the set $\mathfrak{W}_I(\Gamma)$ of realizable vector payoffs is not necessarily convex. As an example we may mention the game Γ, as in (3.2), in which $\mathfrak{x}_1 = \mathfrak{x}_2 = [0, 1]$ and

$$H_1(x_1, x_2) = x_1 x_2 + 2(1 - x_1)(1 - x_2),$$

$$H_2(x_1, x_2) = 2x_1 x_2 + (1 - x_1)(1 - x_2).$$

Here the sets of the player's strategies are line segments, hence convex, and the payoff functions are bilinear and therefore concave in each variable when the values of the others are held constant. The set $\mathfrak{W}_I(\Gamma)$ has the form sketched in Figure 1.9. The curvilinear side of the triangle $\mathfrak{W}_I(\Gamma)$ is an arc of a parabola.

As we can readily imagine, the proof of the existence of equilibrium situations in quasiconcave games is naturally based on fixed-point theorems of the type of Kakutani's theorem.*)

3.10. Theorem. *If the noncooperative game* Γ *as in* (3.1) *is quasiconcave, and the set* $\mathfrak{x}_i$ *of the players' strategies is compact (in some topology) and the payoff functions* H_i *are continuous (in the corresponding topology) on* $\mathfrak{x}$, *then there is an equilibrium situation in* Γ.

Proof. We choose, in any way, a situation x in Γ and consider the set $c_i(x) = \underset{\mathfrak{x}_i}{\operatorname{argmax}}\, H_i(x\|x_i)$ of the strategies x_i^* of player i for which

$$H_i(x\|x_i^*) = \max_{x_i \in \mathfrak{x}_i} H_i(x\|x_i).$$

The set $c_i(x)$ is not empty, since $\mathfrak{x}_i$ is compact and H_i is continuous. Since H_i is quasiconcave in x_i, the set is convex and, again by the continuity of H_i, closed.

The correspondence between the situation x and the corresponding set $c_i(x)$ of strategies is closed, i.e., if we have convergent sequences of situations

$$x^{(1)}, \dots, x^{(m)}, \dots \to x^{(0)}$$

and of strategies of player i,

$$y_i^{(1)}, \dots, y_i^{(m)}, \dots \to y_i^{(0)},$$

and if for $m = 1, 2, \dots$

$$y_i^{(m)} \in c_i(x^{(m)}), \tag{3.5}$$

then

$$y_i^{(0)} \in c_i(x^{(0)}). \tag{3.6}$$

For the proof, we first select an arbitrary $x_i \in \mathfrak{x}_i$. Relation (3.5) shows that

$$H_i(x^{(m)}\|y_i^{(m)}) \geqq H_i(x^{(m)}\|x_i).$$

Since this inequality holds for every m, we may pass to the limit on m, and from the continuity of H_i we obtain

$$H_i(x^{(0)}\|y_i^{(0)}) \geqq H_i(x^{(0)}\|x_i).$$

Since this inequality holds for every $x_i \in \mathfrak{x}$, we obtain (3.6).

We now introduce the mapping $c_i : \mathfrak{x}_i \to 2^{\mathfrak{x}}$, which establishes a correspondence between each situation x and the set

$$c(x) = \prod_{i \in I} c_i(x)$$

*) See, for example, L.V. Kantorovich and F.P. Akilov, Functional Analysis, Pergamon, Oxford and New York, 1982.

of situations. As a Cartesian product of nonempty convex compact sets, $c(x)$ is also nonempty, convex and compact. Since the c_i are closed, c is also closed.

As a result, we have obtained a mapping of the set of situations on itself, which satisfies the hypotheses of Kakutani's theorem. According to this theorem the mapping $c : x \to 2^x$ has a fixed point, i.e., a situation $x^* \in c(x^*)$. But in such situations we will have $x_i^* \in c_i(x^*)$, i.e.

$$H_i(x^*) = \max_{x_i \in x_i} H_i(x^* \| x_i) \text{ for } i \in I,$$

or in other words

$$H_i(x^*) \geqq H_i(x^* \| x_i) \text{ for all } x_i \in x_i \text{ and } i \in I. \qquad \square$$

3.11. If Γ is not only quasiconcave, but actually concave, we can establish the existence of equilibrium situations in Γ, under certain modifications of the continuity of the payoff functions H_i. Here we can use, in place of Kakutani's theorem, Brouwer's theorem, which is more elementary.

Theorem. *Let a game Γ as in (3.1) be compact and concave, let each payoff function H_i be continuous on $x\|x_i$ for each given x_i (i.e., in the notation of 1.6, continuous on $x^{(i)}$), and let the sum $\sum_{i\in I} H_i(x)$ be continuous in x. Then $\mathscr{C}(\Gamma) = \mathfrak{R}(\Gamma) \neq \emptyset$.*

Proof. Since the set x of situations in Γ is a Cartesian product of compact convex sets x_i, this set is convex and compact.

Consider the function

$$\Phi(x, y) = \sum_{i\in I} H_i(y \underset{i}{\|} x_i)$$

on $x \times x$. Since the functions H_i were assumed to be concave and continuous, $\Phi(x, y)$ is concave for each y and continuous in each of the arguments x and y.

If now the situation y^* has the property that the maximum of $\Phi(x, y^*)$, with respect to x, is attained at y^*, then, in particular,

$$\Phi(y^*, y^*) \geqq \Phi(y^* \| x_i, y^*) \text{ for all } i \in I \text{ and } x_i \in x_i,$$

i.e.,

$$\sum_{i\in I} H_i(y^*) \geqq \sum_{\substack{j\in I \\ j\neq i}} H_j(y^*) + H_i(y^* \underset{i}{\|} x_i),$$

so that

$$H_i(y^*) \geqq H_i(y^* \underset{i}{\|} x_I) \text{ for all } i \in I \text{ and } x_i \in x_i,$$

which means that $y^* \in \mathscr{C}(\Gamma)$. It remains only to establish the existence of such a situation.

Suppose that there is no such situation. This means that for every situation y there is a situation x such that $\Phi(x, y) > \Phi(y, y)$. Let $\mathfrak{x}_x$ denote, for a given x, the set of situations y that satisfy this inequality. Since, by hypothesis, some x corresponds to *each* $y \in \mathfrak{x}$, we must have $\bigcup_{x \in \mathfrak{x}} \mathfrak{x}_x = \mathfrak{x}$. Since, in addition, all the sets $\mathfrak{x}_x$ are open, the family of all $\mathfrak{x}_x$ constitutes an open covering of $\mathfrak{x}$. Since $\mathfrak{x}$ is compact, there is a finite subcovering $\mathfrak{x}_{x^{(1)}} \ldots, \mathfrak{x}_{x^{(r)}}$.*)

For $k = 1, \ldots, r$, consider the functions

$$f_k(y) = \max\{0, \Phi(x^{(k)}, y) - \Phi(y, y)\}. \tag{3.7}$$

Evidently $f_k(y) \geqq 0$; by our choice of $x^{(1)}, \ldots, x^{(r)}$ we can find, for each $y \in \mathfrak{x}$, an index k such that $f_k(y) > 0$, so that

$$f(y) = \sum_{k=1}^{r} f_k(y) > 0,$$

and, since $\mathfrak{x}$ is convex, we may consider the situation

$$\sum_{k=1}^{r} \frac{f_k(y)}{f(y)} x^{(k)},$$

which we denote by $\psi(y)$. Since the functions f_k are continuous, ψ is also continuous. In particular, it maps the convex hull of the points $x^{(1)}, \ldots, x^{(r)}$ on itself. Since this convex hull is homeomorphic to a simplex, Brouwer's fixed-point theorem is applicable to ψ, and therefore there is a point y^* for which

$$y^* = \sum_{k=1}^{r} \frac{f_k(y^*)}{f(y^*)} x^{(k)}.$$

Consequently, since Φ is convex, we must have

$$\Phi(y^*, y^*) \geqq \sum_{k=1}^{r} \frac{f_k(y^*)}{f(y^*)} \Phi(x^{(k)}, y^*). \tag{3.8}$$

On the other hand, from $f_k(y^*) > 0$ it follows by (3.7) that $\Phi(x^{(k)}, y^*) > \Phi(y^*, y^*)$, so that

$$f_k(y^*)\Phi(x^{(k)}, y^*) > f_k(y^*)\Phi(y^*, y^*).$$

If we sum this inequality on k (if necessary, adding some zero terms), we obtain, after normalization,

$$\sum_{k=1}^{r} \frac{f_k(y^*)}{f(y^*)} \Phi(x^{(k)}, y^*) > \sum_{k=1}^{r} \frac{f_k(y^*)}{f(y^*)} \Phi(y^*, y^*) = \Phi(y^*, y^*),$$

which, however, contradicts (3.8). □

---

*) Here notations like $x^{(k)}$ are used for enumerating situations, rather than in the sense of 1.6.

3.12. With the same plan as for Theorem 3.10, but using more delicate techniques and more refined methods, we can establish stronger results. We shall prove the following theorem.

Theorem. *Let Γ be a noncooperative game as in 3.12, and let $\mathfrak{K}$ be a star-shaped family of coalitions, consisting of the maximal stars $\mathfrak{K}_1, \dots, \mathfrak{K}_k$, where $\cap \mathfrak{K}_l = A_l$ and $\cup \mathfrak{K}_l = L_l$ for $l = 1, \dots, k$.*

Then if all the sets $\mathfrak{x}_i$ are compact convex subsets of Banach spaces, and, for every $l = 1, \dots, k$, each function

$$H^{(l)}(x) = \sum_{i \in A_l} H_i(x)$$

is continuous and quasiconcave with respect to every coalitional strategy x_K for $K \in \mathfrak{K}_l$, we have $\mathcal{R}_{\mathfrak{K}}(\Gamma) \neq 0$.

3.13. We precede the proof of this theorem by the following discussion.

Let $\mathcal{A}$ and $\mathcal{B}$ be metric spaces, ρ the metric in the Cartesian product $\mathcal{A} \times \mathcal{B}$ generated by the metrics in $\mathcal{A}$ and $\mathcal{B}$, and $\rho_H : 2^{\mathcal{A} \times \mathcal{B}} \times 2^{\mathcal{A} \times \mathcal{B}} \to \mathbf{R}$ the Hausdorff metric for subsets of $\mathcal{A} \times \mathcal{B}$, i.e., for $\mathcal{C}, \mathcal{D} \subset \mathcal{A} \times \mathcal{B}$,

$$\rho_H(\mathcal{C}, \mathcal{D}) = \max\{\sup_{d \in \mathcal{D}} \inf_{c \in \mathcal{C}} \rho(c, d), \sup_{c \in \mathcal{C}} \inf_{d \in \mathcal{D}} \rho(c, d)\}.$$

For the mapping $\varphi : \mathcal{A} \to \mathcal{B}$ we denote by $\operatorname{Gr}\varphi$ the graph of φ, i.e.

$$\operatorname{Gr}\varphi = \{(a, b) : a \in \mathcal{A},\ b \in \mathcal{B},\ b \in \varphi(a)\}.$$

We may describe the following lemma as establishing the existence of "ε-selectors".

Lemma. *Let $\varphi : \mathcal{A} \to \mathcal{B}$ be an upper semicontinuous mapping of a metric space $\mathcal{A}$ into a Banach space $\mathcal{B}$, such that all sets of the form $\varphi(a)$ $(a \in \mathcal{A})$ are nonempty, convex, and closed.*

Then for every $\varepsilon > 0$ there is a single-valued mapping (ε-selector) $f_\varepsilon : \mathcal{A} \to \mathcal{B}$ for which

$$\rho_H(\operatorname{Gr} f_\varepsilon, \operatorname{Gr}\varphi) < \varepsilon.$$

Proof. We choose $\varepsilon > 0$, and for all $z \in \operatorname{Gr}\varphi$ we construct open balls $u_\varepsilon(z) \subset \mathcal{A} \times \mathcal{B}$, with centers at z and radii ε. Set

$$u_\varepsilon(\operatorname{Gr}\varphi) = \bigcup_{z \in \operatorname{Gr}\varphi} u_\varepsilon(z)$$

and consider the mapping $\varphi_\varepsilon : \mathcal{A} \to \mathcal{B}$ for which

$$\varphi_\varepsilon(a) = \overline{\Pr_a(u_\varepsilon(\operatorname{Gr}\varphi))|a}$$

(the horizontal bar denotes, as usual, the closure; see Figure 1.10). None of the sets $\varphi_\varepsilon(a)$ is empty, since they contain $\varphi(a)$; they are closed by construction, and convex, as closures of convex sets.

Let us show that the mapping φ_ε is upper semicontinuous. We choose $(a, b) \in \operatorname{Gr}\varphi_\varepsilon$ arbitrarily. This means that $b \in \overline{\Pr_a(u_\varepsilon(\operatorname{Gr}\varphi))}$, i.e., there is a sequence $b_i, \dots, b_n, \dots$ in $\Pr_a(u_\varepsilon(\operatorname{Gr}\varphi))$, and therefore also a sequence of pairs $(a, b_1), \dots, (a, b_n), \dots$ from $u_\varepsilon(\operatorname{Gr}\varphi)$ that converges to (a, b). But the set $u_\varepsilon(\operatorname{Gr}\varphi)$ is open; hence the mapping φ_ε is upper semicontinuous at each point (a, b_n), and for every sequence $a_m \to a$ there is a sequence $b_{n(m)} \to b_n$ such that $b_{n(m)} \in \varphi_\varepsilon(a_m)$.

An application of the diagonal process produces a sequence $b_{n(m),m} \to b$ for whose terms $(a_m, b_{n(m),m}) \in \varphi_\varepsilon(a_m)$. This means that φ_ε is upper semicontinuous at (a, b).

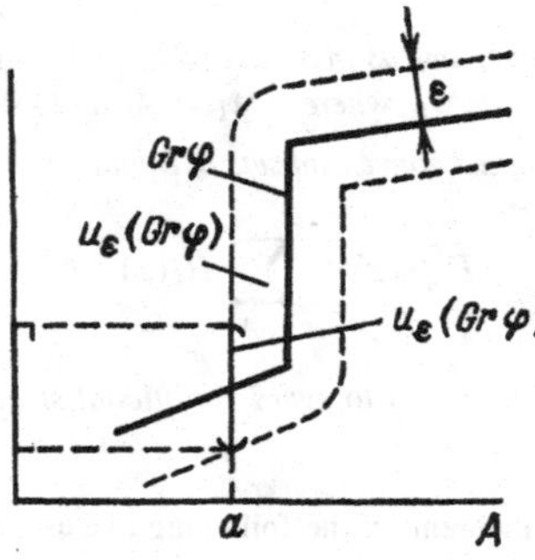

Fig. 1.10

As a result, we have the hypotheses of Michael's theorem,*) according to which a mapping φ_ε has a continuous selector $f_\varepsilon : \mathcal{A} \to \mathcal{B}$ (i.e., a single-valued mapping f_ε for which $f_\varepsilon(a) \in \varphi_\varepsilon(a)$). A direct calculation of distances shows that the selector so constructed is the required mapping.

3.14. We now turn to the proof of the theorem stated in 3.12. For convenience, we break the proof into steps.

1°. Choose a number l from $1, \ldots, k$, and let $K_1, \ldots, K_{r(l)}$ be the maximal coalitions in the star $\mathfrak{K}_l$. Their order is arbitrary, but will be kept fixed.

In what follows, we shall need to use repeatedly the substitution operator $\|$ on coalitional strategies. Consequently, to simplify the notation within this proof we shall denote a situation $(\ldots(x\|y_1)\|\ldots\|y_t)$ simply by $(x, y_1, \ldots, y_t)$.

For any situation $x \in \mathbf{x}$, we set

$$\mathbf{y}_{K_1}(x) = \underset{y_{K_1}}{\operatorname{argmax}}\, H^{(l)}(x, y_k) \subset \mathbf{x}_{K_1},$$

and take this as the first step in our construction. Now let t steps of the construction have been completed and let the sets

$$\mathbf{y}_{K_1}(x),$$

$$\mathbf{y}_{K_2}(x, y_{K_1})$$

....

$$\mathbf{y}_{K_t}(x, y_{K_1}, \ldots, y_{K_{t-1}})$$

of coalitional strategies have been constructed, corresponding to all combinations of coalitional strategies

$$y_{K_1} \in \mathbf{y}_{K_1}(x), \ldots, y_{K_{t-1}} \in \mathbf{y}_{K_{t-1}}(x, y_{K_1}, \ldots, y_{K_{t-2}}).$$

Then the $(t+1)$th step will consist of the construction of

$$y_{K_{t+1}}(x, y_{K_1}, \ldots, y_{K_t}) = \underset{y_{K_{t+1}}}{\operatorname{argmax}}\, H^{(l)}(x, y_{K_1}, \ldots, y_{K_t}).$$

*) See T. Parthasarathy, Selection theorems and their applications, Springer, Berlin, 1972.

With each player $i \in L_l$ we associate the coalition $K_{s(i)}$ from the set $K_1, \ldots, K_{r(l)}$ such that $i \in K_{s(i)}$ and $i \notin K_j$ for $j < s(i)$, and construct the mapping $\mathscr{C}_l : y \to 2^{x^{L_l}}$ $(l = 1, \ldots, k)$ by setting

$$(\mathscr{C}_l(x))_i = \begin{cases} y_{K_1}(x), & \text{if } s(i) = 1, \\ \cup y_{K_{s(i)}}(x, y_{K_1}, \ldots, y_{K_{s(i)-1}}), & \text{if } s(i) > 1, \end{cases}$$

where the union is taken over all combinations of the coalitional strategies $y_{K_1}, \ldots, y_{K_{s(i)-1}}$ that were encountered in the course of the construction. We note that the mapping $\mathscr{C}_l$ is not defined on the whole set x_{L_l} of coalitional strategies, but only on $x_{L_l \setminus K_1}$, i.e., $\mathscr{C}_l(x, y_{K_1}) = \mathscr{C}_l(x, z_{K_1})$ on x. This fact will be used in step 3°.

All the mappings $\mathscr{C}_l$ can be combined into the single mapping

$$\mathscr{C}(x) = \prod_{l=1}^{k} \prod_{i \in L_l} (\mathscr{C}_l(x))_i.$$

We see immediately that

$$\max_{y_{K_t}} H^{(l)}(x, y_{K_1}, \ldots, y_{K_t}) \leqq \max_{y_{K_{t+1}}} H^{(l)}(x, y_{K_1}, \ldots, y_{K_{t+1}}).$$

Therefore, if x^* is a fixed point of the mapping $\mathscr{C}$, i.e. if $x^* \in \mathscr{C}(x^*)$, then successive deviations of the coalitions $K_1, \ldots, K_{r(l)}$ from x^* do not lead to an increase in $H^{(l)}$:

$$H^{(l)}(x^*) = \max_{y_{K_t}} H^{(l)}(x^*, y_{K_1}, \ldots, y_{K_t}), \quad t = 1, \ldots, r(l),$$

so that

$$H^{(l)}(x^*) = \max_{y_{K_{r(l)}}} H^{(l)}(x^*, y_{K_{r(l)}}) \geq \max_{y_{K_{r(l)-1}}} H^{(l)}(x^*, y_{K_{r(l)-1}}$$

$$\geqq \ldots \geqq \max_{y_{K_1}} H^{(l)}(x^*, y_{K_1}).$$

Hence x^* is a K_t-equilibrium situation for each $t = 1, \ldots, r(l)$. Moreover, if a coalition $K \in \mathfrak{K}_l$ is not maximal in $\mathfrak{K}_l$, i.e., $K \subset K_t$, then evidently

$$\max_{y_K} H^{(l)}(x^*, y_K) \leqq \max_{y_{K_t}} H^{(l)}(x^*, y_{K_t}) \leqq H^{(l)}(x^*).$$

Therefore the situation x^* turns out to have the K-equilibrium property. As a result, we find that $x^* \in \mathscr{R}_{\mathfrak{K}_l}(\Gamma)$ for $l = 1, \ldots, r$, i.e. $x^* \in \mathscr{R}_{\mathfrak{K}}(\Gamma)$.

The mapping $\mathscr{C}$ does not satisfy the hypotheses of the usual fixed point theorems (one cannot always establish the contractibility of $\mathscr{C}(x)$). Consequently we are going to establish the existence of ε-$\mathfrak{K}$-equilibrium situations for every $\varepsilon > 0$ as fixed points of mappings $\mathscr{C}_\varepsilon : x \to 2^x$ which are close to $\mathscr{C}$ in the sense of the distance between their graphs.

2°. From the continuity of $H^{(l)}$ in all the variables it follows that for every $\varepsilon > 0$ there is a $\delta(\varepsilon) > 0$ such that $y \in \mathscr{C}(x)$ and $\rho(x, y) < \delta(\varepsilon)$ imply that $y \in \mathscr{R}^{\varepsilon}_{\mathfrak{K}}(\Gamma)$.

It will follow from this proposition that if, for $\delta(\varepsilon) > 0$, the mapping $\mathscr{C}_{\delta(\varepsilon)} : x \to 2^x$ has the property

$$\rho_H(\mathrm{Gr}\,\mathscr{C}, \mathrm{Gr}\,\mathscr{C}_{\delta(\varepsilon)}) < \delta(\varepsilon)/2, \tag{3.9}$$

then a situation that satisfies the condition $x^*_{\delta(\varepsilon)} \in \mathscr{C}_{\delta(\varepsilon)}(x^*_{\delta(\varepsilon)})$ is an ε-$\mathfrak{K}$-equilibrium situation.

In fact, it follows from (3.9) that for every point $(x, y) \in \operatorname{Gr}\mathscr{C}_{\delta(\varepsilon)}$ there is a point $(x', y') \in \operatorname{Gr}\mathscr{C}$ such that $\rho((x, y), (x', y')) < \delta(\varepsilon)/2$ in the topology of $x \times x$.

Let $\rho((x^*, y^*), (x' y')) < \delta(\varepsilon)/2$. Then $\rho(x^*_\delta, x') < \delta(\varepsilon)/2$ and $\rho(x^*_\delta, y') < \delta(\varepsilon)/2$, and consequently $\rho(x', y') < \delta(\varepsilon)$, $y' \in \mathscr{C}(x')$, i.e. y' is an ε-$\mathfrak{K}$-equilibrium situation.

3°. To keep the exposition simple, we consider the case $k = 1$, $\mathfrak{K} = \mathfrak{K}_1 = \{K_1, K_2\}$, $K_1 \cap K_2 \neq \emptyset$. In this case we have $r(i) = 1$ for $i \in K_1 \setminus K_2$ and $r(i) = 2$ for $i \in K_2$,

$$(\mathscr{C}(x))_i = \begin{cases} (y_{K_1}(x_{K_1 \setminus K_2})_i & i \in K_1 \setminus K_2, \\ \cup(y_{K_2}(y_{K_1 \setminus K_2}))_i, & i \in K_2, \end{cases}$$

where the last union is taken over all $y_{K_1} \in y_{K_1}(x_{K_1 \setminus K_2})$ and $y_{K_1 \setminus K_2}$ is the strategy of the coalition $K_1 \setminus K_2$ which is part of the strategy y_{K_1}.

The sets

$$y_{K_1}(x_{K_2 \setminus K_1}) = \operatorname*{argmax}_{y_{K_1}} H^{(1)}(x_{K_2 \setminus K_1}, y_{K_1})$$

are convex, closed, and compact, and the mapping $y_{K_1} : x_{K_1 \setminus K_2} \to 2^{x_{K_1}}$ is upper semicontinuous. By the lemma, for every $\varepsilon > 0$, there is a continuous single-valued mapping $f^1_\varepsilon : x_{K_2 \setminus K_1} \to x_{K_1}$ such that

$$\rho_H(\operatorname{Gr} f^1_\varepsilon, \operatorname{Gr} y_{K_1}) \leqq \varepsilon. \tag{3.10}$$

Since $X_{K_2 \setminus K_1}$ is compact, the sets $f^1_\varepsilon(X_{K_2 \setminus K_1})$ are also compact for every $\varepsilon > 0$.

Choose $\delta(\varepsilon) > 0$ as in step 2°, and consider the mapping

$$y_{K_2} : x_{K_1 \setminus K_2} \to y_{K_2}(y_{K_1 \setminus K_2}) = \operatorname{argmax} H^{(1)}(y_{K_1 \setminus K_2}, y_{K_2})$$

restricted to the set $(f^{(1)}_{\delta(\varepsilon)}, (x_{K_2 \setminus K_1}))_{K_1 \setminus K_2}$. We denote this restriction of y_{K_2} by $y^\varepsilon_{K_2}$. It also satisfies the hypotheses of the lemma, and for each $\delta(\varepsilon)$ there is a single-valued continuous mapping

$$f^2_{\delta(\varepsilon)} : (f^1_\varepsilon(x_{K_2 \setminus K_1}))_{K_1 \setminus K_2} \to x_{K_2},$$

such that

$$\rho_H(\operatorname{Gr} f^2_{\delta(\varepsilon)}, \operatorname{Gr} y^\varepsilon_{K_2}) \leqq \delta(\varepsilon)$$

and consequently

$$\rho_H(\operatorname{Gr} f^2_{\delta(\varepsilon)}, \operatorname{Gr} y_{K_2}) \leqq \delta(\varepsilon). \tag{3.11}$$

Now set

$$(\mathscr{C}_\varepsilon(x))_i = \begin{cases} (f^1_{\delta(\varepsilon)/2}(x_{K_2 \setminus K_1}))_i, & i \in K_1 \setminus K_2, \\ (f^2_{\delta(\varepsilon)/2}(f^1_{\delta(\varepsilon)/2}(x_{K_2 \setminus K_1}))_{K_1 \setminus K_2})_i, & i \in K_2. \end{cases}$$

From the definitions of the mappings $\mathscr{C}$ and $\mathscr{C}_\varepsilon$ and inequalities (3.10) and (3.11), it follows that $\rho_H(\operatorname{Gr}\mathscr{C}, \operatorname{Gr}\mathscr{C}_\varepsilon) < \varepsilon$. Moreover, for every $\varepsilon > 0$ it follows from Schauder's theorem that there is a fixed point of $\mathscr{C}_\varepsilon$, since $\mathscr{C}_\varepsilon$ is a single-valued continuous mapping of the compact convex set x on itself. It follows from proposition 2° that $x^*_\varepsilon \in \mathscr{C}_\varepsilon(x^*_\varepsilon)$ and is an ε-$\mathfrak{K}$-equilibrium situation.

By the continuity of $H^{(l)}$, every limit point of x^*_ε as $\varepsilon \to 0$ is a $\mathfrak{K}$-equilibrium situation. □

§4 Realizability of $\mathfrak{R}$-principles in metastrategies

4.1. Metastrategies. In § 2 we indicated the possibility of interpreting situations in a noncooperative game as agreements among players. In this connection, the optimality, in the sense of some $\mathfrak{R}$-principle, becomes a version of the stability of the agreement.

At the same time, as we know, the stability of agreements can be looked at not only in a categorical version of the form "player i will choose (or must choose) the strategy x_i" but also in the conditional form, for example, "*if* player i chooses the strategy x_i, *then* player j will choose (or must choose) the strategy x_j". There are also more complicated conditional versions: "*if* player i *responds* to the strategy x_j of player j by choosing the strategy x_i, *then* player k must choose the strategy x_k."

Such formulations of a conditional nature will be defined later as "metastrategies". They can be considered as functions that map the set of strategies of one player (or of a whole coalition) to the set of strategies of another player, or as superpositions of such functions. Just such a set of functions has in fact occurred as the set of strategies of player 2 in the game with an informed player that we discussed in 3.7 and 3.8. The natural and fruitful character of the concept of metastrategies is what leads to their formal use in game theory.

4.2. Hierarchical games. The game with an informed player in 3.7 is an especially simple example of metastrategical considerations. Here is a more complicated example.

Example. A hierarchical game. As an introduction, we start from the game

$$\Gamma = \langle I, \{x_i\}_{i\in I}, \{H_i\}_{i\in I}\rangle \tag{4.1}$$

with $I = \{0, 1, \ldots, n\}$, where player 0 is said to be *distinguished* and the payoff functions of the other players have the simplifying property that

$$H_i : x_0 \times x_i \to \mathbf{R} \quad \text{for } i = 1, \ldots, n$$

(i.e., the payoff to player 1 depends only on the strategies of player 1 and of the distinguished player).

We can interpret this game in an interesting way by thinking of player 0 as a *control center* and the remaining players $1, \ldots, n$ as *producers of output*. It is clear that, practically speaking, this interpretation applies not only to the game Γ but also to the following extension ("meta-extension") of it.

The control center chooses "guiding principles", consisting of the determination of its strategy $x_0 \in x_0$ on the basis of actual knowledge of the strategies $x_i \in x_i$ of the producers $1, \ldots, n$. Therefore these principles are, formally speaking, functions of the form

$$\bar{x}_0 : x^{(0)} = \prod_{i\neq 0} x_i \to x_0,$$

i.e., elements of the power $\bar{x}_0 = x^{x^{(0)}}$.

The producers, in turn, choose strategies on the basis of knowledge of the guiding principles of player 0 (provided, let us say, by information of a normative nature). Consequently the behaviors of players $i = 1, \dots, n$ are actually functions

$$\overline{x}_i : \mathfrak{x}_0 \to \mathfrak{x}_i$$

i.e., elements of the sets $\overline{\mathfrak{x}}_i = \mathfrak{x}_i^{\mathfrak{x}_0}$.

Finally, it is natural to set, for each $\overline{x}_i \in \overline{\mathfrak{x}}_i$ $(i = 0, 1, \dots, n)$

$$\overline{\overline{H}}_i(\overline{x}_0, \overline{x}_1, \dots, \overline{x}_n) = \overline{H}_i(\overline{x}_0, \overline{x}_1(\overline{x}_0), \dots, \overline{x}_n(\overline{x}_0))$$
$$= H_i(\overline{x}_0(\overline{x}_1(\overline{x}_0), \dots, \overline{x}_n(\overline{x}_0)), \overline{x}_1(\overline{x}_0), \dots, \overline{x}_n(\overline{x}_0))$$
$$= H_i(\overline{x}_0(\overline{x}_1(\overline{x}_0), \dots, \overline{x}_n(\overline{x}_0)), \overline{x}_i(x_0))$$

(notice that here $\overline{x}_i(\overline{x}_0) \in \mathfrak{x}_i$ $(i \neq 0)$, and therefore $\overline{x}_0(\overline{x}_i(\overline{x}_0), \dots, \overline{x}_n(\overline{x}_0)) \in \mathfrak{x}_0$) and to consider the game

$$\overline{\Gamma} = \langle I, \{\overline{\mathfrak{x}}_i\}_{i \in I}, \{\overline{H}_i\}_{i \in I} \rangle,$$

which it is natural to call *hierarchical*.

4.3. Metagames. Let us introduce the necessary concepts formally.

Definition. A sequence of pairs

$$(a_1, K_1), (a_2, K_2), \dots, (a_r, K_r) \tag{4.2}$$

is said to be a *metastrategic construction* in the game Γ in (4.1), if each K_i is a coalition in Γ, and each a_i is either the set x_{K_i} of the coalitional strategies of K_i in this game, or has the power form (see 3.7)

$$a_{i_0}^{a_{i_1} \times \dots \times a_{i_k}}, \tag{4.3}$$

where $i_0, i_1, \dots, i_k < i$, the corresponding coalitions $K_{i_0}, K_{i_1}, \dots, K_{i_k}$ are pairwise disjoint, and $K_{i_0} = K_i$.

The set a is called the *closed set of metastrategies* of the coalition K if the pair (a, K) is the last element in a metastrategic construction. □

Since it is clear from what we have said that every closed set a_i completely determines the corresponding coalition K_i, we shall occasionally describe a metastrategic construction, not as a sequence (4.2) of pairs, but as a sequence

$$a_1, \dots, a_r \tag{4.4}$$

of sets.

The power (4.3), as a set of functions mapping $a_{i_1} \times \dots \times a_{i_k}$ to a_{i_0}, can be thought of informally as the collection of the various *plans of reaction* by the coalition K_{i_0} to the combinations arising from the previously determined coalitional strategies, from $a_{i_1}, \dots, a_{i_k}$, of the respective coalitions $K_{i_1}, \dots, K_{i_k}$.

It follows directly from the definition (and is established by an uncomplicated inductive proof) that if a_K is the closed set of metastrategies of the coalition K, and $K = K_1 \cup K_2$, with $K_1 \cap K_2 = \emptyset$, then we have the representation

$a_K = a_{K_1} \times a_{K_2}$, where a_{K_1} and a_{K_2} are the closed sets of the coalitions of K_1 and K_2.

In particular, it is always possible to pass from the closed set a_K of metastrategies of K to the closed set a_L of metastrategies for any coalition $L \subset K$, and in particular, to the closed sets a_i of metastrategies of the players belonging to K. The converse transition, generally speaking, is not always possible, since a product of sets of individual metastrategies can have "different origins".

It is easily seen that coalitional strategies, being elements of closed sets of metastrategies, i.e., of powers of sets, are, generally speaking, functions that make a correspondence between a set of metastrategies of other coalitions and a characteristic metastrategy of a, in some sense, "lower" order. Consequently, the choice of metastrategies by a coalition consists of its decision to respond to the metastrategies of other coalitions by one of its own "shorter" metastrategies. The examples given in 4.1 of possible formulations in terms of agreements are just such metastrategies.

Example. Let Γ be a two-person game, where the sets of the players' strategies are x and y.

Here metastrategic constructions are, in particular,

$$x,\ y,\ x^y,\ y^{(x^y)},\ (x^y)^{(y^{(x^y)})} \tag{4.5}$$

and

$$x,\ y,\ x^y,\ y^{(x^y)},\ x^{(y^{(x^y)})}. \tag{4.6}$$

4.4. Not all sets of metastrategies lend themselves to association with situations.

For example, if we take

$$f \in y^{(x^y)} \text{ and } g \in (x^y)^{(y^{(x^y)})},$$

in the metastrategic construction (4.5) we evidently have

$$g(f) \in x^y, f(g(f)) \in y,$$

so that the pair of functions f and g generate the situation

$$((g(f))\,(f(g(f))),\ f(g(f))) \in x \times y.$$

On the contrary, if in the metastrategic construction (4.6) we take the pair of functions

$$f \in y^{(x^y)} \text{ and } g \in x^{(y^{(x^y)})},$$

we can obtain from them the strategy $g(f) \in x$, but we cannot define a definite strategy in y.

What we have just said gives us a reason for introducing the following concept.

Definition. Let $K_1, \ldots, K_k$ be a partition of the set I of players (i.e., a set of pairwise disjoint coalitions whose union is I), and

$$(a_1, K_1), (a_2, K_2), \ldots, (a_k, K_k) \tag{4.7}$$

the set of pairs in which each a_i is the closed set of metastrategies of the corresponding K_i. We call such a set a (*metastrategic*) *basis*. It is clear that the same game can have several metastrategic bases.

A metastrategic development of the basis (4.7) is a sequence

$$(a_1, K_1), \ldots, (a_k, K_k), (a_{k+1}, K_{k+1}), \ldots, (a_n, K_n), \tag{4.8}$$

in which, for $i > k$, each a_i is a closed set of the metastrategies of the coalition K_i, and in addition

a) either there is an $i' < i$ such that

$$a_{i'} = a_i \times a_{i''}, \tag{4.9}$$

where $a_{i''}$ is a closed set of the metastrategies of the coalition $K_{i''} = K_{i'} \setminus K_i$;

b) or there are $i_0, i_1, \ldots, i_r < i$ such that

$$a_{i_0} = a_i^{a_{i_1} \times \ldots \times a_{i_r}}. \tag{4.10}$$

It is easily seen that all metastrategic developments of a given basis differ from each other only in the order of their terms.

A metastrategic development is called *regular* if it contains a set of pairs

$$(a_{j_1}, L_1), \ldots, (a_{j_l}, L_l), \tag{4.11}$$

where $L_1, \ldots, L_l$ form a partition of I, and all $a_{j_r} = x_{L_r}$ $(r = 1, \ldots, l)$ (i.e. it is simply the set of all strategies of the corresponding coalitions).

A basis (4.9) is said to be *regular* if it has a proper metastrategic development.

Thus, for example, if we write the terms of the metastrategic development (4.5) in the opposite order, we obtain a regular metastrategic development; but if this is done with the metastrategic development (4.6), we do not.

For a proper metastrategic basis, the value of the index i in (4.8) for which (4.10) holds is called the *degree* of the metastrategic development (4.8), and also the *degree* of the metastrategic basis (4.7). □

From the uniqueness (up to the order) of a metastrategic development, it follows that each proper basis uniquely determines the set of pairs of the form (4.8).

4.5. The following definition covers all methods of obtaining situations by means of metastrategies.

Definition. Let

$$(a_1, K_1), \dots, (a_k, K_k) \tag{4.12}$$

be a regular metastrategic basis in the game Γ, and let $f_l \in a_i$ for $l = 1, \dots, k$. The system $f = (f_1, \dots, f_k)$ is called a *metasituation* in the basis (4.12).

Let

$$(a_1, K_1), \dots, (a_k, K_k), (a_{k+1}, K_{k+1}), \dots, (a_n, K_n)$$

be a proper metastrategic development of the basis (4.12). Here a sequence

$$f_1, \dots, f_k, f_{k+1}, \dots, f_n$$

of metastrategies is called a *development of the metasituation* f if:

for $i \leqq k$ we must have $f_i \in a_i$;

for $i > k$, under conditions (4.9), $f_{i'} \in a_i$ and $f_{i''} \in a_{i''}$ imply $f_{i'} = (f_i, f_{i''})$;

for $i > k$, under conditions (4.10), $f_{i_t} \in a_{i_t}$ $(t = 0, 1, \dots, r)$ imply $f_i = f_{i_0}(f_{i_1}, \dots, f_{i_r})$.

The terms of the development of a metasituation that correspond to the terms of (4.11) are coalitional strategies in Γ and form a situation in it. This situation is called the *projection* of f *on* Γ and is denoted by $\mathrm{Pr}_\Gamma(f)$. □

From the preceding comments on the uniqueness of a metastrategic development, it follows that every metasituation in a noncooperative game uniquely determines the projection on the game. At the same time, it is to be understood that different metasituations can have the same projection.

Definition. Let Γ be a noncooperative game of the form (4.1), in which $I = \{1, \dots, n\}$; let

$$(a_1, K_1), \dots, (a_k, K_k) \tag{4.13}$$

be one of its metastrategic bases; and let a_i $(i \in I)$ be the closed sets of metastrategies of the players that appear as Cartesian factors in the product a_l. For each metasituation f in this metastrategic basis and for each player i, we set $H_i(f) = H_i(\mathrm{Pr}_\Gamma(f))$.

The system

$$\mathrm{M}_\Gamma = \langle I, \{a_l\}_{l=1,\dots,k}, \{H_i\}_{i \in I} \rangle \tag{4.14}$$

is called the *metastrategic extension* of Γ (or, for short, the *meta-extension* or *metagame* over Γ), generated by the metastrategic basis (4.13). The degree of the metastrategic basis (4.13) will also be called the *degree of the metastrategic extension*, or the degree of the metagame (4.14). In particular, we may speak of the "first", "second", etc., metagames over Γ. □

We emphasize that the metagame M_Γ depends not only on the original game Γ but also on a metastrategic basis for it.

We also observe that the relation of the metagame to the game is transitive: if Δ is a metagame over Γ, and E is a metagame over Δ, then E is a metagame over Γ. To establish this, it is sufficient to the metastrategic basis of Δ, that leads to E adjoin the metastrategic basis of Γ that leads to Δ.

As a very simple example of a (first) metagame, we may take the game with an informed player that was discussed in 3.7. It may be interpreted as the metagame

$$\mathrm{M}_\Gamma = \langle \{1,2\}, \{(x,1),(y^x,2)\}, \{H_1,H_2\}\rangle$$

over the game

$$\Gamma = \langle \{1,2\}, \{x,y\}, \{H_1,H_2\}\rangle.$$

4.6. The course of a game Γ can be represented in terms of metastrategies in the following way. The players in the game choose, in accordance with a given (proper) metastrategic basis of Γ, their metastrategies, which form a metasituation. Then a development of this metasituation is constructed, resulting in a situation in Γ, after which the players obtain the payoffs due to them in this situation.

Here the construction of a development of the metasituation can be thought of in two ways.

On the one hand, one may suppose that the players have "different states of information" and interpret the values of the metastrategy functions as responses to the revealing values of the strategy-arguments of the other players.

On the other hand, the development of a metasituation and the formation of its projection can be interpreted as the activity of an intermediary who "collects" the metastrategies of the players and issues payoffs to the players after a logical analysis of the metasituations implicit in the construction of the development and the resulting projection.

4.7. Stability in meta-extensions of a game. Stability (or $\mathfrak{K}$-stability) of metagames can be predicted from the stability (or $\mathfrak{K}$-stability) of the basic game. Furthermore, until the end of the section we shall confine statements and proofs of propositions about the stability of situations to *finite games*. The transition to the corresponding propositions about ε-stability in general games requires special consideration, on which we shall not dwell.

Let us first consider the two following special propositions which will be useful in what follows.

4.8. Theorem. *Let* Γ *be a finite noncooperative game of the form* (4.1) *in which* $I = \{1,\dots,n\}$.

Choose K *and* $L \subset I$ $(K \cap L = \emptyset)$ *and consider the metagame over* Γ,

$$\mathrm{M}_\Gamma = \langle I, \{\{x_i\}_{i\in I\setminus K}, \{x_i^{x_L}\}_{i\in K}\}, \{H_i\}_{i\in I}\rangle,$$

where we have set, for arbitrary $i \in I, x_L \in x_L$ *and* $f_K : x_L \to x_K$,

$$H_i(f_K, x_L, x_{I\setminus(K\cup L)}) = H_i(f_K(x_L), x_L, x_{I\setminus(K\cup L)}). \tag{4.15}$$

Also let $x^* = (x_K^*, x_L^*, x_{I\setminus(K\cup L)}^*)$ be a situation in Γ and let f_K^* be the metastrategy of K for which $f_K^*(x_L) = x_K^*$ for every $x_L \in \mathfrak{x}_L$ (i.e., a metastrategy which is a function that turns out to be constant).

Finally, let $R \subset I$ and $x^* \in \mathscr{C}_R^\varepsilon(\Gamma)$.

Then, if $R \subset K$ or $R \subset L$, or $R \cap (K \cup L) = \emptyset$, then

$$(f_K^*, x_L^*, x_{I\setminus(K\cup L)}^*) \in \mathscr{C}_R^\varepsilon(\mathrm{M}_\Gamma).$$

Proof. First let $R \cap (K \cup L) = \emptyset$. In this case the transition to the metagame simply does not affect the strategy x_R^*, and the ε-R-optimality of the situation is trivially preserved.

Consequently we shall suppose that $R \cap (K \cup L) \neq \emptyset$. This gives us the possibility of passing (however, mainly to preserve the notation) to the factorization $\bar{\Gamma} = \Gamma / x_{I\setminus(K\cup L)}^*$ (as usual, payoff functions are denoted by $\bar{H}_i$) and $\bar{\mathrm{M}}_\Gamma = \mathrm{M}_\Gamma / x_{I\setminus(K\cup L)}^*$. Evidently $\bar{\mathrm{M}}_\Gamma$ is a metagame over $\bar{\Gamma}$. According to what was said in 2.27, the equivalence of ε-R-stability in Γ and $\bar{\Gamma}$, and also in M_Γ and $\bar{\mathrm{M}}_\Gamma$, provides a basis for reducing the problem to deducing $(f_K^*, x_L^*) \in \mathscr{C}_R(\bar{\mathrm{M}}_\Gamma)$ from $\bar{x}^* \in \mathscr{C}_R^\varepsilon(\bar{\Gamma})$.

Suppose that $R \subset K$ but $(f_K^*, x_L^*) \notin \mathscr{C}_R^\varepsilon(\bar{\mathrm{M}}_\Gamma)$. The latter means that there is a metastrategy $\tilde{f}_R \in \mathfrak{x}_R^{\mathfrak{x}_L}$ such that

$$\bar{H}_r(f_R^*, f_{K\setminus R}^*, x_L^*) \overset{\varepsilon}{\leq} \bar{H}_r(\tilde{f}_R^*, f_{K\setminus R}^*, x_L^*). \tag{4.16}$$

We find from (4.15) that $\bar{H}_r(f_R^*, f_{K\setminus R}^*, x_L^*) = \bar{H}_R(\bar{x}^*)$ and similarly, setting $\tilde{f}_r(x_L^*) = \tilde{x}_r$, that

$$\bar{H}_r(\tilde{f}_R, f_{K\setminus R}^*, x_L^*) = \bar{H}_R(\bar{x}^* \| \tilde{x}_R).$$

Therefore the inequality (4.16) can be rewritten as

$$\bar{H}_R(\bar{x}^*) \overset{\varepsilon}{\leq} \bar{H}_R(\bar{x}^* \| \tilde{x}_R),$$

which contradicts $\bar{x}^* \in \mathscr{C}_R^\varepsilon(\bar{\Gamma})$.

Finally, suppose that $R \subset L$ and $(f_K^*, x_L^*) \notin \mathscr{C}_R^\varepsilon(\bar{\mathrm{M}}_\Gamma)$. This means that there is a strategy $\tilde{x}_R \in \mathfrak{x}_R$ for which

$$\bar{H}_R(f_K^*, x_R^*, x_{L\setminus R}^*) \overset{\varepsilon}{\leq} \bar{H}_R(f_K^*, \tilde{x}_R, x_{L\setminus R}^*).$$

But then, by the definition of the metastrategy f^*, it will follow that

$$\bar{H}_R(\bar{x}^*) \overset{\varepsilon}{\leq} \bar{H}_R(\bar{x}^* \| \tilde{x}_R),$$

and this contradicts $\bar{x}^* \in \mathscr{C}_R^\varepsilon(\bar{\Gamma})$. □

4.9. Corollary. *Let* Γ *be a noncooperative game of the form* (4.1),

$$(a_1, K_1), \dots, (a_k, K_k) \tag{4.17}$$

one of its proper metastrategic bases, and M_Γ *the extension of this basis to a metagame over* Γ. *If* R *is a coalition in* Γ *for which, for each* $i = 1, \dots, k$, *we have* $R \subset K_i$ *or* $R \cap K_i = \emptyset$, *then for a situation* X *in the metagame* M_Γ *it follows from* $\Pr_\Gamma X \in \mathscr{C}^\varepsilon_R(\Gamma)$ *that* $X \in \mathscr{C}^\varepsilon_R(M_\Gamma)$.

The proof is based on the fact that the coalition L, by the hypothesis of the preceding theorem, is one of the sets K_i in the hypothesis of the corollary that is to be proved. After this remark, the proof is carried out by a natural induction on the length of the metastrategical basis (4.17) and reduced to an application of the preceding theorem. □

4.10. It is clear that the use of metastrategies enriches the possibilities for the players. In particular, we can expect that we may find among the metasituations one more stable than among the original situations. The following proposition may serve as a an appropriate illustration.

Theorem. *Consider a finite noncooperative game* Γ *as in* (4.1), *where* $I = K \cup L$, *and* $K \cap L = \emptyset$. *Let* $\mathfrak{K}$ *and* $\mathfrak{L}$ *be quasipartitions of the sets* K *and* L *of players. Let:*

1) *there be in* Γ *for all* $x_L \in \boldsymbol{x}_L$ *a* $\mathfrak{K}$*-stable situation of the form* $(x^*_K(x_L), x_L)$;
2) *there be in the game* $\Gamma^* = \bar{\Gamma}/x^*_K(x_L)$:

$$\Gamma^* = \langle L, \{\boldsymbol{x}_i\}_{i \in L}, \{H^*_i\}_{i \in L} \rangle,$$

where

$$H^*_i(x_L) = H_i(x^*_K(x_L), x_L),$$

an $\mathfrak{L}$*-stable situation* x^*_L.

Then in the metagame

$$M_\Gamma = \langle I, \{\{\boldsymbol{x}_l\}_{l \in L}, \{\boldsymbol{x}_k^{\boldsymbol{x}_L}\}_{k \in K}\}, \{\bar{H}_i\}_{i \in I} \rangle$$

there is a $\mathfrak{K} \cup \mathfrak{L}$*-stable situation.*

Proof. To each strategy $x_L \in \boldsymbol{x}_L$ let there correspond the strategy $x^*_K(x_L)$. This correspondence is a function in $\boldsymbol{x}_K^{\boldsymbol{x}_L}$, i.e. a metastrategy f^*_K of the coalition K, which is a strategy in the metagame M_Γ. It follows from hypothesis 1) and the second concluusion of the theorem in 4.6 that for every $x_L \in \boldsymbol{x}_L$ the situation (x^*_L, f^*_K) is $\mathfrak{K}$-stable in M_Γ. Consequently the situation (x^*_L, f^*_K) is a $\mathfrak{K} \cup \mathfrak{L}$-stable situation.

Next, from hypothesis 2) and the third conclusion of the theorem in 4.6, there follows the $\mathfrak{L}$-stability of (x^*_L, f^*_K) in M_Γ. □

4.11. Equilibrium situations in meta-extensions of games. As an immediate corollary, we establish the following theorem.

Theorem. *In every finite noncooperative n-person game there is an equilibrium situation in the $(n-1)$th meta-extension.*

Proof. We consider a finite noncooperative game Γ as in (4.1), in which, as usual, $I = \{1, \dots, n\}$. For $k = 1, \dots, n$, set

$$\{1, \dots, k\} = K_k,$$
$$\{\{1\}, \dots, \{k\}\} = \mathfrak{K}_k,$$
$$\{k+1\} = L_k, \ \{L_k\} = \mathfrak{L}_k.$$

Evidently, here $K_k \cup L_k = K_{k+1}, K_k \cap L_k = \emptyset$, and K_k and L_k are quasipartitions of the coalitions K_k and L_k, respectively.

We carry out the following inductive argument.

First let $k = 1$. Then $\mathfrak{K}_1 = \{1\}$, and it is evident that a 1-stable equilibrium situation exists in a finite game. To make this formal, see the corollary to the theorem of 3.2.

Now suppose that, whatever the strategies $x_{I \setminus K_k} \in \boldsymbol{x}_{I \setminus K_k}$ and $x^0_{L_k} \in \boldsymbol{x}_{L_k} = \boldsymbol{x}_{k+1}$ may be, there exists, in the factorization of the $(k-1)$th metagame $M_{k-1}/x_{I \setminus (K_k \cup L_k)}$, a $\mathfrak{K}_k$-stable situation of the form $(x^0_{L_k}, x_{K_k})$, where $x_{K_k} = x_{K_k}(x^0_{L_k})$ depends on $x^0_{L_k}$.

In addition, in the subsequent factorization $(M_{k-1}/x_{I \setminus K_K \cup L_k})/x_{K_k}(x^0_{L_k})$ of this metagame, there evidently exists (again, by the corollary to the theorem of 3.2, and in our present notation) an $\mathfrak{L}$-stable situation $x^*_{L_k}$. Therefore, by the preceding theorem, in the metagame of the game we have constructed, which has a factorization of the form $M_k/x_{I \setminus (K_{k+1}, L_{k+1})}$, there is a $\mathfrak{K}_{k+1}$-stable situation of the form $(x^0_{L_{k+1}}, x_{K_{k+1}})$. Since the strategy $x_{I \setminus K_{k+1}} \in \boldsymbol{x}_{I \setminus K_{k+1}}$ was arbitrary, the induction is established.

4.12. Example. Let us apply the preceding theorem to the "odd-even" game. For this purpose, we take, in the zero-sum two-person game $\Gamma = \langle \boldsymbol{x}, \boldsymbol{y}, H \rangle$, the metastrategic basis

$$(\boldsymbol{x}, 1), (\boldsymbol{y}^{\boldsymbol{x}}, 2) \tag{4.18}$$

and its metastrategic extension,

$$(\boldsymbol{x}, 1), (\boldsymbol{y}^{\boldsymbol{x}}, 2), (\boldsymbol{y}, 2),$$

which, as we see at once, is regular. Consequently, the basis (4.18) is also regular.

Here the metasituations are pairs of the form (x, f), where $x \in \boldsymbol{x}$ and $f \in \boldsymbol{y}^{\boldsymbol{x}}$; and

$$\bar{H}(x, f) = H(x, f(x)).$$

We thus arrive at the metagame

$$\mathrm{M}_\Gamma = \langle \{1, 2\}, \{\boldsymbol{x}, \boldsymbol{y}^{\boldsymbol{x}}\}, H \rangle.$$

In our case, $\boldsymbol{x} = \boldsymbol{y} = \{e, o\}$, since the set $\boldsymbol{y}^{\boldsymbol{x}}$ consists of the following four response tactics:

1) $e \to e, \quad o \to e;$

2) $e \to e, \quad o \to o;$

3) $e \to o, \quad o \to e;$

4) $e \to o, \quad o \to o.$

Consequently the values of the payoff function have a simple representation by the following table:

$$\begin{array}{c} \\ \\ e \\ o \end{array} \begin{array}{cccc} e \to e & e \to e & e \to o & e \to o \\ o \to e & o \to o & o \to e & o \to o \end{array}$$
$$\begin{matrix} e \\ o \end{matrix} \begin{pmatrix} -1 & -1 & 1 & 1 \\ +1 & -1 & 1 & -1 \end{pmatrix}$$

As in the original game Γ, all situations are Pareto optimal for the coalition $\{1,2\}$.

It is easily seen that here the equilibrium situations are

$$\left(e, \begin{matrix} e \to e \\ o \to o \end{matrix} \right) \quad \text{and} \quad \left(o, \begin{matrix} e \to e \\ o \to o \end{matrix} \right)$$

and in these situations player 1 *loses* (and player 2 gains) a unit amount.

Notice that if we implement the metastrategic construction starting from the metastrategic basis $(x, 2)$, $(x^y, 1)$ we obtain a metagame with the equilibrium situations

$$\left(\begin{matrix} e \to o \\ o \to e \end{matrix}, e \right) \quad \text{and} \quad \left(\begin{matrix} e \to o \\ o \to e \end{matrix}, o \right)$$

in each of which player 1 *wins* (and player 2 loses) a unit amount.

Therefore, projections of equilibrium metasituations in metagames and the payoffs to the players in them are not uniquely determined by the game and the application of an optimality principle to it, but also depend on a supplementary factor: the initial choice of a metastrategical basis. We shall see later that this multiplicity of solutions is a general characteristic of game theory.

4.13. The lack of realizability of an optimality principle for certain games can be informally thought of as the absence from the game of a situation that is satisfactory in the corresponding sense. An example of such an unsatisfactoriness is the absence of equilibrium situations from the game. A manifestation of it, and also a method for overcoming it in the metastrategical way, were illustrated by the theorem in 4.11 and the specific example of the odd-even game. As another, qualitatively different, example, we may use the lack of a *profitable* equilibrium situation in the game (or, equivalently, of an unstable profitable situation in it) as this occurs in, for example, prisoner's dilemma. An unsatisfactoriness of this kind can also be overcome in a metastrategical way.

Theorem. *Every finite noncooperative two-person game* (i.e., $I = \{1,2\}$) *has a* $\mathfrak{K}$*-stable situation for* $\mathfrak{K} = \{1, 2, \{1,2\}\}$ *in its third metagame.*

Proof. Let us set

$$\Delta = \langle x, y^x, \overline{H}_1, \overline{H}_2 \rangle,$$

where, as usual, for $f \in y^x$ we take

$$\overline{H}_i(x, f) = H_i(x, f(x)), \quad i = 1, 2,$$

and in addition

$$\mathrm{E} = \langle x^{(y^x)}, y^x, \overline{\overline{H}}_1, \overline{\overline{H}}_2 \rangle,$$

where we also take, for $g \in x^{y^x}$ and $f \in y^x$,

$$\overline{\overline{H}}_i(g, f) = \overline{H}_i(g(f), f) = H_i(g(f), f(g(f))), \quad i = 1, 2.$$

According to the theorem of 3.2, there is a situation $(x^0, y^0) \in \mathscr{C}_{\{1,2\}}(\Gamma)$ for which

$$H_1(x^*, y^*) \leqq H_1(x^0, y^0), \tag{4.19}$$

$$H_2(x^*, y^*) \leqq H_2(x^0, y^0). \tag{4.20}$$

Let us show that the desired $\{1, 2, \{1, 2\}\}$-stable situation in E is the situation (g^0, f^0) for which

$$f^0(x) = \begin{cases} y^0, & \text{if } x = x^0, \\ y^*, & \text{if } x \neq x^0, \end{cases}$$

$$g^0(f) = \begin{cases} x^0, & \text{if } f = f^0, \\ x^*, & \text{if } f \neq f^0. \end{cases}$$

In fact,

$$(g^0, f^0) \to (g^0(f^0), f^0) = (x^0, f^0) \to (x^0, f^0(x^0)) = (x^0, y^0),$$

i.e., $(x^0, y^0) = \Pr_\Gamma(g^0, f^0)$, and consequently, by the theorem of 4.8, we must have $(g^0, f^0) \in \mathscr{C}_{\{1,2\}}(\mathrm{E})$. Moreover, for $f \neq f^0$ we have

$$\begin{aligned} \overline{\overline{H}}_2(g^0, f) &= \overline{H}_2(g^0, (f), f) = \overline{H}_2(x^*, f) = H_2(x^*, f(x^*)) \\ &\leqq H_2(x^*, y^*) \quad [\text{since } (x^*, y^*) \in \mathscr{C}_2(\Gamma)] \\ &\leqq H_2(x^0, y^0) \quad [\text{in view of (4.20)}] \\ &= \overline{\overline{H}}_2(g^0, f^0), \end{aligned}$$

so that $(g^0, f^0) \in \mathscr{C}_2(\mathrm{E})$).

Finally, let us take $g \neq g^0$. Here we distinguish two cases:

a) $g(f^0) = x^0$. In this case

$$\overline{\overline{H}}_1(g, f^0) = \overline{H}_1(g(f^0), f^0) = \overline{H}_1(x^0, f^0) = H(x^0, y^0) = \overline{\overline{H}}_1(g^0, f^0);$$

b) $g(f^0) \neq x^0$. In this case

$$\begin{aligned} \overline{\overline{H}}_1(g, f^0) &= \overline{H}_1(g(f^0), f^0) = H_1(g(f^0), f^0(g(f^0))) = H_1(g(f^0), y^*) \\ &\leqq H_1(x^*, y^*) \quad [\text{since } (x^*, y^*) \in \mathscr{C}_1(\Gamma)] \\ &\leqq H_1(x^0, y^0) \quad [\text{in view of (4.19)}] \\ &= \overline{H}_1(g^0, f^0). \end{aligned}$$

In either case we have obtained $H_1(g, f^0) \leqq H_1(g^0, f^0)$, so that $(g^0, f^0) \in \mathscr{C}_1(\mathrm{E})$.

Combining these results, we obtain $(g^0, f^0) \in \mathscr{C}_{\mathfrak{R}}(\mathrm{E})$. □

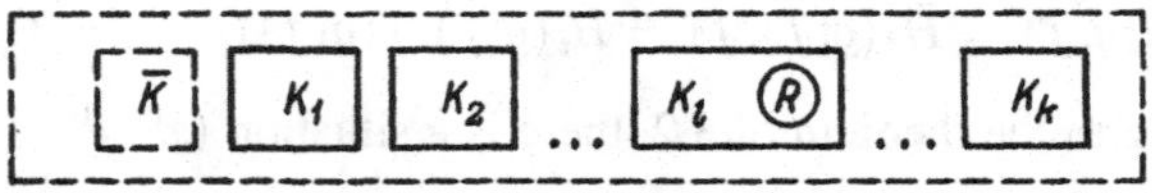

Fig. 1.11

4.14. The theorems established in 4.11 and 4.13 are special cases of the following general proposition.

Theorem. *Let* Γ *in* (4.1) *be a finite noncooperative game, and* $\mathfrak{K}$ *a quasipartition of the set* I *of players.*

Then we can find a regular metastrategic basis

$$(\mathfrak{a}_1, K_1), \ldots, (\mathfrak{a}_k, K_k),$$

depending only on I *and* $\mathfrak{K}$, *in which* $K_1, \ldots, K_k \in \mathfrak{K}$, *such that the metagame over* Γ *generated by it has a* $\mathfrak{K}$*-stable situation.*

Proof. We shall prove the theorem by induction on the length of the quasipartition $\mathfrak{K}$. For the case when the length is unity, the corresponding proposition is the content of the theorem of 3.2.

Let us suppose that the conclusion of the theorem has been established for all quasipartitions of lengths not exceeding m, and consider a quasipartition $\mathfrak{K}$ of length $m+1$.

Let K be a maximal coalition in $\mathfrak{K}$. We denote by $K_1, \ldots, K_k$ the maximal coalitions in $\mathfrak{K}$ that are contained in K (i.e., those such that $K_j \subset K' \subset K$ and $K' \in \mathfrak{K}$ imply either $K_j = K'$ or $K' = K$), and set $I \setminus K = \bar{K}$. We may suppose without loss of generality that $K_1 \cup \ldots \cup K_k = K$, i.e. that each player belonging to K is in some coalition.

Both in the coalitions K_i and in the set $\bar{K}$ of players there may (or may not) be a smaller coalition from $\mathfrak{K}$. The elements of $\mathfrak{K}$ that are contained in $\bar{K}$ evidently form a quasipartition of $\bar{K}$. Let us denote this by $\bar{\mathfrak{K}}$. In addition, we set $\mathfrak{K} \setminus (\bar{\mathfrak{K}} \cup \{K\}) = \mathfrak{K}^*$. As a result we obtain a diagram like the one sketched in Figure 1.11.

Now select any strategy $x_{\bar{K}}$ from $\bar{K}$. Since the length of the quasidecomposition $\mathfrak{K}^*$ does not exceed m, we may construct, by the inductive hypothesis, a metagame

$$\Delta = \langle I, \{y_i\}_{i \in I}, \{H_i\}_{i \in I} \rangle$$

generated by a suitable metastrategic basis, in which situations of the form $(y_K, x_{\bar{K}})$ are $\mathfrak{K}^*$-stable for every $x_{\bar{K}} \in x_{\bar{K}}$. Let us transform Δ metastrategically so that it becomes K-stable.

We now subject the metagame Δ to further metastrategic extensions. In order to simplify the notation, we shall denote the sets y_{K_j} of coalitional strategies of the coalitions K_j simply by y_j, and the strategies themselves by y_j (instead of x_{K_j}).

The dependence of each y_j on $x_{\bar{K}}$ will not always be indicated in the notation. Correspondingly, a $\mathfrak{K}^*$-stable situation in which we are interested will be denoted by

$$y^* = (y_1^*, \ldots, y_k^*, x_{\bar{K}}).$$

According to the theorem of 3.2, the game Δ contains a K-stable situation dominating y^*, with respect to K. We denote it by

$$y^0 = (y_1^0, \ldots, y_k^0, x_{\bar{K}}).$$

(Like strategies appearing in the situation y^*, strategies belonging to y^0 also depend on $x_{\bar{K}}$, but we shall not indicate this dependency explicitly.)

If it happens that $y_K^* = y_K^0$, this situation will be simultaneously $\mathfrak{K}$-stable and K-stable, i.e., $\mathfrak{K}^* \cup \{K\}$-stable, and we will have achieved our objective. We shall therefore suppose that $y_K^* \neq y_K^0$.

We form a metastrategic structure from $2k$ closed sets of metastrategies:

$$\begin{aligned} &\mathbf{y}_1, \mathbf{y}_2, \ldots, \mathbf{y}_k, \\ &\mathbf{f}_j = \mathbf{y}_j^{\mathbf{f}_1 \times \ldots \times \mathbf{f}_{j-1} \times \mathbf{y}_{j+1} \times \ldots \times \mathbf{y}_k}, \quad j = 1, \ldots, k \end{aligned} \tag{4.21}$$

(here $\mathbf{f} \times \ldots \times \mathbf{f}_{j-1} \times \mathbf{y}_{j+1}$ is to be understood as $\mathbf{y}_2$ when $j = 1$; and $\mathbf{f}_{j-1} \times \mathbf{y}_{j+1} \times \ldots \times \mathbf{y}_k$, as $\mathbf{f}_{k-1}$ when $j = k$).

We denote by E the metagame over Δ generated by the metastrategic basis corresponding to (4.21).

We also introduce metastrategies

$$f_j^0 \in \mathbf{f}_j, \qquad j = 1, \ldots, k,$$

by setting inductively, for $j = 1, \ldots, k$, and for arbitrary $y_j \in \mathbf{y}_j$ and $f_j \in \mathbf{f}_j$

$$f_j^0(f_1, \ldots, f_{j-1}, y_{j+1}, \ldots, y_k) = \\ = \begin{cases} y_j^0, & \text{if } f_u = f_u^0 \text{ for } u = 1, \ldots, j-1, \text{ and } y_u = y_u^0 \text{ for } u = j+1, \ldots, k, \\ y_j^* & \text{in all other cases.} \end{cases} \tag{4.22}$$

Let us show that the metasituation

$$f^0 = (f_1^0, \ldots, f_k^0, x_{\bar{K}}) \tag{4.23}$$

is $\mathfrak{K}^* \cup \{K\}$-stable in the game E.

For this purpose we form the projection f of f^0 on Δ.

The construction of the metastrategic development of this metasituation in accordance with (4.22) consists of k steps, each of which in turn replaces f_j^0 by y_j^0 in the list of strategies of the situation. Therefore

$$\mathrm{Pr}_{\Delta}(f^0) = (y_1^0, \ldots, y_k^0, x_{\bar{K}}).$$

The K-stability of a situation t^0 in Δ means that we will not have

$$H_K(\bar{y}) \geq H_K(y^0) \tag{4.24}$$

for any situation $\bar{t}$ of the form $(\bar{t}_K, x_{\bar{K}})$.

But for every metastrategy $\bar{f}_K \in \mathbf{f}_K$ the projection $\mathrm{Pr}_{\Delta}(\bar{f}_K, x_{\bar{K}})$ evidently has the form $(\bar{y}_K, x_{\bar{K}})$, and $y^0 \in \mathrm{Pr}_{\Delta}(f^0)$. Therefore the impossibility of (4.24) implies the impossibility of

$$H_K(\mathrm{Pr}_{\Delta}(\bar{f}_K, x_{\bar{K}}) \geq H_K(\mathrm{Pr}_{\Delta}(f^0)),$$

or, if we pass to the game E, the impossibility of

$$H_K(\bar{f}_K, \bar{x}_K) \geq H_K(f^0).$$

But this means that f^0 is stable in E.

We now prove the R-stability of the situation f^0 for an arbitrary coalition $R \in \mathfrak{K}^*$. For definiteness, we suppose that $R \subset K_1$. Consider an arbitrary coalitional metastrategy $f_R \in \mathbf{f}_R$.

If $f_R = f_R^0$, we trivially have

$$H_R(f^0 \| f_R) = H_R(f^0)$$

and in particular it is impossible to have

$$H_R(f^0 \| f_R) \geq H_R(f^0).$$

Now let $f_R \neq f_R^0$. We shall construct and analyze the projection of the metasituation $f^0 \| f_R$ on the game Δ. For $j = k, k-1, \ldots, l+1$, we construct inductively

$$f_j^0(f_1^0, \ldots, f_{l-1}^0, (f_R, f_{K_l \setminus R}^0), f_{l+1}^0, \ldots, f_{j-1}^0, \bar{y}_{j+1}, \ldots, \bar{y}_k) = \bar{y}_j$$

(evidently there are $k - l$ such formulas).

Here we have assumed that $f_R \neq f_R^0$. Consequently, by (4.22) we must have

$$y_j = y_j^* \text{ for } j = k, k-1, \ldots, l+1.$$

We also set

$$(f_R, f_{K_l \setminus R})(f_1^0, \ldots, f_{l-1}^0, y_{l+1}^*, \ldots, y_k^*) = \bar{y}_l.$$

According to 4.3, the coalitional strategy $\bar{y}_l$ must have the form $(\bar{y}_R, y_{K_l \setminus R}^0)$. If $\bar{y}_j = y_j^* = y_j^0$ for $j = k, k-1, \ldots, l+1$ and in addition $\bar{y}_l = y_l^0$, then by (4.22) the process of metastrategic development continues just as for f^0, and we obtain

$$\mathrm{Pr}_\Delta(f^0 \| f_R) = \mathrm{Pr}_\Delta(f^0).$$

Therefore our question has a trivial answer.

Consequently we suppose that one of the coalitional strategies $\bar{y}_j$ $(j = k, k-1, \ldots, l+1)$ is different from y_j^0. Then by (4.22) we will have

$$f_j^0(f_1^0, \ldots, f_{l-1}^0, y_{l+1}^*, \ldots, y_{l-1}^*, \bar{y}_l, y_{l+1}^*, \ldots, y_k^*) = y_j^*$$

for $j = l-1, l-2, \ldots, 1$, and also

$$f_{K_l \setminus R}^0(f_1^0, \ldots, f_{l-1}^0, y_{l+1}^*, \ldots, y_k^*) = y_{K_l \setminus R}^*.$$

Thus, in this case

$$\mathrm{Pr}_\Delta(f^0 \| f_R) = (y_1^*, \ldots, y_{l-1}^*, \bar{y}_R, y_{K_l \setminus R}^*, y_{l+1}^*, \ldots, y_K^*) = y^* \| \bar{y}_R. \tag{4.25}$$

Because of the assumed R-stability of y^* we cannot have

$$H_R(y^* \| \bar{y}_R) \geq H_R(y^*),$$

and because of the stipulation of K-domination, which includes R-domination,

$$H_R(y^*) \leqq H_R(y^0)$$

is impossible, and

$$H_R(y^* \| y_R) \geq H_R(y^0).$$

Therefore in view of (4.25) we cannot have

$$H_R(f^0 \| f_R) \geq H_R(f^0),$$

and f^0 is R-stable.

We now use the inductive hypothesis again: this time, applied to the game E and a quasipartition $\mathfrak{K}$ whose length does not exceed k. Here we construct a metagame Z over E, in which there is a $\mathfrak{K}$-stable situation of the form $(f_K^0, g_{\bar{K}}^0)$ (or, more precisely, of the form $(f_K^0, g_{\bar{K}}^0(f_K^0))$). Here all the coalitions in the corresponding terms of the metastrategic basis are contained in K, so that the $\mathfrak{K}^* \cup \{K\}$-stability, established above, of the situation

$$(f_K^0, \mathrm{Pr}_\mathrm{E}(g_{\bar{K}}^0(f_K^0))$$

in E is preserved by the transition to the situation $(f_K^0, g_{\bar{K}}^0)$ in Z. Therefore the situation $(f_K^0, g_{\bar{K}}^0)$ is $\mathfrak{K}^* \cup \{K\}$-stable and simultaneously $\bar{\mathfrak{K}}$-stable, hence $\mathfrak{K}$-stable, and the induction is complete. □

§5 Realizability of equilibrium situations in mixed strategies

5.1 Mixed extensions of noncooperative games. In 3.8 we established the wider realizability of $\mathfrak{R}$-optimality in games with complete information, as compared to arbitrary noncooperative games. This led to the introduction of metastrategies, which, in essence, are formally equivalent to providing the players with the possibility of a certain amount of information. As we saw in the preceding section, in this way we succeeded essentially in extending the realizability of $\mathfrak{R}$-principles to arbitrary finite games.

In a similar way, for concave (or quasiconcave) games the theorems in 3.10–3.12 give us a basis for supposing that the introduction of generalized strategies that are convex combinations of the original strategies, and are said to be *mixed*, also leads to new possibilities for realizing $\mathfrak{R}$-optimality principles. This conjecture will be established, just for equilibrium situations.

As it turns out, equilibrium situations in mixed strategies are possessed by games in a fundamental class of noncooperative games that has already been discussed: games with compact strategy spaces and continuous payoff functions (see 1.9 and 3.10).

5.2. We shall construct an extension of the noncooperative game

$$\Gamma = \langle I, \{\mathfrak{x}_i\}_{i\in I}, \{H_i\}_{i\in I}\rangle \tag{5.1}$$

and, more precisely, of the measurable noncooperative game (see 1.10)

$$\Gamma = \langle I, \{\langle \mathfrak{x}_i, \Xi_i\rangle\}_{i\in I}, \{H_i\}_{i\in I},\rangle \tag{5.2}$$

which we shall describe as *mixed*.

For this purpose we consider a set X_i of probability (i.e., countably additive normalized) measures on $\langle \mathfrak{x}_i, \Xi_i\rangle$. We shall call measures belonging to X_i *mixed strategies* of player i in the noncooperative game Γ from (5.1) (and more precisely, in the measurable game from (5.2)).

For our purposes it is natural to suppose that the set X_i is convex, i.e., to suppose that together with arbitrary $X_i', X_i'' \in X_i$ the measure defined by

$$X_i(\mathfrak{y}_i) = \lambda X_i'(\mathfrak{y}_i) + (1-\lambda)X_i''(\mathfrak{y}_i) \text{ for an arbitrary } \mathfrak{y}_i \in \Xi_i,$$

with an arbitrary $\lambda \in [0,1]$, also belongs to X_i; we also set

$$X_i^{(\lambda)} = \lambda X_i' + (1-\lambda)X_i''.$$

In addition, let us assume, what is by no means as natural, that the set X_i of mixed strategies for the player i contains *all degenerate measures* on $\mathfrak{x}_i$. This actually means that we are assuming that $\mathfrak{x}_i \in X_i$. When the $\mathfrak{x}_i$ are thought of as elements of X_i, we call them *pure strategies* of player i, and mixed strategies which are not pure are said to be that player's *strictly mixed* strategies.

As we know*), for any collectively independent countably additive measures X_i on the measure spaces $\langle x_i, \Xi_i \rangle$ $(i \in I)$, it is possible to construct a countably additive product measure $X = X_1 \times \ldots \times X_n$ on $\langle x, \Xi \rangle$ (see 1.10) connected in a specific way with the original measures X_i $(i \in I)$. We emphasize (since this turns out to be essential for the extensions developed in § 3, Chapter 2) that, for the determination of the value of X in individual subsets of Ξ, its countable additivity is used in an essential way.

Measures X on $\langle x, \Xi \rangle$ are known as *situations of* Γ *in mixed strategies*. The set of situations of Γ in mixed strategies is denoted by $\boldsymbol{X}$.

By the definition of a measurable game, each payoff function H_i in (5.2) is measurable with respect to Ξ, i.e. is a real random variable on $\langle x, \Xi \rangle$. Therefore it is integrable with respect to each measure $X \in \boldsymbol{X}$, and we may set

$$\widetilde{H}_i(X) = \int_x H_I(x) dX(x) = \int_{x_1 \times \ldots \times x_n} H_i(x_1, \ldots, x_n) d(X_1 \times \ldots \times X_n), \quad (5.3)$$

or, using the independence of measures $X_1, \ldots, X_n$ and Fubini's theorem,

$$\widetilde{H}_i(x) = \int_{x_1} \ldots \int_{x_n} H_i(x_1, \ldots, x_n) dX_1(x_1) \ldots dX_n(x_n). \quad (5.4)$$

For situations x_I in pure strategies we have

$$\widetilde{H}_i(x_I) = \int_x H_i(x) dx_I(x) = H_i(x_I),$$

i.e., the functions $\widetilde{H}_i$ are extensions of the corresponding functions H_i. Consequently, we shall denote them simply by H_i whenever no ambiguity will result.

We are therefore led to the following definition.

Definition. *A mixed extension* of a measurable noncooperative game Γ as in (5.2) is defined to be the noncooperative game

$$\bar{\Gamma} = \langle I, \{\boldsymbol{X}_i\}_{i \in I}, \{H_i\}_{i \in I} \rangle,$$

where $\boldsymbol{X}_i$ is a convex family of all the probability measures on $\langle x_i, \Xi_i \rangle$ $(i \in I)$, and the payoff functions $H_i : \boldsymbol{X} \to \mathbf{R}$ are defined as mathematical expectations in accordance with (5.3). □

We emphasize that in the definition of a measurable noncooperative game, and hence implicitly in the definition of its mixed strategies, the mixed strategies are "fitted" to the available payoff functions, which must remain measurable.

In practice, we usually construct the σ-algebras Ξ_i (and the mixed strategies) independently of the payoff functions. In this case, of course, we have to verify

*) See, for example, P.R. Halmos, Measure Theory, van Nostrand, New York, 1950.

the measurability of the payoff functions separately. A test for this measurability will be discussed in § 6.

As is easily seen, if two noncooperative games are affinely equivalent their similar mixed extensions (i.e., those having the same sets of mixed strategies) are also affinely equivalent. On the other hand, the similar mixed extensions (in the same sense) of strategically equivalent games are not necessarily strategically equivalent.

5.3. It follows from (5.4) that, for every $x_j^0 \in x_j$,

$$H_i(X\|x_j^0) = \int\limits_{x_1} \cdots \int\limits_{x_{j-1}} \int\limits_{x_{j+1}} \cdots \int\limits_{x_n} H_i(x_1 \ldots, x_{j-1}, x_j^0, x_{j+1} \ldots, x_n) \times dX_1(x_1) \ldots dX_{j-1}(x_{j-1}) dX_{j+1}(x_{j+1}) \ldots dX_n(x_n). \tag{5.5}$$

Together with (5.4), this yields

$$H_i(X) = \int\limits_{x_j} H_i(X\|x_j) dX_j(x_j). \tag{5.6}$$

5.4. Existence of equilibrium situations in mixed strategies. The following proposition is evident.

Theorem. *A mixed extension of a measurable noncooperative game is a concave game.*

For the proof, it is enough to observe that the set of mixed strategies of each player in a measurable game is convex, and the values of the payoff functions, in accordance with (5.6), depend linearly on the mixed strategies, so that they are concave. □

So, for instance, the concave game in the example of 3.9 is a mixed extension of the "battle of the sexes" bimatrix game in 2.16.

The following proposition is an immediate consequence of the preceding theorem and the theorems of 3.10 or 3.11.

Theorem. *If compact convex sets of mixed strategies for the players in a noncooperative game* Γ *can be chosen so that the payoff functions in the resulting situations are continuous, then* Γ *will have equilibrium situations in the mixed strategies.*

This theorem has the serious deficiency that to infer the existence of equilibrium situations in a noncooperative game does not require the verification of specific properties of the game itself, but the construction of a mixed extension that satisfies the necessary conditions.

5.5. It is possible to find theorems that are free of this deficiency. The simplest one reads as follows.

Theorem. *Every finite noncooperative game has equilibrium situations in the mixed strategies.*

Proof. If the game Γ in (5.1) is finite, the sets $\boldsymbol{X}_i$ of mixed strategies of the players in it:

$$\boldsymbol{X}_i = \{X_i : X_i(x_i) \leqq 0 \text{ for } x_i \in \boldsymbol{x}_i \text{ and } \sum_{x_i \in \boldsymbol{x}_i} X_i(x_i) = 1\}$$

are simplexes in finite-dimensional Euclidean spaces, and the topologies in these spaces are compact. Here the spaces are linear, and their subsets $\boldsymbol{X}_i$ are convex.

Besides, for every situation X in the mixed strategies we will have

$$H_i(X) = \sum_{x \in \boldsymbol{x}} H_i(x)X(x) = \sum_{x_1 \in \boldsymbol{x}_1} \cdots \sum_{x_n \in \boldsymbol{x}_n} H_i(x_i, \dots, x_n)X_1(x_1) \dots X_n(x_n).$$

Consequently the functions $H_i : \boldsymbol{X} \to \mathbf{R}$ are linear in each argument, and therefore concave (and especially quasiconcav).

The conclusion of the theorem follows from the theorems of 3.10 or from those of 3.11. □

5.6. The preceding result can be developed in a natural way in two directions.

The first is a matter of pure logic and is connected with the elementary proofs of the theorems in 5.5 which were obtained above. For example, in particular, it is appropriate to raise the question of a more direct application of Brouwer's theorem, or even of the complete rejection of the use of fixed-point theorems here. This line will be developed in 1.1, Chapter 2.

The second direction, however, leads to the consideration of more general results. First of all, there is the question of extending the theorems we have proved to compact games with continuous payoff functions. There are two ways of doing this. In the following subsections the existence of equilibrium situations in mixed strategies for these games will be established in a rather traditional functional-theoretic way, by using Glicksberg's very powerful fixed point theorem, which was proved specifically for game-theoretic applications. In addition, the same theorem on the existence of equilibrium situations will be obtained in § 6 by the exploration of questions that are more natural for this circle of ideas, questions of purely game-theoretic approximate constructions based on the concepts of ε-semihomorphism and ε-homomorphism (see 1.18).

5.7. By way of preparation we introduce some concepts pertaining to functional analysis.

Let $\mathscr{A}$ be a set which is partially ordered with respect to $\leqq$ (i.e., $\leqq$ is reflexive, antisymmetric, and transitive on $\mathscr{A}$). It is said to be *directed* if, for every α_1 and $\alpha_2 \in \mathscr{A}$, there is an $\alpha \in \mathscr{A}$ such that $\alpha_1 \leqq \alpha$ and $\alpha_2 \leqq \alpha$. Every mapping of a directed set $\mathscr{A}$ on an arbitrary set X with $\alpha \mapsto x_\alpha$ is said to be *directed* in X. If X can be topologized, we may speak of the *convergence* of a directed set $\{x_\alpha\}$ to its *limit* x if, for every neighborhood ω of x, there is an α_ω such that $x_\alpha \in \omega$ for all $\alpha \geqq \alpha_\omega$. We denote the convergence of $\{x_\alpha\}$ to x by $x_\alpha \to x$.

We call x a *limit point* of $\{x_\alpha\}$ in X if, for every neighborhood ω of x and every $\alpha_0 \in \mathscr{A}$, there is an $\alpha \geqq \alpha_0$ such that $x_\alpha \in \omega$. It is clear that every limit of a directed set is a limit point, but the converse does not hold.

Let X be a linear Hausdorff space and $S \subset X$. A multivalued mapping $\Phi : S \to X$ that carries points of S to nonempty convex subsets of X is said to be *closed* if its graph (i.e., the set of all ordered pairs of the form (x, y), where $x \in S$ and $y \in \Phi(x)$) is closed in $X \times X$.

In terms of directed sets, the property of being closed in accordance with this definition means for Φ that

$$x_\alpha \to x,\ y_\alpha \in \Phi(x_\alpha),\ \text{and}\ y_\alpha \to y\ \text{imply}\ y \in \Phi(x). \qquad (*)$$

It turns out that the property of being closed is preserved if instead of requiring that $y_\alpha \to y$ we require only that y is a limit point of $\{y_\alpha\}$.

In fact, let $x_\alpha \to x$, $y_\alpha \in \Phi(x_\alpha)$, and let y be a limit point of $\{y_\alpha\}$. Consider a directed family $\{U\}$ of neighborhoods of y, ordered by inclusion (i.e., we set $U_1 \leqq U_2$ if and only if $U_1 \subset U_2$; the family is directed because together with any two neighborhoods it also contains their intersection). We now form the set $\Delta = \{(\delta, U) : y_\delta \in U\}$ of pairs, and introduce the relation $\leqq$ on it by setting

$$(\delta_1, U_1) \leqq (\delta_2, U_2) \text{ if and only if } \delta_1 \leqq \delta_2,\ U_1 \leqq U_2.$$

The family Δ is ordered by $\leqq$. In fact, for all $y_{\delta_1} \in U_1$ and $y_{\delta_2} \in U_2$ we can find $\delta_0 \geqq \delta_1, \delta_2$ and then $\delta > \delta_0$ such that $y_\delta \in U_1 \cap U_2 = U$, from which we obtain $(\delta_1, U_1), (\delta_2, U_2) \leqq (\delta, U)$.

Now set $x_{(\delta,U)} = x_\delta$ and $y_{(\delta,U)} = y_\delta$. Then by hypothesis we will have $x_{(\delta,U)} \to x$ and $\Phi(x_{(\delta,U)}) \ni y_{(\delta,U)}$, and by what has been proved $y_{(\delta,U)} \to y$. Consequently it follows from $(*)$ that $y \in \Phi(x)$.

5.8. We now turn to Glicksberg's fixed point theorem, which is of interest here.

Theorem. *Let Φ be a closed mapping on a locally convex Hausdorff space X, carrying points to nonempty convex sets and transforming a compact convex set $S \subset X$ to itself. Then Φ has a fixed point.*

Proof. We select an arbitrary symmetric convex neighborhood V of the zero element in the linear space X. The family of sets of the form $x + V$, where $x \in S$, evidently covers S, and since S is compact we may select a finite subcovering $\{x_1 + V, \dots, x_n + V\}$. Let us denote the convex hull of the points $x_1, \dots, x_n$ by S_V and set

$$\Phi_V(x) = (\Phi(x) + V) \cap S_V. \qquad (5.7)$$

It follows from $f \in \Phi(x_i) \subset S \subset U_i(x_i + V)$ that f has the form $x_i + v$, where $v \in V$, so that $x_i = f - v$. Since V is symmetric, we have $-v \in V$, so that $x_i \in f + V \subset \Phi + V \subset \Phi + \bar{V}$. Since, in addition, $x_i \in S_V$, we have $x_i \in \Phi_V(x)$, so that $\Phi_V(x) = \emptyset$. The convexity of $\Phi_V(x)$ follows from the convexity of $\Phi(x)$, V, and S_V. Finally, by (5.7), for $x \in S_V$ we have $\Phi_V(x) \subset S_V$.

Let us show that Φ_V is closed. To establish this, we suppose that $x_\alpha \to x$, $y_\alpha \in \Phi_V(x_\alpha)$, and $y_\alpha \to y$. Then $y_\alpha \in (\Phi(x_\alpha) + V) \cap S_V$, i.e. there are $z_\alpha \in \Phi(x_\alpha)$ and $v_\alpha \in V$ such that $y_\alpha = z_\alpha + v_\alpha \in S_V$. But $z_\alpha \in \Phi(x_\alpha) \subset S$, and since S is compact the directed set $\{z_\alpha\}$ has a limit point $z \in S$. Since Φ is a closed mapping, it follows from what we have said that $z \in \Phi(x)$.

Furthermore, since $v_\alpha = y_\alpha - z_\alpha$ and $y_\alpha \to y$, the directed set has the limit point $y - z$. But since $v_\alpha \in V$ and V is closed, we must also have $v \in V$, so that $y = z + v \in \Phi(x) + V$. On the other hand, $y \in S_V$ since S_V is closed. Consequently $y \in (\Phi(x) + V) \cap S_V = \Phi_V(x)$, and we have established that Φ_V is closed.

The set S_V is a convex combination of a finite number of points. Hence we may consider S_V to be a closed bounded convex polygon in a finite-dimensional Euclidean space. Accordingly, with the Euclidean topology, we may apply Kakutani's theorem to the mapping $\Phi_V : S_V \to S_V$, obtaining an $x_V \in S_V$ for which $x_V \in \Phi_V(x_V)$.

The family $\{V\}$ of neighborhoods of the origin that we are considering, and hence also the collection $\{x_V\}$ of points, forms a directed set V, ordered by inclusion. Since all the $x_V \in S$, and S is compact, the directed set $\{x_V\}$ has a limit point x.

As in the criterion for a mapping to be closed, we form a family $\Delta = \{(V,U) : x_V \in U\}$ of pairs, where U is a neighborhood of x. The family of sets U and V is directed by inclusion. Let us set $x_{(V,U)} = x_V$, so that

$$x_{(V,U)} \to x, \tag{5.8}$$

and for every V_0 we can find U and $V \geqq V_0$ such that

$$(V,U) \in \Delta. \tag{5.9}$$

Now choose $z_{(V,U)}$ so that $x_{(V,U)} - z_{(V,U)} \in V$. At the same time, $x_{(V,U)} = x_V \in \Phi(x_V) + V$. It follows from these inclusions that $z_{(V,U)} \in \Phi(x_V)$, so that by (5.9) we must have $z_{(V,U)} \to x$. From this and the hypothesis that Φ is a closed mapping, it follows that $x \in \Phi(x)$. □

5.9. We now turn our attention to the existence of equilibrium situations.

Theorem. *Suppose that, in the noncooperative game* Γ *in* (5.1), *all the sets* x_i *of strategies are (Hausdorff) compact and that all the payoff functions* H_i *are continuous with respect to* x *in the componentwise topology.*

Then Γ *has equilibrium situations in the mixed strategies.*

Proof. For each player $i \in I$, consider the mapping

$$\mathfrak{C}_i : \boldsymbol{X} \to 2^{\boldsymbol{X}_i},$$

where, for each $X \in \boldsymbol{X}$, we set

$$\mathfrak{C}_i(X) = \{X_i^0 : H_i(H\|X_i^0) = \sup_{X_i} H_i(X\|X_i)\},$$

i.e.,

$$\mathfrak{C}_i(X) = \operatorname*{argmax}_{X_i} H_i(X\|X_i),$$

and

$$\mathfrak{C}(X) = \prod_{i\in I} \mathfrak{C}_I(X).$$

Consequently we consider the mapping $\mathfrak{C} : \boldsymbol{X} \to 2^{\boldsymbol{X}}$. Let us verify that $\boldsymbol{X}$ and $\mathfrak{C}$ satisfy the hypotheses of Glicksberg's fixed point theorem. To do this we must find a topology on $\boldsymbol{X}$ that makes it compact, and makes $\mathfrak{C}$ closed.

The probability measures imposed on each $\boldsymbol{X}_i$ are linear functionals on the Banach spaces $C(x_i)$ of continuous functions, i.e. $\boldsymbol{X}_i \subset C^*(x_i)$. As a suitable topology on $C^*(x_i)$ we choose the w^*-topology, in which a fundamental system of neighborhoods of the origin has the form

$$V_{f_1,\dots f_k;\varepsilon} = \{\mu \in C^*(x_i) : |\mu(f_l)| < \varepsilon, \quad l = 1,\dots,k\}$$

(here $f_1,\dots,f_k$ is any finite set of elements of $C(x_i)$, and $\varepsilon > 0$). Convergence of a sequence of measures in the w^*-topology means the convergence of the corresponding sequence of integrals for every given continuous function on x_i.

From the definition of the topology it follows immediately that each of the spaces $C^*(x_i)$ is, in this topology, a locally convex Hausdorff topological linear space. It is known that every norm-bounded w^*-closed set in $C^*(x_i)$ is compact in the w^*-topology. Hence the sets $\mathbf{X}_i$ are w^*-compact sets (since $\|\mu\| = \mu(x_i) = 1$ this set is bounded, and it follows directly that it is closed). Therefore the set $\mathbf{X}$, as the Cartesian product of the $\mathbf{X}_i$, is a w^*-compact subset of the locally convex linear Hausdorff space

$$\prod_{i\in I} C^*(x_i) = \left(\prod_{i\in I} C(x_i)\right)^*.$$

We now turn our attention to the mappings $\mathfrak{C}_i$ and $\mathfrak{C}$. By the very definition of the w^*-topology, the functions H_i are continuous on $C^*(x_i)$ and therefore attain their maxima on a w^*-compact subset of $\mathbf{X}_i$. Consequently the images $\mathfrak{C}(X\|\mathbf{X}_i)$, and therefore $\mathfrak{C}(X)$, are nonempty for each $X \in \mathbf{X}$. Since H_i is linear over $\mathbf{X}_i$ it follows that the images $\mathfrak{C}_i(X)$ are convex; hence so is $\mathfrak{C}(X)$; and since H_i is w^*-continuous, $\mathfrak{C}(X)$ is also w^*-closed.

Consequently $\mathfrak{C}$ carries the points of $\mathbf{X}$ to nonempty convex w^*-closed subsets of $\mathbf{X}$. Finally, since H_i is continuous on x, the mapping $\mathfrak{C}$ is closed.

We see that the hypotheses of Glicksberg's theorem are satisfied. Application of that theorem gives us the existence of a situation $X^* \in \mathbf{X}$ such that $X^* \in \mathfrak{C}(X^*)$, i.e.

$$H_i(X^*) = \sup_{X_i} H_i(X^*\|X_i)$$

for all $i = 1, \ldots, n$, or equivalently

$$H_i(X^*) \geqq H_i(X^*\|X_i) \text{ for all } X_i \in \mathbf{X}_i. \quad \square$$

5.10 Spectra of mixed strategies. We introduce the following useful concept, which will be needed later.

Definition. The *spectrum* of the mixed strategy X_i of the player $i \in I$ is the smallest closed measurable subset supp X_i for which $X_i(\operatorname{supp} X_i) = 1$.

Lemma. *If $x_i \in \operatorname{supp} X_i$, and ω is any X_i-measurable neighborhood of x_i, then* $X_i(x_i) > 0$.

To establish this it is enough to notice that, in the first place, $\operatorname{supp} X_i \setminus \omega$ is closed and not equal to $\operatorname{supp} X_i$; and, in the second place,

$$X_i(\operatorname{supp} X_i \setminus \omega) + X_i(\omega) = X_i(\operatorname{supp} X_i) = 1.$$

Therefore if $X_i(\omega) = 0$ then $X_i(\operatorname{supp} X_i \setminus \omega) = 1$, which contradicts the definition of the spectrum as the *smallest* set of full measure. $\square$

It follows immediately from the definition of the spectrum that, for every X_i-measurable function f, we have

$$\int_{x_i} f(x_i)dX_i(x) = \int_{\operatorname{supp} X_i} f(x_i)dX_i(x_i).$$

Definition. A mixed strategy is called *finite* if its spectrum is finite.

A mixed extension of a noncooperative game Γ in which the sets X_i of all the strategies of the players are the sets of all their finite strategies is said to be a *finite mixed extension* of the game.

It is clear that, with respect to situations in finite mixed strategies, all payoff functions are measurable, so that the problem mentioned at the end of 5.2 has a trivial solution.

5.11. It is clear, by the way, from what we said in 3.10 and 3.11, that mixed strategies play no role in connection with the existence of equilibrium strategies in compact concave continuous games. This observation leads to the following useful generalization.

Let us modify the definition in 3.9: A noncooperative game as in (5.1) is said to be *(strictly) concave* for player i if the set x_i of that player's strategies is a convex compact subset of a Banach space, and the player's payoff function $H_i(x\|x_i)$ is a (strictly) concave function of x_i for every situation x.

The concavity property of the payoff function can be transferred from pure to mixed strategies.

Lemma. *If $H_i(x\|x_i)$ is (strictly) concave in x_i for every situation x (in pure strategies), then $H_i(X\|x_i)$ will be (strictly) concave in x_i for every situation X (in mixed strategies).*

Proof. Choose x_i' and $x_i'' \in x_i$ and $\lambda \in [0,1]$. Then (5.4) gives us

$$H_i(X\|(\lambda x_i' + (1-\lambda)x_i'')) =$$

$$= \int_{x_1} \dots \int_{x_n} H_i(x\|(\lambda x_i' + (1-\lambda)x_i''))dX_1(x_1)\dots dX_{i-1}(x_{i-1})dX_{i+1}(x_{i+1})\dots dX_n(x_n).$$

Here the integrand is a (strictly) concave function of the distribution of the vertical components of the argument. Replacing it by the number

$$\lambda H_i(x\|x_i') + (1-\lambda)H_i(x\|x_i''),$$

which is no smaller (and strictly larger in the case of strict concavity and $\lambda \in (0,1)$), and then summing the integrals and again applying (5.4) to each of them, we obtain the required conclusion. □

For further use of the concept of dominating strategies, see 2.25.

Theorem. *If the game Γ is compact and concave for player i then every mixed strategy of player i is dominated by one of that player's pure strategies. If also Γ is strictly concave for i, then every mixed strategy for i which is not pure is strictly dominated by one of i's pure strategies.*

Proof. We first notice that the compact set $\boldsymbol{X}_i$ is metrizable. Let $X_i \in \boldsymbol{X}_i$; choose any positive numbers ε' and ε'' and form the finite measurable ε'- and ε''-partitions

$$\mathfrak{A}' = \{A'_1, \ldots, A'_{m'}\}, \mathfrak{A}'' = \{A''_1, \ldots, A''_{m''}\}$$

of the set $\operatorname{supp} X_i$ (with the understanding that the diameter of each element of $\mathfrak{A}'$ does not exceed ε', and the diameter of each element of $\mathfrak{A}''$ does not exceed ε''). Further choose

$$x'_{ik} \in A'_k,\ k = 1, \ldots, m', \text{ and } x''_{il} \in A''_l, l = 1, \ldots, m'',$$

arbitrarily, and form the sums

$$s'_i = \sum_{k=1}^{m'} x'_{ik} X_i(A'_{ik}) \text{ and } s''_i = \sum_{l=1}^{m''} x''_{il} X_i(A''_l).$$

Finally, set, for $k = 1, \ldots, m$ and $l = 1, \ldots, m$,

$$A'_k \cap A'_l = A_{kl},$$

choose $y_{ijk} \in A_{kl}$ arbitrarily, and form

$$s_i = \sum_{k=1}^{m'} \sum_{l=1}^{m''} y_{ikl} X_i(A_{kl}).$$

All the sums s'_i, s''_i, and s_i are elements of x_i because this set is convex. We have

$$\begin{aligned} \rho(s'_i, s_i) &= \left\| \sum_{k=1}^{m'} x'_{ik} \sum_{l=1}^{m''} X_i(A_{kl}) - \sum_{k=1}^{m'} \sum_{l=1}^{m''} y_{ikl} X_i(A_{kl}) \right\| \\ &= \left\| \sum_{k=1}^{m'} \sum_{l=1}^{m''} (x'_{ik} - y_{ikl}) X_i(A_{kl}) \right\| \leqq \sum_{k=1}^{m'} \sum_{l=1}^{m''} \rho(x'_{ik}, y_{ikl}) X_i(A_{kl}). \end{aligned}$$

But since x'_{ik} and $y_{ikl} \in A'_k$, we must have $\rho(x'_{ik}, y_{ikl}) < \varepsilon'$, so that

$$\rho(s'_i, s_i) \leqq \varepsilon' \sum_{k=1}^{m'} \sum_{l=1}^{m''} X_i(A_{kl}) = \varepsilon'.$$

Similarly we obtain

$$\rho(s''_i, s_i) \leqq \varepsilon'',$$

whence

$$\rho(s'_i, s''_i) \leqq \varepsilon' + \varepsilon''. \tag{5.10}$$

If we now take arbitrary measurable ε-partitions with successively smaller $\varepsilon > 0$, we obtain a sequence of values of sums s_i in which, by (5.10), each sequence converges, and which all belong to $\boldsymbol{x}_i$. Therefore, since $\boldsymbol{x}_i$ is complete (a consequence of its compactness), these sequences have a common limit. Let us denote this limit by

$$\int_{\boldsymbol{x}_i} x_i dX_i(x_i).$$

Therefore, to each mixed strategy X_i^* there corresponds a pure strategy x_i^*:

$$x_i^* = \int_{\boldsymbol{x}_i} x_i dX_i^*(x_i). \tag{5.11}$$

If now $H_i(x\|x_i)$ is concave in x_i, for each mixed strategy X_i^* we have, choosing x_i^* according to (5.11),

$$H_i(X\|x_i^*) = H_i\left(X\| \int_{\boldsymbol{x}_i} x_i dX_i^*(x_i^*)\right) \leqq \int_{\boldsymbol{x}_i} H_i(X\|x_i)dX_i(x_i) = H_i(X\|x_i).$$

If H_i is strictly concave in x_i and X_i is "strictly" mixed, there is strict inequality in the preceding relation. □

Corollary. *If, in a game that is concave for the player i, this player has a mixed equilibrium strategy, then the player has a pure equilibrium strategy. In a game that is strictly concave for the player i that player does not have any (strictly) mixed equilibrium strategies.*

5.12 A property of the set of equilibrium situations. The structure of the set $\mathscr{C}(\Gamma)$ of equilibrium situations in noncooperative games can be, in general, rather complicated. One of the few properties that belong to all such sets in all noncooperative games is described in the following theorem.

Theorem. *Let X be a situation, and i a player, in the game Γ. Let $\mathcal{M}_i(X) = \{X_i \in \boldsymbol{X}_i : (X\|X_i) \in \mathscr{C}(\Gamma)\}$.*

*The set $\mathcal{M}_i(X)$ is a (possibly empty) closed bounded subset of $\boldsymbol{X}_i$. If Γ is finite, $\mathcal{M}_i(X)$ is a convex polygon.**)

Proof. The following inequalities are necessary and sufficient for $X_i' \in \mathcal{M}_i(X)$:

$$H_i\left(X \underset{i}{\|} x_i\right) \leqq H_i\left(X \underset{i}{\|} X_i'\right) \qquad \text{for all } x_i \in \boldsymbol{x}_i,$$

$$H_j\left(X \underset{i,j}{\|} (X_i, x_j)\right) \leqq H_j\left(X \underset{i}{\|} X_i'\right) \qquad \text{for all } j \neq i \text{ and } x_j \in \boldsymbol{x}_j$$

*) As usual, by a convex polygon in a linear space we mean the convex hull of a finite set of points.

(of course, the first inequality can be formally considered to be a special case of the second). All these inequalities are linear in X_i, so that each one describes a closed convex subset of the linear space X_i. Consequently the set of points of X_i that satisfy all the inequalities is also closed and convex (of course, it might be empty).

If Γ is finite, the number of inequalities defining $\mathcal{M}_i(x)$ is also finite, so that $\mathcal{M}_i(x)$ is a polygon. □

5.13 A necessary condition for equilibrium. We present a test for the equilibrium of a situation in a noncooperative game, which sometimes facilitates the discovery of such situations.

Lemma. *A necessary and sufficient condition for a situation X^* in the mixed strategies of a noncooperative game Γ to be i-optimal is that the inequality*

$$H_i(X^* \| x_i) \leqq H_i(X^*) \tag{5.12}$$

is satisfied for every pure strategy $x_i \in \boldsymbol{x}_i$.

The necessity follows because every pure strategy is also a mixed strategy. For the proof of the sufficiency, we integrate (5.12) with respect to x_i with an arbitrary mixed strategy X_i as the integrator measure. As a result, we obtain

$$H_i(X^* \| X_i) \leqq H_i(X^*). \qquad \square$$

5.14 Properties of the players' payoffs. We present some simple but useful propositions on payoffs to players who apply mixed strategies.

Lemma. *For every situation X and player j each player i has pure strategies x_i^+ and x_i^- which are points of the spectrum of the mixed strategy X_i and for which*

$$H_j(X \| x_i^+) \geqq H_j(X),$$
$$H_j(X \| x_i^-) \leqq H_j(X).$$

Proof. Suppose that

$$H_j(X \| x_i) > H_j(X) \tag{5.13}$$

for all strategies $x_i \in \operatorname{supp} X_i$.

We choose a sequence $\varepsilon_n \downarrow 0$ and denote by ω_n the set of x_i for which

$$H_j(X \| x_i) - \varepsilon_n > H_j(X). \tag{5.14}$$

Evidently $\omega_1 \subset \omega_2 \subset \ldots$. Set

$$\bigcup_{n=1}^{\infty} \omega_n = \omega.$$

By hypothesis, $\omega \supset \operatorname{supp} X_i$, so that by the lemma in 5.10 we must have $X_i(\omega) = 1$. Consequently, by the continuity of X_i (this continuity is known to be equivalent to the countable additivity of the measures), there is an n for which $X_i(\omega_n) > 0$.

By (5.6), we have

$$H_j(X) = \int_{\boldsymbol{x}_i} H_j(X \| x_i) dX_i(x_i)$$

and furthermore

$$H_j(X) = \int_{\omega_n} H_j(X \| x_i) dX_i(x) + \int_{\boldsymbol{x}_i \setminus \omega_n} H_j(X \| x_i) dX_i(x);$$

on the other hand, by estimating $H_j(X \| x_i)$ in the first integral by (5.14), and in the second integral, by (5.13), we obtain

$$H_j(X) \geqq \int_{\omega_n} (H_j(X) + \varepsilon_n) dX_i(x_i) + \int_{\boldsymbol{x}_i \setminus \omega_n} H_j(X) dX_i(x_i)$$

$$= \int_{\boldsymbol{x}_i} H_j(X) dX_i(x_i) + \varepsilon_n \int_{\omega_n} dX_i(x_i) = H_j(X) + \varepsilon_n X_i(\omega_n),$$

which, since $\varepsilon_n > 0$ and $X_i(\omega_n) > 0$, cannot be valid.

This contradiction shows that (5.13) is not satisfied for some strategy $x_i^- \in \boldsymbol{x}_i$, i.e. x_i^- is the required strategy.

The existence of x_i^+ is established similarly. □

Corollary. *For every situation X*

$$\sup_{x_i \in \boldsymbol{x}_i} H_i(X \| x_i) = \sup_{X_i \in \boldsymbol{X}_i} H_i(X \| X_i).$$

Proof. Evidently the right-hand side is not less than the left-hand side. If we assume that it is strictly greater, there is a strategy $X_i^0 \in \boldsymbol{X}_i$ for which

$$\sup_{x_i \in \boldsymbol{x}_i} H_i(X \| x_i) < H_i(X \| X_i^0),$$

and this contradicts the theorem just proved. □

5.15 "Complementary slackness." The following proposition has the same basis as the familiar property of complementary slackness in a standard linear programming problem.

Theorem. *In a noncooperative game as in* (5.1), *let the situation X^* be i-optimal. Then:*

1) *If x_i^0 is a point of the spectrum of X_i^*, and the payoff function H_i is continuous with respect to x_i at the point $X^* \| x_i^0$, then*

$$H_i(X^* \| x_i^0) = H_i(X^*); \tag{5.15}$$

2) *If $X_i^*(x_i^0) > 0$ then* (5.15) *holds.*

Proof. 1) To say that X^* is i-optimal means that

$$H_i(X^* \| x_i) \leqq H_i(X^*) \text{ for all } x_i \in \boldsymbol{x}_i. \tag{5.16}$$

Suppose that, with the hypotheses of the theorem, (5.15) is not valid. Because of (5.16), this means that necessarily

$$H_i(X^* \| x_i^0) < H_i(X^*).$$

Choose $\varepsilon > 0$ so that

$$H_i(X^* \| x_i^0) < H_i(X^*) - \varepsilon.$$

Since we have assumed that H_i is continuous, there is a neighborhood ω of x_i^0 such that

$$H_i(X^* \| x_i) < H_i(X^*) - \varepsilon \text{ for all } x_i \in \omega. \tag{5.17}$$

Since x_i^0 is a point of the spectrum of X_i^* and $\omega \ni x_i^0$, by the lemma in 5.10 we must have $X_i^*(\omega) > 0$. It remains only to repeat the final part of the proof of the lemma in the preceding subsection.

2) If $X_i^*(x_i^0) > 0$, then if we take ω to be the set consisting of the single strategy x_i^0 we may repeat the proof of part 1), this time without using the continuity of H_i. □

Corollary. *In a noncooperative game with continuous payoff functions, a strictly dominated pure strategy of player i cannot appear in the spectrum of any i-optimal strategy.*

Both the theorem of this section and its corollary will be used frequently in various circumstances.

5.16. The following development of the content of the preceding subsection will be useful.

Definition. A coalitional strategy $X_{I\setminus i}^*$ is said to be *equalizing* if the payoff $H_i(X_{I\setminus i}^*, x_i)$ is independent of the strategy $x_i \in \boldsymbol{x}_i$.

In particular, if there are only two players in the game we may speak of equalizing the strategies of the players. □

It follows from the definition that if a coalitional strategy $X^{*(i)}$ is equalizing, then the situation $(X_{I\setminus i}^*, x_i)$ must be i-optimal for every $x_i \in \boldsymbol{x}_i$.

Therefore if a situation X^* has the property that, for every $i \in I$, it contains an equalizing strategy $X_{I\setminus i}^*$, it is an equilibrium situation.

For all its simplicity, this reason is sometimes useful in actually finding equilibrium situations (see, for example, §5, Chapter 3).

In turn, it follows from the theorem in 5.15 that if, in an i-optimal situation, we have $\operatorname{supp} X_i^* = \boldsymbol{x}_i$, and the payoff function H_i is continuous with respect to x_i, then the strategy $X_{I\setminus i}^*$ is equalizing.

§6 Natural topology in games

6.1 On a method of defining a topological structure. Later, we shall be interested in introducing, in an intrinsic way, useful topological structures on sets of strategies of players in a noncooperative game. These structures have rather transparent informal interpretations. They allow us to introduce some natural semi-endomorphisms and endomorphisms of games; these will reduce the study of games of a rather general kind to the study of finite games.

As a preliminary, we describe a general method of defining topological structures on abstract sets.

Let there be given a set $\boldsymbol{y}$, and let there correspond to each pair (y, ε), where $y \in \boldsymbol{y}$ and $\varepsilon > 0$, a subset (ε-neighborhood) $\mathcal{Y}(y, \varepsilon) \subset \boldsymbol{y}$, where the correspondence has the following properties:

1°. Reflexivity: for every $y \in \boldsymbol{y}$ and $\varepsilon > 0$ we have $y \in \mathcal{Y}(y, \varepsilon)$;

2°. Universality: for every y and $y' \in \boldsymbol{y}$ there is an $\varepsilon > 0$ such that $y' \in \mathcal{Y}(y, \varepsilon)$;

3°. Monotonicity: for every $y \in \boldsymbol{y}$ it follows from $\varepsilon < \varepsilon'$ that $\mathcal{Y}(y, \varepsilon) \subset \mathcal{Y}(y, \varepsilon')$;

4°. Transitivity: if y, y', and $y'' \in \boldsymbol{y}$ and ε and $\varepsilon' > 0$, then $y' \in \mathcal{Y}(y, \varepsilon)$ and $y'' \in \mathcal{Y}(y', \varepsilon')$ imply $y'' \in \mathcal{Y}(y, \varepsilon + \varepsilon')$.

The topology on $\boldsymbol{y}$ defined by a neighborhood system of the form $\mathcal{Y}(y, \varepsilon)$ as basis is called the *$\mathcal{Y}$-topology*.

We notice that the definition of the concept of a neighborhood does not include the axiom of symmetry: it does not necessarily follow from $y' \in \mathcal{Y}(y, \varepsilon)$ that $y \in \mathcal{Y}(y', \varepsilon)$. Consequently the $\mathcal{Y}$-topology in itself does not define any metric on $\boldsymbol{y}$, and a neighborhood $\mathcal{Y}(y, \varepsilon)$ is not an "ε-neighborhood" in the usual sense of the word (i.e., is not an ε-ball).

Let us consider a $\mathcal{Y}$-topology on $\boldsymbol{y}$, generated by ε-neighborhoods of points $y \in \boldsymbol{y}$. If $\varepsilon > 0$ is given, the subset $\boldsymbol{y}^\varepsilon \subset \boldsymbol{y}$ for which

$$\bigcup_{y \in \boldsymbol{y}^\varepsilon} \mathcal{Y}(y, \varepsilon) = \boldsymbol{y}$$

is called a $\mathcal{Y}$-ε-net on $\boldsymbol{y}$.

A space $\boldsymbol{y}$ with the $\mathcal{Y}$-topology is said to be *totally bounded* if it has a finite ε-net for every $\varepsilon > 0$.

If $\boldsymbol{y}$ is a subset of a complete metric space and $\mathcal{Y}$ is the topology induced by the enveloping space, then the total boundedness of $\boldsymbol{y}$ is equivalent to its $\mathcal{Y}$-precompactness (i.e., the compactness of the closure of $\mathcal{Y}$ in the enveloping space).*) We also recall that the space $\boldsymbol{y}$ with the $\mathcal{Y}$-topology is said to be $\mathcal{Y}$-separable if it contains a countable subset which is everywhere dense in the $\mathcal{Y}$-sense. It is clear that a $\mathcal{Y}$-precompact space is $\mathcal{Y}$-separable.

---

*) See Kolmogorov and Fomin, Elements of the theory of functions and functional analysis, Moscow, 1976, p. 109; also published as Introductory real analysis.

6.2 Semi-intrinsic topology on spaces of strategies in noncooperative games. In the noncooperative game

$$\Gamma = \langle I, \{\mathfrak{x}_i\}_{i\in I}, \{H_i\}_{i\in I}\rangle \tag{6.1}$$

we select a strategy $x_i' \in \mathfrak{x}$ and an arbitrary $\varepsilon > 0$ and consider the set $\mathcal{S}_i(x_i', \varepsilon)$ of all $x_i'' \in \mathfrak{x}_i$ for which

$$H_i(x\|x_i'') < H_i(x\|x_i') + \varepsilon \tag{6.2}$$

for all $x \in \mathfrak{x}$.

Definition. The set $\mathcal{S}_i(x_i, \varepsilon)$ is called an *$\mathcal{S}_i$-ε-neighborhood* of x_i' in $\mathfrak{x}_i$.

As is easily verified, the system $\mathcal{S}_i(x_i, \varepsilon)$ $(x_i \in \mathfrak{x}, \varepsilon > 0)$ so defined has the properties 1°–4° of 6.1.

The topology on $\mathfrak{x}$ whose basis is the set of all neighborhoods of the form $\mathcal{S}_i(x_i, \varepsilon)$ (for arbitrary $x_i \in \mathfrak{x}$ and $\varepsilon > 0$) is known as the *$\mathcal{S}_i$-topology* on $\mathfrak{x}_i$ or the *semi-intrinsic* topology on $\mathfrak{x}_i$.*)

The $\mathcal{S}_i$-topology on the sets $\mathfrak{x}_i$ generates the "componentwise" topology on the Cartesian product $\mathfrak{x}_K = \prod_{i\in K} \mathfrak{x}_i$ (see 1.9), called the $\mathcal{S}_K$-topology on $\mathfrak{x}_K$ or the semi-intrinsic topology on $\mathfrak{x}_K$. In particular, we may consider the $\mathcal{S}_I$-topology (or simply the $\mathcal{S}$-topology) on the set $\mathfrak{x}_I = \mathfrak{x}$ of all situations.

Consequently, even in the most abstract noncooperative games, we have a basis for considering sets of players' strategies and coalitions in topological spaces.

In the terminology of 2.21, we may say that $\mathcal{S}_i(x_i, \varepsilon)$ is the set of strategies of player i that are ε-dominated by a given strategy x_i. Hence the semi-intrinsic topology, by its very essence, has optimizational content.

In the end, it is the semi-intrinsic topology and the intrinsic topology (to be introduced later) that determine the existence and many of the basic properties of the solutions of noncooperative games. However, the *exterior* topology on spaces of player's strategies, which is "physically" inherent in the spaces, as a rule appears in questions of the solvability of games only to the extent that it predetermines the character of the semi-intrinsic or intrinsic topologies.

The semi-intrinsic topology is connected in a very simple way with the properties of the players' payoff functions.

Theorem. *The payoff function $H_i(x \underset{i}{\|} \cdot) : \mathfrak{x}_i \to \mathbf{R}$ is upper semicontinuous in the $\mathcal{S}_i$ sense. The semicontinuity is uniform in $x \in \mathfrak{x}$.*

The proof follows directly from the definition of the semi-intrinsic topology. Choose $x_i \in \mathfrak{x}_i$ and $\varepsilon > 0$ arbitrarily. Then (6.2) will be satisfied for every x_i' in the neighborhood $\mathcal{S}_i(x_i', \varepsilon)$ of the strategy x_i', and since (6.2) is uniform in x, the choice of the neighborhood is independent of x. □

---

*) S is the first letter of "semi-intrinsic". Later on we shall introduce a similar topology, the $\mathcal{J}$-topology, where the $\mathcal{J}$ stands for "intrinsic".

6.3 ε-equilibrium situations in totally bounded games. Definition. A game Γ as in (6.1) is said to be *totally bounded in the $\mathcal{S}$ sense* if every set $\mathfrak{x}_i$ of strategies is totally bounded in the $\mathcal{S}_i$-topology (i.e., there is a finite $\mathcal{S}_i$-ε-net for every $\varepsilon > 0$). □

The optimality meaning of the $\mathcal{S}_i$-ε-net $\mathfrak{x}_i^\varepsilon$ in the space $\mathfrak{x}$ of strategies of player i is that every strategy in $\mathfrak{x}_i$ leads to ε-i-domination by at least one strategy from $\mathfrak{x}_i^\varepsilon$, i.e., for every pure strategy x_i of player i there is a (pure) strategy $x_i^{(\varepsilon)}$ of that player in the $\mathcal{S}_i$-ε-net such that for arbitrary actions of the other players a change by player i from $x_i^{(\varepsilon)}$ to x_i increases the payoff to the player by at most ε. In this sense, an $\mathcal{S}_i$-ε-net can be said to have "exterior ε-stability".

The union of the mappings $x_i \mapsto x_i^{(\varepsilon)}$ for all $i \in I$ is evidently an ε-semi-endomorphism (see 1.14). Consequently the general theorem of 2.21 is applicable here, and as a consequence we have the following proposition. However, we shall present a proof that does not depend on the theorem of 2.21.

Theorem. *Every game Γ as in* (6.1) *which is totally bounded in the $\mathcal{S}$ sense has, for every $\varepsilon > 0$, ε-equilibrium situations in the finite* (see 5.6) *mixed strategies.*

Proof. We choose an arbitrary $\varepsilon > 0$, construct a finite $\mathcal{S}_i$-ε-net $\mathfrak{x}_i^\varepsilon$ for each $i \in I$ in the set $\mathfrak{x}_i$ of pure strategies, and consider the $\mathfrak{x}_1^\varepsilon \times \ldots \times \mathfrak{x}_n^\varepsilon$-subgame Γ^ε of Γ.

The game Γ^ε is finite, so that, according to the theorem of 5.5 there is an equilibrium situation X^ε for which, by definition,

$$H_i(X^\varepsilon \| x_i^\varepsilon) \leqq H_i(X^\varepsilon) \text{ for all } i \in I \text{ and } x_i^\varepsilon \in \mathfrak{x}_i^\varepsilon. \tag{6.3}$$

Now choose an arbitrary $x' \in \mathfrak{x}$. Denote by x_i^ε the strategy of player i in $\mathfrak{x}_i^\varepsilon$ whose $\mathcal{S}_i$-ε-neighborhood contains x_i'. Then it follows from the definition of an $\mathcal{S}_i$-ε-neighborhood that for every situation x we must have $H_i(x\|x_i') \leqq H_i(x\|x_i^\varepsilon)+\varepsilon$ (in fact there is strict inequality here, but that is not important for us). If, in this inequality, we pass from the situation x to a situation X in the mixed strategies, we find that it is also satisfied for all situations in the mixed strategies, in particular for X^ε:

$$H_i(X^\varepsilon \| x_i) \leqq H_i(X^\varepsilon \| x_i^\varepsilon) + \varepsilon.$$

With (6.3) this gives us

$$H_i(X^\varepsilon \| x_i) \leqq H_i(X^\varepsilon) + \varepsilon \tag{6.4}$$

for all $i \in I$ and $x_i \in \mathfrak{x}_i$. □

6.4. A criterion for a game to have "intrinsic topological" properties can be obtained from the "exterior topological" properties of the sets of strategies and the analytic properties, in this exterior topology, of the payoff functions.

Theorem. *In a noncooperative game Γ as in* (6.1), *let the set $\mathfrak{x}_i$ be a totally bounded subset of a subspace of a topological space with respect to an exterior topology $\mathcal{Y}$, and let the function $H_i(x \underset{i}{\|} \cdot) : \mathfrak{x}_i \to \mathbf{R}$ be defined on the compact $\mathcal{Y}$-closure $\bar{\mathfrak{x}}_i$ of $\mathfrak{x}_i$ and upper semicontinuous on $\bar{\mathfrak{x}}_i$ in this topology for each given x.*

Then the space $\mathfrak{x}_i$ is totally bounded in the semi-intrinsic topology $\mathcal{S}_i$.

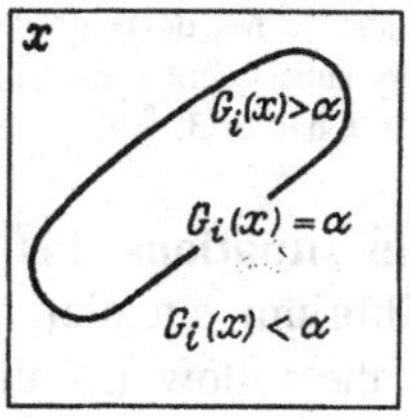

Fig. 1.12

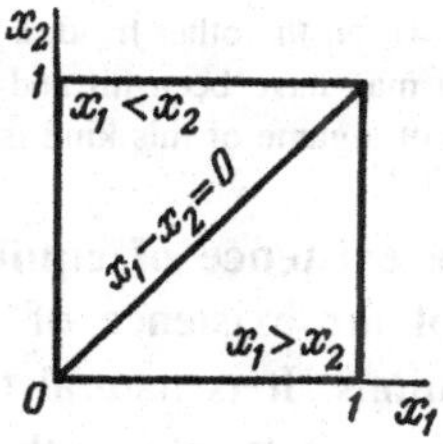

Fig. 1.13

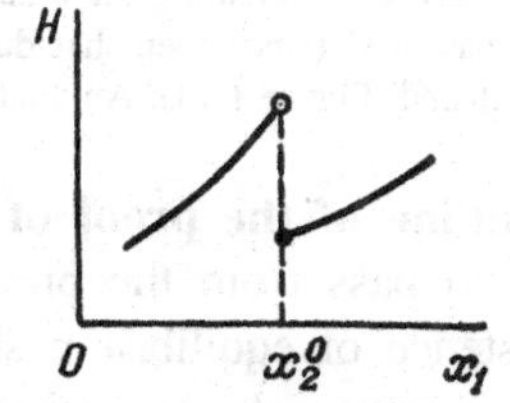

Fig. 1.14

Proof. Choose an arbitrary $\varepsilon > 0$. By the definition of upper semicontinuity we may construct, for every strategy $x' \in x_i$, a neighborhood $\mathcal{S}_i(x_i', \varepsilon)$ such that we have, for all $x_i'' \in \mathcal{S}_i(x_i', \varepsilon)$,

$$H_i(x\|x_i'') \leqq H_i(x\|x_i') + \varepsilon \text{ for all } x \in x.$$

These neighborhoods form a covering of the compact closure $\bar{x}_i$ of the space x_i. From this covering we may extract a subcovering, which is, of course, also a covering of x_i. The corresponding set of strategies x_i in this covering is evidently an $\mathcal{S}_i$-ε-net for x_i. □

Corollary. *If, in the noncooperative game Γ in (6.1), each set x_i is totally bounded (in the topology of the enveloping space) and each function $H_i(x\|\cdot)$ is upper semi-continuous, then Γ is totally bounded in the semi-intrinsic topology (and therefore there are ε-equilibrium situations for every $\varepsilon > 0$).*

6.5. As an example of the application of this theorem we consider a game Γ in (6.1) in which the set x_i of strategies is precompact (for example, is a bounded subset of a finite-dimensional Euclidean space). To specify the conditions of the payoff functions, we suppose that real continuous functions G_i are given on the set x_i, as well as real numbers α_i. We suppose that the payoff functions H_i are continuous on the sets of situations x for which $G_i(x) < \alpha_i$ and $G_i(x) \geqq \alpha_i$ (Figure 1.12), and also that if $\lim_{m\to\infty} x^{(m)} = x^{(0)}$, where $G_i(x^{(0)}) = \alpha_i$, then

$$\lim_{m\to\infty} H_i(x^{(m)}) \leqq H_i(x^{(0)}).$$

It is easily seen that in this case the payoff functions H_i are upper semicontinuous on x. Consequently Γ satisfies the hypotheses of the preceding corollary and therefore has, for every $\varepsilon > 0$, ε-equilibrium situations in the mixed strategies.

6.6. A special case of the game in this example is the game Γ in which $I = \{1, 2\}$ and $x_1 = x_2 = [0, 1]$ (the set of situations in this game is evidently the unit square; representing the graphs of the payoff functions separately, we have two squares, on each of which the corresponding function is given; consequently this game is sometimes called the "two squares" game), $G_1 = -G_2 = x_1 - x_2$, $\alpha_1 = \alpha_2 = 0$ (Figure 1.13); the payoff functions H_i satisfy the hypotheses of 6.5, and in addition, when $x_1 < x_2$ and when $x_1 \geqq x_2$ the function H_1 increases in x_1 and decreases in x_2, whereas when $x_2 < x_1$ and when $x_2 \geqq x_1$ the function H_2 increases in x_2 and decreases in x_1.

It is clear that this game is really a subcase of the preceding game.

This game can usefully be interpreted as a duel between the players: each player chooses an instant $x_i \in [0, 1]$ for firing a shot at the other. On one hand, perhaps the shot is fired as late as possible (for example, because if the duelists are closer, the accuracy of the shot is increased, and so is the probability of defeating the antagonist); on the other hand, a duelist should not delay until the opponent has fired (since then that duelist may have been hit and have his subsequent possibilities sharply reduced, Figure 1.14). An analysis of a game of this kind is given in Chapter 3, § 9.

6.7 Outline of the proof of the existence of equilibrium situations. Let us attempt to pass from the proof of the existence of ε-equilibrium situations to the existence of equilibrium situations. It is natural to use the following rather traditional approach. Assuming, in some topology, the compactness of the space of strategies (and hence — in the componentwise topology — the compactness of the situations) and the continuity, in this topology, of the payoff functions, we have to establish the convergence as $\varepsilon \to 0$ of the sequence of the situations X^ε from the theorem of 6.3 and take the limit in (6.4) as $\varepsilon \to 0$. However, carrying out this quite plausible program with complete rigor encounters many difficulties and the introduction of a number of reservations.

In the first place, in (6.4) we have to take limits not on strategies but on situations; consequently we have to choose a topology on sets of strategies that guarantees appropriate analytic properties not only of functions $H_i(x \underset{i}{\|} \cdot)$ of strategies x_i but also of functions $H_i(\cdot)$ of situations x as well as of situations X in the mixed strategies. Such a topology appears to be reasonable, and should be sufficiently close, in the sense of the topology of strategies, to apply not only for the player using these strategies, but also for each other participant in the game.

In the second place, because of the presence on both sides of (6.4) of the variable X^ε with respect to which we need to carry out the termwise approach to a limit, it is necessary for the function H_i, which depends on this variable, to be not merely upper semicontinuous, but continuous (and moreover continuous on the space of situations in the mixed strategies). This, in particular, will require not just one-sided closeness of the values of the payoff functions, as in (6.2), but closeness on both sides.

In the third place, the existence of the required limits of the situations requires the compactness of the set of situations in the mixed strategies.

Finally, in the fourth place, the limit of a sequence of situations in the finite mixed strategies may turn out to be, in general, a situation in the mixed strategies belonging to a rather wide class. But then we must establish the measurability of the payoff functions with respect to the situations constituting all such strategies.

Almost the entire remainder of this section will be devoted to carrying out this program.

6.8 The intrinsic topology on spaces of strategies and situations. We begin with the choice of a suitable topology on sets x_i of strategies. In the noncooperative game (6.1) we select a strategy $x_i' \in x_i$ and an arbitrary $\varepsilon > 0$, and consider the set $\mathcal{J}_i(x_i', \varepsilon)$ of the $x_i'' \in \boldsymbol{x}_i$ for which

$$|H_j(x \underset{i}{\|} x_i') - H_j(x \underset{i}{\|} x_i'')| < \varepsilon \tag{6.5}$$

for all $x \in \boldsymbol{x}$ and $j \in I$. In other words, $\mathcal{J}_i(x_i', \varepsilon)$ is the set of $x_i'' \in \boldsymbol{x}$ for which

$$\max_{j \in I} \sup_{x \in \boldsymbol{x}} |H_j(x \| x_i') - H_j(x \| x_i'')| < \varepsilon. \tag{6.6}$$

Definition. The set $\mathcal{J}_i(x_i'', \varepsilon)$ is called an $\mathcal{J}_i$-ε-*neighborhood* of the strategy x_i' in $\boldsymbol{x}_i$.

The topology on $\boldsymbol{x}_i$ that has as its basis the family of all sets of the form $\mathcal{J}_i(x_i, \varepsilon)$ (for arbitrary $x_i \in \boldsymbol{x}_i$ and $\varepsilon > 0$) is called the $\mathcal{J}_i$-*topology* on $\boldsymbol{x}$ or the *intrinsic* topology on $\boldsymbol{x}_i$. Sometimes it is also called the *natural* topology on $\boldsymbol{x}_i$. □

As in the case of the semi-intrinsic topology, the intrinsic topology on $\boldsymbol{x}_i$ induces on the Cartesian product

$$\boldsymbol{x}_K = \prod_{i \in K} \boldsymbol{x}_i$$

the componentwise topology $\mathcal{J}_K$; and on the set $\boldsymbol{x}$ of situations, a corresponding topology $\mathcal{J}$.

Since both the $\mathcal{J}_i$- and $\mathcal{J}_K$-topologies are defined by the game Γ as a whole, we shall refer to them simply as $\mathcal{J}$-topologies (on the spaces $\boldsymbol{x}_i$ of individual strategies, $\boldsymbol{x}_K$ of coalitional strategies, and $\boldsymbol{x}$ of situations).

The connection of the intrinsic topology with the properties of the payoff functions of the players is described by the following theorem, which is an analog of the theorem in 6.2.

Theorem. *The payoff functions $H_i(x \underset{i}{\|} \cdot)$ are, in the sense of the $\mathcal{J}_i$-topology, uniformly continuous on $\boldsymbol{x}_i$. They are equicontinuous for $x \in \boldsymbol{x}$.*

The proof follows from the definition of the intrinsic topology. Let $x_i' \in \boldsymbol{x}_i$ and $\varepsilon > 0$. Choose a neighborhood $\mathcal{J}_i(x_i', \varepsilon)$. Then (6.6) is satisfied for every strategy x_i'' in this neighborhood, and the characteristic ε of the neighborhood coincides with the originally chosen ε, i.e. is independent of the choice of x_i'. In general, the neighborhood is independent of x. □

Corollary. *Every payoff function H_j is, in the sense of the component-wise topology $\mathcal{J}$ on $\boldsymbol{x}$, a uniformly continuous function of situations.*

Indeed, let $\varepsilon > 0$ and $x' = (x_1', \ldots, x_n') \in \boldsymbol{x}$. Choose any situation $x'' = (x_1'', \ldots, x_n'')$ for which

$$x_i'' \in \mathcal{J}_i(x_i', \varepsilon/n) \text{ for arbitrary } i \in I.$$

This means that, for each player $i \in I$,

$$|H_j(x \| x_i') - H_j(x \| x_i'')| < \varepsilon/n \text{ for all } x \in \boldsymbol{x} \text{ and } j \in I. \tag{6.7}$$

Setting

$$(x_1'', \dots, x_k'', x_{k+1}', \dots, x_n') = y^{(k)} \text{ for } k = 0, \dots, n$$

(the last component of $y^{(n)}$ is x_n'') we obtain

$$|H_j(x') - H_j(x'')| \leqq \sum_{k=0}^{n} |H_j(y^{(k)}) - H_j(y^{(k+1)})|.$$

$$= \sum_{k=0}^{n} |H_j(y^{(k)} \underset{k}{\|} x_k') - H_j(y^{(k)} \underset{k}{\|} x_k'')|,$$

or, using (6.7),

$$|H_j(x') - H_j(x'')| \leqq n\varepsilon/n = \varepsilon. \quad \square$$

We have not yet found a possibility for a precise statement of the continuity of the payoff functions with respect to mixed strategies or of their continuity as functions of the situations; however, as soon as this has been done, the required continuity will be clear from the preceding proof.

6.9 Definition. A game Γ as in (6.1) will be said to be *totally bounded* in the $\mathcal{J}$ sense if every space $\mathfrak{x}_i$ of strategies is totally bounded in the $\mathcal{J}$-topology. □

Evidently the $\mathcal{J}$-topology is stronger than the $\mathcal{S}$-topology: for every $i \in I$, $x_i \in \mathfrak{x}$, and $\varepsilon > 0$, we have

$$\mathcal{J}_i(x_i, \varepsilon) \subset \mathcal{S}(x_i, \varepsilon).$$

Therefore every $\mathcal{J}_i$-ε-net for $\mathfrak{x}_i$ is also an $\mathcal{S}_i$-ε-net for the same space. Consequently a game that is totally bounded in the $\mathcal{J}$ sense has the same property in the $\mathcal{S}$ sense, and the following proposition follows from the theorem in 6.3.

Theorem. *Every noncooperative game which is totally bounded in the $\mathcal{J}$ sense has, for every $\varepsilon > 0$, an ε-equilibrium situation in the finite mixed strategies.*

This theorem does not present any logical advantage for establishing the total boundedness of noncooperative games in the $\mathcal{J}$ sense, as compared with establishing this in the $\mathcal{S}$ sense. As we shall see in 6.11, a supplementary logical possibility included in the total boundedness of a game in the $\mathcal{J}$ sense is realized in a somewhat different way.

6.10 A natural metric on spaces of strategies. As we noticed in 6.1, the semi-intrinsic topology on spaces of the players' strategies does not immediately define a metric on these spaces. On the other hand, the intrinsic topology does this, and in a rather natural way.

Take x_i' and $x_i'' \in \mathfrak{x}_i$ and set

$$\rho_i(x_i', x_i'') = \max_{j \in I} \sup_{x} |H_j(x \| x_i') - H_j(x \| x_i'')|. \tag{6.8}$$

It is not hard to verify that the function $\rho_i : x_i \times x_i \to \mathbf{R}$ so defined has the following properties:

a) *Nonnegativity:*

$$\rho_i(x'_i, x''_i) \geqq 0;$$

b) *Indivisibility of points:*

$$\rho_i(x_i, x_i) = 0;$$

c) *Symmetry:*

$$\rho_i(x'_i, x''_i) = \rho_i(x''_i, x'_i);$$

d) *Triangle property:*

$$\begin{aligned}\rho_i(x'_i, x''_i) + \rho_i(x''_i, x'''_i) &= \max_{j\in I} \sup_{\bar{x}} |H_j(\bar{x} \parallel x'_i) - H_j(\bar{x} \parallel x''_i)| \\ &\quad + \max_{j\in I} \sup_{\bar{x}} |H_j(\bar{x} \parallel x'') - H_j(\bar{x} \parallel x''')| \\ &\geqq \max_{j\in I} \sup_{\bar{x}} (|H_j(\bar{x} \parallel x'_i) - H_j(\bar{x} \parallel x''_i)| \\ &\quad + |H_j(\bar{x} \parallel x''_i) - H_j(\bar{x} \parallel x'''_i)|) \\ &\geqq \max_{j\in I} \sup_{\bar{x}} |H_j(\bar{x} \parallel x'_i) - H_j(\bar{x} \parallel x'''_i)|)\rho_i(x'_i, x'''_i).\end{aligned}$$

A function that has all these properties is called a *pseudometric*. For a pseudometric to be a *metric*, it must also have the following property:

e) *Separation of points:*

$$x'_i \neq x''_i \text{ implies } \rho(x'_i, x''_i) > 0.$$

Generally speaking, the payoff functions may appear in the game in such a way that the construction of the fundamental function based on them will not have the last property. However, $\rho_i(x'_i, x''_i) = 0$ means that

$$\max_{j\in I} \sup_{\bar{x}} |H_j(\bar{x} \| x'_i) - H_j(\bar{x} \| x''_i)| = 0,$$

i.e., that

$$H_j(\bar{x} \| x'_i) = H_j(\bar{x} \| x''_i) \tag{6.9}$$

identically for every $j \in I$ and in all situations $\bar{x} \in x$.

But this means, in turn, that the use by player $i \in I$ of the strategies x'_i and x''_i leads, in the game Γ, to exactly the same consequences *for every* player $j \in I$. Hence we may suppose that, under the given circumstances, x'_i and x''_i are not so much different strategies of player i as different specimens of one of that player's strategies (compare what is said about this in 2.8, Chapter 3). Formally, this amounts to identifying all the strategies for which (6.9) is satisfied. From now on, we shall suppose in such situations, unless the contrary is specified, that such strategies have already been identified. This means that we shall assume that the pseudometric ρ_i is a metric, and that the set $\mathfrak{x}_i$ of strategies is a metric space.

From the point of view of the metric ρ_i the set $\mathcal{J}_i(x_i, \varepsilon)$ is an ε-ball with center at x_i. Therefore the metric ρ_i on $\mathfrak{x}_i$ precisely generates the $\mathcal{J}$-topology on this set.

In terms of this topology, we may speak of ε-nets on sets of strategies of the players in the game Γ in (6.1). A mapping in which there corresponds, to each strategy x_i of a player $i \in I$, a strategy $x_i^{(\varepsilon)}$ from the ε-net $\mathfrak{x}_i^{(\varepsilon)}$, no farther from it than ε in the metric, is evidently an ε-endomorphism of Γ.

As soon as spaces of players' strategies turn out to be metrizable, we may speak, instead of their *total boundedness* in the $\mathcal{J}$-topology, of their $\mathcal{J}$-*precompactness*.

Considering spaces of players' strategies as metric spaces with the natural metric is convenient in many circumstances.

6.11 Precompact games. There is a certain connection between the $\mathcal{J}$-precompactness of spaces $\mathfrak{x}_i$ of strategies of different players. This is interesting in itself, and in addition allows us to achieve economies in the conditions that have to be imposed on a game in order to establish its solvability.

Theorem. *In terms of the game* Γ *in* (6.1), *let* $i_0 \in I$. *Denote* $I \setminus i_0$ *by* I_0. *Then if all the spaces* $\mathfrak{x}_i^{(\varepsilon)}$ *of strategies with* $i \in I_0$ *are* $\mathcal{J}$-*precompact,* $\mathfrak{x}_{i_0}$ *is also* $\mathcal{J}$-*precompact.*

Proof. To simplify the notation, we suppose that $i_0 = 1$. We take any $\varepsilon > 0$, and for each $\mathfrak{x}_i$ with $i \neq 1$ construct a finite $\varepsilon/(2n+1)$-net $\mathfrak{x}_i^\varepsilon$ in terms of the natural metric ρ_i. Let $\mathfrak{x}_{I_0}^\varepsilon = \prod_{i \in I_0} \mathfrak{x}_i^\varepsilon$. We suppose that the elements of this (finite!) set are enumerated in some specified order.

We also select any $x_1 \in \mathfrak{x}_1$ and associate with this strategy a set of vectors

$$\{H_I(x_I, x_{I_0}^\varepsilon)\}_{x_{I_0}^\varepsilon \in \mathfrak{x}_{I_0}^\varepsilon},$$

where the vectors between the braces are enumerated in the same order as the coalitional strategies $x_{I_0}^\varepsilon$. We suppose that the components of these vectors are the coordinates of the Euclidean space $\mathbf{R}^k$ of the indicated dimension.

In this way we have set up a mapping $\varphi : \mathfrak{x} \to \mathbf{R}^k$. Let us set $\varphi(\mathfrak{x}_1) = \mathfrak{x}_1^*$.

The set $\mathfrak{x}_i^*$ is a bounded subset of $\mathbf{R}^k$ and hence totally bounded in the Euclidean metric ρ_E of this enveloping space. In $\mathfrak{x}_1^*$ we construct a finite ε-net (in terms of the Euclidean metric), which we denote by $\mathfrak{x}_i^{*\varepsilon}$. For each $x_1^{*\varepsilon} \in \mathfrak{x}_1^{*\varepsilon}$ we select an $x_1^\varepsilon \in \varphi^{-1}(x_1^{*\varepsilon})$ and denote the set of such x_1^ε by $\mathfrak{x}_1^\varepsilon$. Let us show that the finite set $\mathfrak{x}_1^\varepsilon$ is an ε-net in terms of the intrinsic metric ρ_1.

In fact, let us choose any $x_1 \in \mathfrak{x}_1$, construct $\varphi(\mathfrak{x}_1) = \mathfrak{x}_1^*$, and find a point $x_1^{*\varepsilon} \in \mathfrak{x}_1^{*\varepsilon}$ for which $\rho_E(\mathfrak{x}_1^*, \mathfrak{x}_1^{*\varepsilon}) < \varepsilon/(2n+1)$. This means that

$$\left(\sum_{i \in I} \sum_{x_{I_0}^\varepsilon \in \mathfrak{x}_{I_0}^\varepsilon} (H_i(x_1, x_{I_0}^\varepsilon) - H_i(x_1^\varepsilon, x_{I_0}^\varepsilon))^2\right)^{1/2} < \frac{\varepsilon}{2n+1}$$

and consequently

$$|H_i(x_1, x_{I_0}^\varepsilon) - H_i(x_1^\varepsilon, x_{I_0}^\varepsilon)| < \varepsilon/(2n+1) \tag{6.10}$$

for all $i \in I$ and $x_{I_0}^\varepsilon \in \mathfrak{x}_{I_0}^\varepsilon$. It then remains only to replace $x_{I_0}^\varepsilon$ in this inequality by an arbitrary $x_{I_0} \in \mathfrak{x}_{I_0}$. The necessary discussion is reminiscent of that used in the proof of the corollary in 6.8. Take an arbitrary $x_{I_0} \in \mathfrak{x}_{I_0}$, and for each individual strategy x_i occurring in x_{I_0} find a strategy $x_i^\varepsilon \in \mathfrak{x}_i^\varepsilon$ 'such that $\rho_i(x_i, x_i^\varepsilon)$ is less than $\varepsilon/(2n+1)$. Set

$$(x_2^\varepsilon, \ldots, x_n^\varepsilon) = x_{I_0}^\varepsilon,$$

$$(x_2^\varepsilon, \ldots, x_k^\varepsilon, x_{k+1}, \ldots, x_n) = y_{I_0}^{(k)} \text{ for } k = 1, \ldots, n$$

(the last component of $y_{I_0}^{(n)}$ is x_n^ε).

We evidently have $y_{I_0}^{(1)} = x_{I_0}$, $y_{I_0}^{(n)} = x_I^\varepsilon$ and

$$(x_1, y_{I_0}^{(k)} \| x_{k+1}^\varepsilon) = (x_1, y_{I_0}^{(k+1)}) \text{ for } k = 1, \ldots, n-1 \tag{6.11}$$

(and similar equations if x_1 is replaced by x_1^ε).

Therefore

$$|H_i((x_1, y_{I_0}^{(k)}) \| x_{k+1}) - H_i((x_1, y_{I_0}^{(k)} \| x_{k+1}^\varepsilon)| < \varepsilon/(2n+1) \tag{6.12}$$

for $k = 1, \ldots, n-1$, and similarly

$$|H_i((x_1^\varepsilon, y_{I_0}^{(n)}) \| x_{k+1}) - H_i((x_1^\varepsilon, y_{I_0}^{(k)}) \| x_{k+1}^\varepsilon)| < \varepsilon/(2n+1) \tag{6.13}$$

for $k = 1, \ldots, n-1$.

If we combine (6.10), (6.12) and (6.13), and then use (6.11), we obtain

$$|H_i(x_1, x_{I_0}) - H_i(x_1^\varepsilon, x_{I_0})| < \varepsilon.$$

We now have only to recall that both $i \in I$ and $x_{I_0} \in \mathfrak{x}_{I_0}$ are arbitrary. Therefore $\rho(x_1, x_1^\varepsilon) < \varepsilon$ and $\mathfrak{x}_1^\varepsilon$ is indeed an $\mathcal{J}$-ε-net. □

6.12 Compact games. Let us consider games that are compact in the $\mathcal{J}$-sense.

Definition. If all spaces x_i of strategies in a noncooperative game Γ as in (6.1) are $\mathcal{J}$-compact, the game Γ itself is said to be *$\mathcal{J}$-compact.* □

Evidently every $\mathcal{J}$-compact game is also $\mathcal{J}$-precompact, i.e. totally bounded, and therefore, for every $\varepsilon > 0$, it contains an ε-equilibrium situation in the finite mixed strategies.

In the case when the game is not only $\mathcal{J}$-precompact but also $\mathcal{J}$-compact, we naturally expect that there exist in it not only ε-equilibrium situations (for every $\varepsilon > 0$) but also equilibrium situations.

We can now give the details of the reasoning, outlined in 6.2, which leads to the existence of equilibrium situations in $\mathcal{J}$-compact games.

First we convert, in a specific way (based on the $\mathcal{J}$-topologies of the spaces of players' strategies), the set of players' strategies into a measurable space, and take the collection of probability measures on them as the sets of the players' mixed strategies. These measurable spaces of strategies generate, in a natural way, the measurable spaces of situations (compare 1.10). In order to have the right to speak of the existence of a mixed strategy in the game, we must establish the measurability of the payoff functions with respect to the measurable space that we have constructed. It turns out that this follows notoriously in case the spaces of players' strategies are $\mathcal{J}$-separable (see 6.2).

Now we deduce from the $\mathcal{J}$-compactness of a game the compactness of the spaces of players' strategies, first in a somewhat different sense that corresponds to the intuitive idea of what the convergence of measures should "really" entail. Later on, we shall call this *weak* convergence. In probability theory it is usually called convergence *in distribution.**) Then we shall establish the compactness of the spaces X_i in the $\mathcal{J}$-topology. We shall also establish the continuous dependence of the payoff functions on the mixed strategies.

When all this has been done, the proof of the existence theorems will be obtained without difficulty.

6.13. We denote by $\mathcal{B}_i$ the Borel σ-algebra of subsets x_i (i.e., the smallest σ-algebra containing all subsets x_i that are open in the sense of the $\mathcal{J}$-topology; see 1.11), and by $\mathcal{B}$ the σ-algebra of subsets x generated by the $\mathcal{B}_i$ $(i \in I)$.

From now on we shall understand by the measurability of a payoff function its measurability with respect to the measurable space $\langle x, \mathcal{B} \rangle$.

Theorem. *If, in a game Γ as in* (6.1), *each space x_i of players' strategies is $\mathcal{J}_I$ separable, then all payoff functions H_i are measurable.*

---

*) See A.N. Shiryayev, Probability, Springer-Verlag, 1984, p. 251.

Proof. Choose a player $i^* \in I$ and a real number h, and denote by $\mathfrak{y}_I^h$ the set of all situations y in Γ for which $H_{i^*}(y) > h$.

Let $\mathcal{S}_i(x_i, \varepsilon)$ be the open sphere in $\mathfrak{x}_i$ of radius ε (in the sense of the metric ρ_i) with center x_i.

Let $y \in \mathfrak{y}^h$ be arbitrary. Set

$$H_{i^*}(y) - h = \varepsilon_y$$

and take $\varepsilon'_y < \varepsilon_y/n$. Then choose a situation

$$y' \in \prod_{i \in I} \mathcal{S}(y_i, \varepsilon'_y)$$

and consecutively form the situations $y = y^{(0)}, y^{(1)}, \ldots, y^{(n)}$, in which, for $k = 1, \ldots, n$,

$$y^{(k)} = y^{(k-1)} \underset{k}{\|} y'_k.$$

In particular, we should evidently have $y^{(n)} = y'$.

We have

$$H_{i^*}(y') = H_{i^*}(y) + \sum_{k=1}^{n} (H_{i^*}(y^{(k)}) - H_{i^*}(y^{(k-1)}))$$

$$= H_{i^*}(y) + \sum_{k=1}^{n} (H_{i^*}(y^{(k-1)}) \underset{k}{\|} y'_k) - H_{i^*}(y^{(k-1)} \| y_k))$$

$$\geqq H_{i^*}(y) - \sum_{k=1}^{n} \sup_{y^*} |H_{i^*}(y^* \underset{k}{\|} y'_k) - H_{i^*}(y^* \| y_k)|$$

$$\geqq H_{i^*}(y) - \sum_{k=1}^{n} \rho_k(y_k, y'_k) \geqq H_{i^*}(y) - n\varepsilon'_y > H_{i^*}(y) - \varepsilon_y = h.$$

Therefore $y' \in \mathfrak{y}^h$, and consequently also

$$\prod_{i \in I} \underline{\mathcal{S}}_i(y_i, \varepsilon'_y) \subset \mathfrak{y}^h \tag{6.14}$$

(the horizontal bar denotes the interior of the set).

Now choose countable dense subsets $\mathfrak{x}'_i$ in each subspace $\mathfrak{x}_i$, and find, for each situation y, a situation y' and a rational number $\varepsilon^*_y < \varepsilon'_y$ such that

$$y \in \prod_{i \in I} \underline{\mathcal{S}}_i(y'_i, \varepsilon^*_y) \subset \prod_{i \in I} \mathcal{S}_i(y_i, \varepsilon'_y). \tag{6.15}$$

It follows from (6.14) and the right-hand inclusion in (6.15) that

$$\bigcup_{\substack{y' \in \mathfrak{x}' \\ \varepsilon^*_y > 0}} \prod_{i \in I} \underline{\mathcal{S}}_i(y'_i, \varepsilon^*_y) \subset \mathfrak{y}^h. \tag{6.16}$$

On the other hand, if $y \in \mathfrak{y}^h$, we have by the left-hand inclusion in (6.15)

$$y \in \prod_{i \in I} \underline{\mathcal{I}}_i(y_i, \varepsilon_y^*) \subset \bigcup_{\substack{y' \in x' \\ \varepsilon_y^* > 0}} \prod_{i \in I} \mathcal{I}_i(y', \varepsilon_y^*). \tag{6.17}$$

From (6.16) and (6.17), we have

$$\mathfrak{y}^h = \bigcup_{\substack{y' \in \mathfrak{y}' \\ \varepsilon_y^* > 0}} \prod_{i \in I} \mathcal{I}_i(y', \varepsilon_y^*),$$

and since we have, on the right-hand side, a union of countably many sets that evidently belong to $\mathcal{B}$, we must also have $\mathfrak{y}^h \in \mathcal{B}$. □

6.14 The natural metric on spaces of mixed strategies. We extend the natural metric, defined on the set x_i of pure strategies of player i, to the set X_i of mixed strategies, by setting, for X_i' and $X_i'' \in X_i$,

$$\rho_i(X_i', X_i'') = \max_{j \in I} \sup_{X} |H_j(X \| X_i') - H_j(X \| X_j'')|. \tag{6.18}$$

Let, in particular, $X_i' = x_i'$ and $X_i'' = x_i''$ be pure strategies of player i. It is clear from (6.8) that

$$\rho_i(x_i', x_i'') \leqq \max_{j \in I} \sup_{X} |H_j(X \| x_i') - H_j(X \| x_i')|. \tag{6.19}$$

On the other hand, it also follows from (6.8) that, for any situation, and for any player j,

$$-\rho_i(x_i', x_i'') \leqq H_j(x \| x_i') - H_j(x \| x_i'') \leqq \rho_i(x_i', x_i'').$$

Passing from this inequality to the mixed strategies of each player other than i, we obtain

$$-\rho_i(x_i', x_i'') \leqq H_j(X \| x_i') - H_j(X \| x_i'') \leqq \rho_i(x_i', x_i'')$$

for every situation X in the mixed strategies and every player j. This means that

$$\max_{j \in I} \sup_{X} |H_j(X \| x_i') - H_j(X \| x_i'')| \leqq \rho_i(x_i', x_i''),$$

which, together with (6.19), yields the equation

$$\rho_i(x_i', x_i'') = \max_{j \in I} \sup_{X} |H_j(X \| x_i') - H_j(X \| x_i'')|.$$

Therefore the distance defined by (6.18) between mixed strategies in the case when X_i' and X_i'' are pure strategies agrees with the previously defined distance ρ_i and is an extension of it.

Definition. The metric ρ_i on the set X_i of the mixed strategies of player i, defined in (6.18) as the distance between the strategies X_i' and X_i'', is called the *intrinsic (or natural)* metric on X_i.

The topology on X_i generated by the intrinsic metric is called the *intrinsic* topology (*$\mathcal{J}$-topology*) on X_i.

The componentwise intrinsic topology (*$\mathcal{J}$-topology*) on the set X of situations in the mixed strategies is defined in the natural way. □

6.15. Our immediate aim is to establish a convenient criterion for the compactness of the space X_i of mixed strategies in the $\mathcal{J}$-topology. We shall see that a sufficient condition is the compactness of the spaces x_i of pure strategies in the $\mathcal{J}$-topology.

Throughout the analysis of games with $\mathcal{J}$-compact spaces of pure strategies for the players we shall be dealing with a topology of weak convergence (see 6.12) in the space of probability measures on a metric space.

Definition. Let $\langle A, \mathcal{A}\rangle$ be a topological measurable space on which the topology is generated by a metric ρ (we consequently call the topology the ρ-topology), in which the open sets are measurable.

Then we shall say that a sequence $\mu_1, \ldots, \mu_n, \ldots$ of probability measures on $\langle A, \mathcal{A}\rangle$ *converges weakly to the measure* μ_0 (see 6.12) if for every open subset $B \subset A$ for which $\mu_0(\bar{B} \setminus B) = 0$ (where $\bar{B}$, as usual, denotes the closure of B) we have

$$\lim_{n\to\infty} \mu_n(B) = \mu_0(B).$$

The topology generated by this kind of convergence is called the *weak* topology. □

The following theorem is not a game-theoretic proposition in the strict sense of the word, but plays an auxiliary, although important, role in what follows.

Theorem. *If the space A is compact in the topology generated by the metric ρ, and all its open subsets are $\mathcal{A}$-measurable, then the space of probability measures on $\langle A, \mathcal{A}\rangle$ is compact in the weak topology.*

Proof. Let A be compact in the ρ-topology, and let $\mu_1, \ldots, \mu_n, \ldots$ be a sequence of measures on $\langle A, \mathcal{A}\rangle$. Choose a sequence, converging to zero, of positive numbers $\varepsilon_1, \ldots, \varepsilon_n, \ldots$.

Let us suppose that:

a) $m_1, \ldots, m_k, \ldots$ are positive integers;

b) for each k there are open subsets

$$A_{l_1,\ldots,l_k}, \quad 1 \leqq l_i \leqq m_i\,; \quad i = 1, \ldots, k, \tag{6.20}$$

pairwise disjoint for a given k, such that

1) $\displaystyle\bigcup_{l_k=1}^{m_k} \bar{A}_{l_1,\ldots,l_k} = \bar{A}_{l_1,\ldots,l_{k-1}}$

(we suppose that $A_{l_1,\ldots,l_k}$ is A when $k = 0$);

2) The ρ-diameter of each set of the form (6.20) is at most ε;

3) $\mu_n(\bar{A}_{l-1,\ldots,l_k} \setminus A_{l_1,\ldots,l_k}) = 0$

for all n, k and $l_1, \ldots, l_k$.

For $B \in \mathcal{A}$, we denote by $\mu_0(B)$ the limit $\lim_{n\to\infty} \mu_n(B)$ (if this limit exists).

By using the standard diagonal process, we may construct a subsequence $\mu_{n_1}, \dots, \mu_{n_l}, \dots$ of the original sequence for which the limits

$$\lim_{j\to\infty} \mu_{n_j}(A_{l_1,\dots,l_k}) = \mu_0(A_{l_1,\dots,l_k}) \tag{6.21}$$

exist for all k and $l_1, \dots, l_k$.

Since

$$\mu_n(A_{l_1,\dots,l_k}) = \mu_n(\bar{A}_{l_1,\dots,l_k})$$

for all n, it is clear that we also have

$$\mu_0(A_{l_1,\dots,l_k}) = \mu_n(\bar{A}_{l_1,\dots,l_k}).$$

In addition, for every open set B we denote by $\mathcal{S}(B)$ the collections of finite unions of closures of sets of the form (6.20) contained in B (possibly for different values of k). Let

$$\sup_{S\in\mathcal{S}(B)} \mu_0(S) = \mu(B).$$

Evidently $\mu(A) = 1$, and μ is nonnegative and finitely additive on ρ-open subsets of A. Let us show that it is countably additive.

For this purpose we chose an infinite sequence

$$B_1, \dots B_n, \dots \tag{6.22}$$

of pairwise disjoint open subsets of A. Set $\bigcup_{r=1}^{\infty} B_r = B$. Let $S \in \mathcal{Y}(B)$. In virtue of its definition, S is compact and contained in B. Since it is compact, $\mathcal{S}$ is contained in the union $\bigcup_{t\in T} B_t$ of a finite number of sets of the form (6.22). We have

$$\mu(S) \leqq \mu\left(\bigcup_{t\in T} B\right) = \sum_{t\in T} \mu(B_t) \leqq \mu(B).$$

Therefore we also have

$$\sup_{S\in\mathcal{S}(B)} \mu(B) = \mu(B) \leqq \sup_T \sum_{t\in T} \mu(B_t) = \sum_{t=1}^{\infty} \mu(B_t),$$

and the countable additivity of μ is established for open sets.

It remains to extend μ to the whole Borel σ-algebra generated by the open subsets of A.

Choose an open set $B \subset A$ for which

$$\mu(\bar{B} \setminus B) = 0. \tag{6.23}$$

Let $d_k(B)$ denote the union of those closed sets of the form (6.23) for a given k which intersect B but are not subsets of B. By (6.23) we have

$$\lim_{k\to\infty} \mu_0(d_k(B)) = 0. \tag{6.24}$$

In addition, let $g_k(B)$ be the union of the closed sets of the form (6.20) which are subsets of B. Evidently

$$\mu_{n_j}(d_k(B) \cup g_k(B)) \geqq \mu_{n_j}(B) \geqq \mu_{n_j}(g_k(B)).$$

Consequently

$$\mu_0(d_k(B) \cup g_k(B)) \geqq \overline{\lim_{j\to\infty}} \mu_{n_j}(B) \geqq \underline{\lim}_{j\to\infty} \mu_{n_j}(B) \geqq \mu_0(g_k(B)). \tag{6.25}$$

But it is clear that

$$\mu_0(d_k(B) \cup g_k(B)) \geqq \mu(B) \geqq \mu_0(g_k(B)), \tag{6.26}$$

and by (6.24) that

$$\lim_{k\to\infty} (\mu_0(d_k(B) \cup g_k(B)) - \mu_0(g_k(B))) = 0. \tag{6.27}$$

It follows from (6.25) and (6.27) that the limit $\lim_{j\to\infty} \mu_{n_j}(B)$ exists, and by (6.26) and (6.27) it equals $\mu(B)$.

Thus we have established the compactness of the space of probability measures in the weak topology. □

6.16 Theorem. *In a noncooperative game* Γ *as in* (6.1), *let the space* $\mathfrak{x}_i$ *of pure strategies be* $\mathfrak{J}_i$*-separable, and let the sequence* $X_i^{(1)}, X_i^{(2)}, \ldots$ *converge weakly to* $X_i^{(0)}$. *Then this convergence is also valid in the* $\mathfrak{J}$*-topology.*

Proof. For an arbitrary $\delta > 0$ we construct a countable family $\{\mathfrak{y}_i^{(1)}, \mathfrak{y}_i^{(2)}, \ldots\}$, consisting of open nonempty pairwise disjoint subsets, for which

$$\bigcup_{k=1}^{\infty} \mathfrak{y}_i^{(k)} = \mathfrak{x}_i,$$

$$X_i^{(0)}(\bar{\mathfrak{y}}_i^{(k)} \setminus \mathfrak{y}_i^{(k)}) = 0,$$

and the diameter of each set $\mathfrak{y}_i^{(k)}$ does not exceed δ. This is possible because the spaces $\mathfrak{x}_i$ are $\mathfrak{J}$-separable.

Let $y_i^{(k)} \in \mathfrak{y}_i^{(k)}$ $(k = 1, 2, \ldots)$ and let $X^{(l*)}$ $(l = 0, 1, \ldots)$ be probability measures on $\mathfrak{x}_i$ such that

$$X_i^{(l*)}(y_i^{(k)}) = X_i^{(l)}(\mathfrak{y}_i^{(k)}).$$

Evidently the sequence $X_i^{(1*)}, X_i^{(2*)}, \ldots$ converges weakly in the ordinary sense to $X_i^{(0*)}$. Consequently, by the boundedness of the functions H_i we have, for every situation $\bar{x}$,

$$\begin{aligned}
\lim_{l\to\infty} H_i(\bar{x} \parallel X_i^{(l*)}) &= \lim_{l\to\infty} \sum_{k=1}^{\infty} X_i^{(l*)}(y_i^{(k)} H_i(x \parallel y_i^{(k)}) \\
&= \sum_{k=1}^{\infty} H_i(x \parallel y_i^{(k)}) \lim_{l\to\infty} X_i^{(l*)}(y_i^{(k)}) \\
&= \sum_{k=1}^{\infty} H_i(x \parallel y_i^{(k)}) X_i^{(0*)}(y_i^{(k)}) = H_i(x \parallel X_i^{(0*)}).
\end{aligned}$$

Here the convergence is uniform for all $x \in \mathfrak{x}$.

Consequently

$$\lim_{l\to\infty} \rho_i(X_i^{(l*)}, X_i^{(0*)}) = 0. \tag{6.28}$$

Moreover, we must have, for all l, and in particular for $l = 0$,

$$|H_i(x \parallel X_i^{(l*)}) - H_i(x \parallel X_i^{(l)})|$$

$$= \left| \sum_{k=1}^{\infty} (X_i^{(l*)}(y_i^{(l*)}) H_i(x \parallel y_i^{(k)}) - \int_{y_i^{(k)}} H_i(x \parallel y_i) X_i^{(l)}(dy_i) \right|$$

$$= \left| \sum_{k=1}^{\infty} \int_{y_i^{(k)}} (H_i(x \parallel y_i^{(k)}) - H_i(x \parallel y_i)) dX_i^{(l)}(y_i) \right|$$

$$\leq \sum_{k=1}^{\infty} \int_{y_i^{(k)}} |H_i(x \parallel y_i^{(k)}) - H_i(x \parallel y_i)| dX_i^{(l)}(y_i).$$

Since $\rho_i(y_i^{(k)}, y_i) < \delta$ for all $y_i \in y_i^{(k)}$,

$$|H_i(x \| X_i^{(l*)}) - H_i(x \| X_i^{(l)})| < \delta \sum_{k=1}^{\infty} \int_{y_i^{(k)}} dX_i^{(l)} = \delta X_i^{(0*)}(x_i) = \delta.$$

Therefore,

$$\rho_i(X_i^{(l*)}, X_i^{(l)}) < \delta. \tag{6.29}$$

Finally, we have

$$\rho_i(X_i^{(l)}, X_i^{(0)}) \leqq \rho_i(X_i^{(l)}, X_i^{(l*)}) + \rho_i(X_i^{(l*)}, X_i^{(0*)}) + \rho_i(X_i^{(0*)}, X_i^{(0)}),$$

and, taking account of (6.26) and (6.27),

$$\lim_{l \to \infty} \rho_i(X_i^{(l)}, X_i^{(0)}) \leqq 2\delta + \lim_{l \to \infty} \rho_i(X_i^{(l*)}, X_i^{(0*)});$$

and in view of (6.28) as well as the arbitrariness of the positive number δ,

$$\lim_{l \to \infty} \rho_i(X_i^{(l)}, X_i^{(0)}) = 0. \quad \square$$

Corollary. *If Γ is $\mathfrak{J}$-compact, its mixed extension is also $\mathfrak{J}$-compact.*

6.17 Existence of equilibrium situations. We have arrived at the objective of our discussion.

Theorem. *If the game Γ in (6.1) is $\mathfrak{J}$-compact, it contains an equilibrium situation.*

Proof. The $\mathcal{J}$-compactness of Γ implies its $\mathcal{J}$-precompactness. Therefore, by the theorem in 6.8, for every $\varepsilon > 0$ there is an ε-equilibrium situation in Γ. Choose $\varepsilon_n \downarrow 0$ and find, for each ε_k, an ε_k-equilibrium situation $X^{(k)}$ in Γ.

If we associate with each player $i \in \Gamma$ that player's strategy $X_i^{(k)}$ corresponding to the situation $X^{(k)}$, then in view of the $\mathcal{J}$-compactness, established above, of the space X_i of mixed strategies of the player, we can successively choose subsequences that converge in the $\mathcal{J}$-topology. Let the limit of such a subsequence for player i be that player's mixed strategy $X_i^{(0)}$. From these limit strategies $X_i^{(0)}$ we form the situation $X^{(0)}$.

Since each situation $X^{(k)}$ is an ε_k-equilibrium situation, we have

$$H_i(X^{(k)} \| x_i) \leqq H_i(X^{(k)}) + \varepsilon_k,$$

for every $i \in I$, $x_i \in \boldsymbol{x}_i$, and $k = 1, 2, \ldots$.

Taking the limit in this inequality as k increases, we obtain

$$\lim_{k\to\infty} H_i(X^{(k)} \| x_i) \leqq \lim_{k\to\infty} H_i(X^{(k)}) - \lim_{k\to\infty} \varepsilon_k.$$

But ε_k tends to zero, and the payoff functions H_i are continuous (since in the proof of the corollary in 6.8 we actually did not make any use of the nature of the strategies; in particular, the strategies might be mixed). Therefore

$$H_i((\lim_{k\to\infty} X^{(k)} \| x_i)) \leqq H_i(\lim_{k\to\infty} X^{(k)}),$$

i.e.

$$H_i(X^{(0)} \| x_i) \leqq H_i(X^{(0)})$$

for all $i \in I$ and $x_i \in \boldsymbol{x}_i$. □

6.18 Special cases and examples. As in the case of the semi-intrinsic topology (see 6.4), in the case considered here we may likewise determine the $\mathcal{J}$-topological properties of the game by its properties in the exterior topology.

Theorem. *In the noncooperative game Γ of (6.1), let each set $\boldsymbol{x}_i$ be compact in the exterior topology $\mathcal{Y}_i$, and let the payoff functions H_i be continuous functions of situations in the sense of the topology $\mathcal{Y}$ generated on $\boldsymbol{x}_i$ by the topologies $\mathcal{Y}_i$.*

Then the game Γ is $\mathcal{J}$-compact.

Proof. For each player $i \in I$ we choose an arbitrary sequence

$$x_i^{(1)}, \ldots, x_i^{(m)}, \ldots \tag{6.30}$$

of that player's pure strategies, and select a subsequence that converges (in the $\mathcal{Y}_i$ sense) to a limit $x_i^{(0)}$; without loss of generality, we may suppose that the whole sequence (6.30) converges. Since the functions H_j are continuous, we have

$$\lim_{m\to\infty} H_j(x \| x_i^{(m)}) = H_i(x \| x_i^{(0)}). \tag{6.31}$$

But continuous functions on a compact set are uniformly continuous, and the number of players is finite. Therefore, using (6.31), we have

$$\lim_{m\to\infty} \rho_i(x_i^{(m)}, x_i^{(0)}) = \lim_{m\to\infty} \max_{j\in I} \sup_{x\in \boldsymbol{x}} |H_j(x \parallel x_i^{(m)}) - H_j(x \parallel x_i^{(0)})|,$$

i.e., the sequence (6.30) converges in the $\mathcal{J}$-topology. □

We notice, as a very important special case, *a continuous game on the unit n-cube*, i.e. a noncooperative game Γ in which $x_i = [0,1]$ $(i \in I)$ (so that the set of situations is the unit n-cube), and all the payoff functions H_i are continuous on the cube.

6.19. The theorem just proved reproduces the result of 5.9 as a corollary of the theorem of 6.17. However, this corollary by no means exhausts the possibilities of that theorem, since one noncooperative game or another, even with very pathological features, can nevertheless be entirely O.K. from the point of view of the intrinsic topology; for example, compact.

As a very simple example, we consider the game Γ on two squares (see the example in 6.6), in which the payoff functions are the superpositions

$$H_i(x_1, x_2) = F_i(\varphi_1(x_1), \varphi_2(x_2)), \quad i = 1, 2,$$

where F_1 and F_2 are continuous, and φ_1 and φ_2 are arbitrary mappings of [0,1] on itself (admitting arbitrarily pathological features).

Evidently no general theorems based solely on the exterior topology are applicable, whereas the theorem of 6.17 allows us to suppose that the "real strategies" of players 1 and 2 are not values of the variables x_1 and x_2 themselves, but those of functions $\varphi_1(x_1)$ and $\varphi_1(x_2)$.

This fact reveals a very deep principle: the topology (and metric) on spaces of strategies (and of situations) has to reflect not only the superficial, the (so to speak) "physical" proximity of strategies as actions, but their intrinsic, "pragmatic", proximity, that is determined by the proximity of the consequences of the choice of the strategies.

From a purely formal point of view, the transition from the strategies x_1 and x_2 of the players to their strategies $\varphi_1(x_1)$ and $\varphi_1(x_2)$ is an exact homomorphism of the corresponding games, and the theorem of 2.21 ensures the possibility of transferring the theorem on the existence of equilibrium situations from one game to another.

6.20. At the same time, the abandonment of the continuity of the payoff functions can immediately lead to a violation of the $\mathcal{J}$-compactness of the game.

For example, suppose that in the game described in 6.6 we have, for every x_1 in an interval $[\alpha, \beta] \subset [0,1]$ and for some $\delta > 0$,

$$H_1(x_1, x_1) - \lim_{x_2 \to x_1 - 0} H_1(x_1, x_2) > \delta. \tag{6.32}$$

Select x_1' and $x_2' \in [\alpha, \beta]$ arbitrarily, and let, for definiteness, $x_1' < x_1''$ (Figure 1.15).

Let us suppose that

$$\rho_1(x_1', x_1'') \leqq \varepsilon < \delta/2. \tag{6.33}$$

Then

$$\sup_{x_2} |H_1(x_1', x_2) - H_1(x_1'', x_2)| \leqq \varepsilon,$$

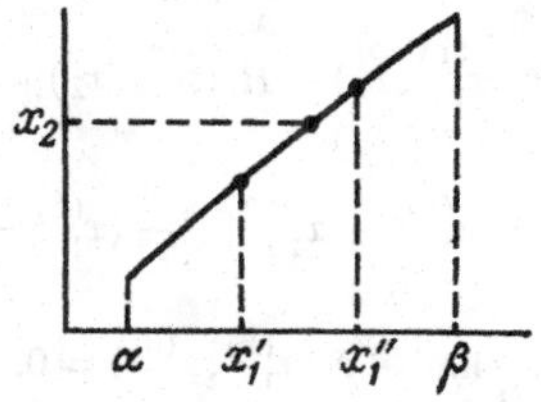

Fig. 1.15

and in particular, with $x_2 = x_1''$,

$$|H_1(x_1', x_1'') - H_1(x_1'', x_1'')| \leqq \varepsilon,$$

and on the other hand, letting $x_2 \to x_1'' - 0$,

$$|H_1(x_1', x_1'') - \lim_{x_2 \to x_1'' - 0} H_1(x_1'', x_2)| \leqq \varepsilon,$$

whence

$$|H_1(x_1'', x_1'') - \lim_{x_2 \to x_1'' - 0} H_1(x_1'', x_2)| \leqq 2\varepsilon,$$

which, together with (6.33), contradicts (6.32).

Consequently *any two points* of $[\alpha, \beta]$ are now separated, in terms of the $\mathcal{J}$-metric, by a distance exceeding a given $\varepsilon > 0$. Here we cannot speak of the $\mathcal{J}$-compactness of the space x_1 of strategies.

However, some inferences on the existence of equilibrium situations in such games can nevertheless be made. For example, this will be discussed for a class of two-person zero-sum games in § 3 of Chapter 3.

6.21. The following example is also instructive. It shows that topologies on the sets of strategies of each player that reflect only the proximity of the values of that player's payoff function are too crude for our purposes (compare the discussion in the second paragraph of 6.7).

Let us introduce a metric $\bar{\rho}_i$ on the set x_i of strategies of player i in the noncooperative game Γ in (6.1), by setting, for $x_i', x_i'' \in x_i$,

$$\bar{\rho}_i(x_i', x_i'') = \sup_{x \in x} |H_i(x\|x_i') - H_i(x\|x_i'')| \tag{6.34}$$

(compare formula (6.8)), and construct an example of a game in which each space x_i is compact in the corresponding metric $\bar{\rho}_i$, but which has no equilibrium situations.

Set

$$I = \{1, 2\}, \ x_1 = x_2 = [0, 1]$$

$$H_1(x_1, x_2) = \begin{cases} -x_1^2 + x_2^2, & \text{if } x_2 \in [0, 1/2], \quad (6.35) \\ -(x_1 - 1)^2 + (x_2 - 1)^2, & \text{if } x_2 \in (1/2, 1], \quad (6.36) \end{cases}$$

$$H_2(x_1, x_2) = \begin{cases} (x_1 - 1)^2 - (x_2 - 1)^2, & \text{if } x_1 \in [0, 1/2], \quad (6.37) \\ x_1^2 - x_2^2, & \text{if } x_1 \in (1/2, 1], \quad (6.38) \end{cases}$$

To analyze this game we select a sequence $x_1^{(1)}, \dots, x_1^{(k)}$ of strategies which converges in the Euclidean topology of the interval [0,1] to the strategy $x_1^{(0)}$. Then, according to (6.34) – (6.36),

$$\bar{\rho}_1(x_1^{(k)}, x_1^{(0)}) = \sup_{x_2} |H_1(x_1^{(k)}, x_2) - H_1(x_1^{(0)}, x_2)|$$

$$= \max\{| - x_1^{(k)2} + x_1^{(0)2}|, | - (x_1^{(k)} - 1)^2 + (x_1^{(0)} - 1)^2|\},$$

so that

$$\lim_{x_1^{(k)} \to x_1^{(0)}} \bar{\rho}_1(x_1^{(k)}, x_1^{(0)}) = 0,$$

and the compactness of x_1 in the Euclidean metric implies the compactness of x_1 in the metric $\bar{\rho}_i$.

The compactness of x_2 in the metric $\bar{\rho}_2$ is proved similarly.

Now suppose that $(X_1^*, X_2^*) \in \mathscr{C}(\Gamma)$, i.e. that

$$H_1(x_1, X_2^*) \leqq H_1(X_1^*, X_2^*) \text{ for } x_1 \in x_1,$$

$$H_2(X_1^*, x_2) \leqq H_2(X_1^*, X_2^*) \text{ for } x_2 \in x_2.$$

It is clear from this and from (6.35) and (6.36) that H_1 is a strictly concave function of x_1 (the second derivative with respect to x_1 is constant, and equal to -2) for all strategies x_2. Therefore, according to the corollary at the end of 5.11, the mixed strategy X_1 (if not pure) is strictly dominated by some pure strategy. Consequently we may set $X_1^* = x_1^*$. Similarly we can deduce from (6.37) and (6.38) that $X_2^* = x_2^*$, so that the game must have an equilibrium situation (x_1^*, x_2^*) in the pure strategies, and this is unique because of the strict concavity that we have established.

Let us suppose that $x_1^* \in [0, 1/2]$. Then by (6.37), $H_2(x_1^*, x_2)$ attains its maximum at $x_2 = 1$, so that necessarily $x_2^* = 1 > 1/2$. But then, by (6.36), $H_1(x_1, x_2^*)$ attains its maximum at $x_1 = 1$, so that we must have $x_1^* = 1 > 1/2$, and we have a contradiction.

Similarly, $x_1^* \in [1/2, 1]$ implies, by (6.38), that $x_2 = 0 \leqq 1/2$, whence by (6.35) we have $x_1^* = 0 < 1/2$, i.e. again a contradiction.

Notes and references for Chapter 1

§1. The concept of a noncooperative game was introduced by J. von Neumann [1]. The matrix arrangement of payoffs for a finite two-person zero-sum game was apparently first used by E. Borel [1]. The term "matrix game" apparently originated with J.B. Dantzig [1]. Bimatrix games were introduced by N.N. Vorob'ev [1]. F. Szidarovszkiĭ [2] discussed a generalization of bimatrix games. For the form of the set of realizable payoff vectors, see N.N. Vorob'ev and G.V. Epifanov [1]. The use of topological properties of the space of the players' strategies (for two-person zero-sum games) was introduced by J. Ville [1]; and of topological structures on the sets of players' strategies, by A. Wald [1,2]. In [2] Wald analyzed some questions connected with the measurability of games, and introduced the notion of a subgame. Automorphisms of noncooperative games were first considered by J. Nash [1]. One can find examples of ε-homomorphisms of games, in the guise of ε-nets for the natural metric, in A. Wald [2]; and for quasi-invariant kernels, in E.B. Yanovskaya [1]. A special construction of ε-homomorphisms was given by N.N.Vorob'ev [8]. Apparently the first use of strategic equivalence was made by J.C.C. McKinsey [1].

§2. O. Cournot [1] described the concept of an equilibrium situation in informal terms for the case of two-person games. In the general case, the corresponding mathematical formulation was given by J. Nash [2]. The axiomatic description of equilibrium situations was introduced by E.I. Vilkas [8] (for the case of two-person games, see also M. Jansen and S. Tijs [1]). Some of their general properties were considered by E.I. Vilkas in [6,7]. For the structure of sets of equilibrium situations see M.M. Chin, T. Parthasarathy, and T.E.S. Raghavan [1]. Wald [1] investigated ε-optimality (for the case of ε-equilibrium in two-person zero-sum games). $\mathfrak{K}$-equilibrium situations were introduced by E.B. Yanovskaya [20,22].

Morgenstern introduced the idea of using literature as source material for game-theoretic analysis (see J. von Neumann and O. Morgenstern [1], p. 199). Methodological problems connected with this were analyzed by N.N. Vorob'ev [9], who suggested the example of Mr. Pickwick. A game theoretical analysis of some biblical stories was undertaken by S. Brams [1,2,3].Detailed analyses of "battle of the sexes" games and of "prisoner's dilemma" are given by R.D. Luce and M. Raiffa [1]. The latter game has inspired many studies, including those of its psychological features. A. Rapoport and A.M. Chammah [1] have devoted a whole monograph to experimental study of this game.

Independence of irrelevant alternatives as a property (or axiom) inherent in optimality goes back to the work of F. Zeuthen [1]. The concept was apparently introduced into game theory by J. Nash [1]. A critical analysis of this concept was carried out by J. Bein [1]. J. Nash [2] raised the question of covariance of equilibrium situations. Special cases of the theorem in **2.22** appear in articles by A. Wald [2], E.B. Yanovskaya [1] and N.N. Vorob'ev [18].

§3. The idea of establishing the existence of equilibrium situations in a two-person game with informed players can actually be found in E. Zermelo [1]. For further discussion see von M. Stackelberg's book [1]. Ju.B. Germeĭer's article [3] and book [4] contain studies of a large number of versions of these games. The existence of equilibrium situations in quasiconcave games (the theorem of **3.10**) was established for two-person zero-sum games by M. Sion [1], and for concave games by H. Nikaidô and K. Isoda [1]. The condition for uniqueness of an equilibrium situation for this class of games was established by J.B. Rosen [1].

The theorem on the existence of $\mathfrak{K}$-equilibrium situations for a star-shaped family $\mathfrak{K}$ of coalitions was proved by E.B. Yanovskaya [22]; subsections **3.12–3.14** were written by her. For the theorem of Michael on the existence of the required selector, see T. Parthasarathy [4].

§4. N. Howard [1] gave the first systematic discussion of metastrategic extensions of noncooperative games with the aim of proving the existence of stable situations. In this connection, see also N.S. Kukushkin [1,2]. The proof of the existence of stable situations in metagames, for the case when $\mathfrak{K}$ is a quasipartition, was given by N.N. Vorob'ev [24].

§5. Mixed strategies as a means of finding a maximin were suggested early in the 18th century by Waldegraf (on the evidence of P. Montmort [1]), but the fact was completely forgotten, and was rediscoverd by E. Borel [1]. (See also R. Fisher's article [1].) The first proof of the existence of equilibrium situations (saddle points) in matrix games (based on fixed-point theory) was given by J. von Neumann [1], who took precisely this result as the starting point of game theory. Subsequently the requirements of game theory served as an important stimulus for many increasingly more general fixed-point theorems. In this connection we çall attention to the papers by M.F. Bohnenblust and S. Karlin [1] and I.L. Glicksberg [1], in which the general theorem of **5.4** was proved. The existence of pure strategies, mentioned in the proof of the theorem in **5.5**, may be seen immediately from proposition 7 of Bourbaki's book.*)

§6. The intrinsic topology and intrinsic metric on spaces of strategies were introduced by A. Wald [2], who used the term natural metric, and also the Helly metric. E. Helly [1] originated this concept, originally for real-valued functions of two variables. See also J. von Neumann [2]. A. Wald [2] proved, for completely bounded (conditionally compact) two-person zero-sum games, the existence of ε-saddlepoints for every $\varepsilon > 0$ in strategies with finite spectrum; and, for compact two-person zero-sum games, the existence of saddlepoints in the mixed strategies. The intermediate results of **6.11–6.16**, for two-person zero-sum games, were also obtained by him. The semi-intrinsic topology on spaces of players' strategies in two-person zero-sum games was proposed Teh-Tjoe Tie [1]. There is no difficulty in the natural extension of the necessary concepts and the existence theorems to general noncooperative games along these lines.

The example in **9.21** is based on E.B. Yanovskaya's ideas.

---

*) N. Bourbaki, Intégration, mesures, intégration des mesures, Actualités Sci. Ind., no. 1244, Hermann, Paris, 1956.

Chapter 2
Finite noncooperative games

§1 Finite noncooperative games

1.1 Fundamental concepts and elementary properties. Let the noncooperative game

$$\Gamma = \langle I, \{x_i\}_{i\in I}, \{H_i\}_{i\in I}\rangle \tag{1.1}$$

be finite (see 1.3, Chapter 1). For each $i \in I$, we set $x_i = \{x_i^1, \ldots, x_i^{m_i}\}$. The system of numbers $\langle m_1, \ldots, m_n\rangle$ is sometimes called the *format* (or the *dimensions*) of the game (1.1). Sometimes the strategies of player i will be identified with their indices $1, \ldots, m_i$.

If we suppose that the sets x_i $(i \in I)$ of strategies are topologized as spaces of isolated points, and their subspaces are assumed to be measurable, then the finite noncooperative game falls under the hypotheses of the theorem of 5.9, Chapter 1, i.e., has equilibrium situations in the mixed strategies. In this section we shall give still another proof of the existence of equilibrium situations for finite noncooperative games and examples of how they are found in certain cases.

Evidently every mixed strategy X_i of player i is given by a set of probabilities of choosing, under its hypotheses, each of the player's pure strategies

$$X_i(x_i^1), \ldots, X_i(x_i^{m_i}), \tag{1.2}$$

where we are to have

$$X_i(x_i^j) \geqq 0, \quad j = 1, \ldots, m_i; \quad \sum_{j=1}^{m_i} X_i(x_i^j) = 1. \tag{1.3}$$

Since the players choose their strategies in a stochastically independent way, we have, for the probability of the situation $x = x_I = (x_1, \ldots, x_n)$, under the conditions of the situation, in the mixed strategies $X = X_I = (X_1, \ldots, X_n)$,

$$X(x) = \prod_{i\in I} X_i(x_i).$$

We note for later use that the description of the mixed strategies of player i in the form of the sequence of numbers (1.2) satisfying (1.3) allows us to represent them as points of an $(m_i - 1)$-dimensional simplex defined by their barycentric coordinates.

In this case, formulas (5.3) and (5.4), Chapter 1, which determine the payoffs in a situation in the mixed strategies, can be written in the form

$$H_i(X) = \sum_{x\in x} H_i(x)X(x) = \sum_{x_1\in x_1} \cdots \sum_{x_n\in x_n} H_i(x_1, \ldots, x_n)X_1(x_1)\ldots X_n(x_n). \tag{1.4}$$

In particular, if we repeat the discussion in Chapter 1, 5.3, we have as an analogue of the formula (5.5), Chapter 1,

$$\begin{aligned} H_i(X\|x_j^0) = \sum_{x_1\in \boldsymbol{x}_1} \cdots \sum_{x_{j-1}\in \boldsymbol{x}_{j-1}} \sum_{x_{j+1}\in \boldsymbol{x}_{j+1}} \cdots \sum_{x_n\in \boldsymbol{x}_n} \\ \times H_i(x_1,\ldots,x_{j-1},x_j^0,x_{j+1},\ldots,x_n) \\ \times X_1(x_1)\ldots X_{j-1}(x_{j-1})X_{j+1}(x_{j+1})\ldots X_n(x_n), \end{aligned} \tag{1.5}$$

and we immediately obtain as in Chapter 1

$$H_i(X) = \sum_{x_j\in \boldsymbol{x}_j} H_i(X\|x_j)X_j(x_j) \tag{1.6}$$

from formulas (1.4) and (1.5).

1.2. In the finite noncooperative game in (1.1), the players' mixed strategies assign certain probabilities to their pure strategies. A pure strategy $x_i \in \boldsymbol{x}_i$ then appears as a point of the spectrum of the mixed strategy $X_i \in \boldsymbol{X}_i$ if $X_i(x_i) > 0$.

The continuity of the payoff functions in the mixed extensions of a finite game assumes the validity of the lemma of Chapter 1, 5.15, which takes the following form under the present conditions.

Lemma. *If, in the finite noncooperative game Γ in (1.1), the situation X^* is i-optimal, and*

$$X^*(x_i^0) > 0, \tag{1.7}$$

then

$$H_i(X^*\|x_i^0) = H_i(X^*). \tag{1.8}$$

The proof would follow from what was said in Chapter 1, 5.15, but the finiteness of Γ allows us to use a more elementary argument.

By the characterization of i-optimality given in Chapter 1, 5.13, we have

$$H_i(X^*\|x_i) \leqq H_i(X^*) \text{ for all } x_i \in \boldsymbol{x}_i.$$

Multiply each of these inequalities by $X_i^*(x_i)$:

$$H_i(X^*\|x_i)X_i^*(x_i) \leqq H_i(X^*)X_i^*(x_i) \tag{1.9}$$

and sum over $x_i \in \boldsymbol{x}_i$:

$$\sum_{x_i\in \boldsymbol{x}_i} H_i(X^*\|x_i)X_i^*(x_i) \leqq H_i(X^*) \sum_{x_i\in \boldsymbol{x}_i} X_i^*(x_i),$$

or, by (1.6), $H_i(X^*) \leqq H_i(X^*)$. Here we actually have strict equality. But a sum of inequalities of the same sense can be an equality only if each of the individual inequalities is an equality. This means that all the terms in (1.9) are equalities. Using (1.9) with $x_i = x_i^0$, and (1.7), we obtain (1.8). □

The conclusion of the lemma is often used in the contrapositive form: if, under its hypotheses,

$$H_i(X^*\|x_i^0) < H_i(X^*)$$

then $X_i^*(x_i^0) = 0$.

1.3. The lemma of the preceding subsection, applied to an important special case, yields the following useful fact.

Definition. A mixed strategy X_i of player i in a finite noncooperative game as in (1.1), with the set $\boldsymbol{x}_i$ of pure strategies for player i, is said to be a *completely mixed* strategy if $X_i(x_i) > 0$ for all $x_i \in \boldsymbol{x}_i$. □

Lemma. 1) *If a situation of the form $X\|\bar{X}_i$ under the completely mixed strategy $\bar{X}_i$ of player i is i-optimal, there is a number h_i such that*

$$H_i(X\|x_i) = h_i \text{ for all } x_i \in \boldsymbol{x}_i \tag{1.10}$$

(i.e., the coalitional strategy $X^{(i)}$ is equalizing).

2) *If* (1.10) *holds, the situation $X\|\bar{X}_i$ is i-optimal no matter what the strategy $\bar{X}_i$ is.*

Proof. 1) The i-optimality of X means that

$$H_i(X\|x_i) \leqq H_i(X), \quad x_i \in \boldsymbol{x}_i, \tag{1.11}$$

and from the completely mixed property of X_i it follows from what precedes that there must be equality here for every $x_i \in \boldsymbol{x}_i$. Taking $H_i(X) = h_i$, we obtain (1.10).

2) Let (1.11) hold. Then, turning to the mixed strategies, we obtain

$$H_i(X\|\bar{X}_i) = H_i(X) = h_i,$$

for every strategy $\bar{X}_i$, whence the i-optimality automatically follows. □

1.4. The statement formulated in 6.3 can be transferred from strategies to situations in the following way.

Definition. A situation is said to be *completely mixed* if all the strategies in it are completely mixed. □

Theorem. *A necessary and sufficient condition for a completely mixed situation X to be an equilibrium situation is that* (1.10) *is satisfied for some numbers h_i* $(i \in I)$.

Proof. *Necessity.* Let X be an equilibrium situation. Therefore it is i-optimal for every $i \in I$; therefore, under our hypotheses, by part 1) of the preceding lemma there must be a number h_i for which (1.10) is satisfied.

Sufficiency. Let (1.10) be satisfied for a given situation X. Then by part 2) of the lemma, the situation $X\|X_i = X$ must be i-optimal. Since $i \in I$ is arbitrary, it follows that X is an equilibrium situation. □

1.5. As another corollary of the lemma in 1.2, we show that the determination of equilibrium situations in bimatrix games (see 1.3 of Chapter 1) is a rational procedure (in the number-theoretical sense of the word).

Let us consider an $m \times n$-bimatrix game $\Gamma_{A,B}$. If X and Y are mixed strategies of players 1 and 2 (see 1.1) and $X(i) = \xi_i$ and $Y(j) = \eta_j$, then, in the situation (X, Y), by (1.4) the respective payoffs to the players are

$$\sum_{i,j} a_{ij}\xi_i\eta_j = XAY^T \text{ and } \sum_{i,j} b_{ij}\xi_i\eta_j = XBY^T.$$

The following theorem provides a potential possibility of an algorithm for determining the set $\mathscr{C}(\Gamma)$ for the bimatrix game Γ.

Theorem. *In every bimatrix game the sets of equilibrium situations can be completely determined by a finite number of rational operations on the elements of the payoff matrices.*

An explicit description of the algorithm for the case of bimatrix games which are, in a certain sense, nondegenerate (and, as it turns out, the overwhelming majority) is given in 3.3–3.7, and, for arbitrary bimatrix games, in Chapter 4, 4.7–4.14. Let us denote by $\boldsymbol{X}$ the set of all mixed strategies of player 1 of the bimatrix game under consideration and by $\boldsymbol{Y}$ that of player 2.

Proof. We choose an arbitrary mixed strategy X of player 1, denote its spectrum by R_X, and set

$$\boldsymbol{X}(R) = \{X : X \in \boldsymbol{X},\ R_X = R\}$$

for each $R \subset x$, and also set

$$\mathscr{C}_1(X) = \{Y : Y \in \boldsymbol{Y}, (X, Y) \in \mathscr{C}_1(\Gamma)\},$$

$$\mathscr{C}_2(Y) = \{X : X \in \boldsymbol{X}, (X, Y) \in \mathscr{C}_2(\Gamma)\},$$

for each $X \in \boldsymbol{X}$ and $Y \in \boldsymbol{Y}$. It is clear that $X \in \mathscr{C}_1(Y)$ is equivalent to $Y \in \mathscr{C}_2(X)$.

According to what was said in 1.2, the inclusion $X \in \mathscr{C}_1(Y)$ (and therefore also $Y \in \mathscr{C}_2(X)$) means that the sets of inequalities

$$A_i \cdot Y^T \leqq XAY^T, \qquad i = 1, \dots, m, \tag{1.12}$$

are satisfied. Besides we must have

$$A_i \cdot Y^T = XAY^T \text{ for } i \in R_X.$$

It then follows that we have, for every strategy $\bar{X} \in \boldsymbol{X}(R_X)$,

$$\bar{X}AY^T = XAY^T,$$

i.e., XAY^T is independent of X on $\boldsymbol{X}(R_X)$ and is a common linear form in Y. Therefore (1.12) is, on $\boldsymbol{X}(R_X)$, a set of linear (not bilinear) inequalities, and $\mathscr{C}_1(X)$ is a convex polyhedron (possibly empty) which depends only on R_X, and each of whose vertices is a value of a rational function of the elements of the matrix A.

If $R_X = R$ we may denote $\mathscr{C}_1(X)$ by $\mathscr{C}_1(R)$. It is clear that $R_1 \subset R_2$ implies $\mathscr{C}_1(R_2) \subset \mathscr{C}_1(R_1)$. Therefore we have

$$\mathscr{C}_1(\Gamma) = \bigcup_{R \subset \boldsymbol{x}} (\mathscr{C}_1(R) \times \boldsymbol{X}(R)) = \bigcup_{R \subset \boldsymbol{x}} (\mathscr{C}_1(R) \times \bar{\boldsymbol{X}}(R)),$$

where the summation is over all subsets of the set of pure strategies of player 1 (the bar denotes closure). Since there are only a finite number of terms in the sum, it can be described in rational terms.

Finally, everything we have said applies equally well to $\mathscr{C}_2(\Gamma)$, so that $\mathscr{C}(\Gamma)$ can be described in rational terms as the intersection $\mathscr{C}_1(\Gamma) \cap \mathscr{C}_2(\Gamma)$. □

1.6 Existence theorem. Finite noncooperative games evidently satisfy the hypotheses of the theorem in Chapter 1, 5.4: all functions are trivially continuous on a finite space of isolated points. Therefore the existence of equilibrium situations in finite noncooperative games can be deduced from Glicksberg's fixed point theorem.

At the same time, the existence of equilibrium situations for the mixed strategies in finite noncooperative games also follows from the theorems of Chapter 1, 3.10 and 3.11, on concave and quasiconcave games, since the simplexes of players' mixed strategies in a finite game are convex subsets of finite-dimensional Euclidean spaces, and the payoff functions on situations in the mixed strategies are, by (1.4), multilinear, i.e., continuous. Consequently, in proving the existence of equilibrium situations in finite noncooperative games it is sufficient to use propositions that are more special than Glicksberg's theorem, for example Brouwer's fixed point theorem. Nash gave an elegant proof depending on Brouwer's theorem (historically the first proof) of the theorem on the existence of equilibrium points in the mixed strategies for finite noncooperative games. This is known as Nash's theorem. Let us present this proof.

Let us choose an arbitrary situation $X = (X_1, \ldots, X_n)$ in the mixed strategies of a finite noncooperative game as in (1.1). For each player $i \in I$ and each of that player's pure strategies $x_i^j \in \boldsymbol{x}_i$, set

$$\varphi_i^j(X) = \max\{0, H_i(X \| x_i^j) - H_i(X)\}$$

and further

$$\bar{X}_i(x_i^j) = (X_i(x_i^j) + \varphi_i^j(X)) \left(1 + \sum_{x_i^j \in \boldsymbol{x}_i} \varphi_i^j(X) \right)^{-1}. \tag{1.13}$$

As is easily verified,

$$\bar{X}_i(x_i^j) \geqq 0 \text{ for all } x_i^j \in x_i,\ i \in I,$$

$$\sum_{x_i^j \in x_i} \bar{X}_i(x_i^j) = 1 \text{ for } i \in I.$$

Therefore the set of numbers $\bar{X}(x_i^j)$, where $x_i^j \in x_i$, can be thought of as a mixed strategy $\bar{X}_i$ of player i, and the system $\bar{X} = (\bar{X}_1, \ldots, \bar{X}_n)$, as a situation in the mixed strategies. Therefore formulas of the form (1.13) make each situation X in the mixed strategies correspond to a situation $\bar{X} = \psi(X)$ in the mixed strategies, i.e. there is a mapping $\psi : \boldsymbol{X} \to \boldsymbol{X}$. But $\boldsymbol{X}$ is a Cartesian product of finite-dimensional simplexes $\boldsymbol{X}_i$, i.e., is itself homeomorphic to a finite-dimensional simplex.

Moreover, since all the functions $H_i(X)$ and $H_i(X \| x_i^j)$ are continuous in X, the continuity of the function $f(t) = \max\{0, t\}$ and the positivity of the denominator in (1.13) imply the continuity of ψ.

Consequently we have the hypotheses of Brouwer's theorem, according to which the mapping ψ has a fixed point, i.e. a situation X^* in the mixed strategies for which

$$X_i^*(x_i^j) = (X_i^*(x_i^j) + \varphi_i^j(X^*)) \left(1 + \sum_{x_i^j \in x_i} \varphi_i^j(X^*)\right)^{-1}.$$

We now show that every point of $\boldsymbol{X}$ that is fixed under the mapping ψ is an equilibrium situation.

According to the lemma in Chapter 1, 5.14, each player $i \in I$ has a pure strategy $x_i^{j_0} \in x_i$ for which

$$X_i^*(x_i^{j_0}) > 0 \text{ and } H_i(X^* \| x_i^{j_0}) \leqq H_i(X^*),$$

i.e. $\varphi_i^{j_0}(X^*) = 0$. We then have

$$X_i^*(x_i^{j_0}) = X_i^*(x_i^{j_0}) \left(1 + \sum_{x_i^j \in x_i} \varphi_i^j(X^*)\right)^{-1},$$

and therefore $\varphi_i^j(X^*) = 0$ for all $x_i^j \in x_i$, i.e.,

$$H_i(X^* \| x_i^j) \leqq H_i(X^*).$$

It remains only to remark that this holds for every player $i \in I$.

1.7 Solution of a game. The process of finding equilibrium situations in a noncooperative game is usually called its solution. A general theory of analytic solutions of finite noncooperative games (and all the more, of noncooperative games in general) does not yet exist, and it is not even clear what such a theory could be.

At the present time the solution of any noncooperative game (especially if it depends on parameters) is an individual problem, usually requiring special consideration. One of the rather general methods of solving such problems is the systematic use of the propositions in 1.2–1.4, which allow the replacement of certain inequalities by equations in definition of equilibrium situations.

1.8 Diagonal games. As an example, we consider the noncooperative game Γ of (1.1), in which

$$x_i = \{1, \dots, m\} \text{ for all } i \in I,$$
$$H_i(x) = \begin{cases} h_{ik} > 0 & \text{if } x_i = k \text{ for all } i \in I, \\ 0 & \text{otherwise.} \end{cases} \tag{1.14}$$

It is natural to call such noncooperative games *diagonal*. They can be considered as struggles for arriving at agreements which are possible only under complete unanimity of the players, but where the players individually evaluate different versions of these agreements. We note that the diagonal matrix games to be described in Chapter 4 (5.7) are not diagonal games in the sense of the present definition.

Let X be any situation in the mixed strategies in the game Γ and let k be a pure strategy of the players. We shall use the term *useability* of the strategy k in the situation X to mean the number of players $i \in I$ for whom $X_i(k) > 0$ (i.e., $k \in \operatorname{supp} X_i$), and denote it by $\operatorname{use}_X k$.

In correspondence with the "diagonality" of games Γ as described in (1.14), we have

$$H_i(X) = \sum_{k=1}^{m} h_{ik} \prod_{j=1}^{n} X_j(k), \tag{1.15}$$

where the only terms that differ from zero are the terms corresponding to the strategies k for which $\operatorname{use}_X k = n$. In particular,

$$H_i\left(X \underset{i}{\|} l\right) = h_{il} \prod_{j \neq i} X_j(l). \tag{1.16}$$

Let us describe at once a wide class of "trivial" equilibrium situations X in which none of the pure strategies appears in the spectrum of equilibrium strategies for more than $n-2$ players, i.e.,

$$\max_k \operatorname{use}_X k \leqq n - 2. \tag{1.17}$$

In such a situation each product in (1.15) has at least two vanishing factors, and in (1.17), at least one. Consequently in the case (1.17)

$$H_i(X) = H_i(X \underset{i}{\|} l) = 0 \text{ for all } i \in I \text{ and } l = 1, \dots, m,$$

and X is an equilibrium situation.

The stability of an equilibrium situation of this type reflects the hopelessness of any deviation from such a situation by any player; in view of the lack of coordination of their actions, individual efforts of individual players cannot improve their positions in this situation.

Let us show further that there is no equilibrium situation X for which

$$\max \operatorname{use}_X k = n - 1. \tag{1.18}$$

In fact, in this case all the terms in (1.15) reduce, as before, to zero. On the other hand, let the maximum in (1.18) be attained for the strategy l, i.e. the pure strategy l is not used in X by *exactly one* player, whom we denote by $i(l)$. Then the right-hand side of (1.16) will be positive, so that

$$H_{i(l)}(X) < H_{i(l)}\left(X \underset{i(l)}{\|} l\right),$$

and this contradicts the equilibrium property of the situation X.

It remains only to consider the case when there is a strategy l for which

$$\operatorname{use}_X l = n \tag{1.19}$$

Let us show that in this case the spectra of all the strategies X_i $(i \in I)$ must coincide. Suppose the contrary, i.e. that

$$s \in \operatorname{supp} X_i \setminus \operatorname{supp} X_j \tag{1.20}$$

for some i and $j \in I$ (evidently in this case $\operatorname{use}_X s < n$), and consider the mixed strategy X^*:

$$X_i^*(x_i) = \begin{cases} X_i(l) + X_i(s) & \text{if } x_i = l, \\ 0 & \text{if } x_i = s, \\ X_i(x_i) & \text{for the remaining strategies } x_i \in x_i. \end{cases} \tag{1.21}$$

Then we write

$$H_i(X \| X_i^*) = \sum_{k=1}^{m} h_{ik} X_i^*(k) \prod_{i \neq j} X_j(k). \tag{1.22}$$

According to (1.20), $X_j(s) = 0$, so that the term in (1.14) for which $k = s$ is zero. In (1.22) the corresponding term is also zero, in view of (1.21). In addition, on the basis of (1.19) the term in (1.15) corresponding to $k = l$ is positive, and by (1.20) $X_i(s) > 0$. Consequently, in passing from (1.15) to (1.22) the coefficient of one of the positive terms is actually increased. Finally, the terms corresponding to indices k that are different from l and s coincide on the right-hand sides of (1.15) and (1.13). As a result, it turns out that $H_i(X \| X_i^*) > H_i(X)$, and this contradicts the equilibrium property of X. Consequently (1.20) is impossible, and therefore the set $\operatorname{supp} X_i$ is actually independent of i. Let us suppose that $\operatorname{supp} X_i = y_0$.

It follows from what has been said that

$$H_i\left(X \underset{i}{\|} l\right) = H_i(X) \text{ for all } l \in y_0$$

or, if we use (1.16),

$$h_{il} \prod_{j \neq i} X_j(l) = H_i(X) \text{ for all } l \in y_0.$$

It is clear that $H_i(X) > 0$. Therefore we naturally obtain

$$X_i(l) = \left(\frac{1}{H_i(X)} \left(\prod_{j \in I} H_j(x) \right)^{1/(n-1)} \right) \left(\frac{1}{h_{il}} \left(\prod_{j \in I} h_{jl} \right)^{1/(n-1)} \right)^{-1} \tag{1.23}$$

for all $l \in y_0$, or, if we sum over $l \in y_0$,

$$1 = \frac{1}{H_i(X)} \left(\prod_{j \in I} H_j(x) \right)^{1/(n-1)} \sum_{l \in y_0} h_{il} \left(\prod_{j \in I} h_{jl} \right)^{-1/(n-1)}.$$

If we denote the sum on the right by A_i we have

$$H_i(X) = A_i \left(\prod_{j \in I} A_j \right)^{-1},$$

and if we insert the expressions for A_i and then the expressions for $H_i(X)$ from (1.22), we obtain explicit expressions for all the probabilities $X_i(l)$.

In this way, to each nonempty subset of the total set of the players' strategies there corresponds an equilibrium situation. Its stability reflects the interest of each player in retaining a mutually advantageous position.

Because of this example we make a special but important remark.

If $n \geqq 3$, and if in (1.14) and in Γ we take the numbers h_{lk} to be rational and also such that the indicated roots in (1.23) are irrational, the probabilities $X_i(l)$ are irrational numbers.

This remark shows that when $n \geqq 3$ there are finite noncooperative games with rational payoffs for which the equilibrium situations are specified by irrational numbers. Combination of this fact with the theorem of 1.5 yields the following fundamental result: there are finite noncooperative games with $n \geqq 3$ players whose solution processes are not reducible to the solution processes of a finite number of finite noncooperative games with two players (i.e., bimatrix games). Informally, this has the meaning that the theory of conflicts of three persons is intrinsically more complex than the theory of games with two persons (even if we allow nonantagonistic games, i.e. those that are not zero sum games). For more about this circle of problems, see § 3.

1.9 Use of dominating strategies. Various methods for the solution of games are based on the constructions of semihomomorphisms and homomorphisms of noncooperative games (see Chapter 1, 1.8). We also discuss the use of the concepts of dominating strategies (see Chapter 1, 2.25), and of symmetry, for solving finite games.

It is clear that, "other things being equal", the complexity of solving a finite noncooperative game is determined by the number of strategies of the participating players. Therefore one should take account of and use every possible method of diminishing this number. One such possibility, already mentioned in Chapter 1, 2.25 and 2.26, consists of discarding dominating strategies.

Of course, some equilibrium situations can be missed in this way, since dominating pure strategies (in the case on non-strict domination) can nevertheless appear in the spectra of i-optimal (and hence equilibrium) mixed situations; and games that admit automorphisms can have equilibrium situations that are not invariant under the automorphisms. (This, of course, does not contradict the covariance, mentioned in Chapter 1, 2.19, of the automorphisms of the set of all equilibrium situations.) Nevertheless, even in this incomplete form the solutions of noncooperative games can be both theoretically and practically interesting.

The precise meaning of the possibility of excluding from consideration pure dominating strategies of players in finite noncooperative games is formulated in the following theorem.

Theorem. *If x_i^0, a pure strategy of player i in the finite noncooperative game Γ, is strictly dominated by one of the player's (pure or mixed) strategies X_i^0, then for every i-optimal situation X we must have $X_i(x_i^0) = 0$ (as usual, X_i is a strategy of player i that appears in the situation X).*

Proof. If $X_i(x_i^0) > 0$, then by the lemma in 1.2 we must have

$$H_i(X\|x_i^0) = H_i(X\|X_i) = H_i(X).$$

But by the assumption of strict domination,

$$H_i(X\|x_i^0) < H_i(X\|X_i^0),$$

so that

$$H_i(X) < H_i(X\|X_i^0),$$

and this contradicts the i-optimality of situation X.

1.10. The case of non-strict domination is somewhat more delicate.

Theorem. *Let the pure strategy x_i^0 of player i in the finite noncooperative game Γ be dominated (not strictly) by one of the player's (pure or mixed) strategies X_i^0, different from x_i^0.*

Then if $X^* \in \mathscr{C}_i(\Gamma)$ there is a strategy $\bar{X}_i \in \mathbf{X}_i$ such that $\bar{X}_i(x_i^0) = 0$ and $X^*\|\bar{X}_i \in \mathscr{C}_i(\Gamma)$.

Proof. If $X_i^*(x_i^0) = 0$ we may take $\bar{X}_i = X_i^*$. Hence we may assume that $X_i^*(x_i^0) > 0$. Since $X_i^0 \neq x_i^0$, we must have $X_i^0(x_i^0) < 1$.

Set

$$X_i'(x_i) = \begin{cases} \frac{X_i^0(x_i)}{1-X_i^0(x_i^0)}, & \text{if } x_i \neq x_i^0, \\ 0 & \text{if } x_i = x_i^0. \end{cases}$$

It is clear that $X_i'(x_i) \geqq 0$ and

$$\sum_{x_i \in \boldsymbol{x}_i} X_i'(x_i) = 1,$$

so that $X_i' \in \boldsymbol{X}_i$.

The stipulated domination of strategies means that

$$H_i(X\|X_i^0) \geqq H_i(X\|x_i^0) \text{ for all } X \in \boldsymbol{X},$$

(where $\boldsymbol{X}$ is the set of all situations in mixed strategies) whence

$$\frac{H_i(X \parallel X_i^0) - H_i(X \parallel x_i^0)X_i^0(x_i^0)}{1 - X_i^0(x_i^0)} \geqq H_i(X\|x_i^0),$$

or

$$\sum_{x \neq x_i^0} (X \parallel x_i)\frac{X_i^0(x_i)}{1 - X_i^0(x_i^0)} = H_i(X\|X_i') \geqq H_i(X\|x_i^0). \tag{1.24}$$

Let us also set

$$\bar{X}_i(x_i) = \begin{cases} X_i^*(x_i) + X_i^*(x_i^0)X_i'(x_i), & \text{if } x_i \neq x_i^0, \\ 0, & \text{if } x_i = x_i^0. \end{cases}$$

It is easy to verify that $\bar{X}_i \in \boldsymbol{X}_i$. Let us show that $X^*\|\bar{X}_i \in \mathscr{C}_i(\Gamma)$. We have

$$\begin{aligned} H_i(X^*\|\bar{X}_i) &= \sum_{x_i \in \boldsymbol{x}_i} H_i(X^*\|x_i)\bar{X}_i(x_i) \\ &= \sum_{x_i \neq x_i^0} H_i(X^*\|x_i)(X_i^*(x_i) + X_i^*(x_i^0)X_i'(x_i)) \\ &= \sum_{x_i \neq x_i^0} H_i(X^*\|x_i)X_i^*(x_i) + X_i^*(x_i^0) \sum_{x_i \neq x_i^0} H_i(X^*\|x_i)X_i'(x_i), \end{aligned}$$

or, remembering that $X_i'(x_i^0) = 0$,

$$H_i(X^*\|\bar{X}_i) = \sum_{x_i \neq x_i^0} H_i(X^*\|x_i)X_i^*(x_i) + X_i^*(x_i^0)H_i(X^*\|X_i'),$$

which means, by (1.24), that

$$H_i(X^*\|\bar{X}_i) \geqq \sum_{x_i \neq x_i^0} H_i(X^*\|x_i)X_i^*(x_i) + H_i(X^*\|x_i^0)X_i^*(x_i^0) = H_i(X^*),$$

and it remains only to refer to the i-optimality of X^*. □

1.11 Use of the symmetry of a game. Let Γ be a noncooperative game and let π be an automorphism of Γ (see Chapter 1, 1.14). We extend it to the mixed extension of Γ, setting, for the mixed strategy $X_i = (X_i(x_i^1), \dots, X_i(x_i^{m_i}))$ of player i,

$$\pi X_i = (X_{\pi_i}(\pi x_i^1), \dots, X_{\pi_i}(\pi x_i^{m_i})) \tag{1.25}$$

and, for the situation $X = (X_1, \dots, X_n)$ in the mixed strategies,

$$\pi X = (\pi X_1, \dots, \pi X_n).$$

A situation X in the mixed strategies is said to be *symmetric* if, for every automorphism π of Γ,

$$X_{\pi_i}(\pi x_i^j) = X_i(x_i^j) \text{ for all } x_i^j \in \boldsymbol{x}_i, \quad i \in I.$$

The set of all symmetric situations in the mixed strategies for a given noncooperative game Γ is denoted by $\mathrm{Sym}(\Gamma)$.

It follows from (1.25) that, if $X \in \mathrm{Sym}(\Gamma)$, we necessarily have

$$H_i(X) = H_{\pi_i}(\pi X). \tag{1.26}$$

1.12. It follows immediately from the definition of an equilibrium situation that, for every automorphism π of Γ, it follows that $\pi X \in \mathscr{C}(\Gamma)$ (also see Chapter 1, 2.19), i.e., that the set of equilibrium situations is invariant under the automorphisms of the game. At the same time, it by no means follows that every equilibrium situation in $\mathscr{C}(\Gamma)$ is invariant under each automorphism. However, it is just the case that equilibrium situations that are invariant under automorphisms are optimal in the full sense of the word: it is natural to want (to speak of demanding would be too rigid) the players participating equally in the game (i.e., equivalent under automorphisms of the game) also to be in the same positions in situations that appear to them to be optimal (in this case, in equilibrium situations). If we suppose that it is the equilibrium situations that it is natural to fix by legislated rules, then their symmetries will correspond to a real equality of the participants with respect to the rules.

The following theorem establishes the existence of at least one such situation in a finite noncooperative game.

Theorem. *Every finite noncooperative game Γ has a symmetric equilibrium situation.*

Proof. We first remark that, for every game Γ, the set $\mathrm{Sym}(\Gamma)$ is not empty. In fact, one such situation is that in which

$$X_i(x_i) = 1/|\boldsymbol{x}_i| = 1/m_i,$$

for all $i \in I$; $x_i \in \boldsymbol{x}_i$, since then under the automorphism π

$$X_{\pi_i}(\pi x_i) = 1/m_{\pi_i} = 1/m_i = X_i(x_i).$$

The set $\mathrm{Sym}(\Gamma)$ is convex, since if X' and $X'' \in \mathrm{Sym}(\Gamma)$ and $\alpha \in [0,1]$,

$$(\alpha X' + (1-\alpha)X'')_{\pi i}(\pi x_i) = \alpha X'_{\pi i}(\pi x_i) + (1-\alpha)X''_{\pi i}(\pi x_i)$$
$$= \alpha X'_i(x_i) + (1-\alpha)X''(x_i) = (\alpha X' + (1-\alpha)X'')_i(x_i).$$

In a similar way we can establish that $\mathrm{Sym}(\Gamma)$ is closed, and consequently this set is homeomorphic to a finite-dimensional simplex.

We now consider the continuous mapping ψ from the proof of Nash's theorem in 1.6 and prove that $\psi\,\mathrm{Sym}(\Gamma) \subset \mathrm{Sym}(\Gamma)$. For this purpose we choose $X \in \mathrm{Sym}(\Gamma)$ and an arbitrary automorphism π. It follows from (1.26) that

$$\varphi_{\pi i}^{\pi x_i^j}(\pi X) = \max\{0, H_{\pi i}(\pi X \underset{\pi i}{\|} \pi x_i^j) - H_{\pi i}(\pi X)\}$$
$$= \max\{0, H_i(X \underset{i}{\|} x_i^j\} - H_i(X)\} = \varphi_i^j(X).$$

Therefore

$$(\pi(\psi X_i))(x_i^j) = X_{\pi i}(\pi x_i^j) + \varphi_{\pi i}^{\pi x_i^j}(\pi X)\left(1 + \sum_{\pi x_i^j \in x\pi i} \varphi_{\pi i}^{\pi x_i^j}(\pi X)\right)^{-1}$$
$$= X_i(x_i^j) + \varphi_i^j(X)\left(1 + \sum_{x_i^j \in x_i} \varphi_i^j(X)\right)^{-1} = (\psi X_i)(x_i^j),$$

i.e., the symmetry of X implies the symmetry of ψX.

Consequently the hypotheses of Brouwer's theorem are satisfied, so that ψ has a fixed point in $\mathrm{Sym}(\Gamma)$, and this is an equilibrium situation. □

1.13. The preceding theorem is the basis for the following simple, but useful, reasoning.

With respect to the set I of players, the automorphisms form a permutation group. Let us suppose that this group is transitive with respect to a coalition K, i.e. that for every pair $i, j \in K$ there is an automorphism π_{ij} for which $\pi_{ij}\, i = j$. Then a symmetric equilibrium situation X^* in Γ will satisfy

$$X^*_{\pi_{ij} i}\left(x_{\pi_{ij} i}^{(\pi_{ij}^k)}\right) = X_i^*(x_i^{(k)})$$

for all players $i, j, \in K$ and every strategy $k \in x_i$ of player i. This means that, in the situation X^*, all the strategies of the players in the coalition K are equidistributed in the stochastic sense (up to the numbering of the strategies in the automorphisms π_{ij}).

1.14 An illustrative example. Consider a game Γ as in (1.1), where

$$I = \{1,2,3\}, \quad \mathfrak{x}_i = \{1,2,3,a\}, \quad i = 1,2,3$$

(in what follows we denote by x_i an arbitrary strategy in $\mathfrak{x}_i$, by y_i an arbitrary strategy in y_i that differs from a , and by p_i the strategy y_i of player i of index p) and where the payoff functions H_i are given by

$$H_i(1_1, 2_2, 3_3) = l, \quad i = 1,2,3, \tag{1.27}$$

$$H_1(1_1, 1_2, 1_3) = H_2(2_1, 2_2, 2_3) = H_3(3_1, 3_2, 3_3) = k, \tag{1.28}$$

$$H_i(y_1, y_2, y_3) = 0 \text{ otherwise}, \tag{1.29}$$

$$H_1(a, y_2, y_3) = -1, \quad H_2(a, y_2, y_3) = m, \quad H_3(a, y_2, y_3) = m, \tag{1.30}$$

$$H_1(x_1, a, y_3) = m, \quad H_2(x_1, a, y_3) = -1, \quad H_3(x_1, a, y_3) = m, \tag{1.31}$$

$$H_1(x_1, x_2, a) = m, \quad H_2(x_1, x_2, a) = m, \quad H_3(x_1, x_2, a) = -1, \tag{1.32}$$

where k, l, m are positive numbers.

This game is readily interpreted as the game of morra*) for three players, where each player simultaneously shows one, two, or three fingers, or, say, a fist (strategy a), and the payoff to each player depends as specified above on the player's position and on the numbers shown by the players.

Let us look for the equilibrium situations in this game.

To this end we first notice that, for arbitrary x_1 and x_2,

$$H_3(x_1, x_2, a) = -1 < 0 \leqq H_3(x_1, x_2, y_3),$$

so that strategy a of player 3 is strictly dominated by that player's other strategies. Consequently it does not appear in the spectrum of any equilibrium situation, and therefore may be disregarded.

Correspondingly we consider the game Γ' that differs from Γ only by omitting the strategy a of player 3; then the equalities (1.32) drop out of the description of the payoff functions in Γ'. In Γ' we have, for all x_1 and y_3,

$$H_2(x_1, a, y_3) = -1 < 0 \leqq H_2(x_1, y_2, y_3),$$

so that strategy a of player 2 is dominated, and by omitting it, we pass to the game Γ'' in which player 2 does not use strategy a, and conditions of the type (1.31) do not appear in the definition of the payoff functions.

In this game, for arbitrary y_2 and y_3,

$$H_1(a, y_2, y_3) = -1 < 0 \leqq H_1(y_1, y_2, y_3),$$

and we are similarly led to consider the game

$$\Gamma''' = \langle I, \{y_i\}_{i \in I}, \{H_i\}_{i \in I} \rangle,$$

where H_i is determined only by conditions (1.27)–(1.29).

Although the payoff functions in this game have a very simple structure (each of them differs from zero in only two situations), a complicated analysis of different versions is needed in order to find the equilibrium situations.

As is easily verified, in Γ''' the permutation π for which

$$\pi_i = i + 1 \qquad \text{for } i \in I,$$

$$\pi_{p_i} = (p+1)_{i+1} \qquad \text{for } p_i \in y_i, i \in I,$$

---

*) The strategies of the players in "dumb morra" are to show some number of fingers of one hand, or ("loud morra") to simultaneously call out a number. The payoffs of course depend on the situation. See P. Mérimée, La Vénus d'Ille, p. 258, Nouvelle Edition de Calmann Lévy, Editeur, Paris, 1888.

is an automorphism, where addition is taken modulo 3. In fact, here the situation $(1_1, 2_2, 3_3)$ is not changed by π, and the situations

$$(1_1, 1_2, 1_3),\ (2_1, 2_2, 2_3),\ (3_1, 3_2, 3_3)$$

are permuted cyclically.

The permutations π and π^2, together with the identity permutation, form the automorphism group of Γ'''.

Let us describe the symmetric equilibrium situations in this game. Let $X^* = (X_1^*, X_2^*, X_3^*)$ be such a situation. Set

$$X_1^*(1_1) = \alpha, \quad X_1^*(2_1) = \beta, \quad X_1^*(3_1) = \gamma.$$

By the symmetry of X^*, we must have

$$X_1^*(1_1) = X_2^*(2_2) = X_3^*(3_3) = \alpha,$$

$$X_1^*(2_1) = X_2^*(3_2) = X_3^*(1_3) = \beta,$$

$$X_1^*(3_1) = X_2^*(1_2) = X_3^*(2_3) = \gamma,$$

By (1.27)–(1.29), we have

$$H_1(X^* \| 1_1) = k\beta\gamma + l\alpha^2, \tag{1.33}$$

$$H_1(X^* \| 2_1) = k\alpha\gamma, \tag{1.34}$$

$$H_1(X^* \| 3_1) = k\alpha\beta. \tag{1.35}$$

If $\alpha = 1$ and $\beta = \gamma = 0$, we obtain the situation $X^* = (1_1, 2_2, 3_3)$ in the pure strategies. This is symmetric, and has

$$H_1(X^* \| 1_1) = l > H_1(X^* \| 2_1) = H_1(X^* \| 3_1) = 0,$$

i.e., it satisfies the condition of acceptability for player 1 (i.e., of 1-optimality). The same applies to players 2 and 3, so that this is an equilibrium situation.

Now let $\alpha = 0$, and let one of β and γ also be 0. For definiteness, suppose that $\beta = 0$. Then $\gamma = 1$, and we have the situation $(3_1, 1_2, 2_3)$. The automorphism π evidently does not change this situation: thus it is symmetric. In this situation, the right-hand sides of (1.33)–(1.35) are zero, and this situation is acceptable for player 1. Similarly it also turns out to be acceptable for the other players, i.e. it is an equilibrium situation. In the same way, the situation with $\beta = 1$, i.e. $(2_1, 3_2, 1_3)$ is also an equilibrium situation.

If $\alpha = 0$, but both β and γ are different from zero, comparison of the right-hand sides of (1.33)–(1.35) shows that the corresponding situation is unacceptable to player 1, and therefore not an equilibrium situation.

We now consider the case when $\alpha > 0$ and $\beta > 0$. Here it follows from (1.34), (1.35), and the conditions for acceptability that $k\alpha\gamma \geqq k\alpha\beta$, i.e. $\gamma \geqq \beta$, and therefore $\gamma > 0$. Similarly, it follows from $\alpha > 0$ and $\gamma > 0$ that $\beta > 0$. Consequently it remains only to consider the case when X^* is a completely mixed situation. But in this case, by (1.33)–(1.34), we must have

$$k\beta\gamma + l\alpha^2 = k\alpha\gamma, \tag{1.36}$$

$$k\alpha\gamma = k\alpha\beta. \tag{1.37}$$

It follows from (1.37) that $\beta = \gamma$ and therefore $\alpha + \beta + \gamma = 1$ implies that $\alpha = 1 - 2\beta$. Therefore (1.36) can be rewritten as

$$k\beta^2 + l(1 - 2\beta)^2 = k(1 - 2\beta)\beta,$$

whence

$$\beta = \gamma = \frac{k + 4l \pm \sqrt{k^2 - 4kl}}{6k + 8l}, \quad \alpha = \frac{4k \mp 2\sqrt{k^2 - 4kl}}{6k + 8l}.$$

If these numbers are real (i.e. if $k \geqq 4l$), they are nonnegative.

Thus when $k > 4l$ we obtain two more symmetric equilibrium situations, so that there are five in all; when $k < 4l$, none (the general number of symmetric equilibrium situations here is three); and in the degenerate case, when $k = 4l$, one (the general number of symmetric equilibrium situations is four).

§2 Dyadic games

2.1 Dyadic games. Definition. A finite noncooperative game

$$\Gamma = \langle I, \{x\}_{i \in I}, \{H_i\}_{i \in I} \rangle \tag{2.1}$$

is called *dyadic* if each player has precisely two pure strategies. □

In a dyadic game with n players there are evidently 2^n situations, and the entire game is specified by $n2^n$ real numbers.

In what follows, we shall take $x_i = \{0_i, 1_i\}$; and if there is no danger of confusing the strategies of different players, $x_i = \{0, 1\}$.

In this notation, each situation in a dyadic game is a sequence of n zeros and ones. Therefore it can be represented as a vertex of the unit cube in an n-dimensional Euclidean space.

The mixed strategies of the players in a dyadic game are therefore represented by points of the interval $[0, 1]$; and the situations in the mixed strategies, by points of the cube introduced above. For the sake of simplicity, we shall identify a mixed strategy of player i in a dyadic game with the corresponding probability of that player's strategy 1.

2.2 Characterization of the equilibrium strategies. Let us describe, for a dyadic game Γ, the sets $\mathscr{C}_i(\Gamma)$ of situations that are acceptable by player i.

For this purpose, we select an arbitrary set $K^i \subset I \setminus i$ of players and denote by (α_i, K^i) (where α_i is 0 or 1) the situation $x = (x_1, \dots, x_n)$ in which

$$x_j = \begin{cases} 1_j & \text{if } j \in K^i, \\ 0_j & \text{if } j \notin K^i \text{ and } j \neq i, \\ \alpha_j & \text{if } j = i. \end{cases}$$

Let X be an arbitrary situation of Γ in the mixed strategies. For the duration of this section, we shall set

$$X_j(1_j) = \xi_j, \quad j = 1, \dots, n.$$

Then to a situation X in our cube there will correspond the point $(\xi_1, \dots, \xi_n)$. Under the hypotheses on X (in the mixed strategies) the probability of the appearance of one of the two situations (in the pure strategies) of the form (α_i, K^i)

with $\alpha_i = 0_i$ or 1_i is determined as

$$X(K^i) = \prod_{j \in K^i} \xi_j \prod_{j \notin K^i} (1 - \xi_j). \tag{2.2}$$

We shall have to use such products of probabilities repeatedly. Consequently, to simplify the notation we write

$$\prod_{i \in K} \xi_i = \xi^{(K)} \text{ and } \prod_{i \in K} (1 - \xi_i) = (1 - \xi)^{(K)}.$$

It is clear that

$$H_i(X) = \xi_i \sum_{K^i} H_i(1, K^i) X(K^i) + (1 - \xi_i) \sum_{K^i} H_i(0, K^i) X(K^i) \tag{2.3}$$

and, in particular,

$$H_i(X \| \alpha_i) = \sum_{K^i} H_i(\alpha_i, K^i) X(K^i), \tag{2.4}$$

where $\alpha_i = 0_i$ or 1_i, and the summations in (2.3) and (2.4) are over all $K^i \subset I \setminus i$.

The inclusion $X \in \mathscr{C}_i(\Gamma)$ means that the inequalities

$$H_i(X \underset{i}{\|} \alpha_i) \leqq H_i(X), \text{ for } \alpha_i = 0, 1, \tag{2.5}$$

are satisfied.

We consider three cases.

a) $\xi_i = 0$. Here $X = X \underset{i}{\|} 0$, so that (2.5) is equivalent to

$$H_i(X \underset{i}{\|} 1) \leqq H_i(X \underset{i}{\|} 0) \tag{2.6}$$

(here the other inequality is trivial).

b) $\xi_i = 1$. Here $X = X \underset{i}{\|} 1$, and (2.5) is equivalent to

$$H_i(X \underset{i}{\|} 1) \geqq H_i(X \underset{i}{\|} 0). \tag{2.7}$$

c) $0 < \xi_i < 1$. In this case both pure strategies of player i are points of the spectrum X_i of that player's mixed strategies, so that by the theorem of Chapter 1, 5.15 (see also 5.16) we must have

$$H_i(X \underset{i}{\|} 0) = H_i(X \underset{i}{\|} 1). \tag{2.8}$$

Relations (2.6)–(2.8) have a rather transparent probabilistic meaning: $H_i(X \| \alpha_i)$ (where $\alpha_i = 0$ or 1) is the conditional expectation of the payoff to player i in the situation X under the hypothesis that the player selects the pure strategy α_i; therefore (2.6) and (2.7) mean that player i does not have any reason for deviating from the already selected strategy α_i, since doing so would not increase the payoff; and (2.8) means that, with the given payoffs, whether to use strategy 0 or 1 is a matter of indifference, since these payoffs have the same probabilities.

We denote by $\boldsymbol{X}_0^i$ the set of those combinations $I \setminus i$ of the players' strategies for which (2.6) is satisfied; by $\boldsymbol{X}_1^i$, the set of combinations for which (2.7) is satisfied; and, finally, by $\boldsymbol{X}_=^i$, the set of combinations for which (2.8) is satisfied.

With this notation, the set $\mathscr{C}_i(\Gamma)$ of i-optimal situations will consist of situations of the form:

$$\begin{aligned} &(0, X^i) \text{ for all } X^i \in \boldsymbol{X}_0^i, \\ &(1, X^i) \text{ for all } X^i \in \boldsymbol{X}_1^i, \\ &(\xi_i, X^i) \text{ for arbitrary } \xi_i \in [0,1] \text{ and } X^i \in \boldsymbol{X}_=^i. \end{aligned} \tag{2.9}$$

This description of the sets of i-optimal situations in dyadic games allows us to find their equilibrium situations in a number of cases.

2.3. From what we have said, there follows a concrete version of the theorem of 1.4 as applied to dyadic games.

Theorem. *A necessary and sufficient condition for a completely mixed situation X in a dyadic game Γ to be an equilibrium situation is that equation* (2.8) *is satisfied for every $i \in \Gamma$.*

2.4 2×2**-bimatrix games.** We now consider some examples. As a first example, we analyze 2×2-bimatrix games.

In order to preserve the usual numbering of the rows and columns of matrices, in this subsection we set

$$1_1 = 1_2 = 1 \text{ and } 0_1 = 0_2 = 2.$$

A 2×2-bimatrix game $\Gamma(A, B)$ is defined by a pair of matrices,

$$A = \begin{bmatrix} a_{11} & a_{12} \\ a_{21} & a_{22} \end{bmatrix}, \qquad B = \begin{bmatrix} b_{11} & b_{12} \\ b_{21} & b_{22} \end{bmatrix}.$$

If, as is usual for situations (X, Y) in the mixed strategies, we set $X = (\xi, 1-\xi)$ and $Y = (\eta, 1-\eta)$, equation (2.4) yields

$$H_1(1, Y) = a_{11}\eta + a_{12}(1-\eta) = a_{12} + \eta(a_{11} - a_{12}),$$

and similarly

$$\begin{aligned} H_1(2, Y) &= a_{21}\eta + a_{22}(1-\eta) = a_{22} + \eta(a_{21} - a_{22}), \\ H_2(X, 1) &= \xi b_{11} + (1-\xi) b_{21} = b_{21} + \xi(b_{11} - b_{21}), \\ H_2(X, 2) &= \xi b_{12} + (1-\xi) b_{22} = b_{22} + \xi(b_{12} - b_{22}). \end{aligned}$$

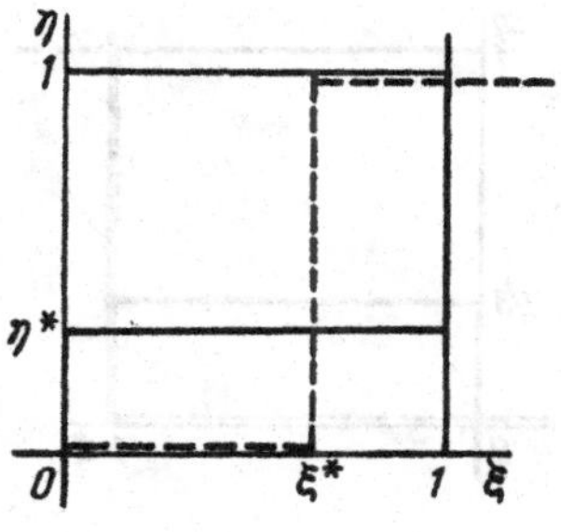

Fig. 2.1

Let us describe the set $\mathscr{C}_1(\Gamma)$, for which we define the sets Y_1, Y_2, and $Y_=$ of strategies of player 2.

For strategies in $Y_=$, equation (2.8) is satisfied, i.e. in this case $H_1(2, Y) = H_1(1, Y)$, or

$$a_{22} + \eta(a_{21} - a_{22}) = a_{12} - \eta(a_{11} - a_{12}),$$

i.e.

$$\eta(a_{11} - a_{12} - a_{21} + a_{22}) = a_{22} - a_{12}.$$

We may rewrite this equation as

$$\eta a = \alpha. \tag{2.10}$$

Here, if $a = 0$ but $\alpha \neq 0$, then $Y_= = \emptyset$, so that the set Y of all strategies is either Y_1 or Y_2, depending on which of a_{12} or a_{22} is larger. Correspondingly, the set $\mathscr{C}_1(\Gamma)$ consists either of all situations of the form $(1, \eta)$, where $\eta \in [0, 1]$, or of all situations of the form $(2, \eta)$.

If $a = 0$ and $\alpha = 0$, then $Y_= = Y$, so that, by (2.9), in general all situations are optimal.

Now consider the nondegenerate case: $a \neq 0$. Then (2.1) implies that $\eta = a/\alpha$ (for simplicity, we denote this value of η by η^*), and the set $Y_=$ consists solely of this strategy η^*.

The strategies in Y_1 satisfy the inequality $\eta a \geqq \alpha$, i.e. the inequality $\eta \geqq \eta^*$ if $a > 0$, and $\eta \leqq \eta^*$ if $a < 0$. In the same way, the strategies in Y_2 satisfy either $\eta \leqq \eta^*$ or $\eta \geqq \eta^*$.

Therefore, in the nondegenerate case the set $\mathscr{C}_1(\Gamma)$ has the shape of a triple zigzag, represented in Figure 2.1 by the heavy solid line when $a > 0$ and $\eta^* = 1/3 \in (0, 1)$. If $a < 0$, the zigzag will have the opposite orientation; if $\eta^* = 0$ or 1, the zigzag will have only two links, and if $\eta^* < 0$ or $\eta^* > 1$, just one (strictly speaking, only two links of the zigzag, or only one, will intersect the situation square).

The set $\mathscr{C}_2(\Gamma)$ can be described similarly. Its form is determined by the numbers

$$b = b_{11} - b_{12} - b_{21} + b_{22} \text{ and } \beta = b_{22} - b_{21}.$$

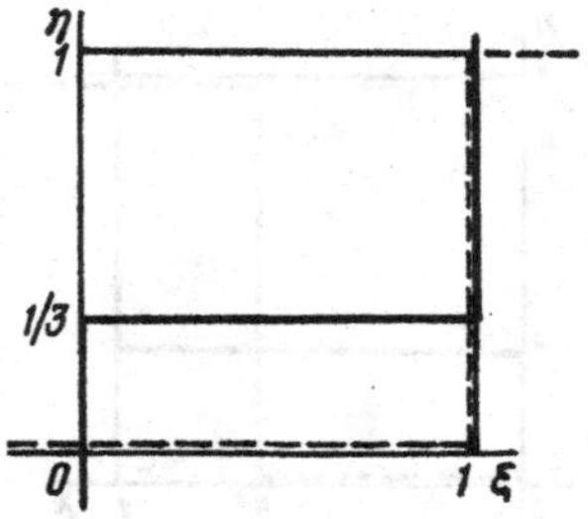

Fig. 2.2

In the nondegenerate case it is also represented by a zigzag (possibly with missing links). This zigzag is represented in Figure 2.1 by the dashed line for the case $b > 0$ and $\xi^* = b/\beta = 2/3 \in (0, 1)$.

It is easily verified that the case represented by Figure 2.1 corresponds to the "battle of the sexes" game which therefore has, in addition to the two equilibrium situations in the pure strategies that were mentioned in Chapter 1, 2.12, an equilibrium situation in the mixed strategies, namely $(2/3, 1/3)$.

It is easy to see that a 2×2-matrix game for which ξ^* and $\eta^* \in (0, 1)$ must have either three equilibrium situations (two in the pure strategies and one in the mixed strategies), if the zigzags $\mathscr{C}_1(\Gamma)$ and $\mathscr{C}_2(\Gamma)$ are oppositely oriented; or a single equilibrium situation (necessarily in the pure strategies) if their orientations are the same. It is easy to verify that the second case occurs for 2×2-matrix games (i.e., two-person zero-sum games).

If we go outside the limits of the 2×2-bimatrix games for which ξ^* and $\eta^* \in (0, 1)$, the set of equilibrium situations may even have a somewhat exotic form. For example, for the game Γ with

$$A = \begin{bmatrix} 1 & 0 \\ 0 & 2 \end{bmatrix}, \qquad B = \begin{bmatrix} 0 & 0 \\ 0 & 1 \end{bmatrix}$$

the set $\mathscr{C}$ consists of a line segment of situations of the form $(1, \eta)$, where $\eta \in [1/3, 1]$, together with the isolated situation (0,0) (Figure 2.2).

2.5 The game "for uniqueness". As a second example, we shall find the equilibrium situations in the dyadic game Γ in which

$$H_i(0, \ldots, 0 \underset{i}{\|} 1) = g_i > 0,$$

$$H_i(1, \ldots, 1 \underset{i}{\|} 0) = h_i > 0, \tag{2.11}$$

$$H_i(x) = 0 \text{ in all other cases.}$$

For $n = 2$ this game is an ordinary bimatrix game of the same type as the "battle of the sexes" game, with three equilibrium situations: $(0, 1)$, $(1, 0)$, and

$(g_2/(g_2+h_2), g_1/(g_1+h_1))$, but for $n > 2$ it can be interpreted as an approach of the players to "uniqueness", i.e., to the choice of pure strategy which is not adopted by any of the other players. Here the "1-uniqueness" of player i is rewarded by the amount g_i; and "0-uniqueness" of the player, by the amount h_i. Such a game admits a natural interpretation in terms of an advertizing campaign.

We first observe that if, in situation X in Γ, any two players choose different pure strategies, the situation will be an equilibrium situation, whatever strategies (pure or mixed) are chosen by the other player. In fact, let, for definiteness, $X_1 = x_1 = 0$ and $X_2 = x_2 = 1$. Then we have, by (2.11),

$$H_1(0,1,\xi_3,\dots,\xi_n) = g_1 \prod_{i=3}^{n} \xi_i \geqq H_1(1,1,\xi_3,\dots,\xi_n) = 0,$$

$$H_2(0,1,\xi_3,\dots,\xi_n) = h_2 \prod_{i=3}^{n}(1-\xi_i) \geqq H_2(0,0,\xi_3,\dots,\xi_n) = 0,$$

and for $j > 2$,

$$H_j(0,1,\xi_3,\dots,\xi_n) = 0 \geqq \begin{cases} H_j(0,1,\xi_3,\dots,\xi_n \underset{j}{\|} 0) = 0, \\ H_j(0,1,\xi_3,\dots,\xi_n \underset{j}{\|} 1) = 0. \end{cases}$$

Now suppose that in situation X two players choose the same strategy. For definiteness, suppose that $X_1 = x_1 = 0$ and $X_2 = x_2 = 0$. Then $H_1(X) = 0$, and a necessary condition for X to be an equilibrium situation is that

$$H_1(X) \geqq H_1(X \underset{1}{\|} 0) = h_1 \prod_{i=3}^{n}(1-\xi_i) = 0$$

for arbitrary $j = 3,\dots,n$ and $\xi_j \in [0,1]$. This is possible if $\xi_i = 1$ for some player i, $i > 2$, i.e. $X_i = x_i = 1$; this means that in the situation X two players (namely, players 1 and i) choose different pure strategies, and we are back in a case that has previously been examined.

The equilibrium situations correspond to points that form $(n-2)$-dimensional cycles that consist of $n(n-1)$ faces of the $(n-2)$-dimensional boundary of the situation cube (there are no other equilibrium situations on the $(n-2)$-dimensional faces of the cube). In particular, if $n = 3$, this set consists of six edges of the three-dimensional cube (Figure 2.3).

If *exactly* one player chooses a pure strategy in a situation, this situation cannot be an equilibrium situation. In fact, let us consider a situation $(0,\xi_2,\dots,\xi_n)$, where $0 < \xi_i < 1$ for $i = 2,\dots,n$. For this to be an equilibrium situation, it is necessary that

$$0 = H_2(0,0,\xi_3,\dots,\xi_n) = H_2(0,1,\xi_3,\dots,\xi_n) = g_2 \prod_{i=3}^{n}(1-\xi_i),$$

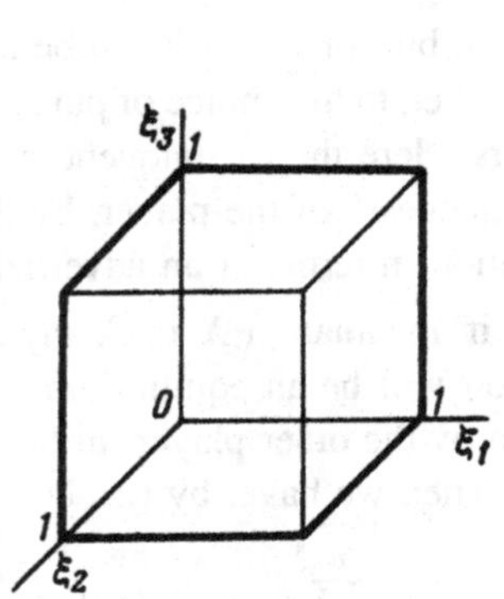

Fig. 2.3

which is possible only if $\xi_i = 1$ for some i, i.e. at least one other player chooses a pure strategy. This means geometrically that there are no points inside the $(n-1)$-dimensional faces of the cube corresponding to equilibrium situations.

We still have to find all the completely mixed equilibrium situations. Let $X = (\xi_1, \ldots, \xi_n)$, where $0 < \xi_i < 1$ for $i = 1, \ldots, n$, be one of them. According to (2.8), the condition for this to be an equilibrium situation can be presented in the form

$$H_i(X \underset{i}{\|} 0) = H_i(X \underset{i}{\|} 1),$$

i.e., by (2.1) we have

$$h_i \prod_{j \neq i} \xi_j = g_i \prod_{j \neq i} (1 - \xi_j).$$

Solving this system of equations, we obtain

$$\xi_i = \left(\frac{g_i}{h_i} \left(\frac{H}{G} \right)^{1/(n-1)} + 1 \right)^{-1},$$

where

$$G = \prod_{i \in I} g_i, \qquad H = \prod_{i \in I} h_i.$$

The game Γ therefore has exactly one completely mixed equilibrium situation. Observe that under such circumstances each probability ξ_i is a decreasing function of g_i (as long as none of the other parameters of the problem is changed). Therefore an increase in the nominal "1-reward" of any player leads to a decrease in the expectation of receiving this reward.

At the same time, if the game is symmetric, i.e. if $g_i = g$ and $h_i = h$ for $i = 1, \ldots, n$, then

$$\xi_i = \left(\left(\frac{h}{g} \right)^{1/(n-1)} + 1 \right)^{-1}.$$

It follows from this that simultaneously increasing the "1-reward" of every player leads to an increase of the players' expectations of receiving this reward. This conflict between individual and collective expediency is a rather typical phenomenon of game theory.

2.6 A model of pollution and protection of the environment. As a more complicated example, let us consider a class of dyadic games which admit an informal interpretation as a scheme for making use of the natural environment.

Let us suppose that there are n factories (players $1, 2, \ldots, n$, the set I), that make use of water drawn from a natural reservoir. Let each of them have two pure strategies: use water-purification equipment (strategy 0), or discharge the water untreated (strategy 1). We suppose that the nature of the reservoir and of the factory equipment is such that, if the set of factories that discharge untreated water is K, then enterprise i incurs a loss $v_i(K)$ from the discharge of untreated water (it is reasonable to suppose that if $K_1 \subset K_2$ we will have $v_i(K_1) \leqq v_i(K_2)$ and that $v_i(\emptyset) = 0$). Let r_i denote the cost to factory i for purifying water.

Under these assumptions we obtain a dyadic game in which

$$H_i(\alpha, K^i) = \begin{cases} -v_i(K^i) - r_i, & \text{if } \alpha = 0, \\ -v_i(K^i \cup i), & \text{if } \alpha = 1. \end{cases}$$

Let X be any given situation in this game. Let S_0 be the set of players that use the pure strategy 0 in situation X; let S_1 be the set of players that use the pure strategy 1 in X; and let S be the set of players that in X use their own (completely) mixed strategies. For X to be an equilibrium situation means that each player $i \in S_0$ uses the inequality (2.6); each player $i \in S_1$ uses (2.7); and each player $i \in S$ uses (2.8). Let us write out these relations to conform with the case in hand.

Inequality (2.6), namely $H_i(X \parallel_i 1) \leqq H_i(X \parallel_i 0)$, for a player $i \in S_0$, now assumes the form

$$-\sum_{i \notin K \subset I} v_i(K \cup i) X(K) \leqq -\sum_{i \notin K \subset I} v_i(K) X(K) - r_i$$

(here the summation is over all coalitions K that do not contain player i), or

$$\sum_{i \notin K \subset I} (v_i(K \cup i) - v_i(K)) X(K) \geqq r_i.$$

However, the set K, considered as an event (and under the assumption of the mixedness of the situation X it is a *random* event) consists of its having its first strategies chosen by the players in K, and only these. Consequently if $K \cap S_0 \neq \emptyset$, or $S_1 \not\subset K$, then $X(K) = 0$. Therefore we may suppose that $K = S_1 \cup (K \cap S)$. Setting $K \cap S = \bar{K}$ and recalling the form of $X(K)$ from (2.2), we have

$$\sum_{\bar{K} \subset S} (v_i(S_1 \cup \bar{K} \cup i) - v_i(S_1 \cup \bar{K})) \xi^{(\bar{K})} (1 - \xi)^{(S \setminus \bar{K})} \geqq r_i. \tag{2.12}$$

In addition, if $i \in S_1$, then $K = (S_1 \setminus i) \cup (K \cap S) = (S_1 \setminus i) \cup \bar{K}$, and inequality (2.7), namely $H_i(X \underset{i}{\|} 1) \geqq H_i(X \underset{i}{\|} 0)$, can be written in the form

$$\sum_{\bar{K} \subset S} v_i(S_1 \cup \bar{K}) - v_1((S_1 \setminus i) \cup \bar{K}))\xi^{(\bar{K})}(1-\xi)^{(S \setminus \bar{K})}) \leqq r_i. \quad (2.13)$$

Finally, if $i \in S$ then $K = S_1 \cup \bar{K}$, and each equation (2.8), namely $H_i(X \underset{i}{\|} 0) = H_i(X \underset{i}{\|} 1)$, can be rewritten in the form

$$\sum_{\bar{K} \subset (S \setminus i)} (v_i(S_1 \cup \bar{K} \cup i) - v_i(S_1 \cup \bar{K}))\xi^{(\bar{K})}(1-\xi)^{(S \setminus (\bar{K} \cup i))} = r_i. \quad (2.14)$$

Relations (2.12)–(2.14) admit specific probabilistic interpretations of the relations (2.6)–(2.8). Thus, the left-hand side of (2.12) is the expectation of the increase in cost to player $i \in S_0$ as the result of entering the set of players who lose in situation X. It is clear that if these costs are not less than the cost r_i of purification, situation X will be i-optimal, and there is no incentive for player i to change to the polluting process. Inequality (2.13) and equation (2.14) have similar interpretations.

2.7. Since the system (2.12)–(2.14) of n relations is a necessary and sufficient condition for X to be an equilibrium situation, we may use it to establish equilibrium conditions for any two situations that are of interest.

Let us first find conditions for equilibrium of the desirable situation in which every participant uses purification. In this case $S_0 = I$ and $S_1 = S = \emptyset$, and $\xi_i = 0$ for all $i \in I$. Here the system (2.12)–(2.14) becomes a set of inequalities of the form (2.12), which reduces to the quite simple form

$$v_i(i) \geqq r_i, \qquad i \in I. \quad (2.15)$$

Consequently, for the use of purification by everyone to be an equilibrium situation, it is not only sufficient but also necessary for the cost to the participants of using pollution to be no less than their use of purification.

The conditions for equilibrium of the situation where every participant uses pollution are "mirror-like". In this case $S_1 = I$, $S_0 = S = \emptyset$, so that $\xi_i = 1$ for all $i \in I$, and then the system (2.12)–(2.14) leads to (2.13), which reduces to

$$v_i(I) - v_i(I \setminus i) \leqq r_i, \qquad i \in I. \quad (2.16)$$

This means that the universal use of pollution is an equilibrium situation if and only if any participant's change from pollution reduces its costs by an amount less than the cost of purification. Regretably, this is the usual state of affairs.

2.8. Relations (2.12)–(2.14) yield, in principle, a realization of a method of determining equilibrium situations in the class of games under consideration, namely the choice, one by one, of a partition $I = S_0 \cup S_1 \cup S$, conforming with the partition in (2.12)–(2.14) and in the determination of sets ξ_S satisfying the system (2.14) followed by the verification that the components of these vectors satisfy (2.12) and (2.13).

In any individual case, the solution of the system (2.14) can present significant difficulties, and an a priori exhaustive analysis of all the variations is hardly possible in general. Consequently we shall restrict ourselves to some particular cases that admit a complete solution or at least promise a sufficiently transparent analysis.

2.9. We first consider the case of "independence of damage", when the increased amount of loss by the factory i from its own participation in the contamination of the reservoir is independent of the set of other participants in the pollution:

$$v_i(K \cup i) - v_i(K) = \varphi_i \text{ for every } K \not\ni i. \tag{2.17}$$

In this case the left hand sides of (2.12)–(2.14) have the form

$$\varphi_i \sum_{K \subset S} \xi^{(K)}(1-\xi)^{(S \setminus K)},$$

where all the sums are equal to unity, as is easily verified. Therefore, here the equilibrium situations are those in which $i \in S_0$ if $\varphi_i > r_i$, $i \in S_1$ if $\varphi_i < r_i$, and $i \in S$ with the choice of an arbitrary probability ξ_i if $\varphi_i = r_i$.

It is also clear that the identity (2.17) corresponds to a very special case of damage from pollution.

2.10. Let us analyze the game with an arbitrary loss function from pollution when there are $n = 3$ players. We shall denote by i, j and k any pairwise different numbers chosen from 1, 2 and 3. In this case (2.12)–(2.14) take the form

$$\begin{aligned}(v_i(i) - v_i(\emptyset))(1-\xi_j)(1-\xi_k) + (v_i(i,j) - v_i(j))\xi_j(1-\xi_k)\\ +(v_i(i,k) - v_i(k))(1-\xi_j)\xi_k + (v_i(I) - v_i(j,k))\xi_j\xi_k \gtreqless r_i,\end{aligned} \tag{2.18}$$

where $\geqq$ corresponds to the case $i \in S_0$, $\leqq$ to $i \in S_1$, and $=$ to $i \in S$. Then (2.18) can be condensed to the form

$$N_i\xi_j\xi_k + P_i\xi_j + Q_i\xi_k + R_i \gtreqless 0, \tag{2.19}$$

where N_i, P_i, Q_i, and R_i are constants that depend on the game only.

The corresponding *equation* which is satisfied by the pairs (ξ_j, ξ_k) describes, in a given case, a set $X^i_=$. In the case $N_i = 0$ this is a line segment; and if $N_i \neq 0$, the part of a hyperbola cut off by the unit square (in this case, if $N_iR_i = P_iQ_i \neq 0$, the hyperbola degenerates to a pair of perpendicular lines). The (closed) domains of the square separated by this curve are the sets X^i_0 (in (2.9) its points correspond to the $\geqq$ sign) and X^i_1 (corresponding to the $\leqq$ sign).

Having clarified the shape of the sets $X^i_=$, X^i_0, and X^i_1, we may, following what was said at the end of 2.2, describe the set $\mathscr{C}_i(\Gamma)$ of i-optimal situations. The appearance of this set can be quite diversified. For example, the case when

$$N_i = 0, \quad R_i > 0, \quad P_i + R_i < 0, \quad P_i + Q_i + R_i > 0$$

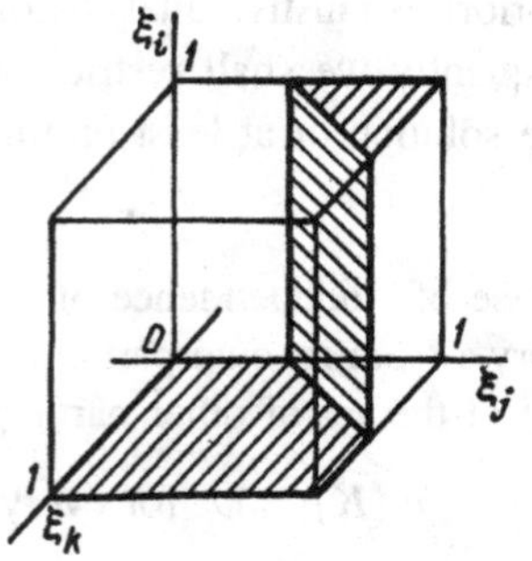

Fig. 2.4

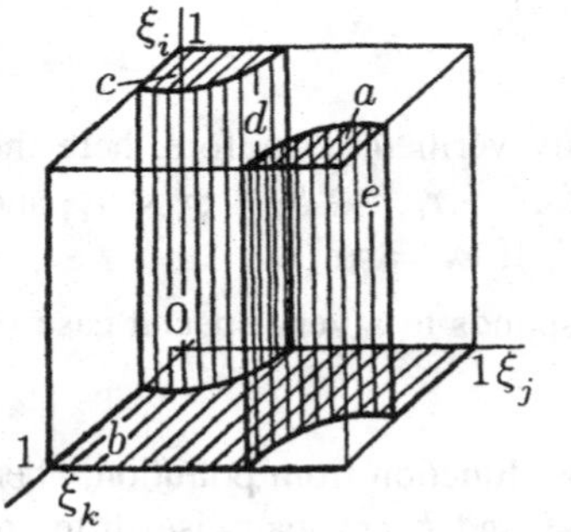

Fig. 2.5

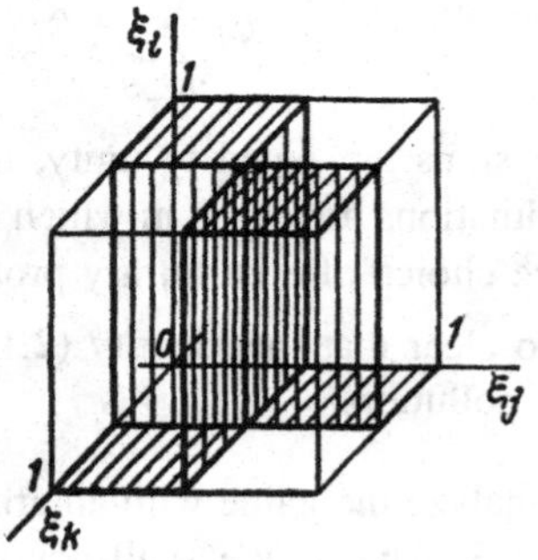

Fig. 2.6

is sketched in Figure 2.4; the case when

$$R_i > \frac{P_i Q_i}{N_i}, \quad R_i < 0, \quad N_i + P_i + Q_i + R_i < 0 \tag{2.20}$$

in Figure 2.5; and the case when

$$N_i R_i = P_i Q_i \neq 0, \quad R_i < 0, \quad N_i + P_i + R_i > 0, \quad Q_i + R_i > 0 \tag{2.21}$$

in Figure 2.6.

Having constructed the sets $\mathscr{C}_1(\Gamma)$, $\mathscr{C}_2(\Gamma)$ and $\mathscr{C}_3(\Gamma)$, we can form $\mathscr{C}(\Gamma) = \mathscr{C}_1(\Gamma) \cap \mathscr{C}_2(\Gamma) \cap \mathscr{C}_3(\Gamma)$. Geometric representations like those sketched in Figures 2.4–2.6 can be very useful here.

2.11. Let us consider a specific example. For $i = 1, 2, 3$, let

$$v_i(K) = \begin{cases} 0 & \text{if } |K| \leq 1 \\ h_i & \text{if } |K| \geq 2. \end{cases} \tag{2.22}$$

As we see from (2.22), in this game it is supposed that contamination produced by only one participant does not cause any danger to any participant; on the other hand, contamination by two participants is already so dangerous that the participation in this contamination by a third participant cannot cause any additional danger.

Here (2.18) assumes the form

$$h_i(\xi_j(1-\xi_k)+(1-\xi_j)\xi_k) \gtreqless 1,$$

and correspondingly (2.19) becomes

$$-2h_i\xi_j\xi_k + h_i\xi_j + h_i\xi_k - 1 \gtreqless 0. \tag{2.23}$$

First let $h_i > 2$. Then we have

$$-1 > \frac{h_i^2}{-2h_i}, \quad -2h_i < 0, \quad -2h_i + h_i + h_i - 1 < 0,$$

i.e. (2.20) is satisfied, and all the sets $\mathscr{C}_i$ have the form sketched in Fig. 2.5, oriented along the corresponding coordinate axes. Let us enumerate the points of intersection of these figures.

First we determine the equilibrium situations in the pure strategies.

Since the vertex (1,1,1) belongs to the hyperbolic sector a in all three sets $\mathscr{C}_i$, this is an equilibrium situation. It can usefully be described as an equilibrium "of hopelessness": any single player who, in conditions of general pollution, starts to apply the pure technology only make his position worse, but does not improve the general situation.

Furthermore, the hexagons b in $\mathscr{C}_i$ and $\mathscr{C}_j$ have common intervals on the ξ_k axis that just reach the triangle c in $\mathscr{C}_k$. Thus we obtain three more equilibrium situations: (0,0,1), (0,1,0) and (1,0,0). These situations can be interpreted as "impudence" of player i in using $\xi_i = 1$ after this player has decisively polluted the reservoir, having thus exhausted the "pollution quota"; nothing remains for the other players to use except their pure technologies.

There are no other equilibrium situations in the pure strategies in this game. Note that the situation (0,0,0) is, so to speak, an "anti-equilibrium" situation: here each player is interested in changing his strategy (the inequality (2.15) is not satisfied for any $i \in I$).

It is easy to verify that there are also no pure equilibrium situations for two players (i.e., those lying within the edges of the situation cube).

Let us look for equilibrium situations in which a pure strategy is used by just one player (i.e., situations lying on the interiors of sides of the cube). Since the interiors of the triangles of types a and c, and of the hexagons of type b, do not intersect (even in pairs), the desired situations must lie on the extreme generators of the cylinders of types d and e. These situations actually occur if the intersections of the generators fall in the corresponding hexagon b.

To find these situations, we write the equations corresponding to (2.23):

$$-2h_1\xi_1\xi_3 + h_1\xi_2 + h_1\xi_3 = 1, \tag{2.24}$$

$$-2h_2\xi_1\xi_3 + h_2\xi_1 + h_2\xi_3 = 1, \tag{2.25}$$

$$-2h_3\xi_1\xi_2 + h_3\xi_1 + h_3\xi_2 = 1. \tag{2.26}$$

Suppose, for definiteness, that

$$h_3 \leqq h_2 \leqq h_1; \tag{2.27}$$

taking $\xi_3 = 0$ in (2.24) and (2.25), we have

$$\xi_1 = 1/h_2, \quad \xi_2 = 1/h_1.$$

Substituting these values into the left side of (2.26), we obtain

$$-2\frac{h_3}{h_1h_2} + \frac{h_3}{h_2} + \frac{h_3}{h_1}. \tag{2.28}$$

If then

$$h_3 \geqq \frac{h_1h_2}{h_1+h_2-2}, \tag{2.29}$$

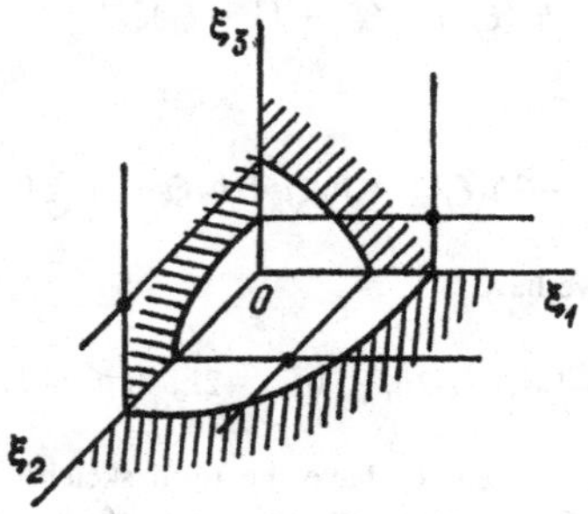

Fig. 2.7

by (2.27) the number (2.28) is at least 1, and we have an equilibrium situation (the point $(1/h_2, 1/h_1, 0)$ lies in the corresponding hexagon). If, however, (2.29) is not satisfied, the number (2.28) is less than 1, and the situation $(1/h_2, 1/h_1, 0)$ is not an equilibrium situation.

If we further set $\xi_2 = 0$ in (2.24) and (2.26), we obtain

$$\xi_1 = -1/h_3, \quad \xi_3 = 1/h_1,$$

and substitution into the left side of (2.25) yields

$$-2\frac{h_2}{h_1 h_3} + \frac{h_2}{h_1} + \frac{h_2}{h_3},$$

which, by (2.27), is always at least 1. Hence $(1/h_3, 0, 1/h_1)$ is certainly an equilibrium situation.

In the same way, $(0, 1/h_3, 1/h_2)$ is also an equilibrium situation. A schematic picture of the case when two of the three situations of the type under consideration are equilibrium situations is shown in Figure 2.7.

Such equilibrium situations can be informally thought of as phenomena that are observed when one of the players deliberately uses the pure technology. It is evident that in this case the other two players are allowed a rather free use of technology.

Let us consider especially the case when (2.29) is satisfied in the opposite sense:

$$h_3 \leqq \frac{h_1 h_3}{h_1 + h_3 - 2}.$$

Then, as is easily verified, the intersections of the generators of two cylinders of type *e* fall in the corresponding triangle of type *a*.

In fact, if we set $\xi_3 = 1$ in (2.24) and (2.25), we obtain

$$\xi_1 = 1 - \frac{1}{h_2}, \quad \xi_2 = 1 - \frac{1}{h_1},$$

and substitution of these expressions into the left side of (2.26) gives us the same expression (2.28), whose size, under our hypotheses, guarantees that the point $(1 - 1/h_2, 1 - /h_1, 1)$ belongs to the corresponding triangle of type *a*. The implication of this possibility is that player 3, who suffers relatively little from the pollution, may then use a cheap technology, even if the other players have a positive probability of polluting the reservoir. It is clear that in essence this situation (if it is an equilibrium situation) recalls the equilibrium of hopelessness (1,1,1), although, of course, it is not as harmful. We also notice that if there is precise equality in (2.29) then both $(1 - 1/h_2, 1 - 1/h_1, 1)$ and $(1/h_2, 1/h_1, 0)$ are equilibrium situations.

Finally, we turn to the determination of completely mixed equilibrium situations. If (ξ_1, ξ_2, ξ_3) is such a situation, then it must, as we know, satisfy equations (2.24)–(2.26).

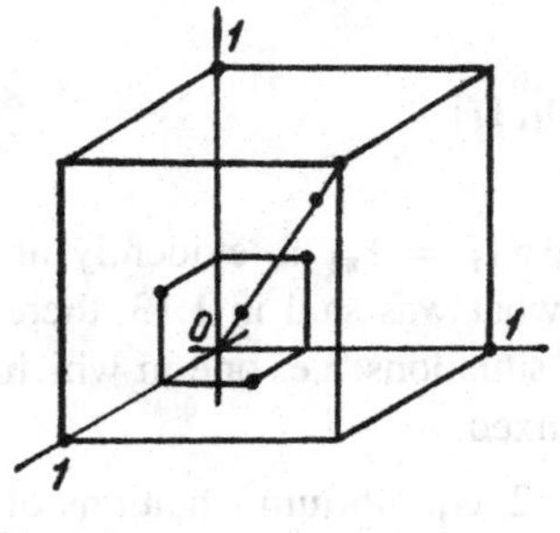

Fig. 2.8

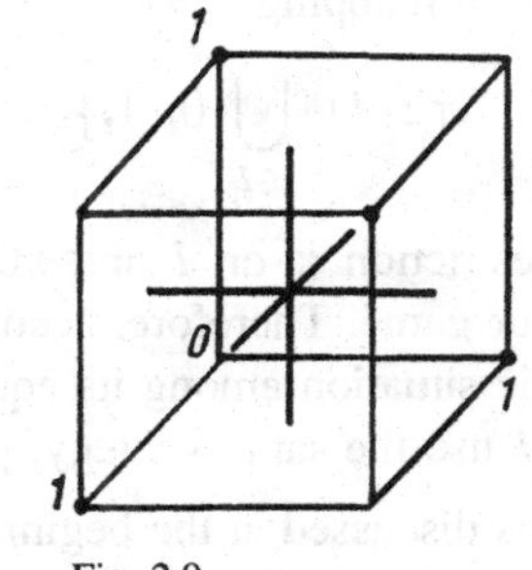

Fig. 2.9

Let us write each of these equations in the form

$$\left(\xi_i - \frac{1}{2}\right)\left(\xi_j - \frac{1}{2}\right) = \frac{1}{4} - \frac{1}{2h_k},$$

from which, setting

$$\frac{1}{4} - \frac{1}{2h_i} = \alpha_i$$

for short ($i = 1, 2, 3$), we can easily obtain two vectors (ξ_1, ξ_2, ξ_3):

$$\left(\frac{1}{2} + \sqrt{\frac{\alpha_2\alpha_3}{\alpha_1}}, \frac{1}{2} + \sqrt{\frac{\alpha_1\alpha_3}{\alpha_2}}, \frac{1}{2} + \sqrt{\frac{\alpha_1\alpha_2}{\alpha_3}}\right),$$

$$\left(\frac{1}{2} - \sqrt{\frac{\alpha_2\alpha_3}{\alpha_1}}, \frac{1}{2} - \sqrt{\frac{\alpha_1\alpha_3}{\alpha_2}}, \frac{1}{2} - \sqrt{\frac{\alpha_1\alpha_2}{\alpha_3}}\right).$$

It is easy to verify that both these vectors are equilibrium situations if $h_i \geqq h_j h_k/(h_j + h_k - 2)$ for all different i, j, k.

When the numbers h_i increase simultaneously, the equilibrium situations in the pure strategies remain in place. The equilibrium situations with a single pure strategy and the completely mixed equilibrium situations are shifted toward the corresponding vertices of the cube. (The general picture has the form sketched in Figure 2.8.)

In conclusion, we turn to the limiting case $h_1 = h_2 = h_3 = 2$ (the case when h_i is less than 2 can be analyzed very simply, and is not of much interest). Here we use (2.21), so that the sets $\mathscr{C}_i$ have the form sketched in Figure 2.5. Elementary geometric considerations show that in this case the equilibrium situations have the form sketched in Figure 2.9. (The bundle of three segments that intersect in the center of the cube, and four vertices of the cube.) This case can be considered degenerate (see 2.13).

2.12. A game of the kind that we are considering, one related to the use of natural resources, is said to be *symmetric* if r_i and $v_i(K)$ are independent of i and in addition $v_i(K)$ does not depend on K itself, but only on the number $k = |K|$ of its elements. This concept of symmetry is appropriate for the applied economic and ecological character of the applications that we consider.

In correspondence with these assumptions, we shall denote $v_i(K)$ by $v(k)$, and r_i by r.

Every one-to-one mapping

$$\pi : I \cup \bigcup_{i \in I} \{0_i, 1_i\} \to I \cup \bigcup_{i \in I} \{0_i, 1_i\}$$

with an arbitrary restriction π on I and $\pi 0_i = 0_{\pi i}$, $\pi 1_i = 1_{\pi i}$ is evidently an automorphism of our game. Therefore, according to what was said in 1.13, there must be a symmetric situation among its equilibrium situations, i.e. one in which all the players $i \in I$ use the same strategy, pure or mixed.

For the reasons discussed at the beginning of 1.12, equilibrium situations of this kind are especially important. We restrict the analysis of our games to these, although there are also asymmetric equilibrium situations (see, for example, Figure 2.8). However, as will be clear later, in "nondegenerate" cases this asymmetry turns out to be rather limited.

We have already considered the case of common pure strategies in 2.7. Here formulas (2.15) and (2.16) can now be written in the forms $v(1) \geqq r$ and $v(n) - v(n-1) \leqq r$.

We turn to the case of common mixed strategies. In this case $\xi_i = \xi \in (0, 1)$, $S = I$, $S_0 = S_1 = \emptyset$, and we have the hypotheses of (2.14), which can be put in the form

$$\sum_{k=0}^{n-1} (v(k+1) - v(k)) \binom{n-1}{k} \xi^k (1-\xi)^{n-k-1} = r \qquad (2.30)$$

by using the symmetry of the game.

In principle, this equation yields a numerical solution for arbitrary losses v and r. Let us consider its qualitative analysis. Setting $\xi = 0$ on the left-hand side, we obtain $v(1)$; and setting $\xi = 1$, we obtain $v(n) - v(n-1)$. We can evidently analyze various possibilities for r, $v(1)$, and $v(n) - v(n-1)$. We examine two of these as examples.

If $r \leqq v(1) \leqq v(n) - v(n-1)$, then among the equilibrium situations there is one for which $\xi = 1$, and there may be no other symmetric equilibrium situations in Γ; or, in view of the continuity of the left-hand side of (2.30) with respect to ξ, there may be an even number of them.

If $v(n) - v(n-1) < r < v(1)$, neither $\xi = 0$ nor $\xi = 1$ yields an equilibrium situation; but, on the other hand, $\xi \in (0, 1)$ yields an odd number (by the same continuity).

Note that in both cases we have as a result an even number of equilibrium situations.

2.13. Let us now show that, except in cases that we may qualify as degenerate, in every equilibrium situation in the games that we consider, all players in S use the same mixed strategy; and describe a processs for finding these equilibrium situations.

Dealing with the partition $I = S_0 \cup S_1 \cup S$, we denote the number of players in each of the sets S_0, S_1, and S by s_0, s_1, and s, and the difference $v(k+1) - v(k)$ by $\Delta^1(k)$. In view of the symmetry of the game, we may replace the partition of the set of players by a partition of their indices: $n = s_0 + s_1 + s$.

The relations (2.6)–(2.8), or equivalently, (2.12)–(2.14) now acquire the following forms:

$$\sum_{k=0}^{s_1} \Delta^1(s_1 + k) \sum_{\substack{K \subset S \\ |K|=k}} \xi^{(K)} (1-\xi)^{(S \setminus K)} \geqq r, \qquad i \in S_0, \tag{2.31}$$

$$\sum_{k=0}^{s} \Delta^1(s_1 + k - 1) \sum_{\substack{K \subset S \\ |K|=k}} \xi^{(K)} (1-\xi)^{(S \setminus K)} \leqq r, \quad i \in S_1, \tag{2.32}$$

$$\sum_{k=1}^{s-1} \Delta^1(s_1 + k) \sum_{\substack{K \subset S \\ |K|=k}} \xi^{(K)} (1-\xi)^{(S \setminus (K \cup i))} = r, \quad i \in S. \tag{2.33}$$

Notice that the inequalities (2.31) for players $i \in S_0$, as well as the inequalities (2.32) for the players $i \in S_1$, are formally identical. Consequently we are concerned here with $s+2$ relationships.

We now multiply out all the terms containing the ξ_j and calculate the sums of the coefficients for each of the resulting products $\xi^{(P)}$. In relations (2.31) and (2.32), this P is any subset of S; and in (2.33), the P is a subset of $S \setminus i$. Later, we shall set $|P| = p$.

The product $\xi^{(P)}$ can be obtained by multiplying any of the products $\xi^{(K)}$ already obtained for a given K by its complement $\xi^{(P \setminus K)}$ obtained as a term by expanding the product

$$\prod_{j \in S \setminus K} (1 - \xi_j) = (1 - \xi)^{(P \setminus K)}.$$

For a given P, the number of sets $K \subset P$ with $|K| = k$ is $\binom{p}{k}$ and each product so obtained has the coefficient $(-1)^{p-k} \Delta^1(s_1 + k)$ under conditions (2.31) and (2.33), and the coefficient $(-1)^{p-k} \Delta^1(s_1 + k - 1)$ under condition (2.32). Therefore the total coefficient of $\xi^{(P)}$ will be either

$$\sum_{k=0}^{p} (-1)^{p-k} \binom{p}{k} \Delta^1(s_1 + k) \text{ or } \sum_{k=0}^{p} (-1)^{p-k} \binom{p}{k} \Delta^1(s_1 + k - 1).$$

By a formula from the calculus of finite differences*) this will be equal to $\Delta^{p+1}(s_1)$ or, respectively, $\Delta^{p+1}(s_1 - 1)$, where Δ^{p+1}, as usual, denotes the $(p+1)$th difference of v.

---

*) I have not found precisely these formulas in books in English, but they are easily derived from, for example, M. Abramowitz and I.A. Stegun, Handbook of Mathematical Functions, Washington, DC., 1964, p. 877. – Translator

We denote by ξ_K the set ξ_i of probabilities for $i \in K$. Then, taking account of what we have said, after the inverse renaming P to K and p to k, and using the usual notation for the fundamental symmetric functions, we can rewrite (2.31)–(2.33) in the form

$$\sum_{k=0}^{s} \Delta^{k+1}(s_1)\sigma_k(\xi_S) \geqq r, \tag{2.34}$$

$$\sum_{k=0}^{s} \Delta^{k+1}(s_1 - 1)\sigma_k(\xi_S) \leqq r, \tag{2.35}$$

$$\sum_{k=0}^{s} \Delta^{k+1}(s_1)\sigma_k(\xi_{S\setminus i}) = r, \quad i \in S. \tag{2.36}$$

The process of enumerating the equilibrium situations in this game, although in principle no different from the process described in 2.6 for solving the system (2.12)–(2.14), can be described in more detail.

We first form all partitions

$$I = S_0 \cup S_1 \cup S, \tag{2.37}$$

and for each partition, the corresponding system (2.34)–(2.36). Every solution of this system is an equilibrium situation corresponding to the partition (2.37). If the system (2.34)–(2.36) is insoluble, this would mean that there is no equilibrium situation corresponding to the partition (2.37).

The process of solving systems of the form (2.34)–(2.36) reduces to finding solutions of (2.36) and verifying that they satisfy (2.34) and (2.35). Here we can use the symmetry property of the game.

We first describe the equilibrium situations in the pure strategies, i.e. those for which $s = 0$. Here the system (2.36) does not occur, and the partition (2.37) reduces to $I = S_1 \cup S_2$, to which there corresponds a unique situation in the pure strategies. A test for this to be an equilibrium situation consists of (2.34) and (2.35), which in the present case become

$$\Delta^1(s_1 - 1) \leqq r \leqq \Delta^1(s_1) \tag{2.38}$$

(evidently only the right-hand inequality applies when $s_1 = 0$, and only the left-hand one when $s_2 = 0$).

Next we consider the case $s = 1$. In this case the system (2.36) reduces to the single equation

$$\Delta^1(s_1) = r. \tag{2.39}$$

If this equation is satisfied, the equilibrium situation corresponding to the choice of (2.37) is a solution of (2.34) and (2.35), which in this case has the form

$$\Delta^2(s_2 - 1)\xi_i + \Delta^1(s_1 - 1) \leqq r \leqq \Delta^2(s_1)\xi_i + \Delta^1(s_1). \tag{2.40}$$

It is clear that if $s_1 = 0$ we may drop the restriction imposed by the left-hand inequality; and when s_0, that imposed by the right-hand side. Otherwise there is no equilibrium situation corresponding to (2.37).

Finally let $s \geqq 2$. Choose any $i' \in S \setminus i$ and notice that

$$\sigma_0 = 1,$$

$$\sigma_k(\xi_{S\setminus i}) = \xi_{i'}\sigma_{k-1}(\xi_{S\setminus\{i',i\}}) + \sigma_k(\xi_{S\setminus\{i',i\}}), \quad 0 < k < s - 1,$$

$$\sigma_{s-1}(\xi_{S\setminus i}) = \xi_{i'}\sigma_{s-2}(\xi_{S\setminus\{i',i\}}).$$

Then we may present the left-hand side of (2.36) as a linear function of $\xi_{i'}$:

$$\xi_{i'} \sum_{k=1}^{s-1} \Delta^{k+1}(s_1)\sigma_{k-1}(\xi_{S\setminus\{i',i\}}) + \sum_{k=0}^{s-2} \Delta^{k+1}(s_1)\sigma_k(\xi_{S\setminus\{i',i\}}) = r \tag{2.41}$$

or, with a natural notation, as

$$\xi_{i'} A_{i'}(\xi_{S\setminus\{i',i\}}) + B_{i'}(\xi_{S\setminus\{i',i\}}) = r.$$

But in the same way, we can write

$$\xi_i A_i(\xi_{S\setminus\{i,i'\}}) + B_i(\xi_{S\setminus\{i,i'\}}) = r,$$

where, as is clear from (2.41), $A_{i'} = A_i$ and $B_{i'} = B_i$. Therefore, in order for ξ_S to be an equilibrium situation either

$$A_{i'}(\xi_{S\setminus\{i',i\}}) = 0 \text{ and } B_{i'}(\xi_{S\setminus\{i',i\}}) = r,$$

and ξ_i and $\xi_{i'}$ are arbitrary, or

$$A_{i'}(\xi_{S\setminus\{i',i\}}) \neq 0 \text{ and } \xi_i = \xi_{i'}.$$

The first case can be considered to be degenerate. Let us suppose that this does not occur for any pair $i, i' \in S$. Then $\xi_i = \xi$ for all $i \in S$. In this case the system (2.36) reduces to the single equation

$$\sum_{k=0}^{s-1} \Delta^{k+1}(s)\binom{s-1}{k}\xi^k = r,$$

which, for clarity, it is useful to transform to a form corresponding to (2.33):

$$\sum_{k=0}^{s-1} \Delta(s_1+k)\binom{s-1}{k}\xi^k(1-\xi)^{s-1-k} = r. \tag{2.42}$$

Similarly we can put (2.34) and (2.35) into the form

$$\sum_{k=0}^{s} \Delta(s_1+k)\binom{s}{k}\xi^k(1-\xi)^{s-k} \geqq r, \tag{2.43}$$

$$\sum_{k=0}^{s} \Delta(s_1+k-1)\binom{s}{k}\xi^k(1-\xi)^{s-k} \leqq r. \tag{2.44}$$

It remains only for us to find, for each triple (s_0, s_1, s), the roots of equation (2.42) (in this connection, see what was said in 2.12 about equation (2.30)) and verify that they satisfy (2.43) and (2.44)).

In conclusion, we notice that, as is clear from, say, the example in 2.11, when $h_1 = h_2 = h_3$, there may be, in the same game, equilibrium situations corresponding to different combinations of s_0, s_1, and s.

2.14 Converse problem for dyadic games. Finally, we discuss an example that is, in a certain sense, fundamental.

Theorem. *Let*

$$f(t) = t^n - a_1 t^{n-1} + \ldots + (-1)^n a_n$$

be a polynomial of any degree n, with rational coefficients, all of whose roots $\alpha_1, \ldots, \alpha_n$ are real and belong to the interval $(0, 1)$.

Then there is a dyadic game Γ with $2n$ players, with rational payoff functions, in which for every completely mixed equilibrium situation

$$X^* = (\xi_1^*, \ldots, \xi_n^*, \xi_{n+1}^*, \ldots, \xi_{2n}^*)$$

the first components $\xi_1^*, \ldots, \xi_n^*$ are a permutation of the roots $\alpha_1, \ldots, \alpha_n$ of f, and the other components $\xi_{n+1}^*, \ldots, \xi_{2n}^*$ are rational numbers in the interval $(0, 1)$, arbitrarily prescribed in advance.

Proof. In fact, we are going to construct a game in which the values of the payoff functions (in situations in the pure strategies) are, first, positive integers; second, the coefficients of the polynomial f; and third, the numbers which we selected to be the values $\xi_{n+1}^*, \ldots, \xi_{2n}^*$.

In fact, consider the dyadic game Γ in which, for $i = 1, 2, \ldots, n$, we take

$$H_i(x \underset{i}{\|} 0) = \begin{cases} 1 & \text{if player } n+i \text{ chooses strategy 1,} \\ 0 & \text{otherwise;} \end{cases} \tag{2.45}$$

$$H_i(x \underset{i}{\|} 1) = \xi_{n+1}^*, \text{ where } \xi_{n+1}^* \in (0, 1); \tag{2.46}$$

$$H_{n+1}(x \underset{n+i}{\|} 0) = \binom{k(x)}{i}, \text{ where } k \text{ is the number of players, among } 1, \ldots, n, \text{ who select strategy 1 in situation } x; \tag{2.47}$$

$$H_{n+i}(x \underset{n+i}{\|} 1) = a_i, \tag{2.48}$$

where $a_i > 0$ $(i = 1, \ldots, n)$ is the absolute value of the corresponding coefficient of f. Let us show that the game Γ so defined is the one required.

Let us select the situation

$$X = (\xi_1, \ldots, \xi_n, \xi_{n+1}, \ldots, \xi_{2n}),$$

and write the equation (2.8) to conform to our case. Its validity is, according to 2.3, a necessary and sufficient condition for equilibrium, and we compute the payoffs $H_i(X \underset{i}{\|} 0)$ and $H_i(X \underset{i}{\|} 1)$, for the players $i = 1, \ldots, 2n$.

For $i = 1, \ldots, n$, in view of (2.4), the payoff $H_i(X \underset{i}{\|} 0)$ is equal to the probability that, in situation X, player $n+i$ chooses strategy 1, i.e.

$$H_i(X \underset{i}{\|} 0) = \xi_{n+1}. \tag{2.49}$$

At the same time, if we turn in (2.46) to the mixed strategies of all the players except i, we obtain

$$H_i(X \underset{i}{\|} 1) = \xi^*_{n+1}. \tag{2.50}$$

Equations (2.49) and (2.50) permit us to rewrite (2.8) as

$$\xi_{n+i} = \xi^*_{n+i}, \qquad i = 1, \ldots, n. \tag{2.51}$$

We now select the player $n+i$ and turn to (2.47). The players $k_1, \ldots, k_i$, that make an arbitrary but fixed combination of i of the players $1, \ldots, n$, choose simultaneously, under the hypotheses of the situation X, their 1-strategies with the probability $\xi_{k_1} \ldots \xi_{k_i}$. Therefore this product is the expectation of the "number of times" with which the players $k_1, \ldots, k_i$ together choose their 1-strategy. But the expectation of the number of times that the first strategy is chosen by an *arbitrary* set of i out of n players is therefore the sum of all such products. In other words,

$$H_{n+1}(X \underset{n+i}{\|} 0) = \sigma_i(\xi_1, \ldots, \xi_n), \tag{2.52}$$

where σ_i is the elementary symmetric function of order i of the variables denoted under the symbol of this function.

It is clear that the transition in (2.48) to the mixed strategies of the players other than $n+i$ leads to

$$H_{n+i}(X \underset{n+i}{\|} 0) = a_i. \tag{2.53}$$

Equations (2.52) and (2.53) give us (2.8) in the form

$$\sigma_i(\xi_1, \ldots, \xi_n) = a_i, \quad i = 1, \ldots, n, \tag{2.54}$$

i.e., the probabilities $\xi_1, \ldots, \xi_n$ form a permutation of the roots of the polynomial f. Together with (2.51), this shows that the game Γ we have constructed is the one required. □

We emphasize that the components $\xi_1, \ldots, \xi_n$ of each completely mixed equilibrium situation in Γ must satisfy (2.51) and (2.54). Consequently there are finitely many of them, and consequently each of these situations is an isolated point in the space of situations. If we are to describe them, we inevitably have to find the roots of the polynomial.

As will be established in section 3, the solution of any finite noncooperative game turns out to be equivalent to the solution of a finite noncooperative game with three players. Hence the fact that we have just established implies the possibility, for every polynomial of arbitrary degree whose roots have the properties specified above, of constructing a finite noncooperative game with three players, such that the description of its equilibrium situations necessarily involves finding the roots of this polynomial. It is clear that this three-player game is, in general, not dyadic.

§3 Solution of general finite noncooperative games

3.1. Obtaining solutions of finite noncooperative games. As is clear from the examples in § § 1 and 2, finding equilibrium situations in finite noncooperative games can present considerable difficulties, even for a small number of players and a small number of strategies for each of them. The example in **2.14** shows that in principle this problem, because of its formal difficulty, is no easier than the problem of finding the zeros of general polynomials with real coefficients.

From the same examples (especially from the problem of the optimal use of the environment, **2.6–2.12**) it follows that the set of equilibrium situations under various values of the parameters that define a game (and ultimately from changing the payoff functions) can have very different forms. When the parameters take their boundary values, there can arise "degeneracies" under which the sets of equilibrium situations can be remarkably varied, and sometimes even fantastic. It is only natural that it is simply impossible to say much about the composition of any algorithm for a complete description of the sets $\mathscr{C}(\Gamma)$ for finite noncooperative games of, say, arbitrary format (or even of a given, but not too small, format).

Nevertheless, it is possible to describe such an algorithm for some special classes of noncooperative games. Moreover, it turns out that in "nondegenerate" cases (in a sense that will be specified later), the total number of equilibrium situations in a finite noncooperative game is finite (and by the way odd), and these situations can be found, more or less effectively, although not completely constructively.

A basis for the reasoning that leads to these results consists of a systematic application of what we said in **1.2**, namely that under the conditions of an equilibrium situation every pure strategy of a player either provides the same (that is, the maximum) payoff, which is also provided in the same situation by a mixed strategy, or does not appear in the spectrum of any mixed strategy.

3.2. In essence, the theorem in **1.5** implicitly contains an algorithm for describing the equilibrium situations in a bimatrix game. However, there are not efficient methods for actually applying this algorithm. This inefficiency has about the same meaning and the same cause as that of the solution of the problem of linear programming (that is, finding an extremum of a linear form on a polyhedron determined by its vertices; a precise formulation of the standard problem of linear programming is given in § 3, Chapter 4) by means of finding all vertices of the polyhedron of admissible solutions and calculating the criterion function for each one. A systematic enumeration of the necessary vertices of the polyhedron by the simplex method makes the problem of linear programming (for reasonable dimensions) solvable in practice.

Something similar can be realized for the determination of equilibrium situations in "nondegenerate" noncooperative games. In the present section we describe a corresponding algorithm, first for bimatrix games, and then for finite noncooperative games with arbitrary finite sets of players. An algorithm for enumerating all equilibrium situations for arbitrary bimatrix games (including the case of degenerate ones) will be presented in § 4, Chapter 4, as a development of a method for finding all solutions of matrix games.

3.3. Let us deal with an $m \times n$-bimatrix game $\Gamma_{A,B}$. The condition that a situation (X^*, Y^*) is an equilibrium situation in this game consists of the satisfaction of the inequalities

$$A_{i\cdot} Y^{*T} \leqq X^* A Y^{*T}, \quad i = 1, \dots, m, \tag{3.1}$$

$$X^* B_{\cdot j} \leqq X^* B Y^{*T}, \quad j = 1, \dots, n. \tag{3.2}$$

Since X^* and Y^* are strategies, it follows that we must have (here and later, $J_p = (1, \dots, 1) \in \mathbf{R}^p$)

$$X^*, Y^* \geq 0, \quad X^* J_m^T = J_n Y^{*T} = 1. \tag{3.3}$$

For the sake of clarity, we transform these conditions for an equilibrium situation. We select an $m \times n$-matrix $E(= E_{m,n})$, consisting only of units, and a number k greater than all the elements of A and B, and consider the system of equations

$$X(kE - B) \geqq J_n, \quad X \geq 0, \tag{3.4}$$

$$(X(kE - B) - J_n) Y^T = 0, \tag{3.5}$$

$$(kE - A) Y^T \geqq J_m^T, \quad Y \geq 0, \tag{3.6}$$

$$X((kE - A) Y^T - J_m^T = 0. \tag{3.7}$$

Let (X, Y) be a solution of this system; we set

$$X^* = \frac{X}{X J_m^T}, \quad Y^* = \frac{Y}{J_n Y^T}. \tag{3.8}$$

Then we obtain

$$X^* B_{\cdot j} \leqq k - \frac{1}{X Y_m^T},$$

from (3.4), and

$$X^* B Y^{*T} = k - \frac{1}{X Y_m^T},$$

from (3.5); these, together, give us (3.2). Similarly, we obtain (3.1) from (3.6) and (3.7). Consequently, the situation (X^*, Y^*) defined by (3.5) is a solution of the system (3.1)–(3.3).

Conversely, let (X^*, Y^*) be a solution of (3.1)–(3.3), and let k be the number of stipulated quantities. Then we set

$$X J_m^T = (k - X^* B Y^{*T})^{-1}, \quad J_n Y^T = (k - X^* A Y^{*T})^{-1},$$

and the pair (X, Y) satisfies (3.4)–(3.7).

In our discussion, the roles of the expressions $kE - A$ and $kE - B$ can be played by arbitrary matrices with positive elements. We shall denote these by C and D. Then (3.4)–(3.7) can be rewritten as

$$XD \geqq J_n, \quad X \geq 0, \tag{3.9}$$

$$(XD - J_n)Y^T = 0, \tag{3.10}$$

$$CY^T \geqq J_m^T, \quad Y \geq 0, \tag{3.11}$$

$$X(CY^T - J_m^T) = 0. \tag{3.12}$$

Every solution (X, Y) of this system will, for the sake of simplicity, also be called an *equilibrium system*.

3.4. We consider the set $\mathfrak{U}$ of m-vectors X that satisfy (3.9). Evidently $\mathfrak{U}$ is a convex set bounded by $n + m$ hyperplanes (of which m are coordinate hyperplanes). If we denote by I_m the unit $m \times m$-matrix, a face of $\mathfrak{U}$ will correspond to the $m \times (n + m)$-matrix (D, I). It is also clear that $\mathfrak{U}$ is the nonnegative orthant except for a bounded set adjacent to the origin.

With each $X \in \mathfrak{U}$ we associate the set S_X of $(m-1)$-dimensional faces of $\mathfrak{U}$ that contain X, or, essentially equivalently, the set of indices i for which $\xi_i = 0$ or $XD_{\cdot j} = 1$; or, finally, the submatrix of (D, I) that consists of the corresponding columns of this matrix.

In connection with the matrix (D, I) we may formulate the following *condition of nondegeneracy*: if $\bar{D}$ is an $m \times r$-submatrix of (D, I) and, for some X, the equality $\bar{D} = S_X$ is satisfied (in the sense of the preceding paragraph), the rank of $\bar{D}$ is equal to r. We shall suppose in addition that the hypothesis of nondegeneracy is satisfied. It follows geometrically from this hypothesis that every vertex of $\mathfrak{U}$ is the intersection of at most m of its faces, and that each edge is the intersection of at most $m - 1$ faces. It also follows from the hypothesis of nondegeneracy that precisely m edges are incident at each vertex of $\mathfrak{U}$. Each of the m edges that goes along the coordinate axes is unbounded. On each unbounded edge of $\mathfrak{U}$ there is a single vertex of $\mathfrak{U}$, but on the other edges of $\mathfrak{U}$ there are two.

3.5. We now consider an analogous n-dimensional polyhedron $\mathfrak{V}$ that consists of n-vectors Y satisfying (3.7), and suppose that they satisfy an analogous condition of nondegeneracy. Then in application to $\mathfrak{V}$, everything we have said about $\mathfrak{U}$ will hold mutatis mutandis for $\mathfrak{V}$. We denote by T_Y the submatrix of the $(m+n) \times n$-matrix $(C, I_n)^T$, whose rows correspond to the equations satisfied by Y.

Finally we set $\mathfrak{W} = \mathfrak{U} \times \mathfrak{V}$. The set $\mathfrak{W}$ is an $(m+n)$-dimensional polyhedron. Its vertices are pairs of vertices of $\mathfrak{U}$ and $\mathfrak{V}$ and its edges consist of pairs of the form "vertex of $\mathfrak{U}$, point of an edge of $\mathfrak{V}$" and "point of an edge of $\mathfrak{U}$, vertex of $\mathfrak{V}$". Since $\mathfrak{U}$ and $\mathfrak{V}$ are not degenerate, it follows that exactly $m + n$ edges are incident at each vertex of $\mathfrak{W}$.

3.6. We can now transform the hypothesis of nondegeneracy of (X,Y) into the property that this pair belongs to $\mathfrak{W}$, plus the validity of the $m+n$ equations

$$\xi_i(C_{i\cdot}Y^T - 1) = 0, \quad i = 1,\ldots,m,$$
$$(XD_{\cdot j} - 1)\eta_j = 0, \quad j = 1,\ldots,n. \tag{3.13}$$

This means that at least one factor on the left-hand side of each equation must vanish.

Some number of these factors depend on X, and the others, on Y. However, by the hypothesis of nondegeneracy no point $X \in \mathfrak{U}$ can satisfy more than m conditions of the form $\xi_i = 0$ or $XD_{\cdot j} = 1$. On the other hand, no $Y \in \mathfrak{B}$ can satisfy more than n analogous conditions. But since, nevertheless, all $m+n$ equations (3.13) must be satisfied, at most one factor of the left-hand side of each equation can vanish. Therefore the condition of nondegeneracy applies once for each i and once for each j. Hence it follows that, first, $Z = (X,Y)$ is a vertex of $\mathfrak{W}$ and, second, that, for each $r = 1,\ldots,m+n$ either the rth column of the matrix (D,I) belongs to S_X, or the rth row of the matrix $(C,I)^T$ belongs to T_Y.

We now specify a value of r and consider the set U_r of the vectors $Z \in \mathfrak{W}$ for which the equations (3.13) are satisfied, except, perhaps,

$$(XD.r - 1)\eta_r = 0. \tag{3.14}$$

By a direct calculation that recalls the one in **3.5**, with the help of the condition of nondegeneracy we verify that all points of $\mathfrak{U}$ lie on the edges of the polyhedron $\mathfrak{W}$ and that among these edges only one is unbounded. It consists of the points that satisfy the maximal number m of equations relating to X and $n-1$ equations relating to Y, and the equations not satisfied by Y originate from the failure to satisfy (3.14) in the part relating to Y, i.e., we must have $\eta_r > 0$ (for points on all the other unbounded edges of $\mathfrak{B}$ we evidently must have $\eta_r = 0$).

3.7. Let us describe all the vertices of $\mathfrak{W}$ that are contained in U_r. We first notice that the points of $\mathfrak{W}$ that satisfy the equations (3.13), that is, the equilibrium situations, automatically belong to *all* the sets U_r.

Let $W = (X,Y)$ be any vertex of $\mathfrak{W}$ that belongs to U_r.

If (3.14) is satisfied for W, it follows that W is an equilibrium situation. As we have made clear, at W (as at every vertex of $\mathfrak{W}$) $m+n$ vertices originate, and in this case just one of the equations $XD_{\cdot r} - 1 = 0$ and $\eta_r = 0$ is satisfied. If $\eta_r = 0$, then among the edges originating at W, only the one at whose points $\eta_r \geqq 0$ will belong to U_r. Similarly, if $XD_{\cdot r} - 1 = 0$, then among the edges originating at W, only the one at whose points $XD_{\cdot r} - 1 > 0$ will belong to U_r. Therefore just one edge belonging to U_r originates at each vertex $W \in U_r$.

Now suppose that (3.14) is not satisfied at the vertex W:

$$(XD_{\cdot r} - 1)\eta_r > 0.$$

Here both factors are different from zero. But since the point Z is a vertex of $\mathfrak{W}$, it must satisfy $m+n$ linear equations. Therefore, in at least one of the $m+n-1$ equations (3.13) for Z which are different from (3.14), both factors must equal zero. Therefore among the $m+n$ edges originating at Z, the two along which both these two factors differ from zero appear in U_r.

A chain of edges of $\mathfrak{W}$ that belong to U_r together with their endpoints is called an *r-chain*. It follows from what was said above that every r-chain contains all the equilibrium situations of the game. It consists of cycles and of linear components whose endpoints, except for one, are equilibrium situations, and this exception is an unbounded edge. In particular, it is clear from this that the number of equilibrium situations in a nondegenerate bimatrix game must be finite and odd.

The method just described makes it rather easy to find one of the equilibrium situations (specifically, one that lies on an r-chain with an associated unbounded edge). To determine all equilibrium situations, the method does not indicate any special procedure except a more or less systematic search through all the vertices of $\mathfrak{W}$.

3.8. It is clear from elementary algebraic considerations that the set of nondegenerate $m \times n$-bimatrix games is everywhere dense (with respect to elementwise convergence of the payoff matrices) in the $2mn$-dimensional space of $m \times n$-bimatrix games.

Consequently, according to the theorem in **2.22**, Chapter 1, in order to find the equilibrium situations in an $m\times n$-matrix game Γ, we may, in principle, use the following plan: form a sequence $\Gamma_1, \Gamma_2 \dots$ of nondegenerate games that converges to Γ; find an equilibrium situation $Z^{*(k)}$ in each $\Gamma^{(k)}$; and determine a limit point of the sequence $Z^{*(1)}, Z^{*(2)}, \dots$. For more detail, see § 4.

It is clear that in this formulation the suggested procedure can hardly be considered as an effective method for finding equilibrium situations in Γ.

3.9. It follows from the definition of equilibrium situations in a finite noncooperative game Γ in mixed strategies that the collection $\mathscr{C}(\Gamma)$ of all equilibrium situations coincides with the set of solutions in R^m of a system of polynomial (in fact, polylinear) inequalities. Such a set is said to be semi-algebraic. Certain results indicate that the equilibrium situations of finite games form a set of semi-algebraic inequalities of an extremely general and sometimes very complicated form; therefore, at present we do not know of any nontrivial geometric description of this set. At the same time, it appears that almost all noncooperative games possess a finite number of equilibrium situations. Moreover, it is a basic result of the present section that this number is always odd (hence, in particular, not zero). The proof is not based on the theorem about the existence of equilibrium situations, so that we shall simultaneously obtain a new proof of Nash's theorem for almost all games (although it is not difficult to deduce Nash's theorem in full generality by means of a limit process). The proof consists of a sequence of searches through certain isolated roots of systems of polynomial equations, the last of which coincides with an equilibrium situation. Consequently, in cases when the system of equations has a simple form (for example, in bimatrix games, where it is linear), it can serve as a basis for the construction of an algorithm for locating equilibrium situations. The algorithm for bimatrix games will be described in **4.2**. (This is the principal effective method for enumerating all the equilibrium situations for arbitrary bimatrix games, as will be described in § 4, Chapter 3.)

3.10. Let us introduce some essential notation. Let

$$\Gamma = \langle I, \{x_i\}_{i\in I}, \{H_i\}_{i\in I}\rangle \tag{3.15}$$

be a finite noncooperative game, and let $(m_1, \dots, m_n)$ be its format (see **1.1**). We set $m_1 + \dots + m_n = m$, and $x_i = \{x_1^{(i)}, \dots, x_{m_i}^{(i)}\}$ for every $i \in I$.

For each $i \in I$ we take some $y_i \subset x_i$ and set

$$\mathcal{F}_y = \mathcal{F}_{y_1,\dots,y_n} = \{X \in \boldsymbol{X} : \operatorname{supp} X_i \cap y_i = \emptyset, i \in I\} \tag{3.16}$$

(in other words, if $X \in \mathcal{F}_y$ and $x^{(i)} \in y_i$, then $X_i(x^{(i)}) = 0$). It is clear that if $\mathcal{F}_y \neq \emptyset$ we will not have $y_i = x_i$ for any $i \in I$. In addition, we denote $|y_i|$ by p_i, and $p_i + \dots + p_n$ by p. We write

$$\mathcal{F}_l^i = \mathcal{F}_{x_l^{(i)}} = \mathcal{F}_{\emptyset,\dots,\{x_l^{(i)}\},\dots\emptyset}.$$

We evidently have

$$\mathcal{F}_y = \bigcap_{i\in I} \bigcap_{x_l^{(i)} \in y_i} \mathcal{F}_{x_l^{(i)}}.$$

Let us choose any $i \in I$ and $x_i^{(l)} \in x_i$ and set

$$\mathcal{G}^{x_l^{(i)}} = \mathcal{G}_i^l = \{X \in \boldsymbol{X} : H_i(X \parallel x_l^{(i)}) = \max_{1\leqq j\leqq m_i} H_i(X \parallel x_j^{(i)}), \tag{3.17}$$

and for each $z_i \subset x_i$ $(i \in I)$,

$$\mathcal{G}^z = \mathcal{G}^{z_1,\dots,z_n} = \bigcap_{i\in I} \bigcap_{x_l^{(i)} \in z_i} \mathcal{G}^{x_l^{(i)}} \tag{3.18}$$

(in other words, if $X \in \mathcal{G}^z$ and $x_l^{(i)} \in z_i$, then the deviation of player i from the situation X resulting from that player's choice of the strategy $x_j^{(i)}$ will be no worse than any other deviation). It is clear that for $\mathcal{G}^z \neq \emptyset$ we cannot have $z_i = \emptyset$ for any $i \in I$. From now on, $|z_i|$ will be denoted by r_i; and $r_1 + \dots + r_n$, by r.

Finally, we set

$$\begin{aligned}\mathcal{H}_y^z = \mathcal{H}_{y_1,\dots,y_n}^{z_1,\dots,z_n} &= \mathcal{F}_y \cap \mathcal{G}^z \\ &= \left(\bigcap_{i\in I} \bigcap_{x_l^{(i)} \in y_i} \mathcal{F}_{x_l^{(i)}}\right) \cap \left(\bigcap_{i\in I} \bigcap_{x_l^{(i)} \in z_i} \mathcal{G}^{x_l^{(i)}}\right).\end{aligned} \tag{3.19}$$

3.11. Let us introduce another series of notations. We set

$$K = \{\mathcal{F}_l^{(i)} : i \in I, 1 \leqq l \leqq m_i\} \cup \{\mathcal{G}_i^{(l)} : i \in I, 1 \leqq l \leqq m_i\}.$$

On the family K, we define a function φ by setting

$$\varphi(\mathcal{F}_l^i) = \varphi(\mathcal{G}_i^l) = \begin{cases} l & \text{if } i = 1, \\ l + \sum\limits_{1 \leqq i \leqq i-1} m_i & \text{if } i > 0. \end{cases}$$

We call a subset $K_1 \subset K$ *normal* if

$$\varphi(K_1) = \{1, 2, \ldots, m\},$$

and call a set $K_2 \subset K$ *distinguished* if

$$\varphi(K_2) \supset \{1, 2, \ldots, m-1\}.$$

Correspondingly, we call the set $\mathcal{H}_y^z$ normal (or distinguished) if the family

$$\{\mathcal{F}_{x_l^{(i)}} : i \in I, x_l^{(i)} \in y_i\} \bigcup \{\mathcal{G}^{x_l^{(i)}} : i \in I, x_l^{(i)} \in z_i\}$$

is normal (or distinguished).

Lemma. *The set $\mathcal{C}(\Gamma)$ of equilibrium situations in the game Γ is the union of all the normal sets $\mathcal{H}_y^z$.*

Proof. Let a situation $X = (X_1, \ldots, X_n) \in \mathbf{X}$, where, for each $i \in I$, the point X_i lies in the interior of the face of the simplex $\mathbf{X}_i$ spanned by the vertices $x_{s_1^{(i)}}^{(i)}, \ldots, x_{s_{k_i}^{(i)}}^{(i)}$. It follows from the definition of the sets $\mathcal{F}_l^i$ that

$$X \in \bigcap_{1 \leqq l \leqq p} \mathcal{F}^i_{r_l^{(i)}},$$

where

$$\{r_i^{(i)}, \ldots, r_{p_i}^{(i)}\} = \{1, 2, \ldots, m\} \setminus \{s_1^{(i)}, \ldots, s_{k_i}^{(i)}\}. \tag{3.20}$$

On the other hand, it follows from the definition of the sets $\mathcal{G}_i^l$ and the property of complementary slackness (Theorem **5.15**, Ch 1) that X is an equilibrium situation if and only if

$$X \in \bigcap_{1 \leqq l \leqq k_i} \mathcal{G}_i^l.$$

By the definition of φ we have

$$\varphi(\mathcal{F}^i_{r_l^{(i)}}) = r_l^{(i)} + \sum_{1 \leqq j \leqq i-1} m_j \quad (1 \leqq l \leqq p_i),$$

$$\varphi(\mathcal{G}_i^{s_l^{(i)}}) = s_l^{(i)} + \sum_{1 \leqq j \leqq i-1} m_j \quad (1 \leqq l \leqq k_i),$$

which, by (3.20), means that the set

$$\{\mathcal{F}^i_{r_l^{(i)}} : i \in I, 1 \leqq l \leqq p_i\} \bigcup \{\mathcal{G}_i^{s_l^{(i)}} : i \in I, 1 \leqq l \leqq k_i\}$$

is normal. □

3.12. In this subsection we define the concept, fundamental for the present section, of the nondegeneracy of a finite noncooperative game.

For an arbitrary situation $X = (X_1, \dots, X_n) \in \boldsymbol{X}$ we set $\xi_l^{(i)} = X_i(x_l^{(i)})$ $(i \in I, 1 \leqq l \leqq m_i)$. We suppose that the mixed strategy X_i of the player $i \in I$ lies in the interior of the face of the simplex $\boldsymbol{X}_i$ spanned by the vertices $x^{(i)}_{s_1^{(i)}}, \dots, x^{(i)}_{s_{k_i}^{(i)}}$. We set

$$\{r_1^{(i)}, \dots, r_p^{(i)}\} = \{1, \dots, m_i\} \setminus \{s_1^{(i)}, \dots, s_{k_i}^{(i)}\}.$$

We now introduce n new variables $\eta_1, \dots, \eta_n$.

Let us choose arbitrarily $\mathcal{H}_y^z \neq \emptyset$ and a situation $X^* \in \mathcal{H}_y^z$; then the point $X^* = (\xi_1^{(i)*}, \dots, \xi_{m_n}^{(n)*})$ satisfies the following system of $r + p + n$ polynomial equations in $m + n$ variables $\xi_1^{(n)}, \dots, \xi_{m_n}^{(n)}, \eta_1, \dots, \eta_n$:

$$\begin{aligned} H_i(X \parallel x_l^{(i)}) &= \eta_i \quad (i \in I, x_l^{(i)} \in z_i), \\ X_i(x_l^{(i)}) &= 0 \quad (i \in I, x_l^{(i)} \in y_i), \\ \sum_{1 \leqq l \leqq m_i} X_i(x_l^{(i)}) &= 1 \quad (i \in I). \end{aligned} \tag{3.7}$$

Definition. A game Γ is said to be nondegenerate if, for an arbitrary set $\mathcal{H}_y^z \neq \emptyset$ and every situation $X = (\xi_1^{(1)}, \dots, \xi_{m_n}^{(n)}) \in \mathcal{H}_y^z$, the Jacobian of the system (3.21) of equations has maximal rank at the point $(\xi_1^{(1)}, \dots, \xi_{m_n}^{(n)}, \eta_1, \dots, \eta_n)$. □

It follows from this definition and the implicit function theorem that the algebraic variety defined on R^{m+n} by the system (3.21) is a smooth $(m - r - p)$-dimensional manifold (in the sense of differential geometry).

Games that are not nondegenerate are called degenerate.

We may observe that, by making small changes (perturbations) in the coefficients of the polynomials $H_i(X \parallel x_l^{(i)})$ $(i \in I,\ x_l^{(i)} \in z_i)$, we can arrange that, for each situation $X \in \mathcal{H}_y^z$, the Jacobian of (3.21) has maximal rank. Consequently the set of all nondegenerate games is dense in the Euclidean space of all games with a given format. Therefore, the typical game is nondegenerate, and the special attention that we give it is justified.

3.13 Lemma. *In a game Γ the set $\mathcal{H}_y^z$ is semi-algebraic.*

If Γ is nondegenerate, the relative interior of $\mathcal{H}_y^z$ is a smooth $(m - r - p)$-dimensional manifold.

Proof. By the definition of the set $\mathcal{H}_y^z$, we have the equation

$$\mathcal{H}_y^z = \mathcal{F}_y \cap \mathcal{G}^z.$$

The set $\mathcal{F}_y$ is evidently semi-algebraic (in fact, a convex polyhedron).

By the definition of $\mathcal{G}^z$, the set $\mathcal{H}_y^z$ coincides with the projection on R^m of the intersection $\tilde{\mathcal{F}}_y \cap A$, where $\tilde{\mathcal{F}}_y = \mathcal{F}_y \times \mathrm{R}^n$ (that is, $\tilde{\mathcal{F}}_y$ is defined on the space R^{m+n}, with coordinates $\xi_1^{(1)}, \ldots, \xi_{m_n}^{(n)}, \eta_1, \ldots, \eta_n$, by the same system of inequalities as $\mathcal{F}_y$), and A is defined by the system of inequalities

$$\begin{aligned} &H_i(X \parallel x_l^{(i)}) = \eta_i \quad (x_l^{(i)} \in z_i), \\ &H_i(X \parallel x_s^{(i)}) \leqq \eta_i \quad (x_s^{(i)} \in (X \setminus z_i), i \in I). \end{aligned} \tag{3.22}$$

By a theorem of Tarski*) the projection of a semi-algebraic set is a semi-algebraic set, so that the set $\mathcal{H}_y^z$ is semi-algebraic.

Moreover, the restriction to A of a projection is injective. In fact, in the contrary case there would be two different points a and $b \in A$ whose projections coincide, that is, for some situation X^1 there would be a player i_0 such that for every strategy $x_l^{(i_0)} \in z_{i_0}$ the equations $H_{i_0}(X^1 \parallel x_l^{(i_0)}) = \eta_{i_0}^{(1)}$ and $H_{i_0}(X^1 \parallel x_l^{(i_0)}) = \eta_{i_0}^{(2)}$ would simultaneously be satisfied, whereas $\eta_{i_0}^{(1)} \neq \eta_{i_0}^{(2)}$, a contradiction. Consequently the projection on A is a homeomorphism.

If a game Γ is nondegenerate, the relative interior M of the set $\tilde{\mathcal{F}}_y \cap A$ is evidently a smooth $(m-p-r)$-dimensional manifold (see (3.21)). Moreover, the projection of M is an immersion (that is, for every point $a \in M$ the projection on R^m of a space tangent to M at the point a is an injection), since for every $i \in I$ and $x_l^{(i)} \in z_i$ the partial derivative

$$\frac{\partial(H_i(X \parallel x_l^{(i)}) - \eta_i)}{\partial \eta_i} = -1$$

is different from zero.

As we know,*) the homeomorphic image of a smooth manifold under an immersion is again a smooth manifold. Therefore the relative interior of $\mathcal{H}_y^z$ is an $m-p-r$-dimensional smooth manifold. □

Up to the end of this section, we shall assume that the game Γ is nondegenerate.

---

*) See, for example, A. Tarski, A decision method for elementary algebra and geometry, University of California Press, Berkeley, 1951.

*) See, for example, M. Hirsch, Differential topology, Springer Verlag, 1976.

3.14. It follows from Lemma 3.13 that if, in a nondegenerate game Γ, we have $p + r = m$ for some set $\mathcal{H}_y^z$, then $\mathcal{H}_y^z$ is a finite set of situations. In particular, every normal set is finite; therefore, by Lemma 3.11, there are only finitely many equilibrium situations in any nondegenerate game. Situations in a zero-dimensional set $\mathcal{H}_y^z$ will be called *vertex* situations.

It also follows from Lemma 3.13 that if, in a nondegenerate game, we have $p + r = m - 1$ for the set $\mathcal{H}_y^z$, then this set is a union of pieces of smooth curves that are disjoint, except possibly for coincident endpoints. We shall call such pieces (for all possible one-dimensional sets $\mathcal{H}_y^z$) *edges*.

Notice that no vertex can lie in the interior of an edge. In fact, let $\mathcal{H}_y^z$ be zero-dimensional and $\mathcal{H}_{y_1}^{z_1}$ one-dimensional, and let the situation $X^1 \in (\mathcal{H}_y^z \cap \operatorname{int}(\mathcal{H}_{y_1}^{z_1}))$, where int denotes the operation of passing into the relative interior. Then X^1 is the projection on R^m of a solution of the system of *equations* (3.21). Let α denote an edge in $\mathcal{H}_{y_1}^{z_1}$ such that $X^1 \in \operatorname{int}(\alpha)$. Then α is contained in the projection on R^m of the manifold $\tilde{\alpha}$ of solutions of the system obtained from (3.21) by deleting one of the first $r + p$ equations. Let us suppose that the deleted equation has the form $(F = 0)$ (i.e., $(X^1, \eta^1) \in \{F = 0\} \cap \tilde{\alpha}$). Since the game is nondegenerate, the sequence of zeros (X^1, η^1) of F, considered on $\tilde{\alpha}$, changes sign. If $(F = 0)$ were one of the first r equations in the system (3.21), this would contradict the definition of the set $\mathcal{G}^{z_1}$; otherwise, we obtain a contradiction with the assumption that the components of the mixed strategy X^1 are nonnegative.

Consequently the vertices are endpoints (i.e., boundary points) of the edges.

3.15. In what follows, we shall be interested only in vertices and edges that are connected components of our sets $\mathcal{H}_y^z$. From what was said in **3.14**, it follows that the endpoints of these edges are distinguished vertices; consequently the distinguished vertices and edges form an unoriented graph, which we call the graph of the game Γ and denote by $G(\Gamma)$.

Notice that each vertex of $G(\Gamma)$ has degree (number of edges incident at the vertex) at most two. In fact, a vertex X of $G(\Gamma)$ coincides with one of the connected components of the finite set $\mathcal{H}_y^z$, and the family

$$K = \{\mathcal{F}_{x_l^{(i)}} : x_l^{(i)} \in y_i, i \in I\} \cup \{\mathcal{G}^{x_l^{(i)}} : x_l^{(i)} \in z_i, i \in I\}$$

is distinguished, that is

$$\varphi(K) = \{1, 2, \ldots, m - 1, q\},$$

where $1 \leqq q \leqq m$. Here $q = m$ if and only if the vertex X is normal. It is evident that there exist only two distinguished proper subsets of K, provided that $q < m$, and consequently no more than two edges containing X. If, however, $q = m$, there is only one such subset; that is, a normal vertex has degree at most 1.

3.16. Let us consider any finite unoriented graph $G = (V, E)$ with set V of vertices and set E of edges, where each vertex $v \in V$ has $\deg(v) \leqq 2$.

From the set V we select a subset $N \subset V$ of vertices which we call *normal.* These vertices satisfy the following condition: if $v \in N$ then $\deg(v) \in \{0, 1\}$; if $v \in V \setminus N$, then $\deg(v) \in \{1, 2\}$. We say that a normal vertex v is *extreme* if $\deg(v) = 0$. We say that a vertex $v \in V \setminus N$ is *extreme* if $\deg(v) = 1$. An example of such a (disconnected) graph is sketched below.

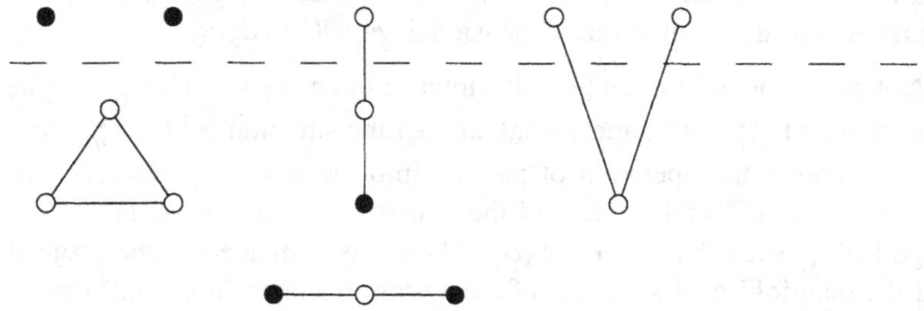

Here the normal vertices are shown in black and the extreme vertices are above the dashed line.

Lemma. *The number of normal vertices in a graph G has the same parity as the number of extreme vertices.*

Proof. Let $V = \{v_j : j = 1, \ldots, p\}$. We set

$$a_j = \begin{cases} 1 & \text{if } v_j \in N; \\ 2 & \text{if } v_j \in V \setminus N. \end{cases}$$

It is clear that the number $|N|$ of normal vertices has the same parity as $a = \sum_{1 \leqq j \leqq p} a_j$.

Let q denote the number of extreme vertices. Then

$$a = \sum_{1 \leqq j \leqq p} \deg(v_j) + q.$$

But the sum of the degrees of all the vertices is even for all graphs; consequently a is odd if q is odd, and conversely. □

For another example of a graph G we may select the graph $G(\Gamma)$ of a nondegenerate game Γ, if we take the normal vertices to be all the equilibrium situations. It follows from **3.14** and **3.15** that $G(\Gamma)$ satisfies all the necessary hypotheses.

3.17 Theorem. *The number of equilibrium situations in a nondegenerate game is odd (in particular, not zero).*

Proof. The proof will be by induction on the number n of players. For $n = 1$, the game Γ has, as is easily seen, a single equilibrium situation, which is a vertex of the polygon $\boldsymbol{X}$.

Let us suppose that every game with $n-1$ players has an odd number of equilibrium situations.

Let Γ be a game with n players. Let Γ^1 denote the game with $n-1$ players that is obtained from Γ if the set X_n is replaced by the single element $x_{m_n}^{(n)}$. Let $X^1 = (X_1, \ldots, X_{n-1})$ be an equilibrium situation in Γ^1.

As in **3.11**, we denote by K the family of sets

$$\{\mathscr{F}_{x_l^{(i)}} : x_l^{(i)} \in \boldsymbol{x}_i, i \in I\} \cup \{\mathscr{G}^{x_l^{(i)}} : i \in I, x_l^{(i)} \in \boldsymbol{x}_i\}.$$

By Lemma 3.11, the situation X^1 lies in the intersection of the elements of a subset $K_1 \subset K$ such that

$$\varphi(K_1) = \left\{1, 2, \ldots, \sum_{1 \leqq i \leqq n-1} m_i\right\}.$$

It follows that the situation $X = (X_1, \ldots, X_{n-1}, x_{m_n}^{(n)}) \in \boldsymbol{X}$ lies, in the first place, in the intersection of the elements of K_1; in the second place, in the intersection $\bigcap_{1 \leqq l \leqq m_{n-1}} \mathscr{F}_l^n$; and in the third place, in one of the sets $\mathscr{G}_n^l$ for which $1 \leqq l \leqq m_n$. Consequently this situation is in the intersection of a distinguished family of sets in K and therefore is a vertex of $G(\Gamma)$.

Let X be a normal vertex. In this case $X \in \mathscr{G}_n^{m_n}$ and also $X \in \mathscr{G}_n^l$ $(1 \leqq l \leqq m_n - 1)$. This means that a sufficiently small neighborhood of X is contained in $\mathscr{G}_n^{m_n}$. Let us suppose that there is an edge α of $G(\Gamma)$ that is incident with X. Then a section of α, of positive length, is contained in $\mathscr{G}_n^{m_n}$, and together with it, the whole edge α lies in $\mathscr{G}_n^{m_n}$. On the other hand, α must satisfy the equation $X = \alpha \cap A$, where $A \in K$, and moreover $\varphi(A) = m$ and $\alpha \not\subset A$. Since $X \notin \mathscr{F}_{m_n}^n$, the set A coincides with $\mathscr{G}_n^{m_n}$ and consequently the inclusion $\alpha \subset \mathscr{G}_n^{m_n}$ is contradictory. Therefore $\deg(X) = 0$ and, by definition, X is an extreme vertex of $G(\Gamma)$.

Now let X not be normal. This means that $X \in \mathscr{G}_n^{l_0}$, $1 \leqq l_0 \leqq m_n - 1$ and $X \notin \mathscr{G}_n^l$ for every $l \neq l_0$, $1 \leqq l \leqq m_n$. Then the only distinguished $(m-1)$-element subset of K, the intersection of whose elements contains X, is

$$K_2 = K_1 \cup \{\mathscr{F}_1^n, \ldots, \mathscr{F}_{l_0-1}^n, \mathscr{F}_{l_0+1}^n, \ldots, \mathscr{F}_{m_{n-1}}^n\} \cup \{\mathscr{G}_n^{l_0}.\}$$

The set K_2 determines the only edge incident with X. Therefore, $\deg(X) = 1$, that is, X is an extreme vertex.

We now prove that the vertices of $G(\Gamma)$, other than situations of the form $X = (X_1, \ldots, X_{n-1}, x_{m_n}^{(n)})$, where $X^1 = (X_1, \ldots, X_{n-1})$ is an equilibrium situation in Γ^1, are not extreme.

Let $X' = (X_1, \dots, X_{n-1}, x_n)$ be a vertex of $G(\Gamma)$ such that $x_n \neq x_{m_n}^{(n)}$. Then $X' \in \bigcap_{1 \leqq l \leqq m_{n-1}} \mathcal{F}_l^n$. Let, for definiteness, $X' \notin \mathcal{F}_{l_0}^n$, where $1 \leqq l_0 \leqq m_n - 1$, which is equivalent to the inclusion $x \in \mathcal{G}_n^{l_0}$.

Let us suppose first that X' is a normal vertex. Then $X' \in \bigcap_{A \in K_3} A$, where $K_3 \subset K$ and $\varphi(K_3) = \{1, 2, \dots, m\}$. If $\varphi(A_0) = m$ for some set $A_0 \in K_3$, then by the inclusion $X' \in \mathcal{G}_n^{l_0} \neq A_0$, the set $\bigcap_{A \in (K_3 \setminus \{A_0\})} A$ has, as is easily verified, the form $\mathcal{H}_y^z$. Moreover, the one-dimensional set $\mathcal{H}_y^z$ is distinguished; consequently X' is not an extreme vertex.

Now let the vertex X' not be normal. This means that there is a number l_n $(1 \leqq l_n \leqq m_n - 1)$ for which $X' \in \mathcal{G}_n^{l_n} \cap \mathcal{F}_{l_n}^n$. It follows that $l_n \neq l_0$, and consequently there are two one-dimensional sets $\mathcal{H}_{y'}^{z'}$ and $\mathcal{H}_{y''}^{z''}$, each containing the point X'. This, in turn, means that X' is not an extreme point.

By the inductive hypothesis, the number of extreme points of $G(\Gamma)$ is odd. By Lemma 3.16, it then follows that the number of normal vertices of $G(\Gamma)$ is odd, that is, the number of equilibrium situations in Γ is odd. □

§4 On the structure of the set of equilibria in finite noncooperative games

4.1. The proposition established in the preceding section can be thought of as an existence theorem for equilibrium situations in nondegenerate games. By using it, we can give another proof of Nash's theorem for general finite noncooperative games, one that does not depend on a fixed point theorem.

In fact, as we noticed in **3.12**, the set of nondegenerate games is dense in the space of all games. Consequently, for an arbitrary (possibly degenerate) game Γ of the form (3.15) and every $\varepsilon > 0$ there is a nondegenerate game

$$\Gamma^{(\varepsilon)} = \langle I, \{x_i\}_{i \in I}, \{H_i^{(\varepsilon)}\}_{i \in I} \rangle$$

that has the same set of strategies x_i as Γ, for which $|H_i^{(\varepsilon)}(x) - H_i(x)| \leqq \varepsilon$ for all $i \in I$ and $x \in \boldsymbol{x}$.

The identity mapping φ_ε of the set of players and of the sets of their strategies on themselves is evidently an ε-isomorphism (see **1.18**) of $\Gamma^{(\varepsilon)}$ onto Γ. This ε-isomorphism can be trivially extended to the mixed extensions of Γ^ε and Γ. Moreover, as $\varepsilon \downarrow 0$ the sequence $\Gamma^{(\varepsilon)}$ converges in φ_ε to Γ, and a mixed extension of $\Gamma^{(\varepsilon)}$ converges to a mixed extension of Γ.

Since each game $\Gamma^{(\varepsilon)}$ is nondegenerate, there are situations (in the mixed strategies) $X^{(\varepsilon)} \in \mathcal{C}(\Gamma^{(\varepsilon)})$ which are elements of $\boldsymbol{X}$. However, since $\boldsymbol{X}$ is compact the set $\{\varphi_\varepsilon(X^{(\varepsilon)})\}_\varepsilon$ has a limit point which, according to **2.22**, belongs to $\mathcal{C}(\Gamma)$.

4.2. The proof of Theorem 3.17 provides a basis for the construction of an algorithm for finding an equilibrium situation in any nondegenerate *bimatrix* game.

Let $\Gamma_{A,B}$ be a nondegenerate bimatrix game. The sets $\mathcal{H}_y^z$ in $\Gamma_{A,B}$ have a simple form: each one of them coincides, as we see from (3.22), with the projection of a convex polyhedron onto $\mathbf{R}^m$, and therefore is again a convex polyhedron, of the same dimension. In particular, every vertex of $G(\Gamma_{A,B})$ is the projection onto $\mathbf{R}^m$ of a single solution of a system of the form (3.20) (for which $p+r=m$) that consists of $m+n$ linear equations in $m+n$ unknowns. Similarly, each edge of $G(\Gamma_{A,B})$ is a subset of the injective projection of a line determined by a system of $m+n-1$ linear equations in $m+n$ variables. The system of equations for the vertices is obtained from the system for the edges by adjoining a single linear equation.

We call a vertex of $G(\Gamma_{A,B})$ a *primary* vertex if it corresponds to an equilibrium situation in the game obtained from $\Gamma_{A,B}$ by replacing the set X_2 by the single strategy $x_{m_2}^{(2)}$. It is clear that a primary vertex coincides with a vertex of the form $(x_{i_0}^{(1)}, x_{m_2}^{(2)})$ of the polyhedron X at which the function $H(x_i^{(1)}, x_{m_2}^{(2)})$ is maximized with respect to $x_i^{(1)}$ over the finite set X_1 and therefore it can easily be found. In particular, we obtain the system of $m+n$ linear equations that determine the vertex.

We see from the proof of Theorem 3.11 that the primary vertex X^1 is an extreme vertex of $G(\Gamma_{A,B})$, and therefore either it is normal (that is, it is an equilibrium situation), or else there is a single edge α^1 incident with it. It is easy to determine which of these alternatives actually occurs. If X^1 is an equilibrium situation, the algorithm has completed its work. In the opposite case, if we delete one of the equations in the system for X^1 we obtain a system of linear equations which yields a line containing the edge α^1. After a search through the linear equations of the form $H_i(X \parallel x_l^{(i)}) = \eta_i$ $(i \in I, 1 \leqq l \leqq m_i)$, the algorithm will reveal the other vertex of α^1.

Further, the algorithm works recursively. At the point when the next step in the algorithm is to be taken, it has found: the vertex X^1, the edge $\alpha^1 \ni X^1$, and the next vertex $X \in \alpha^1$. If X is an equilibrium situation, the algorithm has completed its work. Otherwise, it follows from the discussion in **4.1** that there is a unique edge $\alpha \neq \alpha^1$ incident with the vertex X. After deleting one of the equations for X, the algorithm finds a line containing α, and then (after a search) the other vertex of α. At this point the next step of the algorithm is complete and the algorithm proceeds to the next step. Because the graph $G(\Gamma_{A,B})$ is finite, the process terminates after a finite number of steps.

4.3. The algorithm that we have discussed has a number of deficiencies. In the first place, it is applicable only to nondegenerate matrix games. In the second place, it can find only one equilibrium situation in a game, whereas we know (see **2.16**) that in two-person zero-sum games there exist various equilibrium situations that are not equivalent.

In § 4 of Chapter 3 we shall present an algorithm that finds *all* equilibrium situations in *any* bimatrix game, and consequently is free of these deficiencies. Nevertheless, the procedure described above has an important advantage over the other: it does not require a compulsory search through operations whose number may, depending on the format of the game, increase very rapidly, and consequently the time that it requires is, on the average, appreciably less.

4.4 Stability of a finite noncooperative game. Let $\bar{m} = (m_1, \ldots, m_n)$ be an ordered set of nonnegative integers. We denote by $G_{\bar{m}}$ the class of noncooperative n-player games with the format $\bar{m}$. We shall suppose that, in each game $\Gamma \in G_{\bar{m}}$, both the players and their strategies are indexed, i.e.

$$I = \{1, 2, \ldots, n\}, \quad x_i = \{1, 2, \ldots, m_i\}.$$

This indexing imposes a lexicographical ordering on the set of situations. Then the set $G_{\bar{m}}$ becomes the space of collections $(H_1, \ldots, H_n)$ of real functions defined on the (linearly) ordered finite set x, which we identify with $\mathbf{R}^{nm_1 \ldots m_n}$.

We shall think of the collection of mixed strategies of player i as a standard $(m_i - 1)$-dimensional simplex Δ_{m_i} in $\mathbf{R}^{m_i}$. We set

$$\Delta_{\bar{m}} = \Delta_{m_i} \times \ldots \times \Delta_{m_n} \subset \mathbf{R}^{|\bar{m}|}$$

(here and later, $|\bar{m}| = m_1 + \ldots + m_n$); the polyhedron $\Delta_{\bar{m}}$ is the collection of the situations in the mixed strategies in games of type $\bar{m}$.

We also let F_2 be a two-element field,

$$\Omega_{\bar{m}} = (F_2^{m_1} \setminus \{0\}) \times \ldots \times (F_2^{m_n} \setminus \{0\}), \quad \omega = (\omega_1, \ldots, \omega_n) \in \Omega_{\bar{m}}.$$

We shall say that a situation $\xi = (\xi_1, \ldots, \xi_n) \in \Delta_{\bar{m}}$ has type ω if $\xi_{ij} = 0 \Leftrightarrow \omega_{ij} = 0$. Notice that the nonempty faces of $\Delta_{\bar{m}}$ are in a natural one-to-one correspondence with the elements of the set $\Omega_{\bar{m}}$. If $\Delta_{\bar{m}}^{\omega}$ is the face corresponding to $\omega \in \Omega_{\bar{m}}$, the situations of type ω fill relint $\Delta_{\bar{m}}^{\omega}$. For the set $C(\Gamma)$ of equilibrium situations in Γ, we set

$$C_\omega(\Gamma) = C(\Gamma) \cap \Delta_{\bar{m}}^{\omega}\,; \; s(\Gamma) = |C(\Gamma)|, s_\omega(\Gamma) = |C_\omega(\Gamma)|.$$

Evidently $C(\Gamma)$ is the disjoint union of the sets $C_\omega(\Gamma)$ for all ω.

4.5 Definition. A game $\Gamma \in G_{\bar{m}}$ is ω-stable if

1) the set $C_\omega(\Gamma)$ is finite;

2) the function $s_\omega : G_{\bar{m}} \to Z_+$ is constant in a neighborhood of Γ;

3) for every $\xi \in C_\omega(\Gamma)$ there are a neighborhood U of Γ in $G_{\bar{m}}$, a neighborhood V of ξ in $\mathbf{R}^{|\bar{m}|}$, and an analytic mapping $\varphi :\to \mathbf{R}^{|\bar{m}|}$ for which

$$C_\omega(\Gamma') \cap V = \{\varphi(\Gamma')\}$$

for every game $\Gamma' \in U$.

A game is said to be stable if it is ω-stable for some $\omega \in \Omega_{\bar{m}}$. The set of ω-stable (or stable) games of type m is denoted by $\mathrm{St}_{\bar{m},\omega}$ (or $\mathrm{St}_{\bar{m}}$). It follows from the definition of stability that each of these sets is open in the usual topology of $G_{\bar{m}} \cong \mathbf{R}^{nm_1 \ldots m_n}$. □

4.6 Definition. The subset St of $\mathbf{R}^k$ is said to be extensive if $\mathbf{R}^k - \mathrm{St}$ is a subset of a proper algebraic submanifold of $\mathbf{R}^k$, that is, St has a nonempty interior in the Zariski topology).

The complement of an extensive set has, of course, Lebesgue measure 0. A finite intersection of extensive sets is extensive. Moreover, we shall understand an expression of the form "a property is satisfied almost everywhere" to mean that this property holds on some extensive set of points. Let us show that almost all games are stable.

4.7 Theorem. *For all $\bar{m} \in N^n$ and $\omega \in \Omega_{\bar{m}}$, the set $\mathrm{St}_{\bar{m},\omega}$ is extensive in the space $G_{\bar{m}}$ of games.*

Corollary. *The set of stable games is extensive in the space of games of a given format.*

The theorem just stated can be reduced, by rather simple arguments, to the following special case.

4.8 Theorem. *Denote by* **1** *the element $(1, 1, \dots, 1)$ of the set $\Omega_{\bar{m}}$. The set $\mathrm{St}_{\bar{m},1}$ is extensive in $G_{\bar{m}}$.*

(Notice that a situation of the type **1** is precisely a completely mixed situation.)

4.9. Before we turn to the proof of this reduced theorem, we make some important remarks.

Definition. Two games Γ and $\Gamma' \in G_{\bar{m}}$ will be called equivalent (or weakly equivalent) if there are a set of positive (or respectively nonnegative) numbers $\{a_i\}_{i=1,\dots,n}$ and a set of functions

$$\alpha_i : \mathbf{X}_1 \times \ldots \times \mathbf{X}_{i-1} \times \mathbf{X}_{i+1} \times \ldots \times \mathbf{X}_n \to \mathbf{R}$$

such that for every $i \in I$

$$H_i'(x_1, \dots, x_n) = a_i H_i(x_1, \dots, x_n) + \alpha_i(x_1, \dots, x_{i-1}, x_{i+1}, \dots, x_n).$$

We shall denote equivalence by $\approx$ and weak equivalence by $\sim$. It is easy to prove the following proposition.

Lemma 1. *If $\Gamma \approx \Gamma'$ then $C(\Gamma) = C(\Gamma')$. If $\Gamma \sim \Gamma'$ then $C_1(\Gamma) = C_1(\Gamma')$.*

Let the game in (3.15) have the format $\bar{m}$. We put Γ into correspondence with the equivalent game $\Gamma' = p(\Gamma)$, with payoff function defined by the following equation:

$$H_i'(x_1, \dots, x_n) = H_i(x_1, \dots, x_n) - H_i(x_1, \dots, x_{i-1}, x_{i+1}, \dots, x_n). \tag{4.1}$$

It is clear that

$$H_i'(x_1, \dots, x_{i-1}, 1, x_{i+1}, \dots, x_n) = 0 \tag{4.2}$$

for each $i = 1, \dots, n$. We denote by $H_{\bar{m}}$ the set of games in $G_{\bar{m}}$ with the property (3.10). Then $H_{\bar{m}}$ is a coordinate subspace of $G_{\bar{m}}$ of dimension

$$M = \sum_{i=1}^{n} m_1 \dots m_{i-1}(m_i - 1)m_{i+1} \dots m_n.$$

Formula (4.1) defines a projection $p: G_{\bar m} \to H_{\bar m}$. A completely mixed equilibrium situation in the game $\Gamma' \in H_{\bar m}$ is a solution, with positive coordinates, of the system of equations

$$\begin{cases} \sum_{x_2,\ldots,x_n} H_1(k, x_2, \ldots, x_n)\xi_{2,x_2} \cdots \xi_{nx_n} = 0, \quad k = 2, \ldots, m_1, \\ \ldots\ldots\ldots\ldots\ldots\ldots\ldots\ldots\ldots\ldots\ldots\ldots\ldots\ldots\ldots\ldots \\ \sum_{x_1,\ldots,x_{n-1}} H_n(x_1, \ldots, x_{n-1}, k)\xi_{1,x_1} \cdots \xi_{(n-1)x_{n-1}} = 0, \quad k = 2, \ldots, m_n, \\ \sum_{j=1}^{m_i} \xi_{ij} = 1, \quad i = 1, \ldots, n. \end{cases} \tag{4.3}$$

(Here ξ_{ij} are the coordinates of equilibrium situations in $\mathbf{R}^{|m|}$; the pure strategies $x_i \in X_i$ play the role of integral indices.

For each point $\gamma \in H_{\bar m}$ the equations (4.3) define a manifold $W \in C^{|\bar m|}$ (the set of complex solutions of (4.3) with fixed coefficients determined by γ). Let $W_\gamma(\mathbf{R})$ be the set of real points of the manifold W_γ. Then for every $\Gamma \in G_{\bar m}$ we have

$$C_1(\Gamma) = W_\gamma(\Gamma) \cap \text{relint } \Delta_{\bar m}. \tag{4.4}$$

On the other hand, equations (4.3) determine an algebraic manifold $W_{\bar m} \subset C^M \times C^{|m|}$ (here $H_{\bar m}$ is identified with $\mathbf{R}^M$ and is embedded in C^M in the usual way; the numbers $H(x_1, \ldots, x_n)$ are to be understood in (4.3) as coordinates in C^M).

4.10 Lemma. *The manifold $W_{\bar m}$ is irreducible, and* $\dim W_{\bar m} = \dim H_{\bar m} = M$.

Proof. Let K be any algebraically closed field, X a (Zariski) closed subset of $K^r \times K^s$, π_1 the projection of $K^r \times K^s$ on K^s and $\pi = \pi_1|_X$. Let us suppose that $\pi(X) = K^s$, all the fibers of the projection $\pi^{-1}(y)$ $(y \in K^s)$ are irreducible, and that their dimensions are the same. Then X is also irreducible.

In fact, we can embed K^r in the r-dimensional projective space P^r over the field K in one of the $r+1$ usual ways. Let $\bar X$ denote the closure of X in $P^r \times K^s$. The projection $\hat\pi_1 : P^r \times K^s \to K^s$ carries closed sets to closed sets.*) It is easily verified that $\hat\pi^{-1}(y) \cap X$ is the closure of the fiber $\pi^{-1}(y)$. Consequently all fibers of the mapping $\hat\pi_1|_X$ are irreducible, and their dimensions are the same. Therefore $\bar X$ is irreducible, and consequently so is its open subset X.

Let φ and ψ denote the restrictions to $W_{\bar m}$ of the projections of $C^M \times C^{|\bar m|}$ on its factors. Let Y_i be the hyperplane in C^{m_i} defined by the equation $\sum_{j=1}^{m_i} \xi_{ij} = 1$ and let $Y = Y_1 \times \ldots \times Y_n \in C^{|\bar m|}$. The manifold $W_{\bar m}$ is contained in $C^M \times Y$. The fibers of the morphism $\psi : W_{\bar m} \to Y$ form a linear subspace of C^M with the same dimension (this is easily verified by considering the corresponding system of linear equations: see [1][1])). The conclusion of the lemma is obtained from the preceding discussion, modified for this situation.

Corollary. *For almost all $\gamma \in \mathcal{H}_{\bar m}$, we have $W_\gamma(\mathbf{R}) \cap \partial\Delta_{\bar m} = \emptyset$.*

It follows from Lemma 4.10 that there is a Zariski-open subset of points of C^M whose pre-images are zero-dimensional. (Since $W_\gamma = \psi(\varphi^{-1}(\gamma))$, formula (4.4) lets us assert that almost all games have finitely many equilibrium situations of the type 1.)

For the proof of Theorem 4.8, we need some properties of finite morphisms of algebraic manifolds.[1]) We begin with the following remark.

---

*) [1] I.R. Shafarevich, Theorem 3, §5, Chapter 1, Foundations of algebraic geometry, Nauka, Moscow, 1972.

1) [1], Chapter 1, §5, **3**, and Chapter 2, §5, **3**.

Let $f : V \to W$ be an epimorphism of irreducible algebraic varieties, $\dim V = \dim W$. Then there is a nonempty Zariski-open subset W' of W such that $f|_{f^{-1}(W')}$ is a finite mapping. If W is also normal, and the base field has characteristic 0, then W' contains an open subset W'' over which f is unramified.[2])

4.11 Proof of Theorem 4.8. Let $\mathfrak{f}_1, \dots, \mathfrak{f}_l$ be polynomials with real coefficients, generating an ideal of the manifold $W_{\overline{m}}$ in the ring $C[T_1, \dots, T_{M+|\overline{m}|}]$. Lemma 4.10 and its corollary, together with the preceding remark, ensure the existence of a Zariski-open set St in C^M for which

(a) W_γ is finite for every $\gamma \in \mathrm{St}$;

(b) $\varphi|_{\varphi^{-1}(\mathrm{St})}$ is an unramified finite mapping;

(c) $W_\gamma(\mathbf{R}) \cap \partial\Delta_{\overline{m}} = \emptyset$ for every $\gamma \in \mathrm{St}(\mathbf{R})$.

To prove Theorem 4.8, it is enough to establish that $p^{-1}(\mathrm{St}(\mathbf{R}) \subset \mathcal{S}_{\overline{m},1}$. Let $W'_{\overline{m}} = \varphi^{-1}(\mathrm{St})$, $x \in W'_{\overline{m}}$. It is easily verified[3]) that the point x is not singular on $W_{\overline{m}}$. Consequently, from the generators $\mathfrak{f}_1, \dots, \mathfrak{f}_l$ we can select $|\overline{m}|$ polynomials (we may suppose that they are $\mathfrak{f}_1, \dots, \mathfrak{f}_{|\overline{m}|}$), such that, in a neighborhood of x, the manifold $W_{\overline{m}}$ is defined by the equations $\mathfrak{f}_1 = \mathfrak{f}_2 = \dots = \mathfrak{f}_{|\overline{m}|} = 0$; and moreover, the rank of the matrix

$$A(x) = \begin{pmatrix} \frac{\partial \mathfrak{f}_i}{\partial \gamma_1}(x), \dots, & \frac{\partial \mathfrak{f}_1}{\partial \gamma_M}(x), & \frac{\partial \mathfrak{f}_1}{\partial \xi_1}(x), \dots, & \frac{\partial \mathfrak{f}_1}{\partial \xi_{|\overline{m}|}}(x) \\ & \dots & \dots & \\ \frac{\partial \mathfrak{f}_{|\overline{m}|}}{\partial \gamma_1}(x), \dots, & \frac{\partial \mathfrak{f}_{|\overline{m}|}}{\partial \gamma_M}(x), & \frac{\partial \mathfrak{f}_{|\overline{m}|}}{\partial \xi_1}(x), \dots, & \frac{\partial \mathfrak{f}_{|\overline{m}|}}{\partial \xi_{|\overline{m}|}}(x) \end{pmatrix}$$

is equal[4]) to $|\overline{m}|$. Here $\gamma_1, \dots, \gamma_m$ are coordinates in C^M, and $\xi_1, \dots, \xi_{|\overline{m}|}$ are coordinates in $C^{|\overline{m}|}$. The complex tangent space $T_x(W_{\overline{m}})$ to $W_{\overline{m}}$ at x (considered as a subset of $C^M \times C^{|\overline{m}|}$) is the set of complex solutions of the system of equations whose matrix is $A(x)$. The minor

$$A_1(x) = \left\| \frac{\partial \mathfrak{f}_i}{\partial \xi_j}(x) \right\|_{i,j=1,\dots,\overline{m}}$$

of this matrix has rank $|\overline{m}|$. In fact, if this were not the case the system of equations with the matrix $A_1(x)$ would have the non-zero solution $\xi = (\xi_1, \dots, \xi_{|\overline{m}|})$. Consider the vector

$$\zeta = (0, \dots, 0, \xi_1, \dots, \xi_{|\overline{m}|}) \in T_x(W_{\overline{m}}).$$

Then $\zeta \neq 0$ and $\zeta \in \ker d_x\varphi$, so that $d_x\varphi$ would be an isomorphism for all $x \in W'_{\overline{m}}$.*) The resulting contradiction proves the nondegeneracy of $A_1(x)$.

Now let $x_0 = (\gamma_0, \xi_0)$, $\gamma \in \mathrm{St}(\mathbf{R})$, $\xi_0 \in \mathcal{S}_1(\gamma_0)$. We use the pseudodegeneracy of $A_1(x)$ and apply the implicit function theorem to the mapping

$$f = (f_1, \dots, f_{|\overline{m}|}) : \mathbf{R}^M \times \mathbf{R}^{|\overline{m}|} \leqq \mathbf{R}^{|\overline{m}|}.$$

We conclude that there are a neighborhood U of the point γ_0 in $\mathbf{R}^m$, a neighborhood V of the point (γ_0, ξ) in $\mathbf{R}^M \times \mathbf{R}^{|\overline{m}|}$, and a mapping $g : U \to \mathbf{R}^{|\overline{m}|}$ of class C^∞, such that the conditions

$$(\gamma, \xi) \in V; \qquad f(\gamma, \xi) = 0 \tag{4.5}$$

are equivalent to

$$\gamma \in V, \qquad \xi = g(\gamma).$$

2) [1], Chapter 2, §5, Theorem 7.

3) [1], Chapter 2, §5, Theorem 3, Corollary 3.

4) [1], Chapter 2, §2, 3.

*) I.R. Shafarevich [1], Chapter 2, §5, Corollary 2 of Theorem 8.

In fact, if the neighborhoods U and V are sufficiently small, the validity of (4.5) means that $\xi \in \mathcal{S}_1(\gamma)$ (see (4.4)). Thus games in $p^{-1}(\mathrm{St}(\mathbf{R}))$ possess properties (a) and (c). It remains to prove (b).

Let

$$W_{\gamma_0} \cap \operatorname{relint} \Delta_{\bar{m}} = \{\xi_0, \ldots, \xi_n\}.$$

Let us consider a neighborhood U of the point γ_0 in $\mathrm{St}(\mathbf{R})$; pairwise disjoint neighborhoods V_i of the points (γ_0, ξ_i) in $\mathbf{R}^M \times \mathbf{R}^{|\bar{m}|}$ and a mapping g_i that has the properties described above. Let us suppose that (b) is not satisfied at the point γ_0. Then in every sufficiently small neighborhood there is a point γ_1 for which

$$W_\gamma(\mathbf{R}) \cap \operatorname{relint} \Delta_{\bar{m}} \not\supseteq \{g_0(\gamma), \ldots, g_k(\gamma)\}.$$

Let $\gamma_1, \gamma_2, \ldots \to \gamma_0$ be a convergent sequence of such points. For each γ_i we select from $W_{\gamma_0}(\mathbf{R}) \cap \operatorname{relint} \Delta_{\bar{m}}$ a point ζ_j, different from $g_0(\gamma_j), \ldots, g_k(\gamma_j)$. Since all the ζ_i lie in the compact set $\Delta_{\bar{m}}$, we may suppose that the sequence $\zeta_1, \zeta_2, \ldots$ has a limit $\zeta_0 \in \Delta_{\bar{m}}$. It is clear that $(\gamma_0, \zeta_0) \in W_{\bar{m}}(\mathbf{R})$, i.e., that $\zeta_0 \in W_{\gamma_0} \cap \Delta_{\bar{m}}$. But the set St was chosen so that when $\gamma \in \mathrm{St}(\mathbf{R})$, we have

$$W_\gamma \cap \partial\Delta_{\bar{m}} = \emptyset.$$

Therefore $\zeta \in \mathcal{S}_1(\gamma_0)$, i.e., $\zeta_0 = \zeta_g$ for some $s = 1, \ldots, k$. Consequently, when j is sufficiently large the points ζ_j lie in a neighborhood V_s in which there are no solutions except $g_s(\gamma_i)$. The resulting contradiction completes the proof of Theorem 4.8.

4.12. The theorem of **4.7** shows that the space $\mathcal{G}_{\bar{m}}$ can be constructed in the following way. In $\mathcal{G}_{\bar{m}}$ there is an extensive subset $\mathrm{St}_{\bar{m}}$ of stable games (we recall that $\mathcal{G}_{\bar{m}} - \mathrm{St}_{\bar{m}}$ is contained in an algebraic subset of $\mathcal{G}_{\bar{m}}$ of codimension at least 1). Now $\mathrm{St}_{\bar{m}}$ falls into a finite number of connected components. In each of these the number of equilibrium points (the functions s) is constant, and the coordinates of the equilibrium points depend smoothly on the game. The set $s(\mathrm{St}_{\bar{m}})$ is denoted by $\mathrm{Spec}(\bar{m})$ and is called the spectrum of the games of format $\bar{m}$. We can state that $\mathrm{Spec}(\bar{m})$ is a finite set, and $\min \mathrm{Spec}(\bar{m}) = 1$. A result that we know, that the number of equilibrium situations is odd (see Theorem 3.17), shows that $\mathrm{Spec}(\bar{m})$ consists of odd numbers. The remainder of this section will be devoted to a more detailed study of the properties of the spectrum, and in particular to an upper bound for $\max \mathrm{Spec}(\bar{m})$ as a function of $\bar{m}$.

4.13 On the number of equilibrium situations in stable noncooperative games. What was said at the end of the preceding subsection remains valid for equilibrium situations of a specified type $\omega \in \Omega_{\bar{m}}$. Let us set

$$\mathrm{Spec}_\omega(\bar{m}) = s_\omega(\mathrm{St}_{\bar{m},\omega}).$$

It follows from the theorem in **2.16** that $\mathrm{Spec}_\omega(\bar{m})$ is a finite set of nonnegative integers. Evidently

$$\mathrm{Spec}_\omega(\bar{m}) = \{n \in Z_+ \mid \mathrm{Int}(s_\omega^{-1}(n)) \neq \emptyset\}$$

(as usual, in this section Z_+ denotes the set of nonnegative integers). For $\alpha = (\alpha_1, \ldots, \alpha_n) \in Z_+^n$ we set, as above, $|\alpha| = \alpha_1 + \ldots + \alpha_n$. We denote by $C(\alpha)$ the coefficient of the product $T_1^{\alpha_1} \ldots T_n^{\alpha_n}$ in the polynomial $(T_1 + \ldots + T_n)^{|\alpha|}$. It is easily verified that $C(\alpha) = |\alpha|!/\alpha_1! \ldots \alpha_n!$ (as usual, we set 0! equal to 1). Let $\mathfrak{M}_n$ be the set of $n \times n$ square matrices with elements in Z_+. We denote by $A_{i.}$

the ith row of $A \in \mathfrak{M}_n$, and by $A_{i.}$, its ith column. For $\bar{m} = (m_1, \ldots, m_n) \in Z_+^n$, we set

$$\mathfrak{M}(\bar{m}) = \{A \in \mathfrak{M}_n | A_{i.}, \quad i = 1, \ldots, n ; \operatorname{Tr} A = 0\}$$

(notice that the condition $\operatorname{Tr} A = 0$ means that all the diagonal elements of A are 0). We introduce a function $\mu : Z_+^n \to Z_+$, by setting

$$\mu(m_1, \ldots, m_n) = \sum_{A \in \mathfrak{M}(\bar{m})} C(A_1)C(A_2)\ldots C(A_n).$$

4.14 Theorem. *Let* $\bar{m} = (m_1 + 1, \ldots, m_n + 1)$; $m_i \geqq 0$, $\omega \in \Omega_{\bar{m}}$, *and set* $\nu_i(\omega) = |\{k | \omega_{ik} = 0\}|$. *Then*

$$\max \operatorname{Spec}_\omega(\bar{m}) \leqq \mu(m_1 - \nu_1(\omega), \ldots, m_n - \nu_n(\omega)).$$

The proof is based on the transition from affine algebraic manifolds that describe completely mixed equilibrium situations to projective manifolds, followed by an application of Bezout's theorem.*)

4.15 Remark. It follows from Theorem 4.14 that the number of equilibrium situations of type ω in a stable game of format $\bar{m}$ does not exceed

$$\mu(m_1 - \nu_1(\omega), \ldots, m_n - \nu_n(\omega)).$$

Since the set of stable games of format $\bar{m}$ is extensive, we can say that this number is bounded above for almost all games in $\mathcal{G}_{\bar{m}}$. Let us give some examples.

4.16 Example. Bimatrix games ($n = 2$). In this case the set $\mathfrak{M}(m_1, m_2)$ consists of matrices of the form

$$\begin{pmatrix} 0 & \alpha_1 \\ \alpha_2 & 0 \end{pmatrix}$$

and we must have $\alpha_1 = m_1$, $\alpha_2 = m_2$, $\alpha_1 = m_2$, $\alpha_2 = m_1$. Therefore $\mathfrak{M}(m_1, m_2) \neq \emptyset \Leftrightarrow m_1 = m_1$, $\mu(m_1, m_2) = 1$. Theorem 2 lets us state that almost all bimatrix games have at most one completely mixed equilibrium situation. Moreover, if the players have equal numbers of pure strategies, there are almost always no completely mixed equilibrium situations.

4.17 Example. Dyadic games ($\bar{m} = (2, 2, \ldots, 2)$; see § 2). The matrix A lies in $\mathfrak{M}(1, \ldots, 1)$ if and only if each row and each column contain only one element equal to 1, the rest of the elements are zero, and the 1's are not on the main diagonal. Then

$$|\mathfrak{M}(1, \ldots, 1)| = (n - 1)!.$$

Since $C(0, \ldots, 0, 1, 0, \ldots, 0) = 1$, we must have $\mu(1, \ldots, 1) = (n - 1)!$.

---

*) I.R. Shafarevich [1], Chapter 4, §2.

4.18 Example. Three-player games. Notice that

$$C(\alpha_1, \alpha_2) = \binom{\alpha_1 + \alpha_2}{\alpha_1} = \binom{\alpha_1 + \alpha_2}{\alpha - 2}.$$

We take $\binom{m}{k} = 0$ for $k < 0$ or $k > m$. The matrices

$$A = \begin{pmatrix} 0 & i & m_1 - i \\ j & 0 & m_j - j \\ k & m_3 - k & 0 \end{pmatrix}$$

in $\mathfrak{M}(m_1, m_2, m_3)$ are easily described; they include all nonnegative solutions of the system of equations

$$j + k = m_1, \quad i - k = m_2 - m_3, \quad i + j = m_1 + m_2 - m_3.$$

As a result, we obtain

$$\mu(m_1, m_2, m_3) = \sum_{\substack{i+j=m_2+m_1-m_3 \\ k=m_1-j}} \binom{m_1}{i}\binom{m_2}{j}\binom{m_3}{k}. \tag{4.6}$$

In particular, $\mu(m, m, m) = \sum_{i=0}^{m} \binom{m}{i}^3$.

It follows from (4.6) that $\mu(m_1, m_2, m_3) = 0$ if and only if there is an $i \in \{1, 2, 3\}$ such that

$$m_i > m_1 + m_2 + m_3 - m_i$$

(we recall that the number of strategies of player i is $m_i + 1$).

4.19. The following theorem generalizes the preceding proposition to the case of an arbitrary number of players.

We say that the ω-spectrum of a game of type $\bar{m}$ is trivial if $\mathrm{Spec}_\omega(\bar{m}) = 0$. This means that almost all games in $G_{\bar{m}}$ fail to have an equilibrium situation of type ω. An equivalent condition is that the set $\{\Gamma \in G_{\bar{m}} \,|\, \mathcal{S}_\omega \Gamma \neq \emptyset\}$ has an empty interior in $G_{\bar{m}}$.

Theorem. *Let* $\bar{m} = (m_1, \dots, m_n) \in Z_+^n$, $m_i \geqq 1$, *and let* ν_i *be the number of zeros in the row* $\omega_i \in F_2 \setminus \{0\}$.

The following three propositions are equivalent:

1. $\mathrm{Spec}_\omega(\bar{m}) = 0$,
2. $\mu(m_1 - \nu_1(\omega) - 1, \dots, m_n - \nu_n(\omega) - 1) = 0$,
3. *For some* $i = 1, \dots, n$ *the following inequality holds:*

$$m_i - \nu_i(\bar{m}) > \sum_{j=1, j\neq i}^{n} (m_j - \nu_j(\omega)) - n + 2.$$

In a rather natural way, we can reduce the proof of the theorem to the case $\omega = \mathbf{1}$. The equivalence of the first two propositions is essentially an algebro-geometrical fact, connected with the real solutions of the system (4.3). Condition

2 with $\omega = \mathbf{1}$ is, by the definition of μ, equivalent to the emptiness of the set $\mathfrak{M}(\bar{m}-1)$ (see **4.13**). In turn, a rather complicated combinatorial argument shows that for every multi-index $\bar{k} = (k_1, \dots, k_n)$

$$\mathfrak{M}(\bar{k}) \neq \emptyset \Leftrightarrow \forall i = 1, 2, \dots, n \qquad (2k_i \leqq |\bar{k}|). \qquad \square$$

§5 The complexity of solution of finite noncooperative games

5.1. By a solution of a finite noncooperative game, we mean here a procedure for finding equilibrium situations in the game. In § 3 we described an algorithm for discovering an equilibrium situation in a nondegenerate finite noncooperative (especially) bimatrix game. In this section we discuss an algorithm for the solution of any finite game.

According to one of the equivalent definitions (see **5.13**, Chapter 1) an equilibrium situation in a finite noncooperative game is the solution of a finite system of polynomial (actually polylineal) inequalities in which the components of the mixed strategies are considered as variables. Thus, to find equilibrium situations in finite noncooperative games, it is enough to be able to solve systems of polynomial (polylineal) inequalities.

The principal content of this section is an algorithm for solving an arbitrary system of polynomial inequalities, and its application to noncooperative games. We note that at present it is not known how to make use of the particular (polylineal) form of the system (**5.4**, Chapter 1).

5.2. We begin with some general introductory remarks on the concept of an algorithm. Informally speaking, an algorithm is a finite set of prescriptions (instructions) by means of which an input (conditions of a computational problem) is transformed into an output (solution of the problem). Here it is assumed that the input is "constructive", that is, is given by a finite string of symbols of a finite number of different kinds. The number of symbols in the string is called the length of the input.

The action of the algorithm on a particular input consists of a sequence of elementary steps, each consisting of a single application of some instruction. It is plausible that the longer the input, the larger (generally speaking) the number of elementary steps that will be required. Consequently the effectiveness of the algorithm is naturally determined by the rate of increase of the number of steps as a function of the length of the input. (When the algorithm requires different numbers of steps for inputs of the same length, the value of the function is taken to be the greatest number of steps required.) This measure of effectiveness is also called the working time of the algorithm, or its temporal complexity.

Of course, the working time depends on a more precise definition of what we mean by an algorithm, for example on what is an elementary step, and how elementary it is. At the same time, in many computational problems, including the solution of inequalities, such precision is not very essential. The point is that in these problems the working time is determined up to composition with a polynomial (we may say that the complexity is polynomial in some function of the length of the input) and all reasonable specifications of the notion of the algorithm (any form of Turing machine, RAM programs, etc.) have equivalent working times up to a polynomial factor. Consequently our statements about the temporal complexity of algorithms for the solution of inequalities that relate to noncooperative games will be valid for any computational models.

5.3. We consider a system of polynomial inequalities

$$f_1 \geqq 0, \quad f_2 \geqq 0, \quad \ldots, \quad f_k \geqq 0, \tag{5.1}$$

where $f_1, \ldots, f_k \in Z[X_1, \ldots, X_n]$, that is, $f_1, \ldots, f_k$ are polynomials in n variables, with integral coefficients. Notice that these polynomials can be defined by finite strings of symbols from some finite alphabet, so that the system (5.1) is constructively defined.

Let us suppose that the degrees of the polynomials satisfy inequalities $\deg_{X_1,\ldots,X_n}(f_i) \leqq d$ for all i, $(1 \leqq i \leqq k)$, and that the absolute values of the coefficients do not exceed 2^M, for some integer $M > 0$. Then the length of the system (5.1) has the upper bound $L = Mkd^n$.

Let us take up the construction of an algorithm that will decide for a system (5.1) whether it is compatible in $\mathbf{R}^n$, and, if it is, will find a solution of the system. Here each solution will be given by a set of pairs

$$(B_1, i_1), \ldots, (B_n, i_n),$$

where each $B_j \in Z[X_j]$ is a polynomial in one variable with integral coefficients, and i_j is an open interval with rational endpoints, containing a single root x_j of B_j. The working time of the algorithm is a polynomial in $M(kd)^{c^n}$, i.e. in L^L, with some constant $c \geqq 2$.

5.4. Let $\mathbf{P} \subset \mathbf{R}[X_1, \ldots, X_n]$ be a finite set of polynomials with real coefficients. We denote by $\mathbf{U}(\mathbf{P})$ the family of all connected subsets U of $\mathbf{R}^n$, maximal under inclusion, such that every polynomial $f \in \mathbf{P}$ has constant sign on U (that is, f is, on U, either always positive, always negative, or is the constant zero). It is clear that $\mathbf{U}(\mathbf{P})$ is a partition of $\mathbf{R}^n$.

Let us first describe the algorithm informally. Its essential component is a procedure that constructs, for any finite family of polynomials $\mathbf{P} \subset Z[X_1,\ldots,X_n]$, a family

$$E_{X_i}(\mathbf{P}) \subset Z[X_1, \ldots, X_{i-1}, X_{i+1}, \ldots, X_n]$$

such that, for all sets $U \in \mathbf{U}(\mathbf{P})$, $V \in \mathbf{U}(E_{X_i}(\mathbf{P}))$, and projections

$$\pi_i : (X_1, \ldots X_n) \mapsto (X_1, \ldots, X_{i-1}, X_{i+1}, \ldots, X_n),$$

either $V \subset \pi_i(U)$ or $V \subset \pi_i(U) = \emptyset$. If this procedure is applied, recursively in i $(1 \leqq i \leqq n)$, to the family $\{f_1, \ldots, f_k\}$, the algorithm produces a family $P^{(1)}$ of polynomials in the single variable X_1, and $\mathbf{U}(\mathbf{P}^{(1)})$ consists of isolated points and open intervals. The algorithm finds, in each subset of $\mathbf{U}(\mathbf{P}^{(1)})$ at least one point, and then repeats, for each one, the recursive "descent", but with a somewhat more complicated modification, namely with a given value of X_1. As a result, it obtains a finite set of values of X_2, and so on. At the final step of this kind, it constructs a finite collection $\mathcal{T}$ of points of $\mathbf{R}^n$, and for each set $U \in \mathbf{U}(\{f_1, \ldots, f_k\})$ finds a point $x \in (\mathcal{T} \cap U)$. It then remains only to check for each element of $\mathcal{T}$ whether it satisfies (5.1).

5.5. We define constructively the operations

$$E_{X_i} : 2^{\mathbf{Z}[X_1,\ldots,X_n]} \to 2^{\mathbf{Z}[X_1,\ldots,X_{i-1},X_{i+1},\ldots,X_n]}.$$

For $1 \leqq i \leqq n$,

$$f = \sum_{0 \leqq j \leqq s} a_j X_i^j, \quad g = \sum_{0 \leqq j \leqq r} b_j X_i^j \in Z[X_1, \ldots, X_{i-1}, X_{i+1}, \ldots, X_n][X_i]$$

are polynomials in X_i with polynomial coefficients; we consider the matrix

$$M_{x_i}(f, g, p, q, t) = \underbrace{\begin{pmatrix} a_p & a_{p-1} & \cdots & a_0 & 0 \\ & \ddots & & & \ddots \\ 0 & a_p & a_{p-1} & \cdots & a_0 \\ b_q & b_{q-1} & \cdots & b_0 & 0 \\ & \ddots & & & \ddots \\ 0 & b_q & b_{q-1} & \cdots & b_0 \end{pmatrix}}_{p+q-(t-1)} \begin{matrix} \left.\vphantom{\begin{matrix} a \\ a \\ a \end{matrix}}\right\} q-(t-1) \\ \left.\vphantom{\begin{matrix} a \\ a \\ a \end{matrix}}\right\} p-(t-1) \end{matrix}$$

Let $S_{X_i}(f, g, p, q, t) = \det(MM^T)$, where $M = M_{X_i}(f, g, p, q, t)$.

In addition, for a family $\mathbf{P} \subset Z[X_1, \ldots, X_{i-1}, X_{i+1}, \ldots, X_n]\,[X_i]$ of polynomials, we denote by $C_{X_i}(\mathbf{P})$ the set of all the polynomials $g \in Z[X_1, \ldots, X_{i-1}, X_{i+1}, \ldots, X_n]$ such that g coincides with certain polynomial coefficients of one of the polynomials $f \in \mathbf{P}$.

Finally, we set, for a finite $\mathbf{P} \subset Z[X_1, \ldots, X_n]$:

$$E_{X_i}(\mathbf{P}) = C_{X_i}(\mathbf{P}) \cup \left\{ S_{X_i}(f, g, p, q, t) | f, g \in \left(\mathbf{P} \cup \frac{\partial}{\partial X_i}\mathbf{P}\right), \right.$$

$$\left. \mathbf{P} \leqq \deg_{X_i}(f), q \leqq \deg_{X_i}(g), \ 1 \leqq t \leqq \min\{p, q\} \right\},$$

where $\frac{\partial}{\partial X_i}\mathbf{P}$ denotes the collection of partial derivatives with respect to the variables X_i in all the elements of $\mathbf{P}$.

We have the following proposition.

Lemma. *) *For all sets $U \in \mathbf{U}(\mathbf{P})$, $V \in \mathbf{U}(E_{X_i}(\mathbf{P}))$, one of the following alternatives holds: either $V \in \pi_i(U)$ or $V \cap \pi_i(U) = \emptyset$.*

---

*) Compare the proof in H.R. Wüthrich, Ein Entscheidungsverfahren für die Theorie der reell-abgeschlossenen Körper, Lecture Notes in Comput. Sci. 43 (1976), pp. 138–162

5.6. Let $\mathbf{P} \subset \mathbf{R}[X_1, \ldots, X_n]$ be a finite family of polynomials. A finite set $\mathcal{T}$ of points of $\mathbf{R}^n$ is called a *system of representations* of the family $\mathbf{U}(\mathbf{P})$ if $\mathcal{T} \cap U = \emptyset$ for every $U \in \mathbf{U}(\mathbf{P})$.

We construct an auxiliary procedure which, for every family $\mathbf{P} \subset Z[X_1, \ldots, X_r, Y]$ of polynomials, and, for polynomials $B_j \in Z[X_j]$ $(1 \leqq j \leqq r)$, finds a nonzero polynomial $B_{r+1} \in Z[Y]$ such that the set of its zeros contains a system of representations of $\mathbf{U}(P(x_1, \ldots, x_r, Y))$, where x_i is a root of B_i $(1 \leqq i \leqq r)$. Here $P(x_1, \ldots, x_r, Y)$ denotes the family of polynomials obtained from the family $\mathbf{P}$ of polynomials by substituting x_i for the original variable X_i for all i, $1 \leqq i \leqq r$.

We put the family $\mathbf{P} \subset Z[X_1, \ldots, X_r, Y]$ into correspondence with the family of polynomials

$$\tilde{\mathbf{P}} = \{Y, f, f \pm g, \partial f/\partial y, f \pm 1 | f, g \in \mathbf{P}\}.$$

It is easily seen that, for every vector $(b_1, \ldots, b_r) \in \mathbf{R}^r$, the union of the sets of zeros of the polynomials in $\tilde{\mathbf{P}}(b_1, \ldots, b_r, Y)$ forms a system of representations for $\mathbf{U}(\mathbf{P}(b_1, \ldots, b_r, Y))$. In fact, every element $U \in \mathbf{U}(P(b_1, \ldots, b_r, Y))$ is either a point or an open interval with finite or infinite endpoints. In the first case, U is a root of a polynomial $f \in \mathbf{P}(b_1, \ldots, b_r, Y) \subset \tilde{\mathbf{P}}(b_1, \ldots, b_r, Y)$. Let $U = (\alpha, \beta)$, where $\alpha, \beta \in \mathbf{R}$. Then $f(\alpha) = g(\beta) = 0$ for some $f, g \in \mathbf{P}(b_1, \ldots, b_r, Y)$, so that U contains a root of the polynomial $(\partial f/\partial y)(b_1, \ldots, b_r, Y)$ if $f(b_1, \ldots, b_r, Y) = g(b_1, \ldots, b_r, Y)$, and U contains a root of $f(b_1, \ldots, b_r, Y) \pm g(b_1, \ldots, b_r, Y)$ in the contrary case. To the cases $U = (-\infty, \beta)$, $(\alpha, +\infty)$, $(-\infty, +\infty)$ there correspond roots of the polynomials $f(b_1, \ldots, b_r, Y) \pm 1, Y$.

The procedure operates by recursion on i, $0 \leqq i \leqq r$. First it forms the set $\mathbf{B}_1 = \tilde{\mathbf{P}}$. If $r = 0$, the work is complete. Otherwise, let us suppose that it has constructed the family of polynomials $\mathbf{B}_i \subset Z[X_i, \ldots, x_r]$ $(i \leqq i \leqq r)$. Then the procedure forms the set $B_{i+1} = E_{x_i}(\mathbf{B} \cup \{B_i\})$ and repeats the cycle until it has constructed the family $\mathbf{B}_{r+1} \subset Z[Y]$. Now the desired polynomial is $B_{r+1} = \prod_{f \in \mathbf{B}_{r+1}} f$.

We have not for the present verified the correctness of this procedure. To do this, we establish the following proposition by induction on i, $0 \leqq i \leqq r$: for any vector $(b_{i+1}, \ldots, b_r) \in \mathbf{R}^{r-1}$, the set of roots of the polynomials in $\mathbf{B}_{i+1}(b_{i+1}, \ldots, b_r, Y) \subset Z[Y]$ forms a system of representations for $\mathbf{U}(\mathbf{P}(x_1, \ldots, x_i, b_{i+1}, \ldots, b_r, Y))$. The starting point of the induction, at $i = 0$, was established above. Let us suppose that the proposition has been established for $i = k$. Then, in particular, if we set $b_{i+1} = x_{i+1}$ we find that for every vector $(b_{i+2}, \ldots, b_r) \in \mathbf{R}^{r-i-1}$ the set of roots of the polynomials in $\mathbf{B}_{i+1}(x_{i+1}, b_{i+2}, \ldots, b_r, Y)$ forms a system of representations for $\mathbf{U}(\mathbf{P}(x_1, \ldots, x_i, x_{i+1}, b_{i+2}, \ldots, b_r, Y))$.

Let y be a root of the polynomial $f(x_{i+1}, b_{i+2}, \ldots, b_r, y) \in \mathbf{R}[Y]$, where $f(X_{i+1}, \ldots, X_r, Y) \in \mathbf{B}_{i+1}$. By the Lemma of 5.5, there are sets $V_1, \ldots, V_s$ in the family $\mathbf{U}(B_{i+2})$ whose union $V = \bigcup_{1 \leqq j \leqq s} V_j$ coincides with the projection of the algebraic variety $U = \{(X_{i+1}, \ldots, Y)\} | B_{i+1}(X_{i+1}) = f(X_{i+1}, \ldots, Y) = 0\}$ on the space with coordinates $X_{i+2}, \ldots, X_r, Y$. Here each set is defined by a system of polynomial inequalities. The intersection

$$U \cap \bigcap_{i+2 \leqq i \leqq r} \{(X_{i+1}, \ldots, X_r, Y) | X_j = b_j\}$$

is zero-dimensional, that is, it consists of a finite set of points $(x_{i+1}, b_{i+2}, b_r, y)$. Consequently the projection

$$V \cap \bigcap_{i+2 \leqq j \leqq r} \{(X_{i+2}, \ldots, X_r, Y) | X_j = b_j\}$$

of this intersection is also zero-dimensional and contains the point $(b_{i+2}, \ldots, b_r, y)$. But a zero-dimensional system defined by a system of polynomial inequalities is contained in the set of zeros of one of the polynomials that constitute the system. Therefore y is a zero of a polynomial in $\mathbf{B}_{i+1}(b_{i+2}, \ldots, b_r, Y)$. This completes the proof of the proposition. The consistency of the procedure is assured if we set $i = r$.

5.7. We also require an auxiliary procedure which, starting from a given finite family $\mathbf{P} \subset \mathbf{Z}[X_1, \ldots, X_n]$ of polynomials and a number r, $1 \leqq r \leqq n$, constructs a family of polynomials

$$\mathbf{P}^{(1)} E_{X_n}(E_{x_{n-1}}(\ldots(E_{X_{r+1}}(\mathbf{P})\ldots) \subset \mathbf{Z}[X_1, \ldots, X_r].$$

This procedure acts recursively in correspondence with the formulas determining the operations E_{x_i} (see **5.5**).

5.8. We turn to the construction of the algorithm for the solution of the system (5.1). The algorithm begins by constructing, by recursion on r, $1 \leqq r \leqq n$, a set of polynomials $B_1, \ldots, B_n$, where $B_i \in \mathbf{Z}[X_i]$ $(i \leqq i \leqq n)$ such that the collection of vectors $(x_1, \ldots, x_n)$ forms, for all roots x_i of the polynomials B_i, a system of representations for the family $\mathbf{U}(\{f_1, \ldots, f_k\})$.

As the first step of the recursion, the algorithm finds the polynomial B_1 by first applying the procedure in **5.7** to the family $\{f_1, \ldots, f_k\}$ and the number 1, and then the procedure of **5.6** to obtain as a result a family of polynomials in the single variable X_1.

Let us suppose that at the rth step of the recursion we have constructed polynomials $B_1, \ldots, B_r$ (where $B_i \in \mathbf{Z}[X_i]$, $1 \leqq i \leqq r$), such that the collection of vectors $(x_1, \ldots, x_r)$ for all the roots x_i of B_i $(1 \leqq i \leqq r)$ forms a system of representations for the family $\mathbf{U}(\mathbf{P}^{(r)})$, where

$$\mathbf{P}^{(r)} = E_{x_n}(E_{x_{n-1}}(\ldots(E_{x_{r+1}}(\{f_1, \ldots, f_k\})\ldots).$$

Let us describe the $(r+1)$th step of the recursion. First the algorithm applies the procedure of **5.7** to the family $\{f_1, \ldots, f_k\}$ and the number $r+1$. As a result, it obtains the family of polynomials $\mathbf{P}^{r+1} \subset \mathbf{Z}[X_1, \ldots, X_r, X_{r+1}]$. The algorithm applies the procedure of **5.6** to the family $\mathbf{P}^{r+1}$ and the set of polynomials $B_1, \ldots, B_r$. In this way it constructs the polynomial $B_{r+1} \in \mathbf{Z}[X_{r+1}]$, the set of whose roots forms a system of representations of the family $\mathbf{U}(\mathbf{P}^{(r+1)}(x_1, \ldots, x_r, X_{r+1}))$, where x_i is a root of the polynomial B_i. In addition, the collection of vectors $(x_1, \ldots, x_r, x_{r+1})$ of all roots x_i of the polynomials B_i $(1 \leqq i \leqq r+1)$ is a system of representations for $\mathbf{U}(\mathbf{P}^{(r+1)})$. In fact, let the set $U \in \mathbf{U}(\mathbf{P}^{(r+1)})$; then, by the Lemma of 5.5, we have found a set $V \in U(P^{(r)})$ such that $V \subset \pi_{r+1}(U)$. By the inductive hypothesis, we have found a point $(x_1, \ldots x_r) \in V$, where x_i is a root of B_i $(1 \leqq i \leqq r)$. Consequently, by the fundamental property of B_{r+1}, one of its roots, x_{r+1}, satisfies the inclusion $(x_1, \ldots, x_r, x_{r+1}) \in U$.

At the last step of the recursion we obtain the required set of polynomials $B_1, \ldots, B_n$.

5.9. Furthermore, the algorithm has to verify that for each set $(x_1, \ldots, x_n)$ the roots of the corresponding polynomials $B_1, \ldots, B_n$ satisfy the inequality $f_i(x_1, \ldots, x_n) \geqq 0$ for every i, $1 \leqq i \leqq k$.

It is well known that every root $\alpha \neq 0$ of a polynomial

$$a_0^n Z^n + A_1 Z^{n-1} + \ldots + A_n \in \mathbf{R}[Z]$$

satisfies the inequalities

$$\left(\max_{1 \leqq j \leqq n} \left|\frac{a_j}{a_0}\right| + 1\right)^{-1} \leqq |\alpha| \leqq \max_{1 \leqq j \leqq n} \left|\frac{a_j}{a_0}\right| + 1.$$

Let the point $(x_1, \ldots, x_n) \in \mathbf{R}^n$ not be a root of f_i $(1 \leqq i \leqq k)$, that is, $f_i(x_1, \ldots, x_n) \neq 0$. To estimate the upper and lower bounds of the values of $f_i(x_1, \ldots, x_n)$ it is therefore enough to find a polynomial for which these values are attained. Such a polynomial is, for example, the result of applying the procedure in **5.6** to the family $\{Y - f_i\}$, consisting of a single polynomial, and the set $B_1, \ldots, B_n$. In fact, among the roots of the resulting polynomial B_{n+1} there is a y such that $y - f_i(x_1, \ldots, x_n) = 0$.

By a routine calculation, we can obtain the following estimate: $\deg_Y(B_{n+1}) \leqq (kd)^{c_1^n}$, where the moduli of the coefficients of the polynomial B_{n+1} do not exceed $2^{M(kd)^{c_1^n}}$, where c_1 is some integer, $c_1 \geqq 2$.

Therefore,

$$2^{-M(kd)^{c_2^n}} \leqq |f(x_1,\ldots,x_n)| \leqq 2^{M(kd)^{c_2^n}}$$

for some constant $c_2 \geqq 2$.

Let $\varepsilon = 2^{-M(kd)^{c_2^n}}$. Now the algorithm can, for each i, $1 \leqq i \leqq k$, find rational numbers $q_1,\ldots,q_n$ such that

$$f_i(x_1,\ldots,x_n) - f_i(q_1,\ldots,q_N) < \varepsilon/2. \tag{5.2}$$

In fact, if $f_i(q_1,\ldots,q_n) < \varepsilon/2$, then $|f_i(x_1,\ldots,x_n)| < \varepsilon$ and, consequently, $f_i(x_1,\ldots,x_n) = 0$. If $f_i(q_1,\ldots,q_n) \geqq \varepsilon/2$, then $f_i(x_1,\ldots,x_n) > 0$; and if $f_i(y_1,\ldots,y_n) < -\varepsilon/2$, then $f_i(x_1,\ldots,x_n) < 0$.

It turns out that in order to satisfy (5.2), it is enough to take $q_1,\ldots,q_n$ so that $|x_j - q_j| < 2^{-M(kd)^{c_3^n}}$ for some integer $c_3 \geqq 2$. This follows from the upper bound obtained above for $f_1(x_1,\ldots,x_n)$ (and therefore also for every partial derivative $(\partial f/\partial X_p)(x_1,\ldots,x_n)$, $1 \leqq p \leqq n$), and the n-dimensional version of the mean value theorem.

There are well-known procedures for the rational approximation of roots of polynomials in a single variable. The best of these*) have working time that is polynomial in the length of the input (that is, in the degree of the polynomial and the lengths of the binary representations of its coefficients). By using some such procedure, the algorithm finds, for any vector $(x_1,\ldots,x_n)$ with $B_j(x_j) = 0$ $(1 \leqq j \leqq n)$, corresponding rational numbers $q_1,\ldots,q_n$, as well as pairs $\{b_1^{(1)}, b_1^{(2)}\},\ldots,\{b_n^{(1)}, b_n^{(2)}\}$ of rational numbers, such that x_j belongs to $(b_j^{(1)}, b_j^{(2)})$ but $x_{j'} \notin (b_j^{(1)}, b_j^{(2)})$ for $j' \neq j$. In addition, it calculates $f_i(q_1,\ldots,q_n)$ for each i, $1 \leqq i \leqq k$, and then compares the resulting numbers with $\varepsilon/2$. Depending on the result of the comparison, the algorithm determines the sign of $f_i(x_1,\ldots,x_n)$. If it has found a point $(x_1,\ldots,x_n)$ such that $f_i(x_1,\ldots,x_n) \geqq 0$ for every i, $1 \leqq i \leqq k$, the system (5.1) is consistent, $(x_1,\ldots,x_n)$ is a solution, and the algorithm prints out the set of pairs $(B_1,\{b_1^{(1)}, b_1^{(2)}\}),\ldots,(B_n,\{b_n^{(1)}, b_n^{(2)}\})$. In the opposite case, the algorithm states that the system is inconsistent. We note that if the system is consistent the algorithm finds at least one point in each connected component of the set.

The working time of the algorithm is a polynomial in $M(kd)^{c^n}$ with some constant $c \geqq 2$. This verifies the direct computation. The basic reason for the double exponential in the bound for the time is as follows. Each application of the operation E_{X_i} to a recurrent family of polynomials increases the length of the polynomials in the next family, a polynomial number of times. At the same time, the number of applications of the operation E_{X_i} is at least the number n of the variables. The successive superposition of n polynomials becomes a double exponential.

5.10. We now apply the algorithm to the solution of finite noncooperative games.

In the noncooperative game

$$\Gamma = \langle I, \{X_i\}_{i\in I}, \{H_i\}_{i\in I}\rangle$$

with integral payoffs, let there be n players and not more than m pure strategies for each player, and let the absolute values of the payoffs in situations in the pure strategies not exceed 2^M.

The equilibrium situations in the mixed strategies are solutions of a system of polynomial inequalities (**5.8**). In this system the number of variables, and also the degrees of the polynomials, do not exceed 2^M. If we apply the algorithm of **5.9**

*) See, for example, L.E. Heindel, Integer arithmetic algorithms for polynomial real zero determination, J. Assoc. Comp. Mach. 18, no. 4 (1971), 533–548.

to the system (in **5.8**), we obtain a finite set of equilibrium situations. Moreover, in each connected component $\mathscr{C}(\Gamma)$ of the set of equilibrium situations, there is at least one of the points that are obtained. The equilibrium situations are obtained by the same procedure as for the solution of (5.1).

The working of the algorithm for solving the game is a polynomial in $M(nm)^{c^{nm}}$ with some constant $c \geqq 2$.

§6 Reduction to three-person games

6.1. It is clear, both a priori and from what was said above, that both the complexity and the difficulty of solving nondegenerate games (nothing said about degenerate games) generally increase rapidly as the number of players increases. It is the more surprising that the investigation of equilibrium situations for any finite noncooperative game with any (finite) number of players can be reduced to the study of equilibrium situations for certain associated finite noncooperative games for *three* players. Naturally, decreasing the number of players has to be paid for by a substantial increase in the number of strategies in the new game, so that the computational advantages after this reduction turn out to be rather problematical. In principle, the importance of this reduction is that, as it turns out, in the passage from conflicts with three participants to conflicts with any larger (but, of course, finite) number of participants, no essential additional difficulties arise. In particular, all types of equilibrium behavior of the players in noncooperative games already appear in three-person games (sometimes, of course, in disguised form).

The required reduction is based on the following elementary, but rather nontrivial, considerations.

6.2. Let us be given the noncooperative game

$$\Gamma = \langle I, \{x\}_{i\in I}, \{H_i\}_{i\in I}\rangle.$$

Without loss of generality (see **1.13** and **2.19**, Chapter 1), we may suppose that the values of the payoff functions in this game are nonnegative.

We shall consider coalitional strategies of the form

$$(x_1, \ldots, x_{j-1}, x_{j+1}, \ldots, x_i), \text{ where } j = 1, \ldots, i; \quad i = 2, \ldots, n,$$

and each $x_k \in \boldsymbol{x}_k$. We denote such a coalitional strategy by $x(i,j)$; and the set of coalitional strategies of this form (i.e., those with the same i and j), by $\boldsymbol{x}(i,j)$.

We now associate with Γ the three-person game

$$\Delta = \Delta(\Gamma) = \langle\{1,2,3,\}, \{u, v, w\}, \{K, L, M\}\rangle,$$

in which

$$u = \bigcup_{i \in I} x_i,$$

$$v = \bigcup_{i=2}^{n} \bigcup_{j=1}^{i} x(i,j) \bigcup \{\alpha\}$$

(where α is any object different from the individual or coalitional strategies in Γ),

$$w = \bigcup_{i=2}^{n} \bigcup_{j=1}^{i} x(i,j) \bigcup I,$$

and the payoff functions of the players in the situations in the game Δ are described as follows:

$$K(u,v,w) = \begin{cases} H_i(x_I), & \text{if } u \in x_i, w \in x(n,i)\,^{*)} \\ 1, & \text{if } u \in x_i, w = i, \\ 0, & \text{in the remaining cases.} \end{cases}$$

(We note that K is independent of the strategy v of the second player.)

If $v \neq \alpha$, we take

$$L(u,v,w) = \begin{cases} 1 & \text{if } v = w; \\ -1 & \text{if } u \in x_i \text{ for } i \neq 1 \text{ and } i \neq n, \\ & \qquad v \in x(i+1, i+1), \quad w \in x(i,i), \\ -1 & \text{if } u \in x_{i+1} \text{ for } i \neq 1, \\ & \qquad v \in x(i+1, i) \text{ for } j \leqq i, \quad w \in x(i,j), \\ -1 & \text{if } u \in x_2, v \in x(2,1), w \in I, \\ -1 & \text{if } u \in x_1, v \in x(2,2), w \in I, \\ 0 & \text{for all other cases when } v \neq \alpha. \end{cases}$$

If, however, $v = \alpha$, we take

$$L(u,v,w) = \begin{cases} 0 & \text{if } w \in I, \\ 0 & \text{if } u \in x_i \text{ for } i \neq 1 \text{ and } i \neq n \text{ and } w \in x(i,i), \\ 0 & \text{if } u \in x_{i+1} \text{ for } i \neq 1 \text{ and } w \in x(i,j), \\ 0 & \text{if } u \in x_i \text{ for } i = 1,2 \text{ and } w \in I, \\ -1 & \text{in the remaining cases.} \end{cases}$$

In what follows, it will be convenient to index the coalitional strategies in $\bigcup_{i,j} x(i,j)$ by positive integers, denoting strategies x by the number $p(x)$ and a strategy with index p by x^p. Let the total number of these strategies be N. Then we ascribe the index $N+1$ to strategy α of player 2. In terms of these notations,

---

*) This notation is to be understood as follows: u is some $x_i \in \mathfrak{x}_i$, and w is the coalitional strategy $x_{I \setminus i} \in \mathfrak{x}_{I \setminus i}$; then u and w together form a situation in Γ, which is just x_I.

we set

$$M(u, v, w) = \begin{cases} 1 & \text{if } p(w) = p(v) = 1, \\ 1 & \text{if } p(v) = 1 \text{ and } w \in I, \\ -1 & \text{if } u \in x_i \text{ and } w = i \text{ for } i \in I, \\ 0 & \text{in the remaining cases.} \end{cases}$$

We note that in the game Δ the payoffs to players 2 and 3 depend only on the situations involved in this game, not on the values of the payoff functions in the original game Γ.

It turns out that the equilibria of the original game Γ would be described by the equilibria of the game Δ and conversely.

6.3. Let us consider an arbitrary equilibrium situation $S = (U, V, W)$ in Δ (generally speaking, in the mixed strategies). The conditions that it is an equilibrium situation will depend on the various strategies of the players and has the following form:

The acceptability hypothesis for player 1:

$$\sum_{x_{I\setminus i} \in \mathbf{x}_{I\setminus i}} H_i(x_I) W(x(n, i)) + W(i) \leqq K(S) \tag{6.1}$$

for all $x_i \in \mathbf{x}_i$ with $i \in I$.

The acceptability hypothesis for player 2,

$$L(S \parallel v) \leqq L(S) \text{ for } v \in \mathbf{v},$$

can have the following forms:

a) If $v \in \mathbf{x}(i+1, j)$ for $j \leqq i$, $i \neq 1$, $i \neq n$, then

$$L(S \parallel v) = W(x(i+1, j)) - W(x(i, j))U(x_{i+1}) \leqq L(S). \tag{6.2}$$

b) If $v \in \mathbf{x}(i+1, i+1)$ for $i \neq 1$, $i \neq n$, then

$$L(S \parallel v) = W(x(i+1)) - W(x(i, i))U(x_i) \leqq L(S). \tag{6.3}$$

c) If $v \in \mathbf{x}(2, 1)$, then

$$L(S \parallel v) = W(x(2, 1)) - \sum_{i \in i} W(i)U(x_2) \leqq L(S). \tag{6.4}$$

d) If $v \in \mathbf{x}(2, 2)$, then

$$L(S \parallel v) = W(x(2, 2)) - W(I)U(x_i) \leqq L(S). \tag{6.5}$$

e) If $v = \alpha$, then

$$\begin{aligned} L(S \parallel v) = -1 + W(I) + \Bigg[& \sum_{i=2}^{n-1} \sum_{x \in \mathbf{x}} W(x(i, i))U(x_i) \\ & + \sum_{i=2}^{n-1} \sum_{j=1}^{i} \sum_{x \in \mathbf{x}} W(x(i, j))U(x_{i+1}) \Bigg] \\ & + W(I)U(x_I) + W(I)U(x_2) \leqq L(S). \end{aligned} \tag{6.6}$$

Finally, the acceptability hypothesis for player 3:

$$M(S \parallel w) \leqq M(S)$$

can have the following forms:

a) if $w = i \in I$, then,

$$M(S \parallel w) = V(v^1) - U(x_i) \leqq M(S); \tag{6.7}$$

b) if $w \notin I$ (say, to be specific, $w = w^{k-1}$), then

$$M(S \parallel w) = V(v^k) \leqq M(S). \tag{6.8}$$

6.4. On the basis of what has been said, we establish the following propositions.

1°. $M(S) > 0$.

In fact, it follows from (6.8) that $M(S) \geqq 0$. If we had $M(S) = 0$, we would have $V(v^k) = 0$ for every $k > 1$, and hence $V(v^1) = 1$, so that by (6.7), for every i,

$$U(x_i) \geqq 1,$$

which is evidently impossible.

2°. $L(S) = 0$.

We can see by direct calculation that the sum of the left-hand sides of (6.2)–(6.6) is zero. Consequently, if not all these left-hand sides are zero, at least one of them is positive (and therefore $L(S) > 0$), and at least one is negative. Let $L(S \parallel v^p) < 0$. Then, as we know, $V(v^p) = 0$. Suppose that $p > 1$. Then, by 1°, the corresponding inequality in (6.8) is strict. From this it follows, in turn (by (5.15), Chapter 1), that $W(w^{p-1}) = 0$. But then the inequality in (6.2)– (6.6), which corresponds to the pure strategy v^{p-1} of player 2, would be strict, and therefore we would have $V(v^{p-1}) = 0$. Repeating this argument, we see that

$$V(v^p) = \ldots = V(v^1) = 0. \tag{6.9}$$

But it follows from $V(v^1) = 0$, (6.7), and 1° that there would be strict inequality in the relations in (6.7) corresponding to the strategies of player 3 that belong to I. Therefore

$$W(i) = 0 \text{ for all } i \in I. \tag{6.10}$$

We now turn to inequality (6.6). By (6.10), this can be rewritten as

$$\sum_{i=2}^{n-1} \sum_{x \in \boldsymbol{x}} W(x(i,j))U(x_i) + \sum_{i=2}^{n-1} \sum_{j=1}^{i} \sum_{x \in \boldsymbol{x}} W(x(i,j))U(x_i) \leqq 1 + L(S).$$

On the left, we have the sum of the probabilities of some pairwise disjoint incompatible events; this cannot exceed 1. Therefore on the basis of our hypotheses this inequality is also strict. But then $V(\alpha) = 0$. It remains only to recall that $\alpha = v^{N+1}$, so that we may take $p = N + 1$ in (6.9). However, this will now mean that *all* the pure strategies of player 2 have probability zero, which is impossible.

Thus all the left-hand sides of the inequalities in (6.2)–(6.6) vanish, and our conclusion is established. Therefore we must have $L(S) = 0$ for every equilibrium situation S in Δ.

3°. We can now rewrite (6.2)–(6.5) as

$$W(x(i+1,j)) = W(x(i,j))U(x_{i+1}),$$
$$W(x(i,j)) = W(x(i-1,j))U(x_i),$$
$$\dots\dots\dots$$
$$W(x(j+1,j)) = W(x(j,j))U(x_{j+1}),$$
$$W(x(j,j)) = W(x(j-1,j-1))U(x_{j-1}),$$
$$\dots\dots\dots$$
$$W(x(3,3)) = W(x(2,2))U(x_2),$$
$$W(x(2,2)) = W(I)U(x_1),$$

from which we have

$$W(x(i+1,j)) = W(I)\prod_{\substack{k=1\\k\neq j}}^{i+1} U(x_k) \tag{6.11}$$

for every $i \leqq n-1$.

4°. It follows from (6.11) that $\max_i W(i) > 0$, since otherwise the first factor in (6.11), and with it the whole right-hand side, would vanish, that is, we would have $W(w) = 0$ for every $w \in \mathbf{w}$, which is impossible.

On this basis, inequality (6.1) can, in turn, be rewritten as

$$W(I)\sum_{x^{(i)}\in \mathbf{x}^{(i)}} H_i(x^i, x_i)\prod_{\substack{k=1\\k\neq i}}^{n} U(x_k) + W(i) \leqq K(S) \tag{6.12}$$

for all $x_i \in \mathbf{x}_i$ and $i \in I$.

5°. We now show that

$$U(\mathbf{x}_i) = 1/n \text{ for all } i \in I. \tag{6.13}$$

If we assume the contrary, then for some $i^* \in I$ we will have $U(\mathbf{x}_{i_*}) > 1/n$, and for some other $i_* \in I$ we will have $U(\mathbf{x}_{i_*}) < 1/n$. But then it will follow from (6.7) that

$$V(v^1) - U(\mathbf{x}_{i^*}) < V(v^1) - U(\mathbf{x}_{i_*}) \leqq M(S)$$

so that there is strict inequality in (6.7) when $i = i^*$. Therefore we must have

$$W(i^*) = 0. \tag{6.14}$$

Let us choose $x_{i^*} \in \mathbf{x}_{i^*}$ so that $U(\mathbf{x}_{i^*}) > 0$. Then, by (6.12), we will have

$$K(S \underset{1}{\|} x_{i^*}) = W(I)\sum_{x^i\in\mathbf{x}^i} H_{i^*}(x^i, x_i)\prod_{k\neq i^*} U(x_k) + W(i^*) \leqq K(S).$$

If we now replace all the occurrences of $H_{i^*}(x_I)$ by 1's (which are larger) and use (6.14), we obtain

$$K(S) \leqq W(I)\frac{1}{n-1}\sum_{k\neq i^*}(U(x_k))^{n-1} < W(I)\frac{1}{n-1}\sum_{k\neq i^*}U(\mathbf{x}_k)$$
$$= W(I)\prod_{k\neq i^*}\sum_{x_k\in\mathbf{x}_k} U(x_k) = W(I)\prod_{k\neq i^*}U(\mathbf{x}_k)$$

or, by Cauchy's inequality,

$$K(S) \leqq W(I)\frac{1}{n-1}\sum_{k\neq i^*}(U(x_k))^{n-1} < \frac{W(I)}{n-1}\sum_{k\neq i^*}U(x_k)$$

$$< W(I)\frac{1}{n-1}\frac{n-1}{n} = \frac{W(I)}{n} \leqq \max_i W(i).$$

Let the last maximum be attained for $i = i_0$. By 4° we have $W(i_0) > 0$. Since all the payoffs in Γ are nonnegative, it follows that

$$K(S) < W(I)\sum_{x^i\in \mathbf{x}_i} H_i(x^i, x_i)\prod_{k\neq i_0} U(x_k) + W(i_0),$$

and this contradicts (6.12).

6.5. Now let each equilibrium strategy U of player 1 in Δ correspond to a situation $X = (X_1,\ldots,X_n)$ in the original game Γ, by setting

$$X_i(x_i) = nU(x_i) \tag{6.15}$$

(recall that the range of U is $\mathbf{x}_1 \cup \ldots \cup \mathbf{x}_n$).

Let us show that this mapping is a one-to-one correspondence between equilibrium strategies of player 1 in Δ and equilibrium situations in Γ. It is just this that describes the reducibility of the determination of equilibrium situations in the n-person game Γ to the determination of equilibrium situations in the three-person game Δ.

We first show that the equilibrium property of a strategy U in Δ implies the equilibrium property of a situation X in Γ.

In fact, the equilibrium property of U in Δ implies the inequality (6.12), i.e., the relation, for every $i \in I$ and $x_i \in \mathbf{x}_i$,

$$W(I)\sum_{x^i\in \mathbf{x}^i} H_i(x^i, x_i)\prod_{k\neq 1} U(x_k) + W(i) \leqq K(S).$$

The transition from $U(x_i)$ to the corresponding $X_i(x_i)$ shows that, for all $i \in I$ and $x_i \in \mathbf{x}_i$,

$$\sum_{x^i\in \mathbf{x}^i} H_i(x^i, x_i)\prod_{k\neq i} X_k(x_k) \leqq \frac{K(S) - W(i)}{W(I)} n^{n-1} \tag{6.16}$$

or, if we denote the right-hand side by v_i,

$$H_i(X \parallel x_i) \leqq v_i. \tag{6.17}$$

Moreover, there is actually equality for at least one $x_i \in \mathbf{x}_i$. Consequently $X \in \mathscr{C}(\Gamma)$.

6.6. Now, conversely, let $X \in \mathscr{C}(\Gamma)$; set $H_i(X) = v_i$. Then (6.17) must be satisfied for all $i \in I$ and $x_i \in \mathbf{x}_i$. We define a situation $S = (U, V, W)$ in Δ by setting

$$U(x_i) = X_i(x_i)/n \text{ for all } i \in I \text{ and } x_i \in \mathbf{x}_i, \tag{6.18}$$

$$V(v^1) = \frac{M+n}{n(M+1)}, \tag{6.19}$$

$$V(v) = \frac{n-1}{n(M+1)} \text{ for all } v \neq v^1, \tag{6.20}$$

$$W(i) = \frac{1}{\Sigma}\left(1 + \frac{1}{n^{n-1}}\sum_{e=1}^{n} v_j - nv_i\right) \text{ for all } i \in I, \tag{6.21}$$

where

$$\Sigma = \sum_{j=1}^{n} jn^{2-j}.$$

For the other values of $w \in \boldsymbol{w}$ we define $W(w)$ by (6.11).

We first show that the system (U, V, W) is a situation.

It is evident that U and V are strategies of players 1 and 2.

The verification of the inequality $W(i) \geq 0$ for $n = 2$ is direct; for $n \geqq 3$ it depends on the inequalities

$$\left| \frac{1}{n^n} \left(\sum_{j=i}^{n} v_j - nv_i \right) \right| \leqq \frac{1}{n^{n-1}}$$

and

$$\Sigma = n^2 \sum_{j=1}^{n} \frac{j}{n^j} \leqq n^2 \sum_{j=1}^{\infty} \frac{j}{n^j} = n^2 \frac{1/n}{(1-1/n)^2} \leqq \frac{9}{4} n,$$

from which it follows that

$$W(i) \leqq \frac{4}{9n} \left(1 - \frac{1}{n^{n-1}} \right) > 0.$$

Finally, by (6.11),

$$W(\boldsymbol{w}) = W(I) + \sum_{w \in \boldsymbol{w} \setminus I} W(w) = W(I) + W(I) \sum_{i=2}^{n} \sum_{j=1}^{n} \prod_{k \neq j} U(x_k)$$

or, from 5°,

$$W(\boldsymbol{w}) = W(I) \left(1 + \sum_{i=2}^{n} \sum_{j=1}^{i} \prod_{k \neq j} \frac{1}{n} \right)$$
$$= \frac{1}{\Sigma} \left(n + \frac{1}{n^{n-1}} \sum_{i=1}^{n} \left(\sum_{j=1}^{n} v_j - nv_i \right) \right) \left(1 + \sum_{i=2}^{n} \frac{1}{n^{i-1}} \right) = 1.$$

It remains to show that the resulting situation $S = (U, V, W)$ is an equilibrium situation.

We return to (6.16), which states that the strategy U in Δ is an equilibrium strategy for player 1. This means that actually $H_i(X) = v_i$. Therefore, for the values x_i for which there is strict inequality in (6.16), we must have $X(x_i) = 0$, and hence, by (6.18), we also have $U(x_i) = 0$. If we replace $X_k(x_k)$ in (6.16) by $nU(x_k)$, we obtain

$$\sum_{x^i \in \boldsymbol{x}^i} H_i(x^i, x_i) \prod_{k \neq i} U(x_i) \leqq \frac{v_i}{n^{n-1}}$$

for all $i \in I$ and $x_i \in \boldsymbol{x}_i$.

If we multiply both sides of the preceding inequality by

$$\sum_{l \in I} W(l) = \frac{n}{\Sigma} > 0,$$

adding

$$W(i) = \frac{1}{\Sigma}\left(1 + \frac{1}{n^{n-1}}\left(\sum_{j=i}^{n} v_j - nv_i\right)\right)$$

shows that, for all $i \in I$ and $x_i \in \boldsymbol{x}_i$,

$$\begin{aligned}
&\sum_{x^i \in \boldsymbol{x}^i} H_i(x^i, x_i) \sum_{l \in I} W(l) \prod_{k \neq i} V(x_k) + W(i) \\
&\leqq \frac{nv_i}{n^{n-1}\Sigma} + \frac{1}{\Sigma}\left(1 + n^{n-1}\left(\sum_{j=i}^{n} v_j - nv_i\right)\right) \qquad (6.22) \\
&= \frac{1}{\Sigma}\left(1 + \frac{1}{n^{n-1}}\left(\sum_{j=i}^{n} v_j - nx_i\right)\right);
\end{aligned}$$

furthermore, if this inequality is strict, it follows that $U(x_i) = 0$. Consequently the right-hand side of (6.22) in view of (6.12) is $K(S)$, so that this inequality can be rewritten as

$$K(S \underset{1}{\|} x_i) \leqq K(S) \text{ for all } x_i \in \bigcup_{i \in I} \boldsymbol{x}_i.$$

The admissibility of S for player 1 is therefore established. Let us show that S is also admissible for player 2.

For this purpose we notice that if in (6.2)–(6.3) we substitute, for the components of the strategies U and W, their expressions by the formulas (6.11), (6.19), and (6.21) we obtain zero on the left. For example, for (6.2) we obtain

$$W(I) \prod_{k \neq j} U(x_k) - W(I) \prod_{k \neq j,n} U(x_k) U(x_n) = 0.$$

The remaining calculations are carried out similarly, the only exception being perhaps the case of (6.6), where the expression in square brackets has to be transformed:

$$\begin{aligned}
&\sum_{i=2}^{n-1} \sum_{x \in \boldsymbol{x}} W(x(i,i)) U(x_i) + \sum_{i=2}^{n-1} \sum_{j=1}^{i} \sum_{x \in \boldsymbol{x}} W(x(i,j)) U(x_{i+1}) \\
&= \sum_{i=2}^{n-1} \sum_{x \in \boldsymbol{x}} \left(W(I) \prod_{k \neq i} U(x_k) \right) U(x_i) \\
&+ \sum_{i=2}^{n-1} \sum_{j=1}^{i} \sum_{x \in \boldsymbol{x}} \left(W(I) \prod_{\substack{k \neq j \\ k=1}}^{i} U(x_k) \right) U(x_{i+1}) \\
&= \sum_{i=2}^{n-1} \sum_{j=1}^{i} \sum_{x \in \boldsymbol{x}} W(I) \prod_{\substack{k=1 \\ k \neq j}}^{i+1} U(x_k) = \sum_{i=2}^{n-1} \sum_{j=1}^{i+1} \sum_{x \in \boldsymbol{x}} W(x(i+1,j)) \\
&= 1 - W(I) - W(\boldsymbol{x}_i) - W(\boldsymbol{x}_2).
\end{aligned}$$

It is clear from this that the left-hand side of (6.6) also turns out to equal zero. But then $L(S)$, being a weighted sum of the left-hand sides of (6.2)–(6.6), is also equal to zero; this verifies the validity of the inequalities that express the admissibility of S for player 2.

Finally, to establish the admissibility of S for player 3, we notice that the left-hand sides of (6.7) and (6.8) are equal to $(n-1)/[n(N+1)]$, by (6.19) and (6.20), from which, as in the preceding case, it follows that S is admissible, and therefore an equilibrium situation.

Notes and references for Chapter 2

§1. The theorem in § 1.5 was proved N.N. Vorob'ev [25]. V.S. Bubyalis [3] gave an example of a finite game for nine players in which the set of equilibrium situations is a circumference.

The first proof of the existence theorem for equilibrium situations (in the mixed strategies) was given by J.F. Nash [2]. The existence of symmetric equilibrium situations is also proved there. For the "constructivization" of fixed point theorems see, for example, H. Skarf [1]. Diagonal games were discussed by N.N. Vorob'ev [16]. Yu. P. Ivanilov and B.M. Mukhamediev [1] described a method for the complete solution of a class of finite noncooperative games. J.C. Harsanyi [1] indicated a method for selecting a unique instance from the set of equilibrium situations in a noncooperative game.

M. Dresher [3] calculated the probability of the existence of an equilibrium situation in the pure strategies, for noncooperative games for players of a given format, for random independent values of the payoff function.

§2. A systematic discussion of dyadic games was begun by N.N. Vorob'ev in the textbook [20] (§ § 10–12, chap. 3). There he also discussed a special case of game-theoretic ecological models. D. Nowak [1] investigated 2×2-bimatrix games with forbidden situations (in the mixed strategies).

An analysis of the inverse problem for dyadic games (subsection **2.18**) was published by N.N. Vorob'ev [26] as a supplement to the theorem of V.S. Bubyalis [1].

§3. Nondegenerate matrix games were introduced by H.F. Bohnenblust, S. Karlin, and L.S. Shapley [1]. The algorithm in 3.3–3.8 for finding equilibrium situations in nondegenerate bimatrix games was found by C.E. Lemke and J.T. Howson [1]. Different versions of the proof of the oddness of the number of equilibrium situations in general finite nondegenerate games were found independently by J. Rosenmüller [2] and R.B. Wilson [1]. A different version of these considerations appears in a paper by J.C. Harsanyi [2]. N.N. Vorob'ev, Jr. found the modification presented here.

§4. The stability of finite noncooperative games, thought of only as the continuous dependence of solutions of a game, was apparently first discussed by Wu Wen-tsun and Jiang Jia-he [1]. They showed that the set of all games that are stable in this sense is dense in G. Subsequently O.A. Malafeev [4] extended this result to a wider class of games [1], and in [2] discussed the smoothness of the stability (see [5]). The results in **4.4–4.19** were obtained by A.G. Chernyakov [1–3]. The case of uniqueness of the equilibrium situation in the Theorem of 4.19 was discussed by V.L. Kreps [1–3].

§5. The basic concepts of the complexity of algorithms are presented in the book by A. Aho, J. Hopcroft and J. Ullman [1]. The first algorithm for the solution of systems of polynomial inequalities in $\mathbf{R}_n$ (as well as more general problems) was proposed by A. Tarski [1]; however, the working time of this algorithm is very large. The procedure presented here for complexity $M(kd)^{c^n}$ was given by H.K. Wütrich [1]. At present, more effective algorithms are known for solving inequalities with working time polynomial in $M(kd)^{n^2}$ (see D.Yu. Grigor'ev and N.N. Vorob'ev, Jr., [1]) and J. Renagar [1]; however, these use rather more complicated techniques. Applied to an n player game with m pure strategies and payoffs bounded in absolute value by 2^M, the latter algorithm makes it possible to calculate equilibrium situations in time that is polynomial in $M(mn)^{m^2n^2}$. This can also be used for games with non-Archimedean payoffs.

§6. The very interesting, both mathematically and philosophically, possibility of uniformly reducing the theory of equilibrium in arbitrary finite noncooperative games to that in finite three-person games was discovered and proved by V.S. Bubyalis [2, 4].

Chapter 3
Two-person zero-sum games*)

§1 Optimality in two-person zero-sum games

1.1 Two-person zero-sum games. In this chapter we discuss two-person zero-sum games, i.e. systems of the form

$$\Gamma = \langle \boldsymbol{x}, \boldsymbol{y}, H \rangle, \tag{1.1}$$

where $\boldsymbol{x}$ and $\boldsymbol{y}$ are arbitrary disjoint sets (cf. 1.1 of Chapter 1), which are called *sets of strategies* of players 1 and 2, together with $H : \boldsymbol{x} \times \boldsymbol{y} \to \mathbf{R}$, the *payoff function.* Here the pairs $(x, y) \in \boldsymbol{x} \times \boldsymbol{y}$ are called *situations* in Γ, and the number $H(x, y)$ is the *payoff* to player 1 (or the loss to player 2) in the situation (x, y). □

For any given $x \in \boldsymbol{x}$ we may consider the function $H(x, \cdot) : \boldsymbol{y} \to \mathbf{R}$, which we shall often denote, for short, by H_x. We use the notation H_y similarly.

As we noticed in 2.4, Chapter 1, in a two-person zero-sum game all situations are, in general, Pareto optimal (i.e., $\{1, 2\}$-optimal). Consequently, the question of Pareto optimality of situations does not arise in two-person zero-sum games. On the other hand, the investigation of sets of equilibrium situations (i.e., I-stable situations, or equilibrium situations in the sense of Nash), and also of sets of ε-equilibrium situations, is the subject of significant and quite nontrivial theories, which will be investigated in this chapter.

1.2 ε-equilibrium situations and strategies. Let us recall from 2.15, Chapter 1, a definition that concerns two-person zero-sum games.

Definition. Let Γ be a two-person zero-sum game as in (1.1), and let $\varepsilon = (\varepsilon_1, \varepsilon_2) \geqq 0$. A situation $(x_\varepsilon, y_\varepsilon)$ in Γ is said to be an *ε-equilibrium situation* in Γ if

$$H(x, y_\varepsilon) - \varepsilon_1 \leqq H(x_\varepsilon, y_\varepsilon) \leqq H(x_\varepsilon, y) + \varepsilon_2$$

for all $x \in \boldsymbol{x}$ and $y \in \boldsymbol{y}$.

The set of situations that are $\varepsilon = (\varepsilon_1, \varepsilon_2)$-equilibrium situations in Γ will be denoted by $\mathscr{C}^\varepsilon(\Gamma)$ or $\mathscr{C}^{(\varepsilon_1, \varepsilon_2)}(\Gamma)$.

If $(x_\varepsilon, y_\varepsilon) \in \mathscr{C}^\varepsilon(\Gamma)$, the strategy x_ε of player 1 and the strategy y_ε of player 2 are called the *ε-equilibrium* strategies of the players.

The set of ε-equilibrium strategies of player 1 in Γ will be denoted by $\mathscr{S}^\varepsilon(\Gamma)$, and the corresponding set for player 2, by $\mathscr{T}^\varepsilon(\Gamma)$. □

---

*) In Russian, usually called antagonistic games. (Translator)

Consequently,

$$\mathscr{C}^{\varepsilon}(\Gamma) \subset \mathscr{S}^{\varepsilon}(\Gamma) \times \mathscr{T}^{\varepsilon}(\Gamma).$$

As in general noncooperative games, an ε-equilibrium situation $(x_\varepsilon, y_\varepsilon)$ in a two-person zero-sum game is characterized by the property that player 1 has no interest in deviating from the strategy x_ε if doing so will incur a penalty of amount ε_1, and player 2 will not be interested in deviating from y_ε at the expense of a penalty ε_2.

The inequality (1.2) characterizes the point $(x_\varepsilon, y_\varepsilon)$ as an $\varepsilon = (\varepsilon_1, \varepsilon_2)$-saddle point for H. Consequently the ε-equilibrium situations of a game are often referred to as its *ε-saddle points*.

1.3. The ε-equilibrium strategies of the players that appear in the various ε-equilibrium situations of a given two-person zero-sum game have a sort of "approximate interchangeability".

Theorem. *If $(x_{\varepsilon'}, y_{\varepsilon'}) \in \mathscr{C}^{\varepsilon'}(\Gamma)$ and $(x_{\varepsilon''}, y_{\varepsilon''}) \in \mathscr{C}^{\varepsilon''}(\Gamma)$ where $\varepsilon' = (\varepsilon_1', \varepsilon_2')$ and $\varepsilon'' = (\varepsilon_1'', \varepsilon_2'')$, then*

$$(x_{\varepsilon'}, y_{\varepsilon''}), (x_{\varepsilon''}, y_{\varepsilon'}) \in \mathscr{C}^{(\varepsilon,\varepsilon)}(\Gamma),$$

where $\varepsilon = \varepsilon_1' + \varepsilon_2' + \varepsilon_1'' + \varepsilon_2''$.

Proof. Our hypotheses mean that

$$H(x, y_{\varepsilon'}) - \varepsilon_1' \leqq H(x_{\varepsilon'}, y_{\varepsilon'}) \leqq H(x_{\varepsilon'}, y) + \varepsilon_2' \tag{1.3}$$

for all $x \in \boldsymbol{x}$ and $y \in \boldsymbol{y}$,

$$H(x, y_{\varepsilon''}) - \varepsilon_1'' \leqq H(x_{\varepsilon''}, y_{\varepsilon''}) \leqq H(x_{\varepsilon''}, y) + \varepsilon_2'' \tag{1.4}$$

for all $x \in \boldsymbol{x}$ and $y \in \boldsymbol{y}$.

Setting $x = x_{\varepsilon''}$ and $y = y_{\varepsilon'}$ in (1.3), and $x = x_{\varepsilon'}$ in (1.4), we obtain respectively

$$H(x_{\varepsilon''}, y_{\varepsilon'}) \leqq H(x_{\varepsilon'}, y_{\varepsilon''}) + \varepsilon_1' + \varepsilon_2',$$
$$H(x_{\varepsilon'}, y_{\varepsilon''}) \leqq H(x_{\varepsilon''}, y) + \varepsilon_1'' + \varepsilon_2'',$$

i.e.

$$H(x_{\varepsilon''}, y_{\varepsilon'}) \leqq H(x_{\varepsilon''}, y) + \varepsilon_1' + \varepsilon_2' + \varepsilon_1'' + \varepsilon_2''$$

for all $y \in \boldsymbol{y}$. Similarly we obtain

$$H(x, y_{\varepsilon'}) - (\varepsilon_1' + \varepsilon_2' + \varepsilon_1'' + \varepsilon_2'') \leqq H(x_{\varepsilon''}, y_{\varepsilon'})$$

for all $x \in \boldsymbol{x}$, and the required property of $(x_{\varepsilon''}, y_{\varepsilon'})$ is established.

The case $(x_{\varepsilon'}, y_{\varepsilon''})$ can be discussed similarly. □

Corollary. $\mathscr{S}^{\varepsilon'}(\Gamma) \times \mathscr{T}^{\varepsilon''}(\Gamma) \subset \mathscr{C}^{\varepsilon'+\varepsilon''}(\Gamma)$.

Proof. If $x_{\varepsilon'} \in \mathcal{S}^{\varepsilon'}(\Gamma)$ there is a strategy $y_{\varepsilon'} \in \boldsymbol{y}$ for which $(x_{\varepsilon'}, y_{\varepsilon'}) \in \mathscr{C}^{\varepsilon'}(\Gamma)$. Similarly, $y_{\varepsilon''} \in \mathcal{T}^{\varepsilon''}(\Gamma)$ implies the existence of a strategy $x_{\varepsilon''} \in \boldsymbol{x}$ for which $(x_{\varepsilon''}, y_{\varepsilon''}) \in \mathscr{C}^{\varepsilon''}(\Gamma)$. It remains only to apply the preceding theorem. □

It follows from this corollary that in a two-person zero-sum game the ε-equilibrium property of a player's strategy emerges without reference to the other player.

1.4. In a two-person zero-sum game the players' payoffs in different ε-equilibrium situations cannot be very different.

Theorem. *Let, as above,* $(x_{\varepsilon'}, y_{\varepsilon'}) \in \mathscr{C}^{\varepsilon'}(\Gamma)$ *and* $(x_{\varepsilon''}, y_{\varepsilon''}) \in \mathscr{C}^{\varepsilon''}(\Gamma)$, *where* $\varepsilon' = (\varepsilon_1', \varepsilon_2')$. *Then*

$$|H(x_{\varepsilon'}, y_{\varepsilon'}) - H(x_{\varepsilon''}, y_{\varepsilon''})| \leqq \max\{\varepsilon_1' + \varepsilon_2', \varepsilon_1'' + \varepsilon_2''\}. \tag{1.5}$$

Proof. Setting $y = y_{\varepsilon''}$ in (1.3) and $x = x_{\varepsilon'}$ in (1.4), we obtain

$$H(x_{\varepsilon'}, y_{\varepsilon'}) \leqq H(x_{\varepsilon''}, y_{\varepsilon''}) + \varepsilon_1'' + \varepsilon_2''.$$

If we also set $x = x_{\varepsilon''}$ in (1.3) and $y = y_{\varepsilon'}$ in (1.4), we have

$$H(x_{\varepsilon''}, y_{\varepsilon''}) \leqq H(x_{\varepsilon'}, y_{\varepsilon'}) + \varepsilon_1' + \varepsilon_2'.$$

The last two inequalities together yield (1.5). □

1.5 Minimaxes. The existence of ε-equilibrium situations in Γ as ε-saddle points of the function $H : \boldsymbol{x} \times \boldsymbol{y} \to \mathbf{R}$, and some of their properties, are connected with the "minimaxes"

$$\sup_{x \in \boldsymbol{x}} \inf_{y \in \boldsymbol{y}} H(x, y) \text{ and } \inf_{y \in \boldsymbol{y}} \sup_{x \in \boldsymbol{x}} H(x, y) \tag{1.6}$$

(here, of course, if the extrema are attained they may be renamed accordingly). The first of these mixed extrema is known as the *lower value* of the game Γ in (1.1), and is ordinarily denoted by $\underline{v}_\Gamma$, and the second is the *upper value* and denoted by $\bar{v}_\Gamma$. It is clear that every two-person zero-sum game has both a lower value (possibly $-\infty$) and an upper value (possibly $+\infty$).

The informal sense of the lower and upper values of a two-person zero-sum game Γ is rather transparent. If player 1 chooses a particular strategy $x \in \boldsymbol{x}$, the payoff cannot be less than $\inf_{y \in \boldsymbol{y}} H(x, y)$. Therefore the player is guaranteed a payoff arbitrarily close (from below) to $\underline{v}_\Gamma = \sup_x \inf_y H(x, y)$. On the other hand, player 2, choosing a particular strategy $y \in \boldsymbol{y}$, loses no more than $\sup_{x \in \boldsymbol{x}} H(x, y)$, and therefore is certain of a loss no worse than $\bar{v}_\Gamma = \inf_y \sup_x H(x, y)$.

Therefore player 1 in Γ surely can obtain every sum less than $\underline{v}_\Gamma$, and player 2 is certain to lose no more than $\bar{v}_\Gamma$. In essence, it now follows that $\underline{v}_\Gamma \leqq \bar{v}_\Gamma$, but we shall also give a formal proof of this fact.

Lemma. *For all sets $\boldsymbol{x}$ and $\boldsymbol{y}$ and functions $H : \boldsymbol{x} \times \boldsymbol{y} \to \mathbf{R}$ the following "minimax inequality" holds:*

$$\sup_{x\in\boldsymbol{x}} \inf_{y\in\boldsymbol{y}} H(x,y) \leqq \inf_{y\in\boldsymbol{y}} \sup_{x\in\boldsymbol{x}} H(x,y). \tag{1.7}$$

For the proof, we consider the inequality

$$H(x,y) \leqq \sup_{x\in\boldsymbol{x}} H(x,y),$$

which holds for all $y \in \boldsymbol{y}$, and take its infimum with respect to y:

$$\inf_{y\in\boldsymbol{y}} H(x,y) \leqq \inf_{y\in\boldsymbol{y}} \sup_{x\in\boldsymbol{x}} H(x,y).$$

Since this is true for all $x \in \boldsymbol{x}$, we may take the supremum on the left, which yields (1.7). □

Corollary. *If some extrema are attained in (1.7), then if they are replaced by the corresponding maxima or minima, we obtain, for example, the inequalities*

$$\max_x \inf_y H(x,y) \leqq \min_y \sup_x H(x,y);$$

$$\max_x \min_y H(x,y) \leqq \min_y \max_x H(x,y).$$

1.6. The connection between ε-equilibrium situations in a two-person zero-sum game, on one hand, and their significance (minimaxes of their payoff functions), on the other, is described by the following theorem.

Theorem. *Let Γ be a two-person zero-sum game as in (1.1). Then*

1) *If $\varepsilon = (\varepsilon_1, \varepsilon_2) \geqq 0$, then $x_\varepsilon \in \mathcal{S}^\varepsilon(\Gamma)$ implies*

$$\inf_{y\in\boldsymbol{y}} H(x_\varepsilon, y) \geqq \bar{v}_\Gamma - (\varepsilon_1 + \varepsilon_2), \tag{1.8}$$

and $y_\varepsilon \in \mathcal{T}^\varepsilon(\Gamma)$ implies

$$\sup_{x\in\boldsymbol{x}} H(x, y_\varepsilon) \leqq \underline{v}_\Gamma + (\varepsilon_1 + \varepsilon_2). \tag{1.9}$$

The informal meaning of this is that an ε-equilibrium strategy of player 1 yields, in the worst case, a payoff less by at most $\varepsilon_1 + \varepsilon_2$ than what could generally be expected; and that there is a symmetric interpretation of an ε-equilibrium strategy for player 2.

2) *If $\mathscr{C}^\varepsilon(\Gamma) \neq \emptyset$ for every $\varepsilon = (\varepsilon_1, \varepsilon_2) > 0$, then*

$$\underline{v}_\Gamma = \sup_{x\in\boldsymbol{x}} \inf_{y\in\boldsymbol{y}} H(x,y) = \inf_{y\in\boldsymbol{y}} \sup_{x\in\boldsymbol{x}} H(x,y) = \bar{v}_\Gamma; \tag{1.10}$$

3) *If (1.8), (1.9) and (1.10) are satisfied, then $(x_\varepsilon, y_\varepsilon) \in \mathscr{C}^{\varepsilon^*}(\Gamma)$ with $\varepsilon^* = (2(\varepsilon_1 + \varepsilon_2), 2(\varepsilon_1 + \varepsilon_2))$;*

4) *If (1.10) holds, for every $\varepsilon = (\varepsilon_1, \varepsilon_2) > 0$ we have $\mathscr{C}^\varepsilon(\Gamma) \neq \emptyset$.*

Proof. 1) Let $x_\varepsilon \in \mathscr{S}^\varepsilon(\Gamma)$. This means that there is a $y_\varepsilon \in \boldsymbol{y}$ for which

$$H(x, y_\varepsilon) - \varepsilon_1 \leqq H(x_\varepsilon, y_\varepsilon) \leqq H(x_\varepsilon, y) + \varepsilon_2 \tag{1.11}$$

for every $x \in \boldsymbol{x}$ and $y \in \boldsymbol{y}$. If we take the supremum on the left, and the infimum on the right, we obtain

$$\sup_{x \in \boldsymbol{x}} H(x, y_\varepsilon) - \varepsilon_1 \leqq H(x_\varepsilon, y_\varepsilon) \leqq \inf_{y \in \boldsymbol{y}} H(x_\varepsilon, y) + \varepsilon_2, \tag{1.12}$$

from which it immediately follows that

$$\bar{v}_\Gamma = \inf_{y \in \boldsymbol{y}} \sup_{x \in \boldsymbol{x}} H(x, y) \leqq \inf_{y \in \boldsymbol{y}} H(x_\varepsilon, y) + \varepsilon_1 + \varepsilon_2. \tag{1.13}$$

We obtain (1.9) in a similar way.

2) We now turn, on the right-hand side of (1.11), to the values of functions of x_ε at suprema: $\bar{v}_\Gamma - (\varepsilon_1 + \varepsilon_2) \leqq \underline{v}_\Gamma$. Since ε_1 and ε_2 are arbitrary, it follows that $\bar{v}_\Gamma \leqq \underline{v}_\Gamma$; and since the converse inequality also holds, by the preceding lemma, (1.10) holds, and therefore so does the second part of the theorem.

3) In addition, it follows from (1.8), (1.9), and (1.10) that

$$\sup_{x \in \boldsymbol{x}} H(x, y_\varepsilon) - (\varepsilon_1 + \varepsilon_2) \leqq \underline{v}_\Gamma = \bar{v}_\Gamma \leqq \inf_{y \in \boldsymbol{y}} H(x_\varepsilon, y) + (\varepsilon_1 + \varepsilon_2), \tag{1.14}$$

i.e., for all x and y,

$$H(x, y_\varepsilon) - (\varepsilon_1 + \varepsilon_2) \leqq H(x_\varepsilon, y) + (\varepsilon_1 + \varepsilon_2). \tag{1.15}$$

In particular, if we set $x = x_\varepsilon$, we obtain

$$H(x_\varepsilon, x_\varepsilon) \leqq H(x_\varepsilon, y) + 2(\varepsilon_1 + \varepsilon_2) \text{ for every } y \in \boldsymbol{y}. \tag{1.16}$$

Similarly, setting $y = y_\varepsilon$ we have

$$H(x, y_\varepsilon) - 2(\varepsilon_1 + \varepsilon_2) \leqq H(x_\varepsilon, y_\varepsilon) \text{ for every } x \in \boldsymbol{x}. \tag{1.17}$$

Inequalities (1.16) and (1.17) give us the $(2(\varepsilon_1 + \varepsilon_2), 2(\varepsilon_1 + \varepsilon_2))$-equilibrium property of the situation $(x_\varepsilon, y_\varepsilon)$, as required.

4) Finally, for any $\varepsilon = (\varepsilon_1, \varepsilon_2) > 0$, we can always find values of x_ε and y_ε for which (1.8) and (1.9) are satisfied. Consequently, if (1.10) holds, we are in the hypotheses of the discussion just above, according to which the function H has a $(2(\varepsilon_1 + \varepsilon_2), 2(\varepsilon_1 + \varepsilon_2))$-saddle point. It remains only to notice that $\varepsilon > 0$ can be chosen arbitrarily. □

1.7 Equilibrium situations and optimal strategies. Definition. A situation (x^*, y^*) in a two-person zero-sum game Γ as in (1.1) is said to be an *equilibrium situation* (see 2.10, Chapter 1) if it is an ε-equilibrium situation with $\varepsilon = 0$, i.e. if

$$H(x, y^*) \leqq H(x^*, y^*) \leqq H(x^*, y) \text{ for all } x \in \boldsymbol{x} \text{ and } y \in \boldsymbol{y}. \tag{1.18}$$

The strategies x^* and y^* used by the players in an equilibrium situation of a two-person zero-sum game Γ are called *optimal strategies* of the players in the game. This means informally, that a player's deviation from an equilibrium situation cannot lead to an increase in the player's payoff.

The set of optimal strategies of player 1 in the two-person zero-sum game Γ will be denoted by $\mathcal{S}(\Gamma)$, and the set of optimal strategies of player 2, by $\mathcal{T}(\Gamma)$. □

Equilibrium situations in a two-person zero-sum game are saddle points of the payoff function. Hence they are also called saddle points of the game itself.

1.8. The basic propositions on equilibrium situations in two-person zero-sum games can be obtained from the preceding discussion, either by just setting $\varepsilon = 0$, or by transforming it so that this can be done.

The optimal strategies in a two-person zero-sum game have a property of interchangeability.

Theorem. *If* (x^*, y^*) *and* $(x^0, y^0) \in \mathcal{C}(\Gamma)$, *then* (x^*, y^0) *and* $(x^0, y^*) \in \mathcal{C}(\Gamma)$.

For the proof, it is enough to set $\varepsilon = 0$ in the theorem of 1.3. □

The content of this proposition can be stated as a "rectangle property" of the set of equilibrium situations in a two-person zero-sum game. Here the property of being a two-person zero-sum game is essential. For example, in a game like the "battle of the sexes" (see 2.16 of Chapter 1) the situations (b,b) and (f,f) are equilibrium situations, but (b,f) and (f,b) are not.

The rectangularity of the set of situations in a two-person zero-sum game Γ can be written as $\mathcal{C}(\Gamma) = \mathcal{S}(\Gamma) \times \mathcal{T}(\Gamma)$.

The values of the payoff function are the same at all saddle points.

Theorem. *If* $(x^*, y^*), (x^0, y^0) \in \mathcal{C}(\Gamma)$, *then* $H(x^*, y^*) = H(x^0, y^0)$.

We obtain this by setting $\varepsilon = 0$ in the theorem of 1.4. □

1.9. For equilibrium situations in a two-person zero-sum game, the minimaxes of the payoff function play roughly the same role as for ε-equilibrium situations.

Theorem. *For a two-person zero-sum game* Γ *as in* (1.1) *to have an equilibrium situation* (x^*, y^*), *it is necessary and sufficient that the outer extrema in*

$$\max_{x \in \mathfrak{x}} \inf_{y \in \mathfrak{y}} H(x, y) \text{ and } \min_{y \in \mathfrak{y}} \sup_{x \in \mathfrak{x}} H(x, y) \tag{1.19}$$

exist, are equal, and attained for the strategies x^* *and* y^*.

A proof of this theorem can be obtained by modifying the proof of the theorem of 1.6. However, it is simpler to give an independent proof.

Necessity. It follows from (1.18) that $(x^*, y^*) \in \mathcal{C}(\Gamma)$, and taking extrema yields

$$\sup_{x \in \mathfrak{x}} H(x, y^*) \leqq \inf_{y \in \mathfrak{y}} H(x^*, y), \tag{1.20}$$

so that

$$\begin{aligned}\bar{v}_\Gamma = \inf_{y \in \boldsymbol{y}} \sup_{x \in \boldsymbol{x}} H(x,y) &\leqq \sup_{x \in \boldsymbol{x}} H(x,y^*) \\ &\leqq \inf_{y \in \boldsymbol{y}} H(x^*,y) \leqq \sup_{x \in \boldsymbol{x}} \inf_{y \in \boldsymbol{y}} H(x,y) = \underline{v}_\Gamma.\end{aligned} \tag{1.21}$$

Therefore $\bar{v}_\Gamma \leqq \underline{v}_\Gamma$ which, together with $\underline{v}_\Gamma \leqq \bar{v}_\Gamma$ (see 1.5) yields $\underline{v}_\Gamma = \bar{v}_\Gamma$. Consequently all inequalities in (1.21) are actually equations and, in particular,

$$\inf_{y \in \boldsymbol{y}} \sup_{x \in \boldsymbol{x}} H(x,y) = \sup_{x \in \boldsymbol{x}} H(x,y^*),$$

i.e., here the infimum on the left is actually attained, at y^*. It can be shown similarly that the supremum of $\inf_{y \in \boldsymbol{y}} H(x,y)$ is attained, in fact at x^*. It remains only to notice that we have already shown that $\underline{v}_\Gamma = \bar{v}_\Gamma$.

Sufficiency. Let the extrema in (1.19) be equal, and attained respectively at x^* and y^*. This means that

$$\inf_{y \in \boldsymbol{y}} H(x^*,y) = \sup_{x \in \boldsymbol{x}} H(x,y^*),$$

from which we obtain

$$H(x,y^*) \leqq \sup_{x \in \boldsymbol{x}} H(x,y^*) = \inf_{y \in \boldsymbol{y}} H(x^*,y) \leqq H(x^*,y)$$

for arbitrary $x \in \boldsymbol{x}$ and $y \in \boldsymbol{y}$. If we first set $x = x^*$ on the left, and then $y = y^*$ on the right, we obtain (1.18). □

In turn, it follows immediately from this theorem that the set $\mathscr{C}(\Gamma)$ is rectangular, and that H is constant on it.

1.10 Value of a two-person zero-sum game.

Definition. If (1.10) holds:

$$\underline{v}_\Gamma = \sup_{x \in \boldsymbol{x}} \inf_{y \in \boldsymbol{y}} H(x,y) = \inf_{y \in \boldsymbol{y}} \sup_{x \in \boldsymbol{x}} H(x,y) = \bar{v}_\Gamma,$$

the common value of these mixed extrema is called the *value of the game* Γ in (1.1). □

The value of a two-person zero-sum game is usually denoted by $v(\Gamma)$ or v_Γ. Consequently $v_\Gamma = \underline{v}_\Gamma = \bar{v}_\Gamma$.

The value of a game Γ can be interpreted as a reasonable ("equitable") payment by the first player for participating in the game Γ. This point of view will be developed further in 2.11 and 2.12.

It is interesting that the value of a two-person zero-sum game, thought of as $\underline{v}_\Gamma = \sup\inf H$, specifies the largest possible payoff to player 1 in the game; and, thought of as $\bar{v}_\Gamma = \inf\sup H$, the smallest (i.e., the most stringent) upper bound for the payoff to player 1 as the payoff increases.

Theorem. *In a two-person zero-sum game as in* (1.1), *let the value* v_Γ *exist. Then:*

1) *If* $(x_\varepsilon, y_\varepsilon)$ *is an* $\varepsilon = (\varepsilon_1, \varepsilon_2)$*-saddle point of* H, then

$$H(x_\varepsilon, y) + \varepsilon_1 + \varepsilon_2 \geqq v_\Gamma \text{ for all } y \in \boldsymbol{y},$$

$$H(x, y_\varepsilon) + \varepsilon_1 + \varepsilon_2 \leqq v_\Gamma \text{ for all } x \in \boldsymbol{x},$$

and, in particular,

$$|H(x_\varepsilon, y_\varepsilon) - v_\Gamma| \leqq \varepsilon_1 + \varepsilon_2;$$

2) *If* $\varepsilon = (\varepsilon_1, \varepsilon_2) \geqq 0$ *and we have, for the strategy* $x_{\varepsilon_1} \in \boldsymbol{x}$

$$H(x_{\varepsilon_1}, y) + \varepsilon_1 \geqq v_\Gamma \text{ for every } y \in \boldsymbol{y}, \tag{1.22}$$

then for every $\varepsilon' > \varepsilon_1$ *and* $\varepsilon'' > 0$ *there is a* y_ε *such that* $(x_{\varepsilon_1}, y_\varepsilon)$ *is an* $(\varepsilon', \varepsilon'')$*-saddle point of* H.

3) *A symmetric proposition holds under the inequality*

$$H(x, y_{\varepsilon_2}) - \varepsilon_2 \leqq v_\Gamma \text{ for every } x \in \boldsymbol{x}. \tag{1.23}$$

Proof. 1) follows immediately from part 1) of the theorem of 1.6.

2) Take an arbitrary $\delta > 0$. From the existence of v_Γ, i.e., from the validity of (1.10), on the basis of part 4) of the theorem of 1.6, there is a (δ, δ)-saddle point (x_δ, y_δ). According to part 1) of the same theorem, at that point we have

$$\sup_{x \in \boldsymbol{x}} H(x, y_\delta) \leqq v_\Gamma + 2\delta, \tag{1.24}$$

$$\inf_{y \in \boldsymbol{y}} H(x_\delta, y) \geqq v_\Gamma - 2\delta. \tag{1.25}$$

From here on, we present a discussion similar to that used in the proof of part 3) of the theorem of 1.6. Inequalities (1.22) and (1.24) yield

$$v_\Gamma - \varepsilon_1 \leqq H(x_\varepsilon, y_\delta) \leqq v_\Gamma + 2\delta \tag{1.26}$$

and it follows from (1.25) that

$$v_\Gamma - 2\delta \leqq H(x_\delta, y) \text{ for every } y \in \boldsymbol{y},$$

which, together with the right-hand side of (1.26), yields

$$H(x_\varepsilon, y_\delta) \leqq H(x_\varepsilon, y) + 4\delta \text{ for every } y \in \boldsymbol{y}. \tag{1.27}$$

In addition, (1.24) implies

$$H(x, y_\delta) \leqq v_\Gamma + 2\delta \text{ for every } x \in \boldsymbol{x},$$

and, with (1.22),

$$H(x, y_\delta) - (\varepsilon_1 + 2\delta) \leqq H(x_\varepsilon, y_\delta) \text{ for every } x \in \boldsymbol{x}.$$

Combining this with (1.27), we see that $(x_\varepsilon, y_\delta)$ is an $(\varepsilon_1 + 2\delta, 4\delta)$-saddle point of Γ, and it remains only to use the arbitrariness of the positive number δ.

3) This is obtained by a symmetric argument. □

Remark. If, in the hypotheses of part 2), we add the requirement that $\mathscr{C}(\Gamma) \neq \emptyset$, the conclusion of this part can be strengthened to the existence of a $y_{\bar{\varepsilon}} \in \boldsymbol{y}$ such that $(x_{\varepsilon_1}, y_{\bar{\varepsilon}}) \in \mathscr{C}^{(\varepsilon_1,0)}(\Gamma)$.

1.11 Test for the ε-equilibrium property and optimality of strategies. We call attention to a rather simple but useful fact.

Theorem. *Let a two-person zero-sum game, as in* (1.1), *have the value* v_Γ, *and let* $x^* \in \boldsymbol{x}$ *and* $y^* \in \boldsymbol{y}$. *Then:*

1) *For* $x^* \in \mathscr{S}^\varepsilon(\Gamma)$ *it is necessary; and for every* $\varepsilon' > \varepsilon$, *for* $x^* \in \mathscr{S}^{\varepsilon'}(\Gamma)$, *it is also sufficient, that*

$$H(x^*, y) \leqq v_\Gamma - (\varepsilon_1 + \varepsilon_2) \text{ for every } y \in \boldsymbol{y}. \tag{1.28}$$

2) *For* $x^* \in \mathscr{S}(\Gamma)$, *it is necessary and sufficient that* $H(x^*, y) \geqq v_\Gamma$ *for every* $y \in \boldsymbol{y}$.

3) *Symmetric propositions are valid for the ε-equilibrium property and optimality of the strategies of player* 2.

Proof. 1) *Necessity*. For $x^* \in \mathscr{S}(\Gamma)$, let us find a strategy $y_{\varepsilon_1} \in \boldsymbol{y}$ such that

$$H(x, y_{\varepsilon_2}) - \varepsilon_1 \leqq H(x^*, y_{\varepsilon_2}) \leqq H(x^*, y) + \varepsilon_2$$

for all $x \in \boldsymbol{x}$ and $y \in \boldsymbol{y}$. Taking the supremum on the left yields

$$\sup_{x \in \boldsymbol{x}} H(x, y_{\varepsilon_2}) - \varepsilon_1 \leqq H(x^*, y) + \varepsilon_2 \text{ for every } y \in \boldsymbol{y},$$

and consequently

$$v_\Gamma - \varepsilon_1 \leqq H(x^*, y) + \varepsilon_2 \text{ for every } y \in \boldsymbol{y}.$$

The *sufficiency* follows by applying the proposition of part 2) of the theorem of 1.10 to (1.28).

2) For the proof it is enough to observe that $\mathscr{S}(\Gamma)$ is the set of strategies of player 1 for which the maximum is attained on the right-hand side of (1.19). This means that

$$v_\Gamma = \inf_{y \in \boldsymbol{y}} H(x^*, y)$$

is a necessary and sufficient condition for x^* to be optimal, as claimed.

3) is proved by a symmetric discussion. □

In what follows, when we consider $\varepsilon = (\varepsilon_1, \varepsilon_2)$-equilibrium situations in two-person zero-sum games, we shall restrict ourselves to the case $\varepsilon_1 = \varepsilon_2$; and when $\varepsilon \in \mathbf{R}_+$, by an ε-equilibrium situation we shall understand an $(\varepsilon, \varepsilon)$-equilibrium situation. With these conventions, we can restate part 1 of the preceding theorem as follows: a necessary and sufficient condition for a strategy x^* of player 1 in the game Γ to be an ε-equilibrium strategy is that the inequality

$$H(x^*, y) \geqq v_\Gamma - 2\varepsilon$$

is satisfied for every $y \in \boldsymbol{y}$.

A strategy of this kind for player 1, and a strategy y^* for player 2 that satisfies the parallel condition

$$H(x, y^*) \leqq v_\Gamma + 2\varepsilon$$

for every $x \in \boldsymbol{x}$, are not infrequently called 2ε-optimal in the literature. We shall also use this terminology, emphasizing once again that the ε-equilibrium property of a strategy (in a two-person zero-sum game) is equivalent to its 2ε-optimality property.

1.12. The following theorem provides a convenient formal test for a number to be the value of a two-person zero-sum game and for a pair of strategies in it to be optimal.

Theorem. *If, in a two-person zero-sum game* Γ *as in* (1.1), *the strategies* $x^* \in \boldsymbol{x}$ *and* $y^* \in \boldsymbol{y}$ *and the number* $v \in \mathbf{R}$ *satisfy the inequalities*

$$H(x, y^*) \leqq v \leqq H(x^*, y) \text{ for all } x \in \boldsymbol{x} \text{ and } y \in \boldsymbol{y}, \tag{1.29}$$

then $x^* \in \mathcal{S}(\Gamma)$, $y \in \mathcal{T}(\Gamma)$, *and* $v = v_\Gamma$.

Proof. It follows from (1.29) that

$$\inf_{y \in \boldsymbol{y}} \sup_{x \in \boldsymbol{x}} H(x, y) \leqq \sup_{x \in \boldsymbol{x}} \inf_{y \in \boldsymbol{y}} H(x, y), \tag{1.30}$$

from which it follows immediately that the extrema are equal, so that Γ has the value v. But then the two-sided inequality (1.30) is equivalent to (1.27) and (1.28), and it remains only to refer to the conclusion of the preceding subsection. □

1.13 Strategic and affine equivalence of two-person zero-sum games. For two-person zero-sum games, the definitions of strategic and affine equivalence (see 1.3, Chapter 1) can be put in the following form.

Definition. Two-person zero-sum games $\Gamma = \langle \boldsymbol{x}, \boldsymbol{y}, H \rangle$ and $\Gamma' = \langle \boldsymbol{x}, \boldsymbol{y}, H' \rangle$ are said to be *strategically equivalent* if there is a strictly monotonic continuous function $\varphi : \mathbf{R} \to \mathbf{R}$ for which

$$H'(x, y) = \varphi(H(x, y)) \text{ for all } x \in \boldsymbol{x} \text{ and } y \in \boldsymbol{y}.$$

The games are *affinely equivalent* if the function φ is increasing and linear:

$$\varphi(z) = kz + a \text{ with } k > 0. \qquad \square$$

The following theorem is an obvious special case of the theorem of 2.19, Chapter 1:

Theorem. *If the games* Γ *and* Γ' *are strategically equivalent, and their payoff functions* H *and* H' *satisfy* $H' = \varphi H$, *then* $\mathcal{S}(\Gamma') = \mathcal{S}(\Gamma)$, $\mathcal{T}(\Gamma') = \mathcal{T}(\Gamma)$, *and* $v_{\Gamma'} = \varphi(v_\Gamma)$.

1.14 Values of two-person zero-sum games in mixed extensions. Among the examples of noncooperative games presented in 2.16, Chapter 1, with no equilibrium situations (in the pure strategies) for sufficiently small $\varepsilon > 0$, there were two-person zero-sum games (example 4). According to the theorem of 1.6, such games cannot have values (in the pure strategies). Consequently there is a rather urgent question of extensions of two-person zero-sum games, in particular of mixed extensions of such games.

In this connection we quote, for two-person zero-sum games, some definitions and facts connected with mixed extensions of games.

Definition. Let a two-person zero-sum game as in (1.1) be measurable, i.e. let there be given, on the sets $\boldsymbol{x}$ and $\boldsymbol{y}$ of strategies, σ-algebras Ξ and H of measurable sets, together with a measurable payoff function H on the product $\boldsymbol{x} \times \boldsymbol{y}$. Then the game $\tilde{\Gamma} = \langle \boldsymbol{X}, \boldsymbol{Y}, H \rangle$ is said to be a *mixed extension* of Γ if $\boldsymbol{X}$ and $\boldsymbol{Y}$ are convex sets of measures on $\boldsymbol{x}$ and $\boldsymbol{y}$, respectively, containing all degenerate measures, and if, for all independent sets $X \in \boldsymbol{X}$ and $Y \in \boldsymbol{Y}$,

$$H(X, Y) = \int_{\boldsymbol{x}} \int_{\boldsymbol{y}} H(x, y) dY(y) dX(x). \qquad \square \tag{1.31}$$

Notice that, according to Fubini's theorem, the order in which the repeated integration is performed in (1.31) is irrelevant. Special cases of (1.31) are

$$H(X, y) = \int_{\boldsymbol{x}} H(x, y) dX(x), \tag{1.32}$$

and

$$H(x, Y) = \int_{\boldsymbol{y}} H(x, y) dY(y). \tag{1.33}$$

According to (1.31)–(1.33) (also see the corollary of 5.13 of Chapter 1), under the hypothesis that Γ is a two-person zero-sum game we must have

$$\sup_{x \in \boldsymbol{x}} H(x, Y) = \sup_{X \in \boldsymbol{X}} H(X, Y) \text{ for every } Y \in \boldsymbol{Y}, \tag{1.34}$$

$$\inf_{y \in \boldsymbol{y}} H(X, y) = \inf_{Y \in \boldsymbol{Y}} H(X, Y) \text{ for every } X \in \boldsymbol{X}, \tag{1.35}$$

and in particular

$$\sup_{x \in \boldsymbol{x}} H(x, y) = \sup_{X \in \boldsymbol{X}} H(X, y) \text{ for every } y \in \boldsymbol{y}, \tag{1.36}$$

$$\inf_{y \in \boldsymbol{y}} H(x, y) = \inf_{Y \in \boldsymbol{Y}} H(x, Y) \text{ for every } x \in \boldsymbol{x}, \tag{1.37}$$

It follows that the mixed extrema

$$\sup_{Y} \inf_{Y} H(X, Y) \text{ and } \inf_{Y} \sup_{X} H(X, Y) \tag{1.38}$$

can be written in the form

$$\sup_X \inf_y H(X, y) \text{ and } \inf_Y \sup_x H(x, Y) \tag{1.39}$$

In particular, a necessary and sufficient condition for the existence of ε-saddle points for every $\varepsilon > 0$ can be written as

$$\sup_X \inf_y H(X, y) = \inf_Y \sup_x H(x, Y), \tag{1.40}$$

and the common value of these mixed extrema is the value v_Γ of Γ.

If, in addition, one of the players has an optimal strategy, then, as follows from 1.7, the corresponding outer extremum is attained, and just at this optimal strategy.

Because of its importance, we record a simple special case as the following theorem.

Theorem. 1) *If player* 1 *in the game* Γ *in* (1.1) *has a pure optimal strategy (i.e.,* $\mathcal{S}(\Gamma) \cap \boldsymbol{x} \neq \emptyset$*), then*

$$v_\Gamma = \max_x \inf_y H(x, y);$$

2) *If player* 2 *in the game* Γ *in* (1.1) *has a pure optimal strategy (i.e.,* $\mathcal{T}(\Gamma) \cap \boldsymbol{y} \neq \emptyset$*), then*

$$v_\Gamma = \min_y \sup_x H(x, y).$$

Proof. 1) Under the hypotheses of the theorem, the supremum in the left-hand expression in (1.38), and therefore that in the left-hand expression in (1.39), are attained for a pure strategy; i.e., this supremum, which by (1.40) is equal to the value v_Γ, has the form $\max_x \inf_y H(x, y)$.

2) This is obtained by considering the right-hand expressions in (1.38) and (1.39). □

The strategies in $\mathcal{S}(\tilde{\Gamma})$ and $\mathcal{T}(\tilde{\Gamma})$ are also known as *optimal mixed strategies* of players 1 and 2 in Γ.

1.15 The value of a game is independent of its mixed extension.

Theorem. 1) *If the game* Γ *has the value* v_Γ*, then every mixed extension* $\tilde{\Gamma}$ *has a value* $v_{\tilde{\Gamma}}$*, and* $v_{\tilde{\Gamma}} = v_\Gamma$.

2) $\mathcal{C}(\Gamma) \subset \mathcal{C}(\tilde{\Gamma})$ (i.e., $\mathcal{S}(\Gamma) \subset \mathcal{S}(\tilde{\Gamma})$ *and* $\mathcal{T}(\Gamma) \subset \mathcal{T}(\tilde{\Gamma})$).

Proof. Let Γ have the value v_Γ. For an arbitrary $\varepsilon > 0$ we determine, in accordance with 4) of the theorem of 1.6, $(x_\varepsilon, y_\varepsilon) \in \mathscr{C}^\varepsilon(\Gamma)$. According to the definition of an ε-equilibrium situation, we have

$$H(x, y_\varepsilon) - \varepsilon \leqq H(x_\varepsilon, y_\varepsilon) \leqq H(x_\varepsilon, y) + \varepsilon \text{ for every } x \in \boldsymbol{x} \text{ and } y \in \boldsymbol{y}.$$

If we integrate on the left with respect to X and on the right with respect to Y, we obtain

$$H(X, y_\varepsilon) - \varepsilon \leqq H(x_\varepsilon, y_\varepsilon) \leqq H(x_\varepsilon, Y) + \varepsilon \text{ for all } X \in \boldsymbol{X} \text{ and } Y \in \boldsymbol{Y},$$

i.e. $(x_\varepsilon, y_\varepsilon) \in \mathscr{C}^\varepsilon(\Gamma)$.

From this, it follows, first, that $\mathscr{C}^\varepsilon(\Gamma) \subset \mathscr{C}^\varepsilon(\tilde{\Gamma}) \neq \emptyset$ for every $\varepsilon > 0$, so that $\tilde{\Gamma}$ has the value $v_{\tilde{\Gamma}}$.

Second, we apply (1.22) and (1.23) to each of the games Γ and $\tilde{\Gamma}$:

$$|H(x_\varepsilon, y_\varepsilon) - v_\Gamma| \leqq 2\varepsilon,$$

$$|H(x_\varepsilon, y_\varepsilon) - v_{\tilde{\Gamma}}| \leqq 2\varepsilon.$$

Then it follows that $|v_\Gamma - v_{\tilde{\Gamma}}| \leqq 4\varepsilon$ and it remains only to appeal to the arbitrariness of the positive number ε.

2) We can prove the second part of the theorem by repeating the preceding discussion with $\varepsilon > 0$. □

In what follows, we shall sometimes, when speaking of the value of a mixed extension $\tilde{\Gamma}$ of a two-person zero-sum game Γ, or of the optimal strategies of the players, disregard the extension, and simply write $\mathscr{S}(\tilde{\Gamma})$, $\mathscr{T}(\tilde{\Gamma})$, and $v_{\tilde{\Gamma}}$ as $\mathscr{S}(\Gamma)$, $\mathscr{T}(\Gamma)$, and v_Γ. This misuse of notation will not lead to any ambiguity.

1.16. As for general noncooperative games (see the remark in 5.2, Chapter 1), affine equivalence extends from two-person zero-sum games to their mixed extensions, but general strategic equivalence does not always do so. In this connection, the theorem in 1.3 can be supplemented by the following evident proposition.

Theorem. *If two two-person zero-sum games Γ and Γ' are affinely equivalent: $H'(x, y) = kH(x, y) + a$, and $\tilde{\Gamma}$ and $\tilde{\Gamma}'$ are mixed extensions of them, then $\mathscr{S}(\tilde{\Gamma}') = \mathscr{S}(\tilde{\Gamma})$ and $\mathscr{T}(\tilde{\Gamma}') = \mathscr{T}(\Gamma)$.*

1.17. As we have repeatedly noticed, the same game can acquire mixed extensions in various ways. It follows from what was just proved that if Γ has the value v_Γ then all its mixed extensions have the same value v_Γ.

This proposition can be extended to the case when the original game Γ does not have a value.

Theorem. *If $\Gamma^1 = \langle \boldsymbol{X}^1, \boldsymbol{Y}^1, H\rangle$ and $\Gamma^2 = \langle \boldsymbol{X}^2, \boldsymbol{Y}^2, H\rangle$ are two mixed extensions of the same game $\Gamma = \langle \boldsymbol{x}, \boldsymbol{y}, H\rangle$, and the values $v(\Gamma^1)$ and $v(\Gamma^2)$ exist, then $v(\Gamma^1) = v(\Gamma^2)$.*

Proof. In the games Γ^1 and Γ^2, let the mixed strategies of the players be defined on the σ-algebras $(\mathfrak{A}^1, \mathfrak{B}^1)$ and $(\mathfrak{B}^1, \mathfrak{A}^2)$. Then the payoff function H of Γ must be measurable with respect to each of the σ-algebras $\mathfrak{A}^1 \times \mathfrak{B}^1$ and $\mathfrak{A}^2 \times \mathfrak{B}^2$. Since all the σ-algebras $\mathfrak{A}^1, \mathfrak{B}^1, \mathfrak{A}^2, \mathfrak{B}^2$ are atomic, the families $\mathfrak{A}^1 \cap \mathfrak{A}^2$ and $\mathfrak{B}^1 \cap \mathfrak{B}^2$ are also atomic σ-algebras of subsets of $\boldsymbol{x}$ and $\boldsymbol{y}$. Since

$$(\mathfrak{A}^1 \times \mathfrak{B}^1) \cap (\mathfrak{A}^2 \times \mathfrak{B}^2) \subset (\mathfrak{A}^1 \cap \mathfrak{A}^2) \times (\mathfrak{B}^1 \cap \mathfrak{B}^2),$$

the payoff function H must be measurable with respect to the σ-algebra on the right.

Therefore it makes sense, for all $X \in \boldsymbol{X}$ and $Y \in \boldsymbol{Y}$, to state that

$$\int_{(\boldsymbol{x}, \mathfrak{A}^1)} H(x, y) dX(x) = \int_{(\boldsymbol{x}, \mathfrak{A}^1 \cap \mathfrak{A}^2)} H(x, y) dX(x). \tag{1.41}$$

In addition, by hypothesis and by what was said in the preceding subsection,

$$v(\Gamma') = \sup_{X \in \boldsymbol{X}} \inf_{y \in \boldsymbol{y}} \int_{(\boldsymbol{x}, \mathfrak{A}^1)} H(x, y) dX(x). \tag{1.42}$$

But it follows from (1.41) that

$$\inf_{y \in \boldsymbol{y}} \int_{(\boldsymbol{x}, \mathfrak{A}^1)} H(x, y) dX(x) = \inf_{y \in \boldsymbol{y}} \int_{(\boldsymbol{x}, \mathfrak{A}^1 \cap \mathfrak{A}^2)} H(x, y) dX(x).$$

Hence, since every measure on $\mathfrak{A}^1$ can be restricted to a measure on $\mathfrak{A}^1 \cap \mathfrak{A}^2$, we can use (1.42) to obtain

$$v(\Gamma^1) \leqq \sup_{X \in \boldsymbol{X}^1 \cap \boldsymbol{X}^2} \inf_{y \in \boldsymbol{y}} \int_{(\boldsymbol{x}, \mathfrak{A}^1 \cap \mathfrak{A}^2)} H(x, y) dX(x).$$

In the same way we can obtain

$$v(\Gamma^2) \leqq \sup_{X \in \boldsymbol{X}^1 \cap \boldsymbol{X}^2} \inf_{y \in \boldsymbol{y}} \int_{(\boldsymbol{x}, \mathfrak{A}^1 \cap \mathfrak{A}^2)} H(x, y) dX(x),$$

i.e.,

$$\max\{v(\Gamma^1), v(\Gamma^2)\} \leqq \sup_{X \in \boldsymbol{X}^1 \cap \boldsymbol{X}^2} \inf_{y \in \boldsymbol{y}} \int_{(\boldsymbol{x}, \mathfrak{A}^1 \cap \mathfrak{A}^2)} H(x, y) dX(x).$$

A parallel discussion leads to the inequality

$$\inf_{Y \in \boldsymbol{Y}^1 \cap \boldsymbol{Y}^2} \sup_{x \in \boldsymbol{x}} \int_{(\boldsymbol{y}, \mathfrak{B}^1 \cap \mathfrak{B}^2)} H(x, y) dY(y) \leqq \min\{v(\Gamma^1), v(\Gamma^2)\}.$$

The preceding two inequalities, together with the minimax inequality (1.5) and what was said in 1.14, yield

$$\max\{v(\Gamma^1), v(\Gamma^2)\} \leqq \min\{v(\Gamma^1), v(\Gamma^2)\},$$

which implies that $v(\Gamma^1) = v(\Gamma^2)$. □

1.18 Spectra of mixed strategies and complementary slackness. The theorems of 5.12–5.16, Chapter 1, restated for a two-person zero-sum game Γ as in (1.1), provide us with the following propositions.

1°. Theorem of 5.12. $\mathcal{S}(\Gamma)$ and $\mathcal{T}(\Gamma)$ are convex subsets of the spaces $\boldsymbol{X}$ and $\boldsymbol{Y}$, respectively.

2°. Theorem of 5.13. A necessary and sufficient condition for $(X^*, Y^*) \in \mathscr{C}(\Gamma)$ is that the double inequality

$$H(x, Y^*) \leqq H(X^*, Y^*) \leqq H(X^*, y)$$

is satisfied for all $x \in \boldsymbol{x}$ and $y \in \boldsymbol{y}$.

3°. Lemma of 5.14. For every situation (X, Y) in a game Γ as in (1.1):

a) player 1 has strategies x^+ and x^- which are points of the spectrum of that player's strategy X, and for which

$$H(x^+, Y) \leqq H(X, Y),$$
$$H(x^-, Y) \leqq H(X, Y);$$

b) player 2 has strategies y^+ and y^- which are points of the spectrum of that player's strategy Y, and for which

$$H(X, y^+) \leqq H(X, Y),$$
$$H(X, y^-) \geqq H(X, Y).$$

4°. Theorem of 5.15 ("complementary slackness"). In a two-person zero-sum game Γ as in (1.1), let $(X, Y^*) \in \mathscr{C}(\Gamma)$. Then

a$_1$) if x^0 is a point of the spectrum X^*, and the payoff functions $H(\cdot, Y^*)$ is continuous at x^0, then

$$H(x^0, Y^*) = v_\Gamma; \tag{1.43}$$

a$_2$) if $X^*(x^0) > 0$, then (1.43) holds;

b$_1$) if y^0 is a point of the spectrum Y^*, and the payoff function $H(X^*, \cdot)$ is continuous at y^0, then

$$H(X^*, y^0) = v_\Gamma; \tag{1.44}$$

b$_2$) if $Y^*(y^0) > 0$, then (1.44) holds.

5°. The equalizing of strategies. If, in a two-person zero-sum game Γ as in (1.1), a strategy X^* of a player (say, of the first player) is equalized, i.e. if $H(X^*, y)$ is independent of $y \in \boldsymbol{y}$, then $X^* \in \mathcal{S}(\Gamma)$.

§2 Basis of the maximin principle

2.1 The maximin principle and a method for establishing it. In a two-person zero-sum game

$$\Gamma = \langle \boldsymbol{x}, \boldsymbol{y}, H \rangle \tag{2.1}$$

player 1, having chosen the strategy $x \in \boldsymbol{x}$, cannot obtain less than $\inf_y H(x, y)$, an amount that is the guaranteed payoff if strategy x is used (cf. 1.4). Consequently the largest guaranteed payoff is $\max_x \inf_y H(x, y)$, and is obtained if the outer maximum in this expression is attained at the strategy x. This approach to the maximal guaranteed payoff is known as the *maximin principle*, and the strategy applied here is known as a *maximin* strategy.

On the other hand, in the same game Γ the largest possible loss by player 2 under strategy y is $\sup_x H(x, y)$, and this will be least when player 2 uses the *minimax* strategy y, for which the quantity $\min_y \sup_x H(x, y)$ attains its minimum.

If the extrema in (1.19) are equal, the maximin strategy of player 1 and the minimax strategy of player 2 form an equilibrium situation, i.e. a saddle point of Γ.

In Chapter 1 we discussed $\mathfrak{K}$-optimality principles in noncooperative games, and some of their variants, but without any formal arguments about these principles. We merely gave some a priori or intuitive arguments about the reasonableness of this approach to optimality. We might say that there we took the optimality principle in a form in which it is encountered "naturally" (i.e., according to human psychology), or at least might be encountered. At the same time, this concept of optimality is not normative: different people have, depending on the particular features of their psychology, their upbringing, their professions, or even their age, very different ideas of what they would consider to be an optimality principle.

Two conclusions follow from what we have just said. In the first place, in connection with one and the same class of games, it is perfectly reasonable for people to have different ideas of optimality (as was already noticed in Chapter 1). In the second place, it is desirable to make the most convincing case possible for the choice of an optimality principle as the norm in games of the kind under discussion. The most natural procedure is to formulate separate partial requirements that ought to enter into an optimality principle. These requirements will be axiomatic in nature, and an optimality principle will appear as a theorem in such an axiomatic system. The proof of such a theorem involves, on one hand, the noncontradictory nature of the axiom system; and on the other hand, the establishment of an optimality principle for games of a given class. To say that an optimality principle is unique is to say that the axiom system is complete.

As we noticed in 2.1, Chapter 1, a natural (and in a sense the only) optimality principle for noncooperative games with a single player, i.e. not really a game-theoretic principle, is to maximize the payoff function for the single player in the game. Consequently it is natural to suppose that an optimality principle for some class of games should consist of some kind of maximization. Here we encounter the open question of *just what* should be maximized. Depending on our answer to this question, we may obtain various optimality principles, or at least different versions of the validity of some one principle.

In the theory of two-person zero-sum games, the most natural (although not at all the only) optimality principle is the maximin principle, which consists of trying to maximize the minimal ("guaranteed") payoff to a player. Various axiomatizations will lead to this principle. All of them depend in one way or another on the idea of choosing the maximal (sometimes appearing as "largest" or "most preferable"), and justifies the expediency of maximizing the minimum.

We are now going to present three axiom systems for maximin principles; these deal with the axiomatization of different concepts of reasonable behavior of the players.

In this connection, let us recall that the payoff function in a two-person zero-sum game does not describe a gain for player 2, but a loss. Consequently, from the point of view of player 2, a maximin principle appears as a minimax principle, i.e. as the minimization of the maximum (possible) loss.

2.2 Axiomatization of the concept of a player's opportunities. Let us formalize the idea of the feasible possibilities for a player in a game and its subsets. It is natural to think of a possibility for a player as the possibility of doing something and at the same time achieving something; realizability of some possibility for a player depends, aside from anything else, on the set of strategies available to the opponent. Consequently we may speak of realization functions of the possibilities of player 1:

$$U : 2^{\boldsymbol{y}} \to 2^{\boldsymbol{x} \times \mathbf{R}}$$

and of player 2:

$$V : 2^{\boldsymbol{x}} \to 2^{\boldsymbol{y} \times \mathbf{R}}$$

in a two-person zero-sum game of the form (2.1).

For the sake of definiteness, we shall present the discussion from the point of view of player 1. The discussion for player 2 is the mirror image of this.

Definition. A realization function U for the possibilities of player 1 in a two-person zero-sum game Γ as in (1.1) is said to be *admissible* if it has the following properties (i.e., axioms):

1°. *Opportunity under definite conditions.* For each strategy $y^0 \in \boldsymbol{y}$,

$$U(y^0) = \{\langle x, h\rangle : x \in \boldsymbol{x},\ h \leqq H(x, y^0)\};$$

2°. *Monotonicity.* If $\boldsymbol{y}' \subset \boldsymbol{y}'' \subset \boldsymbol{y}$ then $U(\boldsymbol{y}'') \subset U(\boldsymbol{y}')$.

3°. *Independence of restrictions.* If $\boldsymbol{y}' \subset \boldsymbol{y}$ then $U(\boldsymbol{y}') \subset U(y)$ for each $y \in \boldsymbol{y}'$. □

These axioms have natural informal interpretations.

Axiom 1° means that, for a given strategy of player 2, it is possible for player 1 to choose any strategy, and to obtain any sum within the limits of gain in the arising situation. In particular, according to this axiom the set of possibilities for player 1, under specific conditions, has the property of (lower) "exhaustion": a player who "can" obtain more "can" also obtain less (for example, by taking the greater and throwing off the difference).

Axiom 2° asserts that an extension of the set of strategies of player 2 can only reduce the possibilities for player 1.

Finally, according to axiom 3°, if each strategy of player 2 that belongs to the set $\boldsymbol{y}'$ admits some opportunity, the whole set $\boldsymbol{y}'$ of that player's strategies admits the same opportunity.

2.3 Definition. An opportunity $\langle x^*, h_* \rangle \in U(y)$ for player 1 in Γ is said to be *optimal* if

$$h_* = \max\{h : \langle x, h\rangle \in U(y)\}. \qquad \square$$

This concept of optimality is justified by the property that player 1, using an optimal possibility, will actually not obtain less than under any other possibility.

The definition just stated can, in essence, also be taken as an axiom, thus reducing the construction of an optimality principle for two-person zero-sum games to the "principle" of optimality in a traditional maximization problem.

There is no difficulty in introducing a concept of ε-optimality of the possibilities open to the players and adapting the discussion to it; but we shall not do this.

2.4. The following theorem establishes the completeness and consistency of the preceding axiomatization.

Theorem. *A necessary and sufficient condition for a possibility $\langle x^*, h_* \rangle$ of player 1 in a two-person zero-sum game as in (2.1) to be optimal is that the strategy x^* is a maximin for player 1 and that $h_* = \underline{v}_\Gamma$.*

Proof. Necessity. Axiom 3° means that

$$\bigcap_{y \in \boldsymbol{y}'} U(y) \subset U(\boldsymbol{y}') \tag{2.2}$$

for every $\boldsymbol{y}' \subset \boldsymbol{y}$, and it follows from Axiom 2° that $U(\boldsymbol{y}') \subset U(y)$ for all $y \in \boldsymbol{y}'$, i.e. that

$$U(\boldsymbol{y}) \subset \bigcap_{y \in \boldsymbol{y}'} U(y). \tag{2.3}$$

It follows from (2.2) and (2.3) that

$$U(\boldsymbol{y}') = \bigcap_{y \in \boldsymbol{y}'} U(y).$$

Furthermore, it follows from the definition of the optimality of a possibility that

$$\begin{aligned} h_* &= \max\{h : \langle x, h\rangle \in \bigcap_{y \in \boldsymbol{y}} U(y)\} \\ &= \max_{x \in \boldsymbol{x}} \max\{h : h \leqq H(x, y),\ x \in \boldsymbol{y}\} \\ &= \max_{x \in \boldsymbol{x}} \max\{h : h \leqq \inf_{y \in \boldsymbol{y}} H(x, y)\} = \max_{x \in \boldsymbol{x}} \inf_{y \in \boldsymbol{y}} H(x, y) = \underline{v}_\Gamma. \end{aligned}$$

Here the outer maximum is attained at x^* and this strategy is a maximin.

For the sufficiency, we shall prove that a maximin principle actually satisfies axioms 1°, 2°, and 3°.

For this purpose we consider the possibility $\langle x^*, h_*\rangle$, where x^* is the maximin strategy 1, and $h_* = \inf_{y \in \boldsymbol{y}} H(x^*, y)$. Then $h_* \leqq H(x^*, y^0)$ for every $y^0 \in \boldsymbol{y}$, so that by Axiom 1° we must have $\langle x^*, h_*\rangle \in U(y^0)$. Since this holds for every $y^0 \in \boldsymbol{y}$, by Axiom 3° we also have $\langle x^*, h_*\rangle \in U(\boldsymbol{y})$.

It remains to show that $h_* \geqq h$ for every possibility $\langle x, h\rangle \in U(\boldsymbol{y})$. But it follows from $\langle x, h\rangle \in U(\boldsymbol{y})$ that $\langle x, h\rangle \in U(y^0)$ for every $y^0 \in \boldsymbol{y}$. Consequently $h \leqq \inf_{y \in \boldsymbol{y}} H(x, y)$, and therefore

$$h \leqq \max_{x \in \boldsymbol{x}} \inf_{y \in \boldsymbol{y}} H(x, y) = h. \qquad \square$$

A parallel discussion establishes the parallel theorem for the optimal opportunities of player 2.

Theorem. *A necessary and sufficient condition for a possibility $\langle y_*, h^*\rangle$ of player 2 to be optimal in a two-person zero-sum game as in* (1.1) *is that the strategy y_* of player 2 is a minimax and $h^* = \bar{v}_\Gamma$.* □

It follows immediately from these two theorems that $h_* \leqq h^*$, *and in the case when there is equality here, for all optimal opportunities* $\langle x^*, h_*\rangle$ and $\langle y_*, h^*\rangle$ *of the players the strategies x^* and y_* form a saddle point in Γ and $h_* = h^* = v_\Gamma$.*

Notice that it does not follow from what we have said that the maximin principle is the only possible optimality principle in the theory of two-person zero-sum games. From the completeness of the axiom system 1°–3° it follows only that there is a *unique* optimality principle satisfying these axioms, and that it turns out to be the *maximin principle.*

Finally we note that the maximin principle is, in a sense, incomplete: in connection with any specific two-person zero-sum game, its realizations turn out to be saddle points, of which there may be more than one in the game. Hence there naturally arises the question of further sharpening (or strengthening) of the maximin principle, in the sense that in realizations of the sharper principle there will be, as before, saddle points (if there are any in the game), but at most one ("principal", in a certain sense) for each game.

2.5 Axiomatization of the concept of preference of strategies. As the basis for a different axiom system, which we now consider and which also leads to the maximin principle, we take the concept of *preference* (more precisely, weak preference) of strategies by one of the players in a two-person zero-sum game. For definiteness, we shall consider preferences in the set of strategies of player 1.

For each two-person zero-sum game Γ as in (2.1) let there be a (weak) binary preference relation R_Γ on the set $\mathfrak{x}$ of strategies of player 1. As usual, if $\mathfrak{x}'$ and $\mathfrak{x}'' \in \mathfrak{x}_\Gamma$, we set, in what follows,

$$\mathfrak{x}' I_\Gamma \mathfrak{x}'' \Leftrightarrow \mathfrak{x}' R_\Gamma \mathfrak{x}'' \text{ and } \mathfrak{x}'' R_\Gamma \mathfrak{x}',$$

(*equivalence* of $\mathfrak{x}'$ and $\mathfrak{x}''$), and

$$\mathfrak{x}' P_\Gamma \mathfrak{x}'' \Leftrightarrow \mathfrak{x}' R_\Gamma \mathfrak{x}'' \text{ but not } \mathfrak{x}'' R_\Gamma \mathfrak{x}'$$

(*strict preference* relation).

We consider the following axioms, which a preference relation may (or may not) satisfy.

With the aim of simplifying both the constructions and the general line of the discussion, we shall use some consequences of the preceding axioms in the formulation of those that follow.

2.6 1°. *Independence of irrelevant alternatives. If $\mathfrak{x}'$ and $\mathfrak{x}'' \in \mathfrak{x}^* \subset \mathfrak{x}_\Gamma$ and Γ^* is an $\mathfrak{x}^* \times \mathfrak{y}$-subgame* (see 1.12, Chapter 1) *of Γ, then*

$$\mathfrak{x}' R_{\Gamma^*} \mathfrak{x}'' \Leftrightarrow \mathfrak{x}' R_\Gamma \mathfrak{x}'' \tag{2.4}$$

(cf. 2.18, Chapter 1).

This axiom of independence of irrelevant alternatives is evidently a stronger statement than the optimality principle of the same name mentioned in 2.18, Chapter 1; here it refers not only to optimality (i.e., to the most preferred) alternatives, but to the comparison of *any* two alternatives.

For all its apparent naturalness, this axiom expresses a very deep and frequently used property of preferences: the choice from a set of alternatives (in the present case from two strategies) of the one most preferred is independent of any other irrelevant alternatives (strategies) that are not in this set (in this pair).

At the same time, this property of independence of irrelevant alternatives is not only nontrivial and not self-evident, but it can even be violated in certain cases. Consider, for example, the results of the, say, hockey tournament described by the following table:

	A	B	C	D	E	points	rank
A	*	0	2	2	2	6	1
B	2	*	0	1	2	5	2
C	0	2	*	1	1	4	3
D	0	1	1	*	1	3	4
E	0	0	1	1	*	2	5

It is clear that if the champion A is disqualified, C will have higher rank than B.

In our case the axiom of independence of irrelevant alternatives implies a trivial corollary: if $\mathfrak{x}_1$ and $\mathfrak{x}_2 \subset \mathfrak{x}$, and Γ_1 and Γ_2 are an $\mathfrak{x}_1 \times \mathfrak{y}$-subgame and an

$\mathscr{x}_2 \times \mathscr{y}$-subgame of Γ, and $\{x', x''\} = \mathscr{x}^* \subset \mathscr{x}_1 \cap \mathscr{x}_2$, then under our preference relations, R_{Γ_1} and R_{Γ_2} over $\mathscr{x}_{\Gamma_1}$ and $\mathscr{x}_{\Gamma_2}$, respectively, we must have

$$x' R_{\Gamma_1} x'' \Leftrightarrow x' R_{\Gamma_2} x''.$$

Therefore the interdependence or independence of the strategies x' and x'' of player 1, with respect to the preference relation R_Γ, depend, by the axiom of independence of irrelevant alternatives, not on the whole game Γ in which the strategies are applied, but only on the strategies x' and x'', on the set $\mathscr{y}$ of strategies of player 2, and on the payoffs to player 1 in the situations in $\{x', x''\} \times \mathscr{y}$. Consequently we may replace the indication of the game Γ in the preference relation R_Γ by the name of the set $\mathscr{y}$ of strategies of player 2, and write $R_{\mathscr{y}}$. Considering the dependence of the relation $x' R_\Gamma x''$ on the strategies x' and x'' may seem rather "formalistic". However, its significance will appear as we use the following two axioms.

2.7 2°. *Monotonicity.* If $H(x', y) < H(x'', y)$ for every $y \in \mathscr{y}$, then $x' R_{\mathscr{y}} x''$.

In our usual terminology, this axiom means that, for a player's strategies, strict domination implies weak preference. This "weakness" is introduced for the sake of uniformity of terminology. We shall have to pay for it by the supplementary axiom 9° (see 2.9).

Indeed, this axiom refers to the fact that the maximin principle consists, above all, of a kind of *maximization*.

3°. *Continuity.* If $x_1, x_2, \ldots$ is a sequence of strategies of player 1, and $x_n R_{\mathscr{y}} x''$ for every $n = 1, 2, \ldots$; and x' is a strategy of player 1 for which

$$\lim_{n \to \infty} H(x_n, y) = H(x', y) \text{ for every } y \in \mathscr{y}, \tag{2.5}$$

then $x' R_{\mathscr{y}} x''$.

The axiom of continuity allows us to strengthen the axiom of monotonicity by stating it in the following form: if $H(x', y) \leqq H(x'', y)$ for every $y \in \mathscr{y}$, then $x' R_{\mathscr{y}} x''$. In fact, let us choose a sequence $\varepsilon_n \downarrow 0$ and introduce strategies x_n for which

$$H(x_n, y) = H(x', y) + \varepsilon_n \text{ for all } y \in \mathscr{y}. \tag{2.6}$$

Here, by Axiom 2°, $x_n R_{\mathscr{y}} x''$; moreover, (2.5) follows from (2.6), so that, by Axiom 3°, we must also have $x' R_{\mathscr{y}} x''$.

In particular, we have shown that under our hypotheses the relation $R_{\mathscr{y}}$ is reflexive: if $H(x', y) = H(x'', y)$ for every $y \in \mathscr{y}$, then $x' R_{\mathscr{y}} x''$. Moreover, it also follows that $x'' R_{\mathscr{y}} x'$, so that we also have $x' I_{\mathscr{y}} x''$.

Therefore all strategies of player 1 to which the same function H_x corresponds (see 1.1) are equivalent. Consequently we may identify them as being the same function. Hence, in what follows, we may speak of preferred functions instead of the corresponding preferred strategies.

2.8. In order to make a further simplification of the preference strategies of player 1, we now introduce two additional axioms concerned with the strategies of player 2.

4°. *Equality of a player's strategies.* If x' and $x'' \in \boldsymbol{x}$, y' and $y'' \in \boldsymbol{y}$, and

$$H(x'',y) = \begin{cases} H(x',y'') & \text{for } y = y', \\ H(x',y') & \text{for } y = y'', \\ H(x',y) & \text{for all other strategies } y \in \boldsymbol{y}, \end{cases}$$

then $x' R_{\boldsymbol{y}} x''$.

Since, under the same hypotheses, this axiom also yields $x'' R_{\boldsymbol{y}} x'$, it also leads to $x' I_{\boldsymbol{y}} x''$. Therefore the interchange of two strategies of player 2 does not change player 1's preferences among the corresponding strategies.

5°. *Multiplying player 2's strategies.* Let $\boldsymbol{y} \cap \boldsymbol{z} = \emptyset$, $\varphi : \boldsymbol{y} \to 2^{\boldsymbol{z}}$, and, for every $y \in \boldsymbol{y}$, $z \in \varphi y$, and $x \in \boldsymbol{x}$, let $H(x,y) = H(x,\varphi y)$. Then, for every x' and $x'' \in \boldsymbol{x}$,

$$x' R_{\boldsymbol{y}} x'' \Leftrightarrow x' R_{\boldsymbol{y} \cup \boldsymbol{z}} x''.$$

For the case of a finite set $\boldsymbol{y}$ this axiom reduces, in essence, to the "possibility" of either doubling a column in the payoff matrix of any $2 \times n$-matrix game, or eliminating duplicate columns.

2.9. The following two axioms are completely standard.

6°. *Transitivity.* If x', x'', and $x''' \in \boldsymbol{x}$, then $x' R_{\boldsymbol{y}} x''$ and $x'' R_{\boldsymbol{y}} x'''$ together imply $x' R_{\boldsymbol{y}} x'''$.

This axiom allows us to combine two already established preference relations.

7°. *Completeness.* If x' and $x'' \in \boldsymbol{x}$, then either $x' R_{\boldsymbol{y}} x''$ or $x'' R_{\boldsymbol{y}} x'$.

As is often the case when preferences are considered, in the present situation the completeness axiom is an exceptionally strong requirement, and is frequently debated from the informal point of view.

A rather common, and not striking, axiom is traditionally called the axiom of *convexity*, although it really ought to be called the axiom of concavity. We shall state it in a quite weak form. However, as will be evident in the course of later developments, this axiom turns out, in a certain sense, to be decisive.

8°. *Convexity.* Let x' and $x'' \in \boldsymbol{x}$. Consider a strategy $\bar{x}$ obtained by setting

$$H(\bar{x},y) = \frac{1}{2}(H(x',y) + H(x'',y)) \text{ for every } y \in \boldsymbol{x}.$$

Then $x' I_{\boldsymbol{y}} x''$ implies $\bar{x} R_{\boldsymbol{y}} x'$.

9°. *Discrimination.* If

$$H(x',y) = c' > H(x'',y) = c''$$

for all $y \in \boldsymbol{y}$, then $x' P_{\boldsymbol{y}} x''$.

The last axiom restricts, so to speak, the actions of all its predecessors and is oriented toward the avoidance of claiming the equivalence of obviously inequivalent strategies.

2.10. The axiom system formulated in 2.6–2.9 is consistent, in the sense that there exists an optimality principle (i.e., a method of attaching a preference relation over the set of strategies of player 1 in any two-person zero-sum game Γ) that satisfies all the axioms in the system. Moreover it is complete in the sense that this optimality principle is unique (i.e., the preference relation is uniquely determined.) In fact, we have the following theorem.

Theorem. *The only optimality principle that satisfies the system* 1°–9° *of axioms is the maximin principle, in the following form: in a game* $\Gamma = \langle \mathfrak{x}_\Gamma, \mathfrak{y}_\Gamma, H_\Gamma \rangle$ *the strategy* x' *of player* 1 *is preferred (weakly) to the strategy* x'' *of the same player if and only if*

$$\inf_{y \in \mathfrak{y}_\Gamma} H_\Gamma(x', y) \geqq \inf_{y \in \mathfrak{y}_\Gamma} H_\Gamma(x'', y). \tag{2.7}$$

Proof. In correspondence with what was said above (the axiom of independence of irrelevant alternatives) we may construct arbitrary strategies for player 1 and include them among that player's strategies. In addition, as we noted at the end of 1.18, we may replace the strategies x of player 1 by the corresponding functions H_x.

In those cases when it is possible, we may suppose that the elements of the sets of strategies of player 2 that we encounter have been arranged in some order; we arrange the values of the functions H_x in the same order.

For the proof of the theorem, we successively form all the more general criteria for the equivalence of these functions. Until the end of the proof, all functions will be given explicitly. Therefore, when we mention a preference relation we will sometimes omit the index that identifies the domain $\mathfrak{y}$ of the function H_x.

We begin by setting $\mathfrak{y} = \{y', y''\}$, $H(x', y') = a$, and $H(x', y'') = b$ in Axiom 4°. Then in the 2×2-matrix game with the payoff matrix

$$\begin{bmatrix} a & b \\ b & a \end{bmatrix}$$

both (pure) strategies of player 1 are equivalent: $(a, b) I (b, a)$.

Let us now suppose, for definiteness, that $a < b$, and apply the multiplying axiom to the strategies of player 2 in all possible ways. We obtain

$$(a, b, a) I_\mathfrak{y} (b, a, b) R_\mathfrak{y} (b, a, a) I_\mathfrak{y} (a, b, b) R_\mathfrak{y} (a, a, b) I_\mathfrak{y} (b, b, a) R_\mathfrak{y} (a, b, a).$$

It follows by transitivity that all the payoff vectors compared here are equivalent to each other, and in particular

$$(a, a, b) I_\mathfrak{y} (a, b, a) I_\mathfrak{y} (a, b, b).$$

Then by the convexity axiom it follows that

$$(a, 2^{-1}(a+b), 2^{-1}(a+b))R_{\boldsymbol{y}}(a, b, b),$$

from which we obtain similarly

$$(a, a, 2^{-1}(a+b))I_{\boldsymbol{y}}(a, 2^{-1}(a+b), a)I_{\boldsymbol{y}}(a, b, b).$$

Iterating this procedure and passing to the limit in accordance with the axiom of continuity, we have

$$(a, a, a)I_{\boldsymbol{y}}(a, b, b). \tag{2.8}$$

Hence it follows by the "inverse part" of the multiplying axiom that $(a, a)I_{\boldsymbol{y}}(a, b)$. Moreover, by the discrimination axiom we must have $(b, b)P_{\boldsymbol{y}}(a, a)$. Consequently our theorem is established in the case when $\boldsymbol{y}$ involves only two strategies of player 2.

Now let us consider a set $\boldsymbol{y}$ of more than two elements. Let y' denote the strategy corresponding to the first components of the vectors in (2.8), and let y'' be the strategy corresponding to the third components; set $\boldsymbol{y}^* = \boldsymbol{y} \setminus \{y', y''\}$, and multiply the second component in (2.8) by the set $\boldsymbol{y}^*$. According to Axiom 4° we obtain

$$(a, \{a\}_{y \in \boldsymbol{y}^*}, a)I(a, \{b\}_{y \in \boldsymbol{y}^*}, b). \tag{2.9}$$

Now select an arbitrary strategy x of player 1, i.e., a function H_x with

$$\min_{y \in \boldsymbol{y}} H(x, y) = a, \qquad \sup_{y \in \boldsymbol{y}} H(x, y) = b.$$

We now assign $\underline{y} \in \boldsymbol{y}$ for which $H(x, \underline{y}) = a$ as the first component in the vectors in (2.8). Then we will have

$$(a, \{b\}_{y \in \boldsymbol{y}^*}, b) \geqq (H(x, \underline{y}), \{H(x, y)\}_{y \in \boldsymbol{y}^*}, H(x, y')) \geqq (a, \{a\}_{y \in \boldsymbol{y}^*}, a),$$

from which it follows, by the strong monotonicity axiom, that

$$(a, \{b\}_{y \in \boldsymbol{y}^*}, b)R(H(x, \underline{y}), \{H(x, y)\}_{y \in \boldsymbol{y}^*}, H(x, y''))R(a, \{a\}_{y \in \boldsymbol{y}^*}, a).$$

It follows from this and (2.9) that

$$(H(x, \underline{y}), \{H(x, y)\}_{y \in \boldsymbol{y}^*}, H(x, y''))I(a, \{a\}_{y \in \boldsymbol{y}*}, a).$$

Consequently all strategy-functions H_x, for which the minimum $\min_{y \in \boldsymbol{y}} H(x, y)$ is attained, are equivalent.

Now let the strategy x'' have the property that the function $H_{x''}$ does not attain its minimum on $\boldsymbol{y}$. Let $\underline{a} = \inf_{y \in \boldsymbol{y}} H(x'', y)$ (because we always suppose that H is bounded, we have $\underline{a} > -\infty$), form a sequence $\varepsilon_1 > \varepsilon_2 \ldots > 0$ of numbers, and introduce the strategies $x_1, x_2, \dots; x'_1, x'_2, \dots;$ and x', that are defined by the functions

$$H_{x_n} = \max\{H_{x''}, \underline{a} + \varepsilon_n\},$$

$$H_{x'_n} = \underline{a} + \varepsilon_n,$$

$$H_{x'} = \underline{a}.$$

Here $x'_n R x_{n+1}$ by what has been proved; also, by the axiom of monotonicity, $x_{n+1} R x''$, so that $x'_n R x''$. Since this holds for every n, the axiom of continuity gives us $x' R x''$. But monotonicity implies $x'' R x'$, and we have established what was required. The fact that the strict inequality of infima in (2.7) implies $x' P x''$ follows immediately from Axiom 9°. □

2.11 Axiomatization of the concept of the value of a game. In parallel with the axiomatic description of the optimal behavior of a player, we may raise the question of an axiomatic description of a justified (or fair) payoff, thought of by a player in a two-person zero-sum game Γ as the number $f(\Gamma)$ received by the recipient of $f(\Gamma)$ as fair compensation for taking part in the game in the capacity of player 1. We present this axiomatization for the class of two-person zero-sum games that have values (possibly in some mixed extension).

Let, as usual, Γ be a two-person zero-sum game as in (2.1). We assign to it a number $f(\Gamma)$ which is to denote the justified payoff to player 1 in this game. The functional f may (or may not) satisfy the following requirements (or axioms).

1°. *Monotonicity*. If, for two games, Γ as in (2.1) and $\Gamma' = \langle \boldsymbol{x}, \boldsymbol{y}, H' \rangle$, we have $H(x, y) \leqq H'(x, y)$ for every $x \in \boldsymbol{x}$ and $y \in \boldsymbol{y}$, then $f(\Gamma) \leqq f(\Gamma')$.

This may be interpreted informally in the following way. If in changing from one game to another with the same structure, the payoffs to player 1 in every situation can only increase, then the player's justified gain cannot decrease.

2°. *Effectiveness for player 1*. Let Y be any mixed strategy of player 2 in a game Γ as in (2.1); let $\bar{x} \notin \boldsymbol{x}$ and $\bar{\Gamma} = \langle \boldsymbol{x} \cup \bar{x}, \boldsymbol{y}, \bar{H} \rangle$; and let

$$\bar{H}(x, y) \begin{cases} = H(x, y) & \text{if } x \in \boldsymbol{x}, \\ \leqq H(x, y) & \text{if } x = \bar{x}. \end{cases}$$

Then $f(\Gamma) \leqq f(\bar{\Gamma})$.

Informally, adjoining to the game a new strategy of player 1, no better than any combination of those already present, cannot change the justified payoff for that player.

3°. *Effectiveness for player 2*. Let Y be any mixed strategy of player 2 in a game Γ as in (2.1); let $\bar{\bar{y}} \notin \boldsymbol{y}$ and $\bar{\bar{\Gamma}} = \langle \boldsymbol{x}, \boldsymbol{y} \cup \bar{\bar{y}}, \bar{\bar{H}} \rangle$; and let

$$\bar{\bar{H}}(x, y) \begin{cases} = H(x, y) & \text{if } y \in \boldsymbol{y}, \\ \geqq H(x, y) & \text{if } y = \bar{\bar{y}}. \end{cases}$$

Then $f(\Gamma) = f(\bar{\bar{\Gamma}})$.

Informally, adjoining to the game a new strategy of player 2, no better than any combination of those already present, cannot change the justified loss for that player.

4°. *Objectivity.* If, in a game Γ as in (2.1), $\boldsymbol{x} = \{x_0\}$ and $\boldsymbol{y} = \{y_0\}$ (i.e., the sets $\boldsymbol{x}$ and $\boldsymbol{y}$ consist of a single strategy each), then $f(\Gamma) = H(x_0, y_0)$.

Informally, the justified payoff to player 1 in a game with only a single situation equals the payoff in this situation.

2.12 The enumerated axioms form a complete system. Namely, the following theorem is valid.

Theorem. *If for every two-person zero-sum game Γ that has a value $v(\Gamma)$ (possibly in a mixed extension) there is given a real number $f(\Gamma)$, and the functional relation $\Gamma \mapsto f(\Gamma)$ satisfies Axioms 1°–4°, then $f(\Gamma) = v_\Gamma$.*

We emphasize that, in view of the theorems of 1.15 and 1.17, it is irrelevant whether there is a value in the game Γ itself or in one of its mixed extensions (or in which one).

Proof. Let Γ be a two-person zero-sum game as in (2.1). Suppose first that Γ has a saddle point (x^*, y^*) (in the pure strategies). In this case $v_\Gamma = H(x^*, y^*)$.

Now consider the games

$$\underline{\Gamma} = \langle \boldsymbol{x}, \boldsymbol{y}, \underline{H} \rangle, \quad \bar{\Gamma} = \langle \boldsymbol{x}, \boldsymbol{y}, \bar{H} \rangle,$$

where, for all $x \in \boldsymbol{x}$ and $y \in \boldsymbol{y}$, we set

$$\underline{H}(x, y) = \min\{H(x^*, y), H(x, y)\},$$
$$\bar{H}(x, y) = \max\{H(x, y^*), H(x, y)\}.$$

Intuitively speaking, $\underline{H}$ is obtained by "lower leveling" of every set H_y of values down to the corresponding number $H(x^*, y)$. The function $\bar{H}$ is obtained similarly by "upper leveling" of every set H_x up to the number $H(x, y^*)$.

It is clear that

$$\underline{H}(x, y) \leqq H(x, y) \leqq \bar{H}(x, y),$$

so that, by Axiom 1°,

$$f(\underline{\Gamma}) \leqq f(\Gamma) \leqq f(\bar{\Gamma}). \tag{2.10}$$

The strategy x^* of player 1 in the game $\underline{\Gamma}$ dominates all of its other strategies x, so that all these strategies can, by Axiom 2°, be discarded without changing the function $f(\underline{\Gamma})$. In the reduced game, player 1 has only the single strategy x^*, and moreover,

$$H(x^*, y^*) \leqq H(x^*, y) \text{ for all } y \in \boldsymbol{y}.$$

Therefore y^*, as a strategy of player 2, dominates all that player's other strategies, which may be discarded by Axiom 3°. As a result, as an obvious consequence of Axiom 4°, we obtain

$$f(\underline{\Gamma}) = f(\langle x^*, y^*, H \rangle) = H(x^*, y^*) = v_\Gamma.$$

By parallel reasoning, we obtain $f(\bar{\Gamma}) = v_\Gamma$. The two preceding equations, together with (2.10), yield $f(\Gamma) = v_\Gamma$, and we have completed the analysis of the case when Γ has a saddle point.

Now suppose that Γ does not have a saddle point. However, by hypothesis, Γ has a value v_Γ (in the corresponding mixed extensions) and therefore, for every $\varepsilon > 0$, has an ε-saddle point $(X_\varepsilon, Y_\varepsilon)$. By definition, for this point

$$H(x, Y_\varepsilon) - \varepsilon \leqq H(X_\varepsilon, Y_\varepsilon) \leqq H(X_\varepsilon, y) + \varepsilon$$

for every $x \in \boldsymbol{x}$ and $y \in \boldsymbol{y}$, and in addition, in view of (1.22),

$$v_\Gamma - 2\varepsilon \leqq H(X_\varepsilon, Y_\varepsilon) \leqq v_\Gamma + 2\varepsilon.$$

From these two inequalities we immediately obtain the estimate

$$H(x, Y_\varepsilon) \leqq v_\Gamma + 3\varepsilon \leqq H(X_\varepsilon, y) + 6\varepsilon. \tag{2.11}$$

Now form the game $\Gamma_\varepsilon^+ = \langle \boldsymbol{x} \cup X_\varepsilon, \boldsymbol{y} \cup Y_\varepsilon, H_\varepsilon^+ \rangle$, by setting

$$H_\varepsilon^+(x, y) = \begin{cases} H(x, y) & \text{if } x \in \boldsymbol{x}, \\ H(X_\varepsilon, y) + 6\varepsilon & \text{if } x = X_\varepsilon \text{ but } y \in \boldsymbol{y}, \\ v_\Gamma + 3\varepsilon & \text{if } x = X_\varepsilon \text{ and } y = Y_\varepsilon. \end{cases}$$

With this notation, (2.11) can be rewritten as

$$H_\varepsilon^+(x, Y_\varepsilon) \leqq H(X_\varepsilon, Y_\varepsilon) \leqq H(X_\varepsilon, y)$$

for every $x \in \boldsymbol{x} \cup X_\varepsilon$ and $y \in \boldsymbol{y} \cup Y_\varepsilon$, i.e. the game Γ_ε^+ has a saddle point (namely $(X_\varepsilon, Y_\varepsilon)$) and the value $v_{\Gamma_\varepsilon^+} = v_\Gamma + 3\varepsilon$. Hence, by what has been proved, $f(\Gamma_\varepsilon^+) = v_\Gamma + 3\varepsilon$.

Let us apply Axioms 2° and 3° successively to this game, after setting $\bar{x} = X_\varepsilon$ and $\bar{\bar{y}} = Y_\varepsilon$. We thus obtain an $\boldsymbol{x} \times \boldsymbol{y}$-subgame $\bar{\Gamma}_\varepsilon^+$ of Γ_ε^+, and according to these axioms we have

$$f(\bar{\Gamma}_\varepsilon^+) = f(\Gamma_\varepsilon^+) = v_\Gamma + 3\varepsilon.$$

On the other hand, for $x \in \boldsymbol{x}$ and $y \in \boldsymbol{y}$ we have $H(x, y) \leqq H_\varepsilon^+(x, y)$. Hence, by Axiom 1° it must be true that

$$f(\Gamma) \leqq f(\bar{\Gamma}_\varepsilon^+) = v_\Gamma + 3\varepsilon. \tag{2.12}$$

Now form the game

$$\Gamma_\varepsilon^- = \langle \boldsymbol{x} \cup X_\varepsilon, \boldsymbol{y} \cup Y_\varepsilon, H_\varepsilon^- \rangle$$

by setting

$$H_\varepsilon^-(x, y) = \begin{cases} H(x, y) & \text{if } y \in \boldsymbol{y}, \\ H(X_\varepsilon, y) + 6\varepsilon & \text{if } y = Y_\varepsilon \text{ but } x \in \boldsymbol{x}, \\ v_\Gamma + 3\varepsilon & \text{if } y = Y_\varepsilon \text{ and } x = X_\varepsilon, \end{cases}$$

and, by parallel reasoning, obtain $f(\Gamma) \geqq v_\Gamma - 3\varepsilon$.

Together with (2.12), this yields

$$v_\Gamma - 3\varepsilon \leqq f(\Gamma) \leqq v_\Gamma + 3\varepsilon,$$

and an appeal to the arbitrariness of $\varepsilon > 0$ completes the proof. □

§3 Minimax theorems

3.1 General and specific minimax theorems. After having discussed, in the preceding section, the axiomatic basing of the maximin principle in two-person zero-sum games, let us turn to questions of its realizability. Such a realizability consists of the existence, for players in a two-person zero-sum game, of optimal (or at least ε-optimal for every $\varepsilon > 0$) strategies. Rather many existence theorems of this kind are known. They are usually called *minimax theorems*. Sometimes they are taken to include information about the form (or structure) of the optimal strategies whose existence is to be established.

The various minimax theorems that are used in game theory can be divided rather naturally (though perhaps somewhat conventionally) into two groups. The first group consists of "general" theorems on the existence of equilibrium (or ε-equilibrium) situations in classes of games described in *structural* terms, i.e. in terms that relate to topological, linear, or general combinatorial structures of the spaces of the players' strategies and payoff functions. These conditions are usually connected with notions of compactness, convexity, and continuity. To the second group we assign "particular" theorems, in which more specific conditions are imposed on classes of games, conditions either of an *analytic* or a *combinatorial* nature. These conditions might, of course, be of various kinds. As a rule, they are connected with the game itself, and the existence of optimal (or ε-optimal) strategies is assured in some extension, most often mixed.

In this section we present some rather general structural minimax theorems, as well as a typical example of a particular theorem.

3.2 Fundamental concepts. We first observe that all the results obtained in Chapter 1 for general noncooperative games remain valid, with corresponding simplifications, for two-person zero-sum games. Here we state the fundamental definitions and theorems in formulations adapted to two-person zero-sum games.

Definition. In a two-person zero-sum game

$$\Gamma = \langle \boldsymbol{x}, \boldsymbol{y}, H \rangle \tag{3.1}$$

the $\mathcal{S}$-*topology* (the *semi-intrinsic topology*; see 6.2, Chapter 1) on $\boldsymbol{x}$ is defined to be the topology in which the basis is the family of all sets of the form $\mathcal{S}(x^0, \varepsilon)$, where $x^0 \in \boldsymbol{x}$ and $\varepsilon \in \mathbf{R}_+$, i.e. the sets of strategies $x \in \boldsymbol{x}$ which are strictly ε-1-dominated by the strategy x^0. In other words, $\mathcal{S}(x^0, \varepsilon)$ consists of the strategies x for which, for every strategy $y \in \boldsymbol{y}$,

$$H(x, y) < H(x^0, y) + \varepsilon.$$

We shall often refer to the ε-1-domination of the strategies of player 1 simply as ε-domination.

The $\mathcal{S}$-topology on the set $\boldsymbol{y}$ is defined correspondingly.

In a two-person zero-sum game Γ as in (3.1), the $\mathcal{J}$-*topology* (*intrinsic topology*; see 6.8, Chapter 1) on $\boldsymbol{x}$ is defined to be the topology whose basis is the family of sets of the form $\mathcal{J}(x^0, \varepsilon)$, i.e. the set of strategies $x \in \boldsymbol{x}$ for which, for every strategy y,

$$|H(x^0, y) - H(x, y)| < \varepsilon.$$

The $\mathcal{J}$-topology on $\boldsymbol{y}$ is defined correspondingly. □

We now turn our attention to the fact that both the $\mathcal{S}$- and $\mathcal{J}$-topologies on the spaces of the players' strategies are actually topologies on the function spaces whose elements are the functions H_x on $\boldsymbol{y}$ with $x \in \boldsymbol{x}$, or H_y on $\boldsymbol{x}$ with $y \in \boldsymbol{y}$.

The $\mathcal{J}$-topologies on $\boldsymbol{x}$ and $\boldsymbol{y}$ are generated in a natural way by the pseudometrics

$$\rho_1(x', x'') = \sup_{y \in \boldsymbol{y}} |H(x', y) - H(x'', y)|,$$

$$\rho_2(y', y'') = \sup_{x \in \boldsymbol{x}} |H(x, y') - H(x, y'')|,$$

which, as we noticed in 6.10, Chapter 1, after appropriate identification of equivalent strategies, can be thought of as metrics.

According to 6.3 and 6.17 of Chapter 1, we have the following theorem.

Theorem. *For every $\varepsilon > 0$, each $\mathcal{S}$-precompact two-person zero-sum game has ε-saddle points in the finite strategies.*

Therefore, according to the theorem of 1.6 on $\mathcal{S}$-precompact two-person zero-sum games, we have the equation

$$\sup_{X \in \boldsymbol{X}} \inf_{Y \in \boldsymbol{Y}} H(X, Y) = \inf_{Y \in \boldsymbol{Y}} \sup_{X \in \boldsymbol{X}} H(X, Y)$$

and each such game has a value.

Theorem. *Every $\mathcal{J}$-compact two-person zero-sum game has a saddle point in the mixed strategies.*

An application of the theorem of 6.18 of Chapter 1 to two-person zero-sum games gives us the following proposition.

Theorem. *If, in a two-person zero-sum game Γ as in (3.1), each of the sets $\boldsymbol{x}$ and $\boldsymbol{y}$ is compact in some extrinsic topology, and the payoff function H is continuous in both variables in the topology of the space of situations, then Γ has saddle points in the mixed strategies.*

In particular, there are saddle points in the mixed strategies for continuous games on the unit square, i.e. for those two-person zero-sum games Γ in which $\boldsymbol{x} = \boldsymbol{y} = [0, 1]$ and H is continuous in both variables.

3.3 Convex and concave games. As we showed in 3.9, Chapter 1, a concave noncooperative game has, under rather general hypotheses, equilibrium situations (or at least ε-equilibrium situations for every $\varepsilon > 0$). Of course, this theorem holds, in particular, for two-person zero-sum games.

It is natural to suppose that, for two-person zero-sum games that are not concave, there is something like concavity that implies that they have a minimax principle. A property of this kind would be the existence of equilibrium situations (saddle points) in the pure strategies, or something resembling them.

Definition. A two-person zero-sum game Γ as in (3.1) is said to be *convex* if $\boldsymbol{y}$ is a convex subset of a real linear topological space, and all partial payoff functions H_x (i.e., functions $H(x, \cdot) : \boldsymbol{y} \to \mathbf{R}$) are convex for $y \in \boldsymbol{y}$.

A *concave* game is defined by imposing a convexity condition on the space $\boldsymbol{x}$ and a concavity condition on the functions $H_y : \boldsymbol{x} \to \mathbf{R}$.

A two-person zero-sum game which is both convex and concave is called *convexo-concave*. □

We note that there is a disagreement, stipulated by tradition, between the preceding definition and the definition of a concave noncooperative game in 3.9, Chapter 1. Here we are thinking of the convexity of a two-person zero-sum game as the convexity of the payoff function of player 1 *from the point of view of player 2*, for whom it is a *loss function*. The payoff function of player 2 is $-H$, which of course is *concave*.

In the following discussion of convex games and their generalizations, we shall consistently use the notation

$$\boldsymbol{y}_\alpha(x) = \{y : H(x,y) \leqq \alpha\}, \quad \boldsymbol{x}_\alpha(y) = \{x : H(x,y) \geqq \alpha\}. \tag{3.2}$$

We immediately note in this connection that it follows at once from $\alpha < \beta$ for every $x \in \boldsymbol{x}$ that $\boldsymbol{y}_\alpha(x) \subset \boldsymbol{y}_\beta(x)$.

Furthermore, if H_x is a convex function, then all sets $\boldsymbol{y}_\alpha(x)$ are convex. If H_x is lower semicontinuous in y, then all the sets $\boldsymbol{y}_\alpha(x)$ are closed.

We write, for every $\boldsymbol{x}' \subset \boldsymbol{x}$,

$$\bigcap_{x \in \boldsymbol{x}'} \boldsymbol{y}_\alpha(x) = \boldsymbol{y}_\alpha(\boldsymbol{x}') \tag{3.3}$$

and set

$$\inf_{y \in \boldsymbol{y}} \sup_{x \in \boldsymbol{x}'} H(x,y) = v_{\boldsymbol{x}'}. \tag{3.4}$$

It is clear that $v_{\boldsymbol{x}} = \overline{v}_\Gamma$. The set of $y \in \boldsymbol{y}$ for which this infimum is attained will be denoted by $\boldsymbol{y}^*_{\boldsymbol{x}'}$. This set might, generally speaking, be empty. We set $\boldsymbol{y}^*_{\boldsymbol{x}} = \boldsymbol{y}^*$.

For concave two-person zero-sum games, we might introduce similar notations. We shall not do this, since in what follows we shall, for the sake of definiteness, always consider just the convex version.

A convex game does not necessarily have a value. We may, for example, consider the mixed extensions of the games in 4.2 and 4.3.

Lemma. *If the game Γ is convex and has a value, then $\bar{v}_\Gamma = \underline{v}_\Gamma = v_\Gamma$.*

Proof. According to 1.10, we have

$$\underline{v}_\Gamma = \inf_{Y \in \boldsymbol{Y}} \sup_{X \in \boldsymbol{X}} H(X, Y).$$

In view of 1.14, this can be rewritten as follows:

$$\underline{v}_\Gamma = \inf_{Y \in \boldsymbol{Y}} \sup_{x \in \boldsymbol{x}} H(x, Y).$$

But the function $\sup_{x} H_x : \boldsymbol{y} \to \mathbf{R}$ is convex. Therefore

$$\underline{v}_\Gamma = \inf_{y \in \boldsymbol{y}} \sup_{x \in \boldsymbol{x}} H(x, y) = \bar{v}_\Gamma = v_\Gamma. \quad \square$$

3.4. We are going to make use of the following proposition, which is also of independent interest.

Lemma. 1) *If the game Γ of* (3.1) *is convex and has the value $v_\Gamma = v$, $\boldsymbol{y}_x^* = \boldsymbol{y}^* \neq \emptyset$, and for some $\varepsilon > 0$ and some $\boldsymbol{c} \subset \boldsymbol{x}$ we have $\boldsymbol{y}_{v-\varepsilon}(\boldsymbol{c}) = \emptyset$, then player* 1 *has, in Γ, an ε-optimal strategy (see* 1.11*) whose spectrum is contained in $\boldsymbol{c}$.*

2) *If, in a convex game Γ as in* (3.1), *the function H is continuous; if, as before, $\boldsymbol{y}^* \neq \emptyset$; and for some $\varepsilon > 0$ and finite $\boldsymbol{c} \subset \boldsymbol{x}$ we have* $\operatorname{int} \boldsymbol{y}_{v-\varepsilon}(\boldsymbol{c}) = \emptyset$, *then for every $\varepsilon' > \varepsilon$ player* 1 *has an ε'-optimal strategy with spectrum contained in $\boldsymbol{c}$.*

Proof. 1) In view of the preceding lemma, the equation $\boldsymbol{y}_{v-\varepsilon}(\boldsymbol{c}) = \emptyset$ means that

$$v_{\boldsymbol{c}} = \inf_{y \in \boldsymbol{x}} \max_{x \in \boldsymbol{c}} H(x, y) \leqq v - \varepsilon.$$

This means that, for every strategy $y^* \in \boldsymbol{y}^*$,

$$v = \min_{y \in \boldsymbol{y}} \sup_{x \in \boldsymbol{c}} H(x, y) = \sup_{x \in \boldsymbol{x}} H(x, y^*) < v_{\boldsymbol{c}} + \varepsilon;$$

i.e., in the first place we must have, for every $x \in \boldsymbol{x}$,

$$H(x, y^*) < v_{\boldsymbol{c}} + \varepsilon; \tag{3.5}$$

and, in the second place, the strategy y^* is optimal in Γ for player 2.

Now consider the $\boldsymbol{c} \times \boldsymbol{y}$-subgame $\Gamma^{\boldsymbol{c}}$ of Γ. The set $\boldsymbol{c}$ is finite, and therefore $\mathcal{J}$-precompact; consequently (see 6.11, Chapter 1) $\Gamma^{\boldsymbol{c}}$ is itself precompact. Therefore, as was already mentioned in 3.2, $\mathcal{S}^\varepsilon(\Gamma) \neq \emptyset$ for every $\varepsilon > 0$. But then, by part 2) of the theorem of 1.6, the game $\Gamma^{\boldsymbol{c}}$ has the value $v_{\Gamma^{\boldsymbol{c}}} = v_{\boldsymbol{c}}$. Then, according to part 1) of the theorem of 1.10 (if we take $\varepsilon_1 = \varepsilon_2 = \varepsilon/2$), if $X_\varepsilon \in \mathcal{S}^{\varepsilon/2}(\Gamma^{\boldsymbol{c}})$ we must have

$$v_{\boldsymbol{c}} - \varepsilon \leqq H(X_\varepsilon, y) \text{ for every } y \in \boldsymbol{y}.$$

The set X^c is compact, and the function $H_y : X^c \to \mathbf{R}$ is linear on X^c and therefore continuous. Consequently, if we let ε tend to zero (if necessary, on a subsequence), we obtain $X_\varepsilon \to$ (say)X^c, from which

$$v_c \leqq H(X^c, y) \text{ for every } y \in \mathcal{y}.$$

In particular, $v_c \leqq H(X^c, y^*)$. But it follows from (3.5) that $H(X^c, y^*) \leqq v_c + \varepsilon$. Consequently we have

$$H(x, y^*) - \varepsilon \leqq H(X^c, y^*) \leqq H(X^c, y) + \varepsilon \text{ for all } x \in \mathcal{x} \text{ and } y \in \mathcal{y},$$

i.e. $X^c \in \mathcal{S}^\varepsilon(\Gamma)$, and moreover supp $X^c \subset \mathcal{c}$.

2) The emptiness of the interior of the intersection $\mathcal{y}_{v-\varepsilon}(\mathcal{c})$ means that for every $y \in \mathcal{y}$ and every neighborhood $\omega(y)$ there is a $y_\omega \in \omega(y)$ which does not belong to the set $\mathrm{y}_{v-\varepsilon}(\mathcal{c})$, so that there is an $x \in \mathcal{c}$ such that $H(x, y_\omega) > v - \varepsilon$. Therefore, by the continuity of H_x, we have $H(x, y) \geqq v - \varepsilon > v - \varepsilon'$. It remains only to repeat the proof of the first part of the lemma with ε replaced by ε'. □

3.5. In what follows, unless the contrary is explicitly stated, we shall consider convex games Γ as in (3.1), for which the space $\mathcal{y}$ is compact and the function H_x is continuous.

Theorem. *In a convex game* Γ *as in* (3.1) *player* 2 *has a pure optimal strategy, and player* 1 *has, for every* $\varepsilon > 0$, *an* ε*-optimal strategy which is a mixture of finitely many pure strategies. Such a game possesses a value.*

Proof. Choose any finite set $\mathcal{c} \subset \mathcal{x}$. Since $\mathcal{c}$ is finite, the function $\max_{x \in \mathcal{c}} H_x$, together with the functions H_x, is continuous in y; and since $\mathcal{y}$ is compact it attains its minimum, so that $\mathcal{y}_c^* \neq \emptyset$. Therefore, by (3.3) and (3.4),

$$\mathcal{y}_{v_c}(e) = \{y : y \in \mathcal{y}, \max_{x \in \mathcal{c}} H(x, y) \leqq v_\varepsilon\} \neq \emptyset,$$

from which it follows that

$$\inf_{y \in \mathcal{y}} \max_{x \in \mathcal{c}} H(x, y) = \max_{x \in \mathcal{c}} H(x, y^*) \text{ for } y^* \in \mathcal{y}_c^*,$$

and similarly $\mathcal{y}_v(x) = \mathcal{y}^*$ (and it has now been shown that the last set is not empty).

Since $v_c \leqq v$, it follows that $\mathcal{y}_{v_c}(x) \subset \mathcal{y}_v(x)$ for every $x \in \mathcal{x}$, so that

$$\emptyset \neq \mathcal{y}_{v_c}(\mathcal{c}) \subset \mathcal{y}_v(\mathcal{c}) \neq \emptyset,$$

i.e. the intersection of any number of sets of the form $\mathcal{y}_v(x)$ is not empty. Therefore, by the compactness of $\mathcal{y}$ we also have $\mathcal{y}^* = \mathcal{y}_v(\mathcal{x}) \neq \emptyset$, i.e. player 2 has a pure optimal strategy.

Now choose any $\varepsilon > 0$. In accordance with the preceding lemma, it is enough, for our purposes, to establish the existence of a finite $\mathcal{c} \subset \mathcal{x}$ for which the intersection $\mathcal{Y}_{v-\varepsilon}(c)$ is empty. However, if such a finite set c cannot be found, i.e., all intersections $\mathcal{Y}_{v-\varepsilon}(c)$ are nonempty, then the compactness of $\mathcal{Y}$ would imply that $\mathcal{Y}_{v-\varepsilon}(\mathfrak{x}) \neq \emptyset$, i.e. we would have found a $y^0 \in \mathcal{Y}$ for which

$$\sup_{x \in \mathfrak{x}} H(x, y^0) < v - \varepsilon,$$

which contradicts the definition of v. □

Corollary. *Every convexo-concave game Γ has an equilibrium situation.*

For the proof, it is enough to notice that, by the theorem just proved, the convexity of Γ implies $\mathcal{T}(\Gamma) \neq \emptyset$, and its concavity implies, by symmetry, $-\mathcal{S}(\Gamma) \neq \emptyset$.

3.6. The following proposition is a further corollary of the theorem.

Theorem. *If, in a two-person zero-sum game Γ as in (3.1), the set $\mathcal{Y}$ is compact, and all functions H_x are continuous, then for every $\varepsilon > 0$ the game Γ contains an ε-equilibrium situation $(X_\varepsilon, Y_\varepsilon)$, in the mixed strategies, for which X_ε has a finite spectrum.*

Proof. According to the theorem of 6.18, Chapter 1, under our present hypotheses the space $\mathcal{Y}$ of pure strategies of player 2 is compact in the natural topology. Consequently, by the theorem of 6.15, Chapter 1, the space Y of mixed strategies is compact (in the weak topology).

Let us form a "mixed semiextension" of Γ: the two-person zero-sum game $\Gamma^* = \langle \mathfrak{x}, Y, H \rangle$. This game satisfies the hypotheses of the theorem of 3.6, and therefore has corresponding ε-equilibrium situations, which also are necessarily equilibrium situations of the original game Γ. □

3.7. If appropriate conditions are imposed on the set $\mathcal{Y}$ of strategies of player 2 in the game Γ, and on the set $\mathcal{Y}^* = \mathcal{T}(\Gamma)$ we can improve the theorem of 3.5. An essential role in the uniformity of the results so obtained is played by Helly's theorem on convex sets.*)

Theorem. *In the convex game Γ let the set $\mathcal{Y}$ of strategies of player 2 be a (compact) subset of $\mathbf{R}^n$. Then, for every $\varepsilon > 0$, player 1 in Γ has an ε-optimal strategy whose spectrum contains at most $n + 1$ points.*

Proof. It is evident that $\mathcal{Y}_{v-\varepsilon}(\mathfrak{x}) = \emptyset$ for the game Γ and for every $\varepsilon > 0$. For Γ, we have $\dim \mathcal{Y} \leqq n$. Therefore, by Helly's theorem (in contrapositive form), there exists $c \subset \mathfrak{x}$ with $|c| = n + 1$, for which $\mathcal{Y}_{v-\varepsilon}(c) = \emptyset$, and we have only to appeal to the lemma of 3.4. □

---

*) See any books on convex sets.

3.8. If player 2 in the convex game Γ has a sufficiently large set y^* of optimal strategies, the property that we established, of the ε-optimality of the strategies of player 1, can be further refined.

Let us prove an auxiliary combinatorial lemma.

Lemma. *Let $\{y(x)\}_{x\in c}$ be any finite family of closed convex subsets of $\mathbf{R}^n$. Suppose that*

$$y(c) = \bigcap_{x\in c} \partial y(x) \neq \emptyset$$

(where ∂, as usual, denotes the operation of forming the boundary of a set).

Then, if $\dim y(c) = p$, there exists $c' \subset c$ with $|c'| \leqq n - p + 1$ such that

$$y(c') = \bigcap_{x\in c'} \partial y(x). \tag{3.6}$$

Proof. Let $\mathbf{R}^p$ denote the affine subspace of $\mathbf{R}^n$ spanned by $y(c)$. Each vector $y \in \mathbf{R}^n$ has a unique representation as a sum $y^p + z$, where $z \perp \mathbf{R}^p$. We carry out the natural factorization γ of $\mathbf{R}^n$ by $\mathbf{R}^p$ (projection of $\mathbf{R}^n$ on the subspace $(\mathbf{R}^p)^\perp$). We may set $\gamma(y^p + z) = z$ and $\gamma\mathbf{R}^p = \theta$.

Suppose that

$$\bigcap_{x\in c} \gamma y(x) \ni \theta' \neq \theta.$$

This means that, for each $x \in c$, there is a point $y_x = y_x + \theta'$.

In $\mathbf{R}^n$, we form the cones with common bases $y(c)$ and vertices y_x. None of these lie in $\mathbf{R}^p$. Therefore they have common points not lying in $\mathbf{R}^p$ (for example, $y_0^p + \varepsilon(\theta' - \theta)$, where y_0 is the centroid of $y(c)$ and $\varepsilon > 0$ is sufficiently small). This, however, contradicts the definition of $y(c)$.

On the other hand, for arbitrary $x_0 \in c$ we must have $y(c) \subset \partial y(x_0)$, since $y(c)$ lies in the supporting hyperplane of $y(x_0)$ that contains $\mathbf{R}^p$. Its intersection with $\gamma\mathbf{R}^n$ is the supporting hyperplane of $\gamma y(x_0)$ that contains θ. Therefore

$$\theta \in \bigcap_{x\in c} \partial\gamma y(x).$$

As a result, we have

$$\theta = \bigcap_{x\in c} \gamma y(x) = \bigcap_{x\in c} \partial\gamma y(x).$$

Hence it follows immediately that

$$\bigcap_{x\in c} \operatorname{int} \gamma y(x) = \emptyset.$$

If we apply Helly's theorem to the finite number of open sets $\operatorname{int}\gamma y(x)$ with $x \in c$, we obtain a family $c' \subset c$ with $|c'| \leq n - p + 1$ for which

$$\bigcap_{x\in c'} \operatorname{int} \gamma y(x) = \emptyset,$$

i.e., $\bigcap_{x\in c'} \gamma y(x) = \bigcap_{x\in c'} \partial y(x)$.

Moreover, evidently,

$$\mathcal{y}(c') \supset \bigcap_{x \in c'} \partial \mathcal{y}(x).$$

If we had in fact found

$$y \in \mathcal{y}(c') \setminus \bigcap_{x \in c'} \partial \mathcal{y}(x),$$

then we would have had $y \in \operatorname{int} \mathcal{y}(x)$ for at least one $x \in c'$. But then we would also have $\gamma y \in \operatorname{int} \gamma \mathcal{y}(x)$, i.e., $\gamma y \notin \partial \gamma \mathcal{y}(x)$, and we would have a contradiction. □

3.9 Theorem. *If, for a convex game* Γ *as in* (3.1), $\dim \mathcal{y}^* = p \leqq n$, *then player* 1 *in* Γ *has, for every* $\varepsilon > 0$, *an* ε*-optimal strategy, whose spectrum contains at most* $n - p + 1$ *points.*

Proof. According to the theorem of 3.7, for every $\varepsilon > 0$ player 1 has an ε-optimal strategy in Γ with a finite spectrum c. Form the $c \times \mathcal{y}$-subgame Γ_c of Γ. Evidently $\mathcal{S}(\Gamma_c) \subset \mathcal{S}^{\varepsilon}(\Gamma)$.

Let c' be the spectrum of an optimal strategy of player 1 in Γ_c and let $v' = v(\Gamma_c)$. Without loss of generality, we may take $c' = c$. From $\mathcal{S}(\Gamma_c) \neq \emptyset$ and the property of complementary slackness, it follows that

$$\mathcal{y}_{v'}(c) = \bigcap_{x \in c} \mathcal{y}_{v'}(x) = \bigcap_{x \in c} \partial \mathcal{y}_{v'}(x) \neq \emptyset. \tag{3.7}$$

If we take the set $\mathcal{y}_{v'}(x)$ as the $\mathcal{y}(x)$ in the hypotheses of the Lemma in 3.8, we can find a $c' \subset c$ for which $|c'| \leqq n - p + 1$ and (3.6) is satisfied. It follows from (3.7) that $\dim \mathcal{y}_{v'}(c) < n$, so that $\operatorname{int} \mathcal{y}_{v'}(c) = \emptyset$.

Consequently, by part 2) of the lemma in 3.4, in the game Γ player 1 has, for every $\varepsilon' > \varepsilon$, an ε-optimal strategy whose spectrum is a subset of c^*. It remains only to use the arbitrariness of ε and ε'. □

3.10. There is still another refinement when the set $\mathcal{y}^* = \mathcal{T}(\Gamma)$ belongs to the boundary of $\mathcal{y}$.

Lemma. *If* c *is a finite family of convex subsets of* $\mathbf{R}^n$ *and* $\mathcal{y}(c) = \bigcap_{x \in c} \partial \mathcal{y}(x) = \{y^*\}$, *then for each* $x_0 \in c_0$ *there is a subsequence* $c \subset c$ *such that* $x_0 \in c_0$, $|c_0| \leqq n + 1$, *and*

$$\mathcal{y}(c_0) = \bigcap_{x \in c_0} \partial \mathcal{y}(x).$$

Proof. Under our hypotheses, there is, by Helly's theorem, a $c' \subset c$ *for which* $|c'| \leqq n + 1$ *and*

$$\mathcal{y}(c') = \bigcap_{x \in c'} \partial \mathcal{y}(x).$$

Let us set $\mathcal{y}(c' \setminus x_i) = \mathcal{y}_i$ for $i = 1, \ldots, n + 1$, and notice that $\mathcal{y}_i \cap \mathcal{y}_j = y^*$ for every $i \neq j$.

It is enough, for our purposes, to prove that, for every hyperplane G passing through y^*, there are sets $\mathcal{Y}_i$ and $\mathcal{Y}_j$ that G separates. In fact, if we take G to be the supporting hyperplane at y^* of the given $\mathcal{Y}(x_0)$, we find that $\mathcal{Y}(x_0)$ and some $\mathcal{Y}_i$ are separated, i.e. $\mathcal{Y}(x_0) \cap \mathcal{Y}_{i_0} = \partial\mathcal{Y}(x_0) \cap \partial\mathcal{Y}_{i_0}$.

Further, it is enough to establish our proposition for the case when all the sets $\mathcal{Y}(x_i)$ with $x_i \in \mathcal{C}$ are closed halfspaces $P(x_i)$. In fact, let us take $P(x_i)$ to be the halfspace determined by the supporting hyperplane of $\mathcal{Y}(x_i)$, containing $\mathcal{Y}(x_i)$, and suppose that, for some hyperplane $G \ni y^*$, there are n-faced cones P_s and P_r that are separated by G. Then it follows from the inclusion $\mathcal{Y}_i \subset P_i$ for every i that G also separates $\mathcal{Y}_r$ and $\mathcal{Y}_s$.

It remains to show that there are at least two such separating cones. We shall suppose, for the sake of clarity, that y^* is the origin in $\mathbf{R}^n$. Because of the projective nature of the statement of the proposition that we are to prove, we may restrict ourselves to the case when n of our hyperplanes are coordinate hyperplanes and the $(n+1)$th intersects them in an arbitrary way (passing, say, through the origin). Let its equation be $\alpha_1 y_1 + \ldots + \alpha_n y_n = 0$. All coefficients α_i must be different from zero. Since the directions of the coordinate axes can be chosen arbitrarily, we may change the orientation of those that correspond to negative coefficients α_i. Finally, if we change the scale of each coordinate axis by a factor α_i, we obtain the hyperplane $y_1 + \ldots + y_n = 0$.

Now we are concerned with the $n+1$ cones

$$K_0 = \{y : y_j \geqq 0,\ j = 1, \ldots, n\},$$
$$K_i = \{y : y_j \geqq 0,\ j \neq i \text{ and } y_1 + \ldots + y_n \leqq 0\},\ i = 1, \ldots, n.$$

Let us show that every hyperplane $\beta y^T = \beta_1 y_1 + \ldots + \beta_n y_n = 0$ separates at least one pair of these cones.

If $\beta > 0$ then $\beta y^T \geqq 0$ for $y \in K_0$. Taking $i_0 = \arg\max \beta_i$, we will have, for $y \in K_{i_0}$,

$$\beta y^T = \sum_{i \neq i_0} (\beta_i - \beta_{i_0}) y_i + \beta_{i_0} \sum_{i=1}^{n} y_i \leqq 0.$$

Now let the coefficients β_i have different signs. Let $\beta_{i+} = \arg\max \beta_i$ and $\beta_{i-} = \arg\min \beta_i$. Then for $y \in K_{i+}$ we will have

$$\beta y^T = \sum_{i \neq i+} (\beta_{i-} - \beta_{i+}) y_i + \beta_{i+} \sum_{i=1}^{n} y_i \leqq 0,$$

and for $y \in K_{i-}$,

$$\beta y^T = \sum_{i \neq i-} (\beta_{i+} - \beta_{i-}) y_i + \beta_{i-} \sum_{i=1}^{n} y_i \geqq 0. \qquad \square$$

3.11 Theorem. *If, in a convex two-person zero-sum game* Γ *as in* (3.1), *the set* $\mathcal{T}(\Gamma) = \mathcal{y}^*$ *has dimension* p *and lies on the boundary of the set* $\mathcal{y}$ *of all strategies, then for every* $\varepsilon > 0$ *player 1 has an* ε*-optimal strategy with a spectrum consisting of at most* $n - p$ *points.*

Proof. If $p > 0$ we may use projections, as was done in the proof of the lemma in 3.8. We may therefore suppose that $p = 0$.

By the same kind of method as in the proof of the theorem in 3.9, we obtain the formula (3.7), which can evidently be modified to read $\mathcal{y}_{v'}(\mathcal{c}) = \bigcap_{x \in \mathcal{c}} \partial \mathcal{y}_{v'}(x) \cap \partial \mathcal{y}$.

Therefore we can apply to the sets $\{\{\mathcal{y}_{v'}(x)\}_{x \in \mathcal{c}}, \mathcal{y} = \mathcal{y}(x_0)\}$ the lemma of the preceding subsection, according to which there is a $\mathcal{c}_0 \subset \mathcal{c}$ such that $|\mathcal{c}_0| \leqq n$, and $\mathcal{c}$ can be replaced by $\mathcal{c}_0$ in the preceding equation. Then, because $\mathcal{y}_{v'}(x) \subset \mathcal{y}$ for every $x \in \mathcal{c}_0$, we must have $\mathcal{y}_{v'}(\mathcal{c}_0) = \bigcap_{x \in \mathcal{c}_0} \partial \mathcal{y}_{v'}(x)$, from which, as in the theorem of 3.9, we obtain what was required.

3.12. In order to pass from the existence, in convex games, of ε-optimal strategies for player 1 for every $\varepsilon > 0$, to the existence of optimal strategies for the same player, it is natural to suppose that the set $\mathcal{x}$ is compact in some topology, and to subject H_y to the hypothesis of lower semicontinuity (in the same topology).

Then the necessary transition is supplied by the following lemma.

Lemma. *Suppose in the game* Γ *in* (3.1), *with the hypotheses that* $\mathcal{x}$ *is compact and* H_y *is upper semicontinuous for every* $\varepsilon > 0$, *there is an* ε*-optimal strategy for player* 1, *whose spectrum contains at most* n *points, where* n *is independent of* ε.

Then player 1 *in* Γ *also has an optimal strategy whose spectrum contains at most* n *points.*

Proof. Choose $\varepsilon_k \downarrow 0$ arbitrarily. Let X_k be an ε-optimal strategy of player 1, with spectrum containing at most n points. We have

$$H(X_k, y) \geqq v_\Gamma - \varepsilon_k \text{ for every } y \in \mathcal{y}.$$

Form the Cartesian product $\mathcal{x}^* = (\mathcal{x} \times [0, 1])^*$. This is a compact space in the natural induced topology on this product. The mixed strategies X_k can be thought of as points of $\mathcal{x}^*$. We can select a convergent subsequence from the sequence of these points; we may suppose as usual that the original sequence converges. We denote its limit by X_0. It is easily verified that the functions H_y are upper semicontinuous. Therefore passage to the limit on k yields

$$H(X_0, y) \geqq v_\Gamma \text{ for every } y \in \mathcal{y},$$

i.e. the optimality of X_0. □

3.13. What has been said leads us to a theorem which, together with the theorem in 2.5, provides a general description of the results of applying the maximum principle for convex games.

Theorem. *If, in a convex game* Γ *as in* (3.1), *the function* H *is upper semicontinuous on* $\boldsymbol{x}$, *then* $\mathcal{T}(\Gamma) \neq \emptyset$ *and, moreover,*

1) *if* $\boldsymbol{y} \subset \mathbf{R}^n$, *then player* 1 *has an optimal strategy with spectrum containing at most* $n+1$ *points;*

2) *if* $\boldsymbol{y} \subset \mathbf{R}^n$, *and* $\dim \mathcal{T}(\Gamma) = p \leqq n$, *then player* 1 *has an optimal strategy with spectrum containing at most* $n-p+1$ *points;*

3) *if* $\boldsymbol{y} \subset \mathbf{R}^n$, $\dim \mathcal{T}(\Gamma) = p \leqq n$, *and* $\mathcal{T}(\Gamma) \subset \partial \boldsymbol{y}$, *then player* 1 *has an optimal strategy with spectrum containing at most* $n-p$ *points.*

The proof of part 1) follows immediately from 3.7 and 3.12; part 2), from 3.9 and 3.12; and part 3), from 3.11 and 3.12. □

3.14 Convexlike games. It follows from the very definition of a convex (or concave) game that these concepts make sense only for games in which the space of strategies has a linear structure. This restriction can be overcome in various ways. The first of these is based on the concept of a concavelike (or convexlike) game.

Definition. A function

$$f : \boldsymbol{z} \to \mathbf{R} \tag{3.8}$$

is said to be *concavelike* if, for all z' and $z'' \in \boldsymbol{z}$ and $\lambda \in [0,1]$, there exists $z^\lambda \in \boldsymbol{z}$ such that

$$f(z^\lambda) \geqq \lambda f(z') + (1-\lambda) f(z'').$$

A function (3.8) is said to be convexlike if, under the same conditions,

$$f(z^\lambda) \leqq \lambda f(z') + (1-\lambda) f(z''). \qquad \square$$

Definition. A two-person zero-sum game as in (3.1) is said to be *concavoconvexlike* if:

the function H_x is convexlike for every $x \in \boldsymbol{x}$; (3.9)

the function H_y is concavelike for every $y \in \boldsymbol{y}$; (3.10)

a function $H : \boldsymbol{x} \times \boldsymbol{y} \to \mathbf{R}$ that satisfies (3.9) and (3.10) is also called *concavoconvexlike.* □

A natural iteration of concavelikeness and convexlikeness leads to the following proposition.

If the function f in (3.8) is concavelike, then for all $z_1, \dots, z_n \in \boldsymbol{z}$ and $\lambda = (\lambda_1, \dots, \lambda_n) \in \mathbf{R}^n$ with $\lambda_i \geqq 0$ and $\sum_{i=1}^n \lambda_i = 1$, there exists $z^\lambda \in \boldsymbol{z}$ such that

$$f(z^\lambda) \geqq \sum_{i=1}^n \lambda_i f(z_i).$$

Correspondingly, if the function f in (3.8) is convexlike, then, with the same notation, there exists $z^\lambda \in \boldsymbol{z}$ such that the opposite inequality is satisfied.

3.15. The following theorem provides a nearly unique (necessary and sufficient) test for the existence of a value for a game with an $\mathscr{S}$-precompact space of strategies for player 1. In addition, it can serve as a basis for obtaining further results.

Theorem. *For a game* Γ *as in* (3.1) *with an* $\mathscr{S}$*-precompact space* $\boldsymbol{x}$ *to have a value*

$$v = \sup_{x \in \boldsymbol{x}} \inf_{y \in \boldsymbol{y}} H(x,y) = \inf_{y \in \boldsymbol{y}} \sup_{x \in \boldsymbol{x}} H(x,y), \tag{3.11}$$

it is necessary and sufficient that, for every $\varepsilon > 0$ *and all finite subsets* $\boldsymbol{x}' \subset \boldsymbol{x}$ *and* $\boldsymbol{y}' \subset \boldsymbol{y}$, *there exist* $x^0 \in \boldsymbol{x}$ *and* $y^0 \in \boldsymbol{y}$ *such that*

$$\max_{x \in \boldsymbol{x}'} H(x, y^0) \leqq \min_{y \in \boldsymbol{y}'} H(x^0, y) + \varepsilon. \tag{3.12}$$

Proof. *Necessity.* It follows from (3.1) that for all finite $\boldsymbol{x}' \subset \boldsymbol{x}$ we must have

$$\inf_{y \in \boldsymbol{y}} \max_{x \in \boldsymbol{x}'} H(x,y) \leqq v,$$

so that there is a $y^0 \in \boldsymbol{y}$ such that

$$\max_{x \in \boldsymbol{x}'} H(x, y^0) \leqq v - \varepsilon/2. \tag{3.13}$$

In a similar way, we find that for every $\boldsymbol{y}' \subset \boldsymbol{y}$ there is an $x^0 \in \boldsymbol{x}$ for which

$$v + \varepsilon/2 \leqq \min_{y \in \boldsymbol{y}'} H(x^0, y). \tag{3.14}$$

Then (3.12) follows from (3.13) and (3.14).

Sufficiency. It follows from the $\mathscr{S}$-precompactness of the space $\boldsymbol{x}$ that, for every $\varepsilon > 0$ and $x \in \boldsymbol{x}$, there are a finite $\boldsymbol{x}_\varepsilon \subset \boldsymbol{x}$ and a strategy $x_\varepsilon \in \boldsymbol{x}_\varepsilon$ such that, for each y,

$$H(x,y) \leqq H(x_\varepsilon, y) + \varepsilon.$$

But then

$$\sup_{x \in \boldsymbol{x}} H(x,y) \leqq \max_{x_\varepsilon \in \boldsymbol{x}_\varepsilon} H(x_\varepsilon, y) + \varepsilon,$$

and consequently

$$\inf_{y \in \boldsymbol{y}} \sup_{x \in \boldsymbol{x}} H(x,y) \leqq \inf_{y \in \boldsymbol{y}} \max_{x_\varepsilon \in \boldsymbol{x}} H(x_\varepsilon, y) + \varepsilon. \tag{3.15}$$

Now choose a finite $\boldsymbol{y}' \subset \boldsymbol{y}$. Under our hypotheses there exist $x^0 \in \boldsymbol{x}$ and $y^0 \in \boldsymbol{y}$ such that (3.12) is satisfied. But then

$$\inf_{y \in \boldsymbol{y}} \sup_{x_\varepsilon \in \boldsymbol{x}_\varepsilon} H(x_\varepsilon, y) \leqq \sup_{x \in \boldsymbol{x}} \min_{y' \in \boldsymbol{y}'} H(x, y') + \varepsilon. \tag{3.16}$$

Let $\mathcal{H}(\boldsymbol{y})$ denote the family of finite subsets of $\boldsymbol{y}$. It follows from (3.16) that

$$\inf_{y \in \boldsymbol{y}} \sup_{x_\varepsilon \in \boldsymbol{x}_\varepsilon} H(x_\varepsilon, y) \leqq \inf_{\boldsymbol{y}' \in \mathcal{H}(\boldsymbol{y})} \sup_{x \in \boldsymbol{x}} \min_{y' \in \boldsymbol{y}'} H(x, y') + \varepsilon. \tag{3.17}$$

Denote the right-hand extremum by h. Then (3.17) and (3.15) yield

$$\inf_{y \in \boldsymbol{y}} \sup_{x \in \boldsymbol{x}} H(x, y) \leqq h + 2\varepsilon. \tag{3.18}$$

Let us show, in addition, that

$$h \leqq \sup_{x \in \boldsymbol{x}} \inf_{y \in \boldsymbol{y}} H(x, y) + 2\varepsilon. \tag{3.19}$$

It is evidently enough to restrict ourselves to the case $-\infty < h$. By the $\mathcal{S}$-precompactness of $\boldsymbol{x}$, we also cannot have $h = +\infty$. Consequently we may suppose that $h \in \mathbf{R}$. Suppose that

$$h > \sup_{x \in \boldsymbol{x}} \inf_{y \in \boldsymbol{y}} H(x, y) + 2\varepsilon.$$

This means that, for every $x \in \boldsymbol{x}$ and, in particular, for every $x_\varepsilon \in \boldsymbol{x}_\varepsilon$, we have

$$\inf_{y \in \boldsymbol{y}} H(x_\varepsilon, y) < h - 2\varepsilon.$$

Therefore, by the $\mathcal{S}$-compactness of $\boldsymbol{x}$, we can find, for each $x \in \boldsymbol{x}$, an $x_\varepsilon \in \boldsymbol{x}_\varepsilon$, and a corresponding $y_\varepsilon \in \boldsymbol{y}$, such that

$$H(x, y_\varepsilon) \leqq H(x_\varepsilon, y_\varepsilon) + \varepsilon < h - 2\varepsilon + \varepsilon = h - \varepsilon.$$

Let $\boldsymbol{y}_\varepsilon$ denote the set of all y_ε so found. The preceding proposition shows that

$$\sup_{x \in \boldsymbol{x}} \min_{y_\varepsilon \in \boldsymbol{y}_\varepsilon} H(x, y_\varepsilon) < h - \varepsilon.$$

But since $\boldsymbol{y}_\varepsilon \in \mathcal{H}(\boldsymbol{y})$, the left-hand side cannot be less than h, i.e. we have $h < h - \varepsilon$, which is impossible. This establishes (3.1). Together with (3.18), this yields

$$\inf_{y \in \boldsymbol{y}} \sup_{x \in \boldsymbol{x}} H(x, y) \leqq \sup_{x \in \boldsymbol{x}} \inf_{y \in \boldsymbol{y}} H(x, y) + 4\varepsilon.$$

Since $\varepsilon > 0$ is arbitrary, omitting the term 4ε on the right does not affect the inequality, and it remains only to apply the minimax inequality.

3.16. Let us apply the preceding theorem to a concavo-convex game.

Theorem. *If the set $\boldsymbol{x}$ in a concavo-convex game Γ, as in* (3.1), *is $\mathcal{S}$-precompact, then Γ has a value.*

Proof. In view of what precedes, it is enough to verify that, given an arbitrary $\varepsilon > 0$, and finite $\mathfrak{x}_\varepsilon \subset \mathfrak{x}$ and $\mathfrak{y}_\varepsilon \subset \mathfrak{y}$, the inequality (3.12) is satisfied for some $x^0 \in \mathfrak{x}$ and $y^0 \in \mathfrak{y}$.

Let us consider a finite $\mathfrak{x}_\varepsilon \times \mathfrak{y}_\varepsilon$-subgame Γ_ε of Γ, and a saddle point $(X_\varepsilon, Y_\varepsilon)$ of Γ_ε (in the mixed strategies). Set

$$\mathfrak{x}_\varepsilon = \{x_1, \dots, x_m\}, \quad X_\varepsilon(x_i) = \xi_i, \quad i = 1, \dots, m,$$

$$\mathfrak{y}_\varepsilon = \{y_1, \dots, y_n\}, \quad Y_\varepsilon(y_j) = \eta_j, \quad j = 1, \dots, n.$$

Then, since H is concavo-convexlike, there are $x^0 \in \mathfrak{x}$ and $y^0 \in \mathfrak{y}$ for which

$$H(X_\varepsilon, y) = \sum_{i=1}^{m} \xi_i H(x_i, y) \leqq H(x^0, y) \text{ for every } y \in \mathfrak{y},$$

$$H(x, Y_\varepsilon) = \sum_{j=1}^{n} \eta_j H(x, y_j) \geqq H(x, y^0) \text{ for every } x \in \mathfrak{x},$$

and since $(X_\varepsilon, Y_\varepsilon)$ is an equilibrium situation,

$$H(x_i, y_0) \leqq H(x_i, Y_\varepsilon) \leqq H(X_\varepsilon, Y_\varepsilon) \leqq H(X_\varepsilon, y_j) \leqq H(x^0, y_j)$$

for every $x_i \in \mathfrak{x}_\varepsilon$ and $y_j \in \mathfrak{y}_\varepsilon$. Thus (3.12) is satisfied (with something to spare). □

3.17. A special case of the preceding theorem has been known for a long time: if a concavo-convexlike game is $\mathcal{J}$-precompact, then it has a value.

Finally, if we strengthen concavo-convexlikeness to concavo-convexity, we obtain the theorem of 3.10, Chapter 1, for the case of a two-person zero-sum game.

3.18. Connectedness of the payoff function. Another way of avoiding the linear structure of the space of strategies is to replace the condition of convexity of the payoff function by the weaker condition of its connectedness.

Definition. A function $H : \mathfrak{x} \times \mathfrak{y} \to \mathbf{R}$ is said to be *α-connected* on $\mathfrak{x} \times \mathfrak{y}$ if the following two conditions are satisfied:

1) For every finite subset $\mathfrak{y}' \subset \mathfrak{y}$, the intersection $\bigcap_{y \in \mathfrak{y}'} \mathfrak{x}_\alpha(y) = \mathfrak{x}_\alpha(\mathfrak{y}')$ (see (3.2)) is connected (or empty).

2) For every y' and $y'' \in \mathfrak{y}$ there is a continuous mapping $u : [0, 1] \to \mathfrak{y}$ for which $u(0) = y'$, $u(1) = y''$, and the inequalities $0 \leqq \mu \leqq \lambda \leqq \nu \leqq 1$ imply

$$\mathfrak{x}_\alpha(u(\lambda)) \subset \mathfrak{x}_\alpha(u(\mu)) \cup \mathfrak{x}_\alpha(u(\nu)). \tag{3.20}$$

The function H is said to be *α^--connected* on $\mathfrak{x} \times \mathfrak{y}$ if there is a sequence of numbers

$$\varepsilon_1 \geqq \varepsilon_2 \geqq \dots \geqq 0, \tag{3.21}$$

that converges monotonically to zero, such that H is $(\alpha - \varepsilon_n)$-connected on $\mathfrak{x} \times \mathfrak{y}$.

The function H is said to be *strictly* α^-*-connected* on $\mathfrak{x} \times \mathfrak{y}$ if there is a strictly monotonic sequence (3.21) that satisfies the conditions for α^--connectedness. □

Later we shall assume that $\mathfrak{x}$ is compact in some topology, and that the payoff function H is upper semicontinuous in x in this topology; we shall not mention this explicitly in each of the lemmas that follow, but state only any supplementary hypotheses.

Let us consider the upper value of Γ:

$$\bar{v} = \bar{v}_\Gamma = \inf_y \sup_x H(x, y) \tag{3.22}$$

and establish some auxiliary propositions.

3.19 Lemma. *Let there exist a sequence* (3.21) *for which*

$$\bigcap_{y \in \mathfrak{y}'} \mathfrak{x}_{\gamma - \varepsilon_n}(y) \neq \emptyset \tag{3.23}$$

for every n *and every finite subset* $\mathfrak{y}' \subset \mathfrak{y}$.

Then Γ *has a value.*

Proof. It follows from (3.22) that, for every strategy $y \in \mathfrak{y}$,

$$\sup_x H(x, y) \geqq \bar{v}_\Gamma,$$

i.e., for every ε_n there is an x_n such that

$$H(x_n, y) \geqq \bar{v} - \varepsilon_n.$$

But this means that for every $y \in \mathfrak{y}$

$$\mathfrak{x}_{\bar{v} - \varepsilon_n}(y) \neq \emptyset.$$

On the other hand, the upper semicontinuity of H_y implies that each set $\mathfrak{x}_{\bar{v} - \varepsilon_n}(y)$ is closed. In addition, by hypothesis, the system of nonempty closed sets $\mathfrak{x}_{\bar{v} - \varepsilon_n}(y)$ is centralized. Consequently, by the compactness of $\mathfrak{x}$, this system has a nonempty intersection.

Let the strategy x_n belong to all the sets $\mathfrak{x}_{\bar{v} - \varepsilon_n}(y)$. This means that $H(x_n, y) \geqq \bar{v} - \varepsilon_n$ for all $y \in \mathfrak{y}$, i.e.

$$\inf_y H(x_n, y) \geqq \bar{v} - \varepsilon_n. \tag{3.24}$$

Since $\mathfrak{x}$ is compact, we can select from the sequence $x_1, x_2, \ldots$ a convergent subsequence with limit x_0. By the upper semicontinuity of H_y, letting $n \to \infty$ in (3.24) yields

$$\inf_y H(x_n, y) \geqq \bar{v},$$

and therefore

$$\sup_x \inf_y H(x, y) \geqq \bar{v}.$$

The proof is completed by referring to the minimax inequality. □

3.20. We are going to present two versions of this discussion, in which we impose different analytic requirements on H with respect to its dependence on y.

Lemma. *If H is lower semicontinuous in y and is $(\bar{v}-\varepsilon)$-connected for some $\varepsilon > 0$, then we must have, for arbitrary y' and $y'' \in \mathcal{y}$,*

$$\mathcal{x}_{\bar{v}-\varepsilon}(y') \cap \mathcal{x}_{\bar{v}-\varepsilon}(y'') \neq \emptyset. \tag{3.25}$$

Proof. Suppose that, contrary to the conclusion of the lemma, we had

$$\mathcal{x}_{\bar{v}-\varepsilon}(y') \cap \mathcal{x}_{\bar{v}-\varepsilon}(y'') = \emptyset. \tag{3.26}$$

Construct, corresponding to the strategies y' and y'', a mapping $u : [0,1] \to \mathcal{y}$ that establishes the $(\bar{v}-\varepsilon)$-connectedness. Here $y' = u(0)$ and $y'' = u(1)$. Hence, by (3.20), we must have

$$\mathcal{x}_{\bar{v}-\varepsilon}(u(\lambda)) \subset \mathcal{x}_{\bar{v}-\varepsilon}(y') \cup \mathcal{x}_{\bar{v}-\varepsilon}(y'') \tag{3.27}$$

for an arbitrary $\lambda \in [0,1]$.

If the set $\mathcal{x}_{\bar{v}-\varepsilon}(u(\lambda))$ had a nonempty intersection with both sets $\mathcal{x}_{\bar{v}-\varepsilon}(y')$ and $\mathcal{x}_{\bar{v}-\varepsilon}(y'')$, then by (3.26) these two intersections would themselves not intersect and, being nonempty closed subsets of $\mathcal{x}_{\bar{v}-\varepsilon}(u(\lambda))$, would separate it, contrary to its connectedness.

Therefore it follows from (3.27) that $\mathcal{x}_{\bar{v}-\varepsilon}(u(\lambda))$ is completely contained in one of the sets $\mathcal{x}_{\bar{v}-\varepsilon}(y')$ or $\mathcal{x}_{\bar{v}-\varepsilon}(y'')$. Let M' be the set of points $\lambda \in [0,1]$ for which the first possibility is realized, and let M'' be the set of points λ for which the second is realized. Clearly $0 \in M'$, $1 \in M''$, $M' \cap M'' = \emptyset$, and $M' \cup M'' = [0,1]$. Moreover, if $0 \leqq \lambda_1 < \lambda_2 \leqq 1$, we have, by (3.20),

$$\mathcal{x}_{\bar{v}-\varepsilon}(u(\lambda_1)) \subset \mathcal{x}_{\bar{v}-\varepsilon}(y') \cup \mathcal{x}_{\bar{v}-\varepsilon}(u(\lambda_2)).$$

Therefore, if $\lambda_2 \in M'$, i.e. if

$$\mathcal{x}_{\bar{v}-\varepsilon}(u(\lambda_2)) \subset \mathcal{x}_{\bar{v}-\varepsilon}(y'),$$

then it must also be true that

$$\mathcal{x}_{\bar{v}-\varepsilon}(u(\lambda_1)) \subset \mathcal{x}_{\bar{v}-\varepsilon}(y'),$$

i.e, $\lambda_1 \in M'$. It follows from what we have said that M' is an interval containing its left-hand endpoint 0. Similarly we find that M'' is also an interval, containing its right-hand endpoint 1.

Let

$$\bar{\lambda} = \sup M' = \inf M'', \tag{3.28}$$

and suppose for definiteness that $\bar{\lambda} \in M'$. This means that

$$\mathcal{x}_{\bar{v}-\varepsilon}(u(\bar{\lambda})) \subset \mathcal{x}_{\bar{v}-\varepsilon}(y').$$

It follows that

$$H(x, u(\lambda')) < \bar{v} - \varepsilon \text{ for all } x \notin \mathcal{x}_{\bar{v}-\varepsilon}(y'). \tag{3.29}$$

Now choose a sequence λ_m that converges to $\bar{\lambda}$ from the right. By the definition of M', we must have, for every such λ_m,

$$\boldsymbol{x}_{\bar{v}-\varepsilon}(u(\lambda_m)) \subset \boldsymbol{x}_{\bar{v}-\varepsilon}(y'').$$

Consequently, for every $x \notin \boldsymbol{x}_{\bar{v}-\varepsilon}(y'')$, and therefore, in view of (3.26), for every $x \in \boldsymbol{x}_{\bar{v}-\varepsilon}(y')$ we must have $H(x, u(\lambda_m)) < \bar{v} - \varepsilon$; and since H is lower semicontinuous in its second argument,

$$H(x, u(\bar{\lambda})) \leqq \bar{v} - \varepsilon \text{ for all } x \in \boldsymbol{x}_{\bar{v}-\varepsilon}(y'). \tag{3.30}$$

It follows from (3.29) and (3.30) that

$$H(x, u(\bar{\lambda})) \leqq \bar{v} - \varepsilon \text{ for all } x \in \boldsymbol{x},$$

whence

$$\inf_{y \in \boldsymbol{y}} \sup_{x \in \boldsymbol{x}} H(x, y) \leq \bar{v} - \varepsilon,$$

which, however, contradicts (3.22). □

3.21 Lemma. *Let H be lower semicontinuous in x and strictly $\bar{v}^-$-connected with a sequence $\varepsilon_1 > \varepsilon_2 > \ldots > 0$. Then, for every ε_n in this sequence and every finite subset $\boldsymbol{y}' \subset \boldsymbol{y}$, we must have*

$$\bigcap_{y \in \boldsymbol{y}'} \boldsymbol{x}_{\bar{v}-\varepsilon_n}(y) \neq \emptyset.$$

The proof is by induction on $|\boldsymbol{y}'|$. The case $|\boldsymbol{y}'| = 2$ is the content of the preceding lemma. Suppose that the conclusion has been established for $|\boldsymbol{y}'| = k - 1$, and consider the case $|\boldsymbol{y}'| = k$ (for definiteness, let $\boldsymbol{y}' = \{y^1, \ldots, y^k\}$).

Now set $\boldsymbol{x}^n = \boldsymbol{x}_{\bar{v}-\varepsilon_n}(y^k)$ and form the $\boldsymbol{x}^n \times \boldsymbol{y}$-subgame Γ^n of Γ. For this game,

$$\boldsymbol{x}^n_\alpha(y) = \{x : H(x, y) \geqq \alpha, x \in \boldsymbol{x}^n\} = \boldsymbol{x}^n \cap \{x : H(x, y) \geqq \alpha\} = \boldsymbol{x}^n \cap \boldsymbol{x}_\alpha(y).$$

Let us show that Γ^n satisfies the hypotheses of the preceding lemma. Evidently we have lower semicontinuity in y and $(\bar{v} - \varepsilon)$-connectedness for the payoff function H. However, we have to show that the number $\bar{v}$ plays the same role for Γ^n as for Γ, i.e., is its upper value.

Evidently, $\bar{v}(\Gamma^n) \leqq \bar{v}(\Gamma) = \bar{v}$. Now choose $r > n$ and observe that, by the preceding lemma, $\boldsymbol{x}_{\bar{v}-\varepsilon_r}(y^k) \cap \boldsymbol{x}_{\bar{v}-\varepsilon_r}(y) \neq \emptyset$ for every $y \in \boldsymbol{y}$; and since

$$\boldsymbol{x}_{\bar{v}-\varepsilon_r}(y^k) \subset \boldsymbol{x}_{\bar{v}-\varepsilon_n}(y^k),$$

we must also have

$$\boldsymbol{x}_{\bar{v}-\varepsilon_n}(y^k) \cap \boldsymbol{x}_{\bar{v}-\varepsilon_n}(y) \neq \emptyset \text{ for every } y \in \boldsymbol{y}.$$

This means that, for every $y \in \boldsymbol{y}$, there exists $x \in \boldsymbol{x}'$ such that $H(x, y) \geqq \bar{v} - \varepsilon_r$, i.e., for every $y \in \boldsymbol{y}$ we have

$$\sup_{x \in \boldsymbol{x}^n} H(x, y) \geqq \bar{v} - \varepsilon_r,$$

and therefore

$$\bar{v}(\Gamma^n) = \inf_{y \in \mathcal{y}} \sup_{x \in \mathcal{x}^n} H(x, y) \geqq \bar{v} - \varepsilon_r .$$

Letting ε_r tend to zero, we obtain $\bar{v}(\Gamma^n) \geqq \bar{v}$, and, by taking account of the remark above, also $\bar{v}(\Gamma^n) = \bar{v}$. This verifies our assertion.

Application of the inductive hypothesis to Γ^n yields

$$\bigcap_{l=1}^{k-1} x^n_{\bar{v}-\varepsilon_n}(y^l) = \bigcap_{l=1}^{k-1} x_{\bar{v}-\varepsilon_n}(y^l) \cap x_{\bar{v}-\varepsilon_n}(y^k) = \bigcap_{l=1}^{k} x_{\bar{v}-\varepsilon_n}(y^l),$$

and the induction is complete. □

3.22. The results of the lemmas in 3.19 and 3.21 constitute the following theorem.

Theorem. *If the set of strategies of player 1 in the game Γ is compact, and the payoff function H is strictly $\bar{v}$-connected and is upper semicontinuous in x and lower semicontinuous in y, then the game has a value.*

For the proof, it is enough to notice that under the hypotheses of the theorem, the hypotheses of Lemmas 3.20 and 3.21 are satisfied, and the conclusion of Lemma 3.21 ensures that the hypotheses of Lemma 3.19 are satisfied; this is just what was required. □

3.23. It turns out that, in the theorem just established, the hypothesis that H is lower semicontinuous in y can be replaced by the hypothesis that H is upper semicontinuous in y, provided that the connectivity condition is somewhat modified.

Lemma. *If the function H is upper semicontinuous in y and is $(\bar{v} - \varepsilon)$-connected for some $\varepsilon \geqq 0$, then (3.26) will hold for all y' and $y'' \in \mathcal{y}$.*

Proof. Suppose, as in the proof of the lemma of 3.20, that (3.27) holds for some y' and $y'' \in \mathcal{y}$. As before, choose sets M' and M'' which are intervals covering [0, 1]. Suppose, for definiteness, that the number $\bar{\lambda}$ in (3.29) belongs to the left-hand interval, namely M'.

It follows that

$$x_{\bar{v}-\varepsilon}(u(\bar{\lambda})) \cap x_{\bar{v}-\varepsilon}(y'') = \emptyset,$$

i.e., $H(x, u(\bar{\lambda})) < \bar{v} - \varepsilon$ for all $x \in x_{v-\varepsilon}$.

Since we have assumed that H_x is upper semicontinuous in $y \in \mathcal{y}$, for every $x \in x_{\bar{v}-\varepsilon}(y'')$ there must be a neighborhood $\mathcal{y}(x)$ of the strategy $u(\bar{\lambda})$ in which all strategies satisfy

$$H(x, y) < \bar{v} - \varepsilon, \qquad y \in \mathcal{y}(x). \tag{3.31}$$

Since u is continuous, the preimage $u^{-1}(\mathcal{y}(x))$ is an open subset of [0,1] that contains $\bar{\lambda}$. Choose points $\lambda_1 = \lambda_1(x)$ and $\lambda_2 = \lambda_2(x)$ for which $\bar{\lambda} \in [\lambda_1, \lambda_2] = I_x$. Since $u(\lambda_i) \in \mathcal{y}(x)$ for $i = 1, 2$, it follows from (3.31) that $H(x, u(\lambda_1)) < \bar{v} - \varepsilon$. Therefore, by the upper semicontinuity of H in x, we can find, for $i = 1, 2$, neighborhoods $w_i(x)$ of x such that

$$H(x', u(\lambda_i)) < \bar{v} - \varepsilon \text{ for all } x' \in w_i(x).$$

Let $w(x) = w_1(x) \cap w_2(x)$. It is clear that for all $x' \in w(x)$ we have $H(x', u(\lambda_i)) < \bar{v} - \varepsilon$ for both $i = 1$ and 2, i.e., $x' \notin x_{\bar{v}-\varepsilon}u(\lambda_i))$ for $i = 1, 2$. But then it follows from (3.20) with $\alpha = \bar{v} - \varepsilon$,

$\mu = \lambda_1$, and $\nu = \lambda_2$ that for any $\lambda \in [\lambda_1, \lambda_2]$ we must have $x \notin x_{\bar{v}-\varepsilon}(u(\lambda))$, and consequently $H(x', u(\lambda)) < \bar{v} - \varepsilon$.

Consequently, to each strategy $x \in x_{\bar{v}-\varepsilon}(y'')$ there correspond a neighborhood $w(x) \ni x$ and an interval $I_x \ni \bar{\lambda}$, such that $H(x', u(\lambda)) < \bar{v} - \varepsilon$ for any choice of $\lambda \in I_x$ and $x' \in w(x)$. Since the space x is compact and $x_{\bar{v}-\varepsilon}(y'')$ is closed in this space, it follows that $x_{\bar{v}-\varepsilon}(y'')$ is compact. Therefore there is a finite set $q \subset x$ such that

$$\bigcup_{x \in q} w(x) \supset x_{\bar{v}-\varepsilon}(y'').$$

This means that for each $x \in x_{\bar{v}-\varepsilon}(y'')$ there is an $x' \in q$ for which $x \in w(x')$. Take $\lambda \in \bigcap_{x \in q} I_x = I$. Here we must have $H(x, u(\lambda)) < \bar{v} - \varepsilon$, i.e., $x \notin x_{\bar{v}-\varepsilon}(u(\lambda))$. Therefore, for every $\lambda \in I$ we must have $x_{\bar{v}-\varepsilon}u(\lambda)) \cap x_{\bar{v}-\varepsilon}(y'') = \emptyset$, i.e., $x_{\bar{v}-\varepsilon}(u(\lambda)) \subset x_{\bar{v}-\varepsilon}(y')$ or, in other words, $I \subset M'$; but this contradicts the fact that $\max I > \bar{\lambda} = \max M'$. □

3.24. By a deduction similar to that of the lemma of 3.21 from the lemma of 3.20 by means of the lemma of 3.19, we can obtain the following proposition from the lemma of 3.23.

Lemma. *Let H be upper semicontinuous in x and y, and also $\bar{v}$-connected for the sequence $\varepsilon_n \downarrow 0$. Then, for an arbitrary n and a finite $y' \subset y$, we have*

$$\bigcup_{y \in y'} x_{\bar{v}-\varepsilon_n}(y) = \emptyset.$$

The following theorem is an immediate consequence of this lemma together with the lemmas of 3.19 and 3.23.

Theorem. *If the set x of strategies of player 1 in the game Γ is compact, and the payoff function H is both $\bar{v}^-$-connected and upper semicontinuous in both x and y, then Γ has a value.*

3.25. The hypothesis of $\bar{v}^-$-connectedness is very general, but difficult to verify and, in any case, insufficiently transparent. Consequently it is natural to introduce a stronger condition that is more intuitive.

Definition. A function $H : x \times y \to \mathbf{R}$ is said to be *strongly connected* if it is α-connected for every real α.

It is clear that the strong connectedness of H implies its $\bar{v}^-$-connectedness, as well as its strict $\bar{v}^-$-connectedness. Consequently the theorems of 3.22 and 3.24 imply the following proposition.

Theorem. *If the payoff function H is strongly connected, upper semicontinuous in x, and lower (or upper) semicontinuous in y, then Γ has a value.*

3.26. The property of strong connectedness of a function consists of the satisfaction, for every real α, of the two hypotheses in 3.18. The second of these can be modified, if x is compact and H is upper semicontinuous in x, by replacing it by the following more powerful hypothesis:

Let $\bar{y}_\alpha(x) = \{y : H(x, y) \geqq \alpha\}$.

For every y' and $y'' \in y$, there is a continuous mapping $u : [0, 1] \to y$ for which $u(0) = y', u(1) = y''$, and the set $u^{-1}(\bar{y}_\alpha(x))$ is connected (or empty) for every $x \in x$.

That the modified hypothesis of strong connectedness of the payoff function is more powerful follows from the following relationships.

Connectedness of the set $u^{-1}(\tilde{y}_\alpha(x))$ means that whenever $0 \leqq \mu \leqq \lambda \leqq \nu \leqq 1$,

$$\mu, \nu \in u^{-1}(\tilde{y}_\alpha(x)) \text{ imply } \lambda \in u^{-1}(\tilde{y}_\alpha(x)),$$

i.e.,

$$u(\mu), u(\nu) \in \tilde{y}_\alpha(x) \text{ imply } u(\lambda) \in \tilde{y}_\alpha(x),$$

i.e.,

$$H(x, u(\mu)) \geqq \alpha \text{ and } H(x, u(\nu)) \geqq \alpha \text{ imply } H(x, u(\lambda)) \geqq \alpha,$$

i.e., arguing from the contrapositive,

$$H(x, u(\lambda)) < \alpha \text{ implies } H(x, u(\mu)) < \alpha \text{ or } H(x, u(\nu)) < \alpha.$$

Consequently we also have

$$H(x, u(\lambda)) < \alpha \text{ implies } H(x, u(\mu)) \leqq \alpha \text{ or } H(x, u(\nu)) \leqq \alpha.$$

Therefore for every $\varepsilon > 0$

$$x \in x_{\alpha-\varepsilon}(u(\lambda)) \text{ implies } x \in x_\alpha(u(\mu)) \text{ or } x \in x_\alpha(u(\nu)),$$

i.e.,

$$x_{\alpha-\varepsilon}(u(\lambda)) \subset x_\alpha(u(\mu)) \cup x_\alpha(u(\nu)).$$

Finally, since x is compact and H is upper semicontinuous in x, we can let ε tend to zero to obtain (3.23).

Consequently, in the statement of the theorem of 3.25, we can replace the original hypothesis of strong connectedness of the payoff function by the modification that we have discussed.

3.27 Quasiconcavo-convex games. Let us think of the two-person zero-sum game of (3.21) as a noncooperative game in the sense of its formal definition, i.e. as the system

$$\langle \{1, 2, \{x, y\}, \{H, -H\} \rangle.$$

Then the quasiconcavity of the game in the sense of 3.9 of Chapter 1 means the quasiconcavity of each function H_x for $x \in x$ and each function $-H_y$ for $y \in y$. The latter is evidently equivalent to the quasiconvexity of each function H_y. Consequently we are led to the following definition.

Definition. If a quasiconcave noncooperative game is a two-person zero-sum game, it is said to be *quasiconcavo-convex* (cf. the remark on the definition in 3.3). □

In just the same way, if we are dealing with a two-person zero-sum game Γ which is, as a noncooperative game, upper semicontinuous (see 1.9, Chapter 1), this implies the upper semicontinuity of H with respect to x and lower semicontinuity with respect to y.

3.28. We can now obtain from the theorem of 3.19 a proposition that is close to the theorem of 3.10, Chapter 1, for two-person zero-sum games.

Theorem. *In a quasiconcavo-convex two-person zero-sum game, as in* (3.1), *let one of the sets x and y of strategies be compact, and let the payoff function H be upper semicontinuous in x and lower semicontinuous in y. Then Γ has a value.*

Proof. In accordance with the theorem of 3.22, it is enough to prove that a quasiconconvexo-concave function is strictly v^--connected.

In fact, by hypothesis every set of the form $\mathfrak{x}_\alpha(y)$, and therefore the intersection of any family of such sets, is convex and therefore connected. This is guaranteed by the first condition of α-connectedness in 3.18.

Now choose y' and $y'' \in \mathfrak{y}$ arbitrarily, and form the interval

$$\{u(\lambda) : \lambda y' + (1-\lambda)y'', \lambda \in [0,1]\},$$

which is a subset of $\mathfrak{y}$ by the convexity of that set. If now $0 \leqq \mu \leqq \lambda \leqq \nu \leqq 1$, then $u(\lambda)$ is in the interval whose endpoints are $u(\mu)$ and $u(\nu)$. Let us verify the validity of (3.20). For this purpose we rewrite it explicitly as

$$\{x : H(x, u(\lambda)) \leqq \alpha\} \subset \{x : H(x, u(\mu)) \leqq \alpha\} \cup \{x : H(x, u(\nu)) \leqq \alpha\},$$

or

$$H(x, u(\lambda)) \leqq \alpha \text{ implies } H(x, u(\mu)) \text{ or } H(x, u(\nu)) \leqq \alpha,$$

$$H(x, u(\lambda)) \geqq \min\{H(x, u(\nu)), H(x, u(\nu))\},$$

and this is the case for the quasiconcave function H_x. □

3.29 A specific minimax theorem. It follows from what was said in 3.2 that a continuous game on the unit square has a saddle point.

In cases when the payoff function in a game on the unit square is not continuous, but is discontinuous only on a finite number of intervals parallel to the sides of the situation square, the game can easily be shown to retain $\mathcal{J}$-compactness, and therefore has saddle points.

The situation is completely different when the set of discontinuities of the payoff function has a more complicated structure. In such cases the spaces of the players' pure strategies have a very meager natural intrinsic topology.

Let us consider, for example, a case when the set of discontinuities of the payoff function H is the diagonal of the unit square:

$$H(x,y) = \begin{cases} 0 & \text{if } x \leqq y, \\ 1 & \text{if } x > y. \end{cases}$$

In such a game, it is clear that $\rho(x_1, x_2) = 1$ for all x_1 and $x_2 \in \mathfrak{x} = [0,1]$, so that, in the intrinsic topology, the space $\mathfrak{x}$ consists of a set of isolated points of the power of the continuum. At the same time, this game has, as is easily verified, an equilibrium situation, even in the pure strategies (namely (0,1)).

Games on the unit square with payoff functions that are discontinuous on the diagonal but continuous off the diagonal occur, for example, in all cases when the players' strategies involve the choice of an instant of time, and it is essential to know which player is ahead. A specimen of problems of this kind will be studied in detail in section 9 of this chapter.

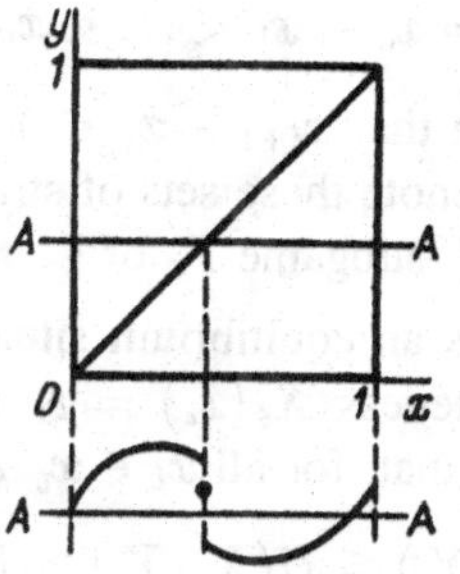

Fig. 3.1

In such cases, the introduction of a natural metric in spaces of strategies turns out to be unpromising. However, for various classes of two-person zero-sum games with discontinuous payoff functions there are a substantial number of theorems on the existence of optimal strategies for the players, or at least on the existence of a value for the game. We shall consider only a relatively elementary example.

3.30 Theorem. *If* Γ *is a game on the unit square with the payoff function*

$$H(x,y) = \begin{cases} L(x,y) & \text{for } x < y, \\ \Phi(x) & \text{for } x = y, \\ M(x,y) & \text{for } x > y. \end{cases},$$

and if L and M are continuous in the closed triangles where they are defined, and, for every $x \in [0,1]$, *the value of* $\Phi(x)$ *lies between* $L(x,x)$ *and* $M(x,x)$:

$$\min\{L(x,x), M(x,x)\} \leqq \Phi(x) \leqq \max\{L(x,x), M(x,x)\} \tag{3.32}$$

(Figure 3.1), *then the players in* Γ *have optimal mixed strategies.*

We call attention to the fact that, in contrast to the hypotheses of the example in 6.6 of Chapter 1, the upper semicontinuity of the payoff function is not assumed in this theorem.

The function H is somtimes called the *kernel* of the game; L and M are its *semikernels*.

Proof. Let us first prove that Γ has a value. For this purpose we choose an arbitrary $\varepsilon > 0$ and use the uniform continuity of L and M to find a δ such that, in the first place, when $|x' - x''| < \delta$ and $y \in [0,1]$ is arbitrary,

$$|L(x',y) - L(x'',y)| < \varepsilon, \tag{3.33}$$

$$|M(x',y) - M(x'',y)| < \varepsilon, \tag{3.34}$$

and, in the second place, similar inequalities are satisfied for $|y' - y''| < \delta$ and arbitrary $x \in [0,1]$.

Now choose any points

$$0 = x_0 < x_1 < \ldots < x_n = 1,$$

subject only to the condition that $x_{i+1} - x_i < \delta$ for $i = 0, 1, \ldots, n-1$, and set $y_i = x_i$ $(i = 0, 1, \ldots, n)$. Denote these sets of strategies of the players by x_ε and y_ε, and consider the $x_\varepsilon \times y_\varepsilon$-subgame Γ_ε of Γ.

The finite game Γ_ε has an equilibrium situation $(X_\varepsilon, Y_\varepsilon)$ (see 1.6, Chapter 2), where we set for definiteness $X_\varepsilon(x_i) = \xi_i$ $(i = 0, \ldots, n)$. That this is an equilibrium situation means that, for all $x_i \in x_\varepsilon$ and $y_i \in y_\varepsilon$,

$$H(x_i, Y_\varepsilon) \leqq H(X_\varepsilon, Y_\varepsilon) \leqq H(X_\varepsilon, y_j). \tag{3.35}$$

Notice that

$$H(X_\varepsilon, y_j) = \sum_{i<j} \xi_i L(x_i, y_j) + \xi_j \Phi(x_j) + \sum_{i>j} \xi_i M(x_i, y_j) \tag{3.36}$$

and similarly

$$H(X_\varepsilon, y_{j+1}) = \sum_{i<j+1} \xi_i L(x_i, y_{j+1}) + \xi_{j+1} \Phi(x_{j+1}) + \sum_{i>j+1} \xi_i M(x_i, y_{j+1}). \tag{3.37}$$

Let us further select any $y \in [0, 1]$. Assuming that $y \notin y_\varepsilon$, let us suppose that $y \in (y_j, y_{j+1})$. We have

$$H(X_\varepsilon, y) = \sum_{i \leqq j} \xi_i L(x_i, y) + \sum_{i>j} \xi_i M(x_i, y). \tag{3.38}$$

Suppose that

$$\Phi(x_j) \leqq L(x_j, y_j). \tag{3.39}$$

In this case, subtracting (3.38) from (3.36) yields

$$\begin{aligned} H(X_\varepsilon, y_j) - H(X_\varepsilon, y) &= \sum_{i<j} \xi_i (L(x_i, y_j) - L(x_i, y)) \\ &+ \xi_j (\Phi(x_j) - L(x_j, x_j)) + \xi_j (L(x_j, x_j) - L(x_j, y)) \\ &+ \sum_{i>j} \xi_i (M(x_i, y_i) - M(x_i, y)). \end{aligned}$$

By inequalities (3.33) and (3.34), the sum of all the terms on the left that do not contain $\Phi(x_j)$ does not exceed ε in absolute value, and by (3.39) the remaining terms are not positive. Therefore

$$H(X_\varepsilon, y) + \varepsilon \geqq H(X_\varepsilon, y_j), \; y \in [y_j, y_{j+1}], \; j = 0, \ldots, n-1. \tag{3.40}$$

Now suppose that

$$\Phi(x_{j+1}) \geqq M(x_{j+1}, x_{j+1}).$$

By an argument similar to that given above, we find that

$$H(X_\varepsilon, y_{j+1}) - H(X_\varepsilon, y) = \sum_{i<j+1} \xi_i (L(x_i, y_{j+1}) - L(x_i, y))$$

$$+ \xi_{j+1}(\Phi(x_{j+1}) - M(x_{j+1}, y_{j+1})) + \xi_{j+1}(M(x_{j+1}, x_{j+1}) - M(x_{j+1}, y))$$

$$+ \sum_{i>j+1} \xi_i (M(x_i, y_{j+1}) - M(x_i, y)),$$

from which, as before, we obtain

$$H(x_\varepsilon, y) + \varepsilon \geqq H(X_\varepsilon, y_{j+1}),\ j = 0, 1, \dots, n-1,\ y \in [y_j, y_{j+1}].$$

What we have said means that

$$\Phi(x_i) \geqq \min\{L(x_i, x_i), M(x_i, x_i)\}, \quad i = 0, 1, \dots, n,$$

implies

$$H(X_\varepsilon, y_j) \leqq H(X_\varepsilon, y) + \varepsilon, \quad j = 1, \dots, n-1, \quad y \in [y_j, y_{j+1}],$$

which, together with (3.40) yields

$$H(X_\varepsilon, Y_\varepsilon) \leqq H(X_\varepsilon, y) + \varepsilon, \quad y \in [0, 1].$$

Consequently Y_ε is an ε-optimal strategy for player 2.

Parallel reasoning shows that

$$\Phi(x_i) \leqq \max\{L(x_i, x_i), M(x_i, x_i)\}, \quad i = 0, 1 \dots, n,$$

implies that X_ε is an ε-optimal strategy for player 1, and the existence of a value for Γ is established. As usual, we denote this value by v_Γ.

Let us now denote the distribution function of the ε-optimal strategy X_ε by F_ε. It is known (by Helly's "first" theorem) that we can find a sequence $\varepsilon > 0$ that converges to zero and for which a sequence of functions F_ε converges to a limit distribution function at each of its points of continuity. During the following proof we may take ε to be a number in this sequence.

Choose any $y \in [0, 1]$, and suppose, as is usual in the Russian-language literature, that the distribution function is continuous on the left. Since the set of discontinuities of F is at most countable, we can, by slightly modifying the proof of Helly's theorem, also require the existence of the limits $\lim_{\varepsilon \to 0} F_\varepsilon(y)$ and $\lim_{\varepsilon \to 0} F_\varepsilon(y + 0)$ at all discontinuities of F (we do not, of course, require that these limits are equal to $F(y)$ and $F(y + 0)$).

Now choose a sequence of positive numbers δ, converging to zero, for which the points $y \pm \delta$ are points of continuity, both of all the functions F_ε and of the limit function F. From now on, we consider only such numbers δ.

For arbitrary numbers ε and δ from those selected, we have

$$F_\varepsilon(y - \delta) \leqq F_\varepsilon(y) \leqq F_\varepsilon(y + 0) \leqq F_\varepsilon(y + \delta).$$

Letting ε tend to zero, we obtain

$$F(y-\delta) \leqq \lim_{\varepsilon\to 0} F_\varepsilon(y) \leqq \lim_{\varepsilon\to 0} F_\varepsilon(y+0) \leqq F(y+\delta),$$

and, letting δ tend to zero,

$$F(y) \leqq \lim_{\varepsilon\to 0} F_\varepsilon(y) \leqq \lim_{\varepsilon\to 0} F_\varepsilon(y+0) \leqq F(y+0). \tag{3.41}$$

By way of abbreviation, we set

$$\begin{aligned} F_\varepsilon(y) - F(y) &= \alpha_\varepsilon(y), \quad \lim_{\varepsilon\to 0} \alpha_\varepsilon(y) = \alpha(y), \\ F_\varepsilon(y+0) - F_\varepsilon(y) &= \beta_\varepsilon(y), \quad \lim_{\varepsilon\to 0} \beta_\varepsilon(y) = \beta(y), \\ F(y+0) - F_\varepsilon(y+0) &= \gamma_\varepsilon(y), \quad \lim_{\varepsilon\to 0} \gamma_\varepsilon(y) = \gamma(y). \end{aligned} \tag{3.42}$$

Let ε and δ be chosen arbitrarily from the numbers previously selected, and let $y \in [0,1]$. Since F_ε is an ε-optimal strategy, we have

$$v_\Gamma - \varepsilon \leqq H(X_\varepsilon, y-\delta) \tag{3.43}$$

or

$$\begin{aligned} v_\Gamma - \varepsilon \leqq & \int_0^{y-\delta} L(x, y-\delta)\,dF_\varepsilon(x) + \int_{y-\delta}^{y} M(x, y-\delta)\,dF_\varepsilon(x) \\ & + \beta_\varepsilon(y) M(y, y-\delta) + \int_{y}^{y+\delta} M(x, y-\delta)\,dF_\varepsilon(y) \\ & + \int_{y+\delta}^{1} M(x, y-\delta)\,dF_\varepsilon(x) \end{aligned}$$

(the term containing Φ is absent here because F_ε is continuous at $y-\delta$). Letting $\varepsilon \to 0$ and using, wherever possible, Helly's "second" theorem (on taking limits under the integral sign in a Stieltjes integral), we obtain

$$\begin{aligned} v_\Gamma \leqq & \int_0^{y-\delta} L(x, y-\delta)\,dF(x) + \lim_{\varepsilon\to 0} \int_{y-\delta}^{y} M(x, y-\delta)\,dF_\varepsilon(x) \\ & + \beta(y) M(y, y-\delta) + \lim_{\varepsilon\to 0} \int_{y}^{y+\delta} M(x, y-\delta)\,dF_\varepsilon(x) \\ & + \int_{y+\delta}^{1} M(x, y-\delta)\,dF(x). \end{aligned} \tag{3.44}$$

Let us estimate the limits that remain. For this purpose, we notice that, since M is continuous, the difference

$$\omega(x, y-\delta) = M(x, y-\delta) - M(y-\delta, y-\delta)$$

with $x \in [y-\delta, y]$, tends to zero with δ. Consequently

$$\lim_{\varepsilon\to 0} \int_{y-\delta}^{y} M(x, y-\delta)\,dF_\varepsilon(x)$$

$$= M(y-\delta, y-\delta) \lim_{\varepsilon\to 0} \int_{y-\delta}^{y} dE_\varepsilon(x) + \lim_{\varepsilon\to 0} \int_{y-\delta}^{y} \omega(x, y-\delta)\,dF_\varepsilon(x)$$

$$= M(y-\delta, y-\delta)\big(\lim_{\varepsilon\to 0} F_\varepsilon(y) - F(y-\delta)\big) + \omega(\delta),$$

where $\omega(\delta)$ tends to zero with δ.

Similarly we obtain

$$\lim_{\varepsilon\to 0} \int_{y}^{y+\delta} M(x, y-\delta)\,dF_\varepsilon(x)$$

$$= M(y-\delta, y-\delta)\big(F(y+\delta) - \lim_{\varepsilon\to 0} F_\varepsilon(y)\big) + \omega'(\delta),$$

where $\omega'(\delta)$ tends to zero with δ. If we now replace the limits on the right-hand side of (3.44) by the expressions we have obtained for them, let δ tend to zero, and use the continuity of L and M, we obtain

$$v_\Gamma \leqq \int_0^v L(x,y)\,dF(x) + M(y,y)\,(\alpha(y) + \beta(y) + \gamma(y)) + \int_y^1 M(x,y)\,dF(x). \tag{3.45}$$

As in (3.43), we write

$$v_\Gamma - \varepsilon \leqq H(x_\varepsilon, y-\delta).$$

Repeating the preceding discussion almost word for word yields

$$v_\Gamma \leqq \int_0^y L(x,y)\,dF(x) + L(y,y)\big(\alpha(y)+\beta(y)+\gamma(y)\big) + \int_y^1 M(x,y)\,dF(x). \tag{3.46}$$

Finally, notice that

$$\begin{aligned} H(F, y) &= \int_0^1 H(x, y) dF(x) \\ &= \int_0^y L(x, y) dF(x) + \Phi(y)\big(F(y+0) - F(y)\big) + \int_y^1 M(x, y) dF(x), \end{aligned} \tag{3.47}$$

where, by (3.42), the coefficient of $\Phi(y)$ is $\alpha(y) + \beta(y) + \gamma(y)$. Therefore, by (3.42), the right-hand side (and therefore the left-hand side) of (3.47) lies between the values of the right-hand sides of (3.45) and (3.46), neither of which is less than $v(\Gamma)$. Consequently

$$H(X, y) \geqq v_\Gamma, \quad y \in (0, 1). \tag{3.48}$$

It remains only to consider the cases when $y = 0$ or $y = 1$. We shall discuss only the case when $y = 1$.

By the ε-optimality of the strategy X_ε we have, for each δ,

$$v_\Gamma - \varepsilon \leqq H(X_\varepsilon, 1),$$
$$v_\Gamma - \varepsilon \leqq H(X_\varepsilon, 1 - \delta),$$

or, correspondingly,

$$v_\Gamma - \varepsilon \leqq \int_0^{1-\delta} L(x, 1) dF_\varepsilon(x) + \int_{1-\delta}^1 L(x, 1) dF_\varepsilon(x) + \Phi(1)\big(1 - F_\varepsilon(1)\big),$$

$$v_\Gamma - \varepsilon \leqq \int_0^{1-\delta} L(x, 1 - \delta) dF_\varepsilon(x) + \int_{1-\delta}^1 M(x, 1) dF_\varepsilon(x).$$

Letting ε tend to zero, we obtain

$$v_\Gamma \leqq \int_1^{1-\delta} L(x, 1) dF(x) + \lim_{\varepsilon \to 0} \int_{1-\delta}^1 L(x, 1) dF_\varepsilon(x) + \Phi(1)\big(1 - \lim_{\varepsilon \to 0} F_\varepsilon(1)\big) \tag{3.49}$$

and correspondingly

$$v_\Gamma \leqq \int_1^{1-\delta} L(x, 1 - \delta) dF(x) + \lim_{\varepsilon \to 0} \int_{1-\delta}^1 M(x, 1 - \delta) dF_\varepsilon(x). \tag{3.50}$$

For the limits that appear here, we have

$$\lim_{\varepsilon \to 0} \int_{1-\delta}^{1} L(x,1)dF_\varepsilon(x)$$
$$= L(1,1)\big(\lim_{\varepsilon \to 0} F_\varepsilon(1) - F(1-\delta)\big) + \omega''(\delta) + \Phi(1)\big(1 - \lim_{\varepsilon \to 0} F_\varepsilon(1)\big)$$
$$= L(1,1)\alpha(1) + \omega''(\delta) + \Phi(1)\beta(1),$$

$$\lim_{\varepsilon \to 0} \int_{1-\delta}^{1} M(x, 1-\delta)dF_\varepsilon(x) = M(1,1)\big(1 - F(1-\delta)\big) + \omega'''(\delta)$$
$$= M(1,1)\big(\alpha(1) + \beta(1)\big) + \omega'''(\delta),$$

where $\omega''(\delta)$ and $\omega'''(\delta)$ tend to zero with δ.

If we substitute these limits in (3.49) and (3.50), and let $\delta \to 0$, we obtain

$$v_\Gamma \leqq \int_0^1 L(x,1)dF(x) + L(1,1)\alpha(1) + \Phi(1)\beta(1), \tag{3.51}$$

$$v_\Gamma \leqq \int_0^1 L(x,1)dF(x) + M(1,1)\big(\alpha(1) + \beta(1)\big). \tag{3.52}$$

Finally,

$$H(F,1) = \int_0^1 L(x,1)dF(x) + \Phi(1)\big(\alpha(1) + \beta(1)\big). \tag{3.53}$$

If, then,

$$L(1,1) \leqq \Phi(1) \leqq M(1,1),$$

we obtain $v_\Gamma \leqq H(F,1)$ by comparing (3.51) and (3.53). If, however,

$$L(1,1) \geqq \Phi(1) \geqq M(1,1),$$

then $v_\Gamma \leqq H(F,1)$ by comparing (3.52) and (3.53).

Comparing these results with (3.48), we find that

$$v_\Gamma \leqq H(X,y) \text{ for all } y \in [0,1],$$

i.e., X is an optimal strategy for player 2; the existence of this can be proved similarly. □

§4 Finitely additive strategies

4.1 Examples of games that are unsolvable in the mixed strategies. As we saw in § 3, in continuous games, as well as in games that differ "only slightly" from continuous games, there are saddle points in the mixed strategies, or at least ε-saddle points for every $\varepsilon > 0$ (the latter case, as we have observed, is equivalent to the existence of a value for the game). Moreover, stronger theorems are known that establish the existence of a maximin principle in two-person zero-sum games for wider classes of games. Further progress in this direction suggests various interesting possibilities. However, the possibilities of solving two-person zero-sum games in the mixed strategies appear to be limited; we can illustrate this by the two following examples.

4.2 Example 1. In the two-person zero-sum game

$$\Gamma = \langle x, y, H \rangle \tag{4.1}$$

let the components be defined as follows: $x = y = \{1, 2, \ldots\}$,

$$H(x,y) = \begin{cases} 1 & \text{if } x > y, \\ 0 & \text{if } x = y, \\ -1 & \text{if } x < y. \end{cases}$$

We can describe this game informally as the simultaneous choice by players 1 and 2 of the numbers x and y; then the player who chose the larger number wins one unit of the other player. If the numbers are equal, the payoffs are zero to both players.

Let us show that this game has no value.

Let X be any mixed strategy of player 1, and $dX(x) = \xi_x$.

Here $\xi_x \geqq 0$ and $\sum_{x=1}^{\infty} \xi_x = 1$. Choose any $\varepsilon > 0$ and find y_ε such that

$$\sum_{x \leqq y_\varepsilon} \xi_x > 1 - \varepsilon.$$

Then

$$\begin{aligned} H(X, y_\varepsilon) &= \sum_{x=1}^{\infty} \xi_x H(x, y_\varepsilon) \\ &= \sum_{x \leqq y_\varepsilon} \xi_x H(x, y_\varepsilon) + \sum_{x > y_\varepsilon} \xi_x H(x, y_\varepsilon) \\ &= -\sum_{x < y_\varepsilon} \xi_x + \sum_{x > y_\varepsilon} \xi_x < -1 + 2\varepsilon, \end{aligned}$$

so that $\inf_Y H(X,Y) < -1 + 2\varepsilon$, and since $\varepsilon > 0$ is arbitrary and $H(x,y)$ takes no values less than -1,

$$\inf_Y H(X,Y) = -1.$$

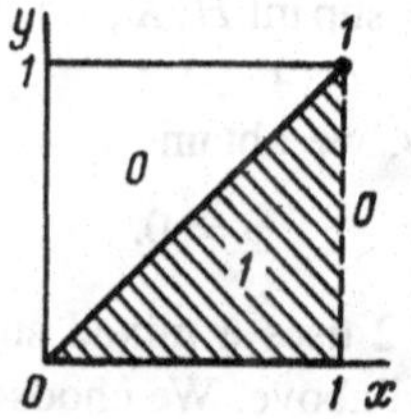

Fig. 3.2

Consequently, since X is arbitrary,

$$\underline{v} = \sup_X \inf_Y H(X,Y) = -1. \tag{4.2}$$

Arguing by symmetry, we obtain

$$\overline{v} = \inf_Y \sup_X H(X,Y) = 1. \tag{4.3}$$

Equations (4.2) and (4.3) show that the game does not have a value (in the mixed strategies).

4.3 Example 2. In a two-person zero-sum game as in (4.1), let $x = y = [0,1]$,

$$H(x,y) = \begin{cases} 1 & \text{if } y \leqq x < 1 \text{ or } x = y = 1, \\ 0 & \text{otherwise.} \end{cases} \tag{4.4}$$

A visual representation of the function H is given in Figure 3.2.

Let us analyze this game. Let player 1 choose a mixed strategy X. Take an arbitrary $\varepsilon > 0$, and then choose δ small enough so that the choice by player 1, under the conditions of the mixed strategy X, of a pure strategy on the interval $[1-\delta, 1)$ will take place with probability less than ε:

$$\int_{1-\delta}^{1-0} dX(x) < \varepsilon. \tag{4.5}$$

Such a choice of δ is possible since the strategy X, considered as a distribution function, is continuous on the left.

We have

$$H(X,y) = \int_0^{1-\delta} H(x,y)dX(x) + \int_{1-\delta}^{1-0} H(x,y)dX(x) + H(1,y)dX(1).$$

If player 2 chooses a pure strategy $y \in (1-\delta, 1)$, the first integral on the left is zero, by (4.4). The second integral does not exceed ε, by (4.5) and the fact that $H(x,y) \leqq 1$; and the third term is zero by (4.3), since $y < 1$. Therefore $H(X,y) \leqq \varepsilon$ for this choice of the strategy y, and consequently $\inf_Y H(X,Y) \leqq \varepsilon$.

Since the preceding argument applies to any strategy X, we must have

$$\underline{v} = \sup_X \inf_Y H(X,Y) \leqq \varepsilon.$$

Finally, since $\varepsilon > 0$ is arbitrary, we obtain

$$\underline{v} = 0. \tag{4.6}$$

Suppose now that player 2 uses a mixed strategy Y. The reasoning in this case is similar to what was used above. We choose an $\varepsilon > 0$ and find a δ so small that under the conditions of Y player 2 will use pure strategies from the interval $[1-\delta, 1)$ with probability less than ε:

$$\int_{1-\delta}^{1-0} dY(y) < \varepsilon. \tag{4.7}$$

Let player 1 use the mixed strategy X that consists of the choice of a pure strategy $x_0 \in [1-\delta, 1)$ and the pure strategy x_1, each with probability one-half. Then

$$\begin{aligned} H(X,Y) &= \frac{1}{2}\left(\int_0^1 H(x_0, y)dY(y) + \int_0^1 H(1,y)dY(y)\right) \\ &= \frac{1}{2}\left(\int_0^{1-\delta} H(x_0,y)dY(y) + \int_{1-\delta}^{1-0} H(x_0,y)dY(y) + H(x_0,1)dY(1)\right) \\ &+ \frac{1}{2}\left(\int_0^{1-0} H(1,y)dY(y) + H(1,1)dY(1)\right). \end{aligned}$$

The first term in the big parentheses on the right reduces to

$$\int_0^{1-\delta} dY(y),$$

the second term is nonnegative, and the third is equal to zero. In addition, the fourth term (in the second set of parentheses) is also equal to zero, and the fifth is $dY(1)$. Consequently,

$$H(X,Y) \geqq \frac{1}{2}\left(\int_0^{1-\delta} dY(y) + dY(1)\right) = \frac{1}{2}\left(1 - \int_{1-\delta}^{1-0} dY(y)\right),$$

i.e., in view of (4.7), $H(X,Y) \geqq \frac{1}{2}(1-\varepsilon)$ and

$$\sup_X H(X,Y) \geqq \frac{1}{2}(1-\varepsilon).$$

Since this inequality holds for all Y and $\varepsilon > 0$, it follows from what we have obtained that

$$\bar{v} = \inf_Y \sup_X H(X,Y) \geqq \frac{1}{2}. \tag{4.8}$$

Combining (4.6) with (4.8), we see that $\underline{v} \neq \bar{v}$ in this game, and hence the game does not have a value.

To complete the analysis of this game, we show that there is actually equality on the right-hand side of (4.8). For this purpose we choose the mixed strategy Y^* in which $y_0 < 1$ and $y_1 = 1$ are used with probability $\frac{1}{2}$ for each. We have

$$H(x,Y^*) = \frac{1}{2}\big(H(x,y_0) + H(x,y_1)\big).$$

But the set of points x for which $H(x,y_0) = 1$ does not intersect the set of points x for which $H(x,y_1) = 1$. Therefore $H(x,Y^*) \leqq \frac{1}{2}$; hence, since x is arbitrary, we find by a standard argument that $\bar{v} \leqq \frac{1}{2}$.

4.4 Statement of the question of finitely additive extensions of games. It follows from the preceding examples that mixed extensions of games are still not broad enough, and for obvious extensions of games with "sufficiently discontinuous" payoff functions, still further extensions are required.

Such an extension is, for example, an analog of the transition from pure to mixed strategies which was discussed in detail in § 5 of Chapter 1. However, an attempt to reproduce it in the present case turned out to be not very successful. A suitable construction can be made in the following way.

Since we expect the outcome of the game Γ in (4.1) to be measurable, i.e. to be representable in the form

$$\Gamma = \langle\langle \boldsymbol{x}, \Xi\rangle, \langle \boldsymbol{y}, \mathrm{H}\rangle, H\rangle\,,$$

we may consider the measure space $\langle \boldsymbol{x} \times \boldsymbol{y}, Z\rangle$ of all situations generated by the spaces of the players' strategies.

We consider the normalized finitely additive measures $\widehat{X}$ and $\widehat{Y}$ on the measure spaces $\langle \boldsymbol{x}, \Xi\rangle$ and $\langle \boldsymbol{y}, \mathrm{H}\rangle$, and we call these measures *finitely additive strategies* of players 1 and 2. We denote by $\widehat{\boldsymbol{X}}$ and $\widehat{\boldsymbol{Y}}$ the sets of the finitely additive strategies of players 1 and 2 that we consider. In general, there is a wide latitude in the choice of these sets, but it is natural to suppose that $\boldsymbol{x} \subset \boldsymbol{X} \subset \widehat{\boldsymbol{X}}$ and $\boldsymbol{y} \subset \boldsymbol{Y} \subset \widehat{\boldsymbol{Y}}$.

If, now, we are given two independent finitely additive measures $\widehat{X} \in \widehat{\boldsymbol{X}}$ and $\widehat{Y} \in \widehat{\boldsymbol{Y}}$, the construction of a finitely additive product measure $\widehat{X} \times \widehat{Y}$, with the desirable intrinsic properties of a product measure, will generally not be possible. (The lack of countable additivity can make it impossible to define the value of the measure $\widehat{X} \times \widehat{Y}$ on certain subsets of Z.) Consequently the integral

$$H(\widehat{X},\widehat{Y}) = \int_{\mathbf{x}\times\mathbf{y}} H(x,y)d(\widehat{X}\times\widehat{Y})(x,y) \tag{4.9}$$

will not necessarily exist.

Consequently a finitely additive extension of a two-person zero-sum game must be introduced more circumspectly.

4.5 The Yosida-Hewitt representation of finitely additive measures. We begin by describing Yosida and Hewitt's idea*) and the representation, which is useful for our purposes, of finitely additive measures as a sort of superposition of countably additive measures of a special kind.

Let $\mathbf{z}$ be any set, $2^{\mathbf{z}}$ the σ-algebra of all its subsets, and Ω the set of all finitely additive measures on $\langle z, 2^{\mathbf{z}}\rangle$ with value 0 or 1. A measure from Ω is known as a 0-1-*measure*. Degenerate measures are special cases of 0-1-measures.

It is easily verified that if ω is a 0-1-measure, and $\omega(\mathbf{v}_1) = \ldots = \omega(\mathbf{v}_k) = 1$, then

$$\omega\left(\bigcap_{i=1}^{k} \mathbf{v}_i\right) = 1$$

(this proposition cannot be extended to infinite intersections because ω is not necessarily countably additive).

For any $\mathbf{v} \subset \mathbf{z}$ we set

$$\Delta_{\mathbf{v}} = \{\omega \in \Omega : \omega(\mathbf{v}) = 1\}.$$

Taking all sets of the form $\Delta_{\mathbf{v}}$ to be open generates a topology on Ω, under which it forms a compact Hausdorff space.

We consider the characteristic functions of sets, setting, for arbitrary $\mathbf{v} \subset \mathbf{z}$ and $\omega \in \Omega$,

$$\chi_{\Delta_v}(\omega) = \begin{cases} 1 & \text{if } \omega \in \Delta_{\mathbf{v}}. \\ 0 & \text{otherwise.} \end{cases}$$

Then, as Yosida and Hewitt showed, corresponding to every finitely additive measure μ on $\langle \mathbf{z}, 2^{\mathbf{z}}\rangle$ there is a countably additive measure $\bar{\mu}$ on the family of Borel subsets of Ω, such that for every $\mathbf{v} \subset \mathbf{z}$

$$\mu(\mathbf{v}) = \int_{\Omega} \chi_{\Delta_v}(\omega)d\bar{\mu}(\omega),$$

and for every bounded function f on $\mathbf{z}$ (evidently this, like all other functions, is $2^{\mathbf{z}}$-measurable)

$$\int_{\mathbf{z}} f(x)d\mu(x) = \int_{\Omega} f(\omega)d\bar{\mu}(\omega),$$

where, naturally,

$$f(\omega) = \int_{\mathbf{z}} f(x)d\omega(x) \tag{4.10}$$

(this function is continuous in the topology of Ω).

---

*) K. Yosida and E. Hewitt. Finitely additive measures, Trans. Amer. Math. Soc. 72 (1952), 46–66.

4.6 A family of auxiliary games. Now let Γ be a game as in (4.1) with *arbitrary* sets x and y of strategies, and let the payoff function H satisfy the natural requirement of boundedness.

Let us consider the σ-algebras 2^x and 2^y of all the subsets of x and y. Evidently H is measurable in x with respect to 2^x and in y with respect to 2^y.

Let us denote by Ω_x and Ω_y the sets of all 0-1-measures on 2^x and 2^y. Consider the σ-algebras of subsets of Ω_x and Ω_y generated by their families of Borel subsets. Let $\widehat{\Omega}_x$ and $\widehat{\Omega}_y$ be the sets of probability measures on these σ-algebras.

The double integral

$$H(\omega_1,\omega_2) = \int\limits_{x\times y} H(x,y)d(\omega_1\times\omega_2). \tag{4.11}$$

exists for certain pairs $(\omega_1,\omega_2)\in\Omega_x\times\Omega_y$. It exists, for example, for all the pairs in which at least one of ω_1 and ω_2 is degenerate.

Let L denote the set of pairs (ω_1,ω_2) for which the integral (4.11) *does not* exist. Defining, for an arbitrary $a\in\mathbf{R}$, the function $H_a:\widehat{\Omega}_x\times\widehat{\Omega}_y\to\mathbf{R}$, let us set, for $F\in\widehat{\Omega}_x$ and $G\in\widehat{\Omega}_y$,

$$H_a(F,G) = \begin{cases} \int\limits_{\Omega_x\times\Omega_y} H(\omega_1,\omega_2)dFdG & \text{if } (F\times G)(L)=0,\\ a, & \text{in the contrary case.}\end{cases} \tag{4.12}$$

Notice that if F is degenerate and concentrated on a degenerate measure ω_1 or if G is degenerate and concentrated on a degenerate measure ω_2 then the integral in (4.12) exists, so that in this case $H_a(F,G)$ is independent of a.

What we have said allows us to consider the family

$$\Gamma_a = \langle\widehat{\Omega}_x,\widehat{\Omega}_y,H_a\rangle,\quad a\in\mathbf{R},$$

of auxiliary measures.

Each such game can be informally thought of in terms of the players' selection of their finitely additive strategies (for example, in the representations described in 4.5), with the payoff to player 1 given by the double integral (4.11), if it exists, and by the number a if this integral does not exist.

Here we actually enter the domain of games with forbidden situations (see 16 in the Introduction).

4.7 Lemma. 1. *If*

$$\underline{v}(\Gamma_a) = \sup_F \inf_G H_a(F,G) < a, \tag{4.13}$$

then

$$\overline{v}(\Gamma_a) = \inf_G \sup_F H_a(F,G) \leqq a. \tag{4.14}$$

2. *If* $\overline{v}(\Gamma_a) > a$ *then* $\underline{v}(\Gamma_a) \geqq a$.

Proof. Consider the iterated integral

$$H^1(\omega_1, \omega_2) = \int_x \int_y H(x,y) d\omega_2 d\omega_1.$$

By (4.10) the function $H^1(\cdot, \omega_2)$ is continuous in ω_1 in the topology of Ω_x for each given $\omega_2 \in \Omega_y$. Consequently the integral

$$H^1(F,G) = \int_{\Omega_y} \int_{\Omega_x} H^1(\omega_1, \omega_2) dF\ dG$$

is continuous in F in the weak* topology of the game $\widehat{\Omega}_x$.

If now $(\omega_1, \omega_2) \notin L$ we have

$$H_a(\omega_1, \omega_2) = H^1(\omega_1, \omega_2).$$

Therefore $(F \times G)(L) = 0$ must imply that

$$H_a(F,G) = H^1(F,G). \tag{4.15}$$

However, (4.13) means that for every $F \in \widehat{\Omega}_x$ there is a $G \in \widehat{\Omega}_y$ such that $H_a(F,G) < a$, and then by (4.15) $H^1(F,G) < a$. Let

$$A_G = \{F \in \widehat{\Omega}_x, H^1(F,G) < a,\ G \in \widehat{\Omega}_y\}.$$

Each of the sets A_G is open, and they constitute a covering $\widehat{\Omega}_x$. The compactness of $\widehat{\Omega}_x$ implies the compactness of $\widehat{\Omega}_x$ in the weak* topology, so that $\widehat{\Omega}_x$ can be covered by a finite number of sets of the form A_G. Let

$$\widehat{\Omega}_x \subset \bigcup_{i=1}^{k} A_{G_i}.$$

Then for every $F \in \widehat{\Omega}_x$ there is a G_i such that $H^1(F, G_i) < a$. Because $H^1(\cdot, G)$ is continuous and linear on F there is a convex linear combination $\sum_{i=1}^k \lambda_i G_i$ such that

$$H^1\left(F, \sum_{i=1}^{k} \lambda_i G_i\right) < a \text{ for all } F \in \widehat{\Omega}_x.$$

But then it follows from (4.13) and (4.15) that

$$H_a\left(F, \sum_{i=1}^{k} \lambda_i G_i\right) \leqq a \text{ for all } F \in \widehat{\Omega}_x,$$

and this implies (4.14).

To establish the second conclusion of the lemma we consider the function

$$H^2(\omega_1, \omega_2) = \int_y \int_x H(x,y) d\omega_1, d\omega_2. \qquad \square$$

4.8 Lemma. *For every game Γ_a we have the alternative:*

$$\text{either } a = \overline{v}(\Gamma_a) \text{ or } a = \underline{v}(\Gamma_a). \tag{4.16}$$

Proof. For every $F \in \widehat{\Omega}_x$, let L_F denote the set of all $G \in \widehat{\Omega}_y$ for which $(F \times G)(L) > 0$, and let M_F denote the set of the remaining $G \in \widehat{\Omega}_y$. Then we have, for every $F \in \widehat{\Omega}_x$ and $a \in \mathbf{R}$,

$$\inf_G H_a(F,G) = \min\{\inf_{G \in L_F} H_a(F,G),\ \inf_{G \in M_F} H_a(F,G)\},$$

or in view of (4.12) and the facts that $y \subset M_F$ for every F and the infimum of the payoff functions over all mixed strategies is equal to the infimum over all pure strategies,

$$\inf_G H_a(F,G) = \min\{a, \inf_{y \in y} H(F,y)\}. \tag{4.17}$$

It follows from this that, in the first place,

$$\inf_G H_a(F,G) \leqq \inf_y H(F,y), \tag{4.18}$$

so that

$$\underline{v}(\Gamma_a) = \sup_F \inf_G H_a(F,G) \leqq \sup_F \inf_y H(F,y). \tag{4.19}$$

In the second place, it follows from (4.17) that

$$\inf_G H_a(F,G) \leqq a,$$

so that

$$\underline{v}(\Gamma_a) \leqq a. \tag{4.20}$$

A parallel discussion yields

$$\overline{v}(\Gamma_a) = \inf_G \sup_F H_a(F,G) \geqq \inf_G \sup_x H(x,G) \tag{4.21}$$

and

$$\overline{v}(\Gamma_a) \geqq a. \tag{4.22}$$

Now suppose that there is strict inequality in (4.20). This means that (4.20) becomes (4.13). But then (4.14) holds; this, with (4.22), yields $\overline{v}(\Gamma_a) = a$. In turn, there is equality in (4.20), i.e. $\underline{v}(\Gamma_a) = a$. □

4.9 Lemma. *The set of solutions of the equation*

$$v(\Gamma_a) = \underline{v}(\Gamma_a) = \overline{v}(\Gamma_a) = a \tag{4.23}$$

is a nonempty closed interval $[\underline{a}, \overline{a}]$ (which may reduce to a single point).

Proof. First consider the equation

$$\underline{v}(\Gamma_a) = a. \tag{4.24}$$

Since $\underline{v}(\Gamma_a)$ is continuous as a function of a, its solution set is closed. By (4.19), this set is also bounded above. Let us show that it is not bounded below.

It is evident that adding a positive amount to the payoff function in a two-person zero-sum game increases the lower value (as well as the upper value) by a number that does not exceed this amount. Hence it follows from the inequality $a_1 < a_2$ that

$$\underline{v}(\Gamma_{a_2}) - \underline{v}(\Gamma_{a_1}) \leqq a_2 - a_1,$$

whence

$$\underline{v}(\Gamma_{a_2}) - a_2 \leqq \underline{v}(\Gamma_{a_1}) - a_1.$$

By what we have just said, a_2 can only equal $\underline{v}(\Gamma_{a_2})$ or $\overline{v}(\Gamma_{a_2})$. If $\underline{v}(\Gamma_{a_2}) = a_2$, then $\underline{v}(\Gamma_{a_1}) - a_1 \geqq 0$ and a_1 cannot equal $\overline{v}(\Gamma_{a_1})$ (if, of course, $\overline{v}(\Gamma_{a_1}) > \underline{v}(\Gamma_{a_1})$). Therefore $\underline{v}(\Gamma_{a_1}) = a_1$. Consequently the solution set of (4.24) cannot be bounded below.

By the preceding discussion, the solution set of (4.24) is the half-open interval $(-\infty, \bar{a}]$.

Similarly the solution set of the equation $\overline{v}(\Gamma_a) = a$ is of the form $[\underline{a}, +\infty)$.

It follows from the preceding lemma that every a must satisfy one of the equations $\underline{v}(\Gamma_a) = a$ or $\overline{v}(\Gamma_a) = a$. Therefore

$$(-\infty, \bar{a}] \cap [\underline{a}, +\infty) \neq \emptyset,$$

i.e. $\underline{a} \leqq \bar{a}$, and $[\underline{a}, \bar{a}]$ is the solution set of (4.23). □

4.10 Finitely additive extensions of a game. As a matter of fact, any game Γ_a with $a \in [\underline{a}, \bar{a}]$ can serve as the desired finitely additive extension of Γ. The multiplicity of choices of the parameter a can be overcome by requiring the choice to satisfy the following condition, of the nature of an axiom.

Definition. *A finitely additive extension* of a two-person zero-sum game Γ is a game Γ_a in which

$$a = f(\underline{a}, \bar{a}) \in [\underline{a}, \bar{a}],$$

where f is some function. □

A priori, the function f is not subject to any restrictions, but actually it turns out to be uniquely defined.

Theorem.

$$f(\underline{a}, \bar{a}) = (\underline{a} + \bar{a})/2. \tag{4.25}$$

Proof. It follows from (4.23) that

$$f(\underline{a}, \bar{a}) = v(\Gamma_a).$$

This means that the affine equivalence of transformations of two-person zero-sum games (see 1.13) must be manifest for the function f just as for the values of the game Γ_a. In particular, we must have

$$f(\underline{a} + \alpha, \bar{a} + \alpha) = f(\underline{a}, \bar{a}) + \alpha \qquad \text{for } a \in \mathbf{R},$$

$$f(k\underline{a}, k\bar{a}) = kf(\underline{a}, \bar{a}) \qquad \text{for } k \geqq 0,$$

$$f(\underline{a}, \bar{a}) = -f(-\bar{a}, -\underline{a}).$$

An obvious application of these formulas yields

$$f(0,1) = -f(-1,0) = -f(0,1) + 1,$$

from which $f(0,1) = \frac{1}{2}$, and therefore

$$f(\underline{a}, \bar{a}) = f(0, \bar{a} - \underline{a}) + \underline{a} = (\bar{a} - \underline{a})f(0,1) + \underline{a}$$
$$= (\bar{a} - \underline{a})/2 + \underline{a} = (\underline{a} + \bar{a})/2. \qquad \square$$

4.11. In the light of the preceding discussion, we can give an equivalent definition of a finitely additive extension of a two-person zero-sum game.

Definition. The game

$$\bar{\Gamma} = \langle \widehat{\Omega}_x, \widehat{\Omega}_y, H_{a^*} \rangle,$$

where $a^* = (\underline{a} + \bar{a})/2$, is said to be a *finitely additive extension* of the two-person zero-sum game Γ.

We may call a^* the *value of the finitely additive extension* $\bar{\Gamma}$ of Γ, and also the value of the original game Γ (in the finitely additive strategies). According to this definition, *every two-person zero-sum game has a value in the finitely additive strategies.*

4.12 Solution of insoluble games in the finitely additive strategies. Let us find solutions in the finitely additive strategies for the two-person zero-sum games in the examples of 4.2 and 4.3.

The game in 4.2 can be analyzed in the following way.

Let ω_1 be any nondegenerate 0-1-measure, and let n be a positive integer. Choose an $A \subset x$ such that $\omega_1(A) = 1$. For this set A we will have

$$\omega_1(A) = \omega_1(A \cap \{1, \ldots, n\}) + \omega_1(A \setminus \{1, \ldots, n\}) = \omega_1(A \setminus \{1, \ldots, n\}) = 1,$$

so that by (4.1)

$$\omega_1(\{x : x > n\}) = \omega_1(\{x : H(x, n) = 1\}) = 1,$$

and therefore

$$H(\omega_1, n) = \int_x H(x, n) d\omega_1(x) = 1. \tag{4.26}$$

Similarly we find that, for every positive integer m and nondegenerate 0-1-measure ω_2 we will have

$$H(m, \omega_2) = \int_{\mathbf{y}} H(m, y) d\omega_2(x) = 1. \tag{4.27}$$

Further, it follows from (4.26) that

$$\int_{\mathbf{x}} \int_{\mathbf{y}} H(x, y) d\omega_2(y) d\omega_1(x) = \int_{\mathbf{x}} H(x, \omega_2) d\omega_1(x) = \int_{\mathbf{x}} (-1) d\omega_1(x) = -1,$$

and from (4.27), that

$$\int_{\mathbf{y}} \int_{\mathbf{x}} H(x, y) d\omega_1(x) d\omega_2(y) = 1.$$

That the last two iterated integrals have different values means that the double integral

$$H(\omega_1, \omega_2) = \int_{\mathbf{x} \times \mathbf{y}} H(x, y) d(\omega_1 \times \omega_2)(x, y)$$

does not exist.

It follows that $(F, G) \in M$ only if F and G are ordinary mixed strategies (probability measures on sets of pure strategies). The payoff function of Γ_a can be described schematically by the diagram

	n	ω_2
m	$H(m, n)$	-1
ω_1	1	a

Here it is evident that when $a \in [-1, 1]$ we have $a = v(\Gamma_a)$. Consequently $\underline{a} = -1$, and $\bar{a} = 1$, so that $a^* = 0$.

Thus the finitely additive extension of Γ consists of having every finitely additive strategy of either player beat every one of the opponent's mixed strategies, and having every situation that consists of finitely additive strategies turn out to be "forbidden" under the original rules of the game, but, under the supplementary (finitely additive) rules, provide zero payoffs. Here every pair of finitely additive strategies is a saddle point of the game.

4.13. Let us turn to the example in 4.3. Here the only 0-1-measures on $\mathbf{x}$ that are of interest are the measures ω_1 for which $\omega_1(\mathbf{x}') = 1$ if and only if $\mathbf{x}'$ contains a sequence $x_1, x_2, \ldots$ that converges to $x_0 = 1$, and the similar measure ω_2 on $\mathbf{y}$ for which $\omega_2(\mathbf{y}') = 1$ if and only if $\mathbf{y}'$ contains a sequence $y_1, y_2, \ldots$ that converges to $y_0 = 1$.

As in the preceding example,

$$H(\omega_1, y) = \int_{\mathbf{x}} H(x, y) d\omega_1(x) = 1$$

for every $y \in [0, 1)$, and

$$H(x, \omega_2) = \int_{\mathbf{y}} H(x, y) d\omega_2(y) = 0,$$

for every $x \in [0, 1)$.

Naturally we find that

$$\int\limits_X \int\limits_Y H(x,y)d\omega_2(y)d\omega_1(x) = 0 \neq \int\limits_Y \int\limits_X H(x,y)d\omega_1(x)d\omega_2(y) = 1.$$

Therefore, for the probability distributions F and G on the combined sets, we consider 0-1-measures and $(F, G) \in M$ in the pure strategies if and only if F and G are ordinary mixed strategies (i.e., there are nondegenerate 0-1-strategies in their spectra).

Therefore the payoff functions in Γ_a are described by the following diagram:

	y	ω_2	1
x	$H(x, y)$	0	0
ω_1	1	a	0
1	0	0	1

(4.28)

There are no saddle points among the situations displayed here. However, it is easily seen that one appears if we combine the rows corresponding to the finitely additive strategy F^* that is a mixture of a 0-1-strategy of type ω_1 with probability $\frac{1}{2}$ and the pure strategy 1, also with probability $\frac{1}{2}$.

It is clear from the diagram (4.28) that in this case

$$H(F^*, y) = H(F, 1) = \frac{1}{2},$$

for $y \in [0, 1)$. However, matters are less trivial for $H(F^*, \omega_2)$. Constructing, according to (4.28), the average

$$H(F^*, \omega_2) = \left(\frac{1}{2}\right)(H(\omega_1, \omega_2) + H(1, \omega_2)) = \frac{a}{2}$$

contradicts the general construction of the games Γ_a, according to which it follows from $(F^*, \omega_2)(L) = 1/2 > 0$ that $H(F^*, \omega_2) = a$.

As a result we should adjoin to the table (4.28) the row

F^*	1/2	a	1/2.

and the situation (F^*, ω_2) appears as a saddle point in Γ_a if $0 \leqq a \leqq \frac{1}{2}$. Therefore we have $\underline{a} = 0$, $\bar{a} = \frac{1}{2}$, so that $a^* = \frac{1}{4}$.

§5 Analytic games on the unit square

5.1 Analytic games on the unit square. As follows from what we said at the end of 6.18 of Chapter 1 (also see 3.2), every continuous game Γ on the unit square has equilibrium situations, i.e. saddle points. If we apply the discussion in 6.12–6.18 to such a game Γ, we see that these saddle points can be obtained as limits of sequences of saddle points of finite subgames of Γ. As a matter of fact, this is all that can be said about continuous games on a product of compact sets and, in particular, about continuous two-person zero-sum games on the unit square. Beyond this, the continuity of the payoff function does not imply any further general properties of two-person zero-sum games on the unit square. The infinite differentiability of the payoff function adds very little to continuity. We now introduce a more restricted class of games on the unit square.

Definition. A game on the unit square is said to be *analytic for player 1* if its payoff function H is analytic in the strategies of player 1, i.e., for every $y \in [0,1]$ the function $H(x,y)$ can be expanded in a convergent Taylor series in a neighborhood of each point $x \in [0,1]$.

There is a symmetric definition of a game that is *analytic for player* 2.

A game on the unit square is said to be *analytic* if it is analytic for both players. □

It is clear that if a game is analytic for player 1, then $H(x,Y)$ is analytic in x for every mixed strategy Y; and if the game is analytic for player 2, then $H(X,y)$ is analytic for every mixed strategy X.

5.2 Equalizing strategies and finiteness of the spectra of analytic games. In connection with analytic games, the discussion in 5.10 of Chapter 1 (also see 1.18) yields the following more precise proposition.

Theorem. *If the game Γ is analytic for player* 1 *and the spectrum of one of the optimal strategies is infinite, then every optimal strategy of player* 2 *is equalizing.*

In the same way, every optimal strategy of player 1 *is equalizing in a game that is analytic for player* 2, *provided that one of the optimal strategies has an infinite spectrum.*

Proof. Under the hypotheses of the first part of the theorem, let Y be an optimal strategy of player 2 in Γ. Then for $x \in \operatorname{supp} X$ we must have

$$H(x,Y) = v_\Gamma. \tag{5.1}$$

Since the set $\operatorname{supp} X$ is infinite and the function $H(\cdot,Y)$ in the preceding equation must be analytic for all $x \in [0,1]$, the conclusion follows.

The second part of the theorem is proved by a parallel argument. □

5.3. From the preceding theorem, there follow some tests for the finiteness of the spectra of the optimal strategies in analytic games.

Theorem. *In a game Γ that is analytic for player* 1 *with payoff function H let there exist an x_0 such that the derivative $\frac{\partial}{\partial x}H(x,y)|_{x=x_0}$ is either nonnegative for all y or nonpositive for all y.*

Then either the spectrum of each optimal strategy of player 1 *is finite, or the spectrum of each optimal strategy of player* 2 *is a subset of the set of y's for which the derivative is zero.*

There is a parallel proposition for games that are analytic for player 2.

Proof. Let some optimal strategy $X \in \mathbf{X}$ have infinite spectrum, and suppose for definiteness that

$$\frac{\partial}{\partial x} H(x,y)|_{x=x_0} \geqq 0. \tag{5.2}$$

Since Γ is analytic for all $x \in [0,1]$, by the theorem of 5.2 the equation (5.1) must be satisfied. This can be rewritten in the form

$$\int_0^1 H(x,y)dY(y) = v_\Gamma,$$

or, if we differentiate with respect to x and set $x = x_0$,

$$\int_0^1 \frac{\partial}{\partial x} H(x,y)v|_{x=x_0} dY(y) = 0.$$

Because of (5.2), this is possible only when the set of y's for which there is strict inequality in (5.2) has Y-measure zero. □

Sometimes it is technically convenient to use, instead of the partial derivative of the payoff function H itself, the partial derivative of some strictly monotonic (increasing or decreasing) differentiable function φ of H. Because of the equation

$$\frac{\partial}{\partial x}\varphi(H(x,y)) = \varphi'(H(x,y))\frac{\partial}{\partial x}H(x,y),$$

the partial derivative on the left has a constant sign if and only if the same is true for the derivative on the right.

As examples, we may use the games with the payoff functions

$$H(x,y) = \left(1 + \lambda(x-y)^2\right)^{-1} \tag{5.3}$$

with $\varphi(t) = 1/t$, and

$$H(x,y) = \exp\left(\left(\frac{3}{2}\right)x - y\right)^2$$

with $\varphi(t) = \log t$.

5.4. The following theorem provides another test for the finiteness of the spectrum of optimal strategies for a player in a two-person zero-sum game.

Theorem. *Let $H(x,y)$ be an analytic function which is positive on the unit square. Furthermore, let z be a complex variable, and let there be a curve C in the z plane, joining 0 and ∞, such that as z varies along C the function $H(z,y)$ remains analytic and bounded. If there is a sequence of points $z_1, z_2, \ldots \to \infty$ on C for which*

$$\lim_{n\to\infty} H(z_n, y) = 0,$$

uniformly for $y \in [0,1]$, the spectrum of every optimal strategy of player 1 is finite.

We prove this by contradiction. If player 1 has an optimal strategy X with an infinite spectrum, by the theorem of 4.2 every optimal strategy Y of player 2 is equalizing:

$$H(x,Y) = \int_0^1 H(x,y)dY(y) = v_\Gamma > 0 \tag{5.4}$$

for all $x \in [0,1]$. But $H(z,Y)$ is analytic for $z \in C$. Therefore it follows from (5.4) that, for all $z \in C$,

$$H(z,Y) = v_\Gamma > 0.$$

But this contradicts

$$v_\Gamma = \lim_{n\to\infty} H(z_n,Y) = \lim_{n\to\infty} \int_0^1 H(z_n,y)dY(n) = \int_0^1 \lim_{n\to\infty} H(z_n,y)dY(y) = 0.$$

□

As examples, we may use the games with the payoff functions (5.3), and also

$$H(x,y) = \left(1 + \frac{\sin(x-y)}{x^2+y^2+2}\right)^{-1}.$$

In both examples the positive real axis can serve as C.

5.5 The inverse problem for analytic games. We see that for any analytic game one of the following possibilities must occur: either the optimal strategy of one of the players has a finite spectrum, or the optimal strategy of the other player is equalizing. This provides a basis for expecting that there is either a "finitary" solution of the game, obtainable by combinatorial and algebraic arguments, or that the solution can be reduced to an analysis of equation (5.1), or, as a matter of fact, to the solution of an integral equation (a Fredholm equation of the first kind).

Unfortunately, the actual realization of these possibilities is successful only in very special cases. One reason is that in analytic games the players' optimal strategies can be quite arbitrary probability measures on unit intervals (including some that are arbitrarily pathological from the analytic point of view). In this subsection we take all strategies to be probability measures on $[0,1]$.

Theorem. *For arbitrary strategies X and Y, there is an analytic function H, defined on the unit square, such that the game on the unit square with this payoff function has the pair (X,Y) of strategies as its only equilibrium situation.*

This theorem has a rather simple proof in the case when each of the strategies X and Y has an infinite spectrum.

We set

$$H(X,Y) = \sum_{n=1}^{\infty} 2^{-n}(x^n - \mu_n)(y^n - \nu_n), \tag{5.5}$$

where

$$\mu_n = \int_0^1 x^n \, dX(x) \text{ and } \nu_n = \int_0^1 y^n \, dY(y)$$

are the nth moments of X and Y.

As is easily verified by integrating (5.5) with respect to x and y, we have, for arbitrary y and x,

$$H(X,y) = H(x,Y) = H(X,Y) = 0,$$

so that (X,Y) is a saddle point. Let us show that it is unique.

If X^* is any optimal strategy of player 1, we must have

$$H(X^*,y) = \sum_{n=1}^{\infty} 2^{-n}(\mu_n^* - \mu_n)(y^n - \nu_n) \tag{5.6}$$

for all $y \in \operatorname{supp} Y$ (here μ_n^* is the nth moment of X^* about zero).

By the theorem in 5.2, equation (5.6) is an identity. But then all the coefficients in the sum must vanish, i.e., $\mu_n^* = \mu_n$ for $n = 1, 2, \ldots$, and, by the theory of the moment problem,*) the optimal strategy X turns out to be unique. The uniqueness of the optimal strategy is established in the same way.

If the spectrum of at least one of the strategies X and Y is finite, the proof of the theorem reduces to an analysis of some constructions that are defined specially for different forms of the spectra of X and Y.

1°. The spectrum of X is infinite; the spectrum of Y is finite and contains at least one point of (0,1).

In this case, let

$0 < y_1 < \ldots < y_{l-1} < 1;$

possibly $y_0 = 0$, and/or $y_l = 1$, are points of the spectrum of Y. (5.7)

Set

$$H(x,y) = \sum_{n=1}^{\infty} 2^{-n}(x^n - \mu_n)\left((y - y_n)^{2n-1} - \bar{\nu}_{2n-1}\right) + \exp\left(-\prod_{j=1}^{l}(y - y_j)^{-2}\right),$$

where, at points of the spectrum of y, the exponent is replaced by its zero limiting value and, as above, μ_n is the nth moment of X and

$$\bar{\nu}_{2n-1} = \int_0^1 (y - y_1)^{2n-1} \, dY(y).$$

*) See, for example, L.V. Kantorovich and G.P. Akilov, Functional Analysis, Nauka, Moscow, 1977, p. 264.

In this case

$$H(X,y) = \exp\left(-\prod_{j=1}^{l}(y-y_j)^{-2}\right),$$
$$H(x,Y) = H(X,Y) = 0, \tag{5.8}$$

so that the strategies X and Y are optimal in Γ, and $v_\Gamma = 0$.

Now let X^* be any optimal strategy of player 1. Then

$$H(X^*,y) = \sum_{n=1}^{\infty} 2^{-n}(\mu_n^* - \mu_n)\big((y-y_1)^{2n-1} - \bar{\nu}_{2n-1}\big)$$
$$+\exp\left(-\prod_{j=1}^{l}(y-y_j)^{-2}\right) \geqq 0,$$

and by 5.5 of Chapter 1,

$$H(X^*,y_1) = \sum_{n=1}^{\infty} 2^{-n}(\mu_n^* - \mu_n)(-\bar{\nu}_{2n-1}) = 0.$$

This means that the sum

$$\sum_{n=1}^{\infty} 2^{-n}(\mu_n^* - \mu_n)(y-y_1)^{2n-1} + \exp\left(-\prod_{j=1}^{l}(y-y_j)^{-2}\right) \tag{5.9}$$

is always nonnegative (and when $y = y_1$, it reduces to zero). But if one of the coefficients $\mu_n^* - \mu_n$ is different from zero, then when $y = y_1$ the expression (5.9), which has the sign of its lowest term, must change its sign. This contradiction shows that $\mu_n^* = \mu_n$, i.e., that the optimal strategy X for player 1 must be unique.

Now consider any optimal strategy Y^* for player 2. It follows from (5.8) that $H(X,y) = 0 = v_\Gamma$ only if y is a point of the spectrum of Y^*. As above, reference to the theorem of 5.2 yields

$$H(x,Y^*) = \sum_{n=1}^{\infty} 2^{-n}(x^n - \mu^n)(\bar{\nu}_{2n}^* - \bar{\nu}_{2n-1}) = 0.$$

Therefore $\bar{\nu}_{2n}^* = \bar{\nu}_{2n-1}$ for all positive integers n, i.e.

$$\sum_{j=0}^{l}(y_j - y_1)^{2n-1}Y^*(y_j) = \sum_{j=0}^{l}(y_j - y_1)^{2n-1}Y(y_j),$$

or

$$\big(Y^*(0) - Y(0)\big)y_1^{2n-1} = \sum_{j=2}^{l}(y_j - y_1)^{2n-1}\big(Y^*(y_j) - Y(y_j)\big). \tag{5.10}$$

Now suppose that $2y_l$ is not a point of the spectrum of Y. Select numbers $y_1, y_2 - y_1, \ldots, y_l - y_1$, which are pairwise different, and pick out the largest one. Let this be $y_{j^*} - y_1$: if the largest number is y_1 the argument will be slightly different. Now divide (5.9) by $(y_{j^*} - y_1)^{2n-1}$:

$$(Y^*(0) - Y(0)) \left(\frac{y_1}{y_{j^*} - y_1} \right)^{2n-1} = \sum_{j=2}^{2n-1} \left(\frac{y_j - y_1}{y_{j^*} - y_1} \right)^{2n-1} (Y^*(y_j) - Y(y_j)). \tag{5.11}$$

Letting n tend to infinity in (5.11) yields zero on the left and $Y^*(y_{j^*}) - Y(y_{j*})$ on the right. Consequently $Y^*(y_{j^*}) = Y(y_{j*})$; if we drop the corresponding zero term in (5.6), we can repeat the process to obtain $Y^*(y_j) = Y(y_j)$ for every j, $j = 0, 1, \ldots, l$, i.e., $Y^* = Y$.

Now suppose that $2y_l = y_j$. Then we replace Γ by a game $\bar{\Gamma}$, also on the unit square, whose payoff function $\bar{H}$ satisfies $\bar{H}(x, y) = H(x, y^r)$ (with $r > 0$). The games Γ and $\bar{\Gamma}$ are isomorphic. Therefore, according to 2.24, Chapter 1, the optimal strategies of player 1 will be the same in Γ and $\bar{\Gamma}$, and there will be a one-to-one correspondence $Y \leftrightarrow \bar{Y}$ between the optimal strategies of player 2 in these games, in which $\bar{Y}([0,1]) = Y([0, Y^r])$. Since $r > 0$ is arbitrary, we can choose it so that y_l is different from any of the numbers $y_2, \ldots, y_{l-1}$. Then $\bar{\Gamma}$ will, as before, be analytic and come under the case already discussed, so that the optimal strategy of player 2 in it will be unique. Consequently the optimal strategy of player 2 in Γ will also be unique.

2°. Let X and Y both have finite spectra, each of which contains points of (0,1). We assume that

$0 < x_1 < \ldots < x_{k-1} < 1$

and possibly $x_0 = 0$, and/or $x_k = 1$ are points of the spectrum of X(5.12) and (5.7) holds. Set

$$H(x,y) = \sum_{n=1}^{\infty} 2^{-n} ((x - x_1)^{2n-1} - \bar{\mu}_{2n-1})((y - y_1)^{2n-1} - \bar{\nu}_{2n-1})$$

$$+ \exp\left(- \prod_{j=1}^{l} (y - y_j)^{-2} \right) - \exp\left(- \prod_{i=1}^{k} (x - x_i)^{-2} \right),$$

where

$$\bar{\mu}_{2n-1} = \int_0^1 (x - x_1)^{2n-1} dX(x),$$

and $\bar{\nu}_{2n-1}$ is defined as above.

In this case we shall have

$$H(X, y) = \exp\left(- \prod_{j=1}^{l} (y - y_j)^{-2} \right) \geqq 0, \tag{5.13}$$

$$H(x,Y) = -\exp\Big(-\prod_{i=1}^{k}(x-x_k)^{-2}\Big) \leqq 0, \tag{5.14}$$

$$H(X,Y) = 0,$$

so that the strategies X and Y are optimal, and $v_\Gamma = 0$.

It follows from (5.13) and (5.14) that the spectra of all optimal strategies of players 1 and 2 must be contained in the spectra of the strategies of X and Y.

Let us choose an arbitrary strategy X^* of player 1. Then, for all $y \in [0,1]$, we shall have

$$H(X^*,y) = \sum_{n=1}^{\infty} 2^{-n}(\bar{\mu}^*_{2n-1} - \bar{\mu}_{2n-1})(y-y_1)^{2n-1} + \exp\Big(-\prod_{j=1}^{l}(y-y_j)^{-2}\Big) \geqq 0,$$

where $\bar{\mu}_{2n-1}$ is defined as above. Since y_1 belongs to the spectrum of Y, we have

$$H(X^*,y_1) = \sum_{n=1}^{\infty} 2^{-n}(\mu^*_{2n-1} - \bar{\mu}_{2n-1})(-\bar{\nu}_{2n-1}) = 0,$$

and comparison with the preceding formula yields

$$\sum_{n=1}^{\infty} 2^{-n}(\mu^*_{2n-1} - \bar{\mu}_{2n-1})(y-y_1)^{2n-1} + \exp\left(-\prod_{j=1}^{l}(y-y_j)^{-2}\right) \geqq 0,$$

which, when $\bar{\mu}^*_{2n-1} \neq \bar{\mu}_{2n-1}$ contradicts the change of sign of this expression in passing from y to y_1. Therefore $\bar{\mu}^*_{2n-1} = \bar{\mu}_{2n-1}$ for all $n = 1, 2 \ldots$. This case can be analyzed as in case 1°.

3°. Let the spectrum of the strategy X be finite and contain at least one point of the interval (0,1) (i.e., let it have the form described in (5.12), and let Y be a strict mixture of strategies 0 and 1. Let us suppose that $Y(0) = \gamma$ ($\gamma \in (0,1)$).

Set

$$H(x,y) = \sum_{n=1}^{\infty} 2^{-n}\big((x-x_1)^{2n-1} - \mu_{2n-1}\big)\left(y^n \sin\frac{1}{y} - (1-\gamma)\sin 1\right)$$

$$+ \exp(-y^{-2}(1-y)^{-2}) - \exp\left(-\prod_{i=1}^{k}(x-x_i)^{-2}\right).$$

Here

$$H(X,y) = \exp(-y^{-2}(1-y)^{-2}) \geqq 0$$

$$H(x,Y) = -\exp\left(-\prod_{i=1}^{k}(x-x_i)^{-2}\right) \leqq 0,$$

$$H(X,Y) = v_\Gamma = 0.$$

As above, the spectra of the optimal strategies of player 1 are contained in supp X; and the spectra of the optimal strategies of player 2, in the set $\{0, 1\}$. If Y^* is any optimal strategy of player 2 with $Y^*(0) = \gamma^*$, we find, by arguing as before, that $(1 - \gamma^*) \sin 1 = (1 - \gamma) \sin 1$, i.e., $\gamma^* = \gamma$.

In addition, for every optimal strategy X^* of player 1 it follows from $\gamma \in (0, 1)$ that

$$\sum_{n=1}^{\infty} 2^{-n}(\bar{\mu}^*_{2n-1} - \bar{\mu}_{2n-1})(-(1-\gamma)\sin 1) = 0,$$

and therefore

$$\sum_{n=1}^{\infty} 2^{-n}(\bar{\mu}^*_{2n-1} - \bar{\mu}_{2n-1})y^n \sin\frac{1}{y} + \exp(-y^{-2}(1-y)^{-2}) \geqq 0. \tag{5.15}$$

Here the first nonzero term of the sum exceeds, in a sufficiently small neighborhood of $y = 0$, the sum of the moduli of the remaining terms. For a sufficiently large m, take

$$y_m = \pi^{-1}\left(2m + \frac{1}{2}\right)^{-1} \quad \text{and} \quad y'_m = \pi^{-1}\left(2m - \frac{1}{2}\right)^{-1}.$$

Then the first nonzero term will equal either

$$2^{-n}(\bar{\mu}^*_{2n-1} - \bar{\mu}_{2n-1})y^n_m \quad \text{or} \quad -2^{-n}(\bar{\mu}^*_{2n-1} - \bar{\mu}_{2n-1})y'^n_m,$$

and if $\bar{\mu}^* \neq \bar{\mu}_{2n-1}$, this term (and with it, the whole sum) must take values of both signs in a neighborhood of zero. However, this contradicts (5.15), and X is consequently the only optimal strategy.

4°. The strategy X has an infinite spectrum, and the spectrum of Y consists of the points 0 and 1. As before, let $Y(0) = \gamma$.

Here we set

$$H(x,y) = \sum_{n=1}^{\infty} 2^{-n}(x^n - \mu_n)\left(y^n \sin\frac{1}{y} - (1-\gamma)\sin 1\right) + \exp(-y^{-2}(1-y)^{-2}).$$

As before, (X, Y) is a saddle point of Γ, $v_\Gamma = 0$, and the spectrum of every different optimal strategy of player 2 contains only the points 0 and 1.

As in the case when X and Y have infinte spectra, here we have $H(x, Y) = 0$ for the optimal strategy Y and arbitrary x. Hence it follows that there is only one optimal strategy for player 2; the uniqueness of the optimal strategy of player 1 follows, as above, from the uniqueness of its moments.

5°. Now let supp X = supp Y = $\{0, 1\}$ with $X(0) = \alpha \in (0, 1)$ and $Y(0) = \gamma \in (0, 1)$.

Take

$$H(x,y) = -\gamma x + (2-\alpha)y + x^2 + (1-\alpha-\gamma)xy - y^2.$$

This game is strictly concave for each player. Consequently the spectra of all the player's optimal strategies must be in the set $\{0,1\}$ and every saddle point of Γ must be a saddle point of the matrix game whose payoff matrix is

$$\begin{bmatrix} H(0,0) & H(0,1) \\ H(1,0) & H(1,1) \end{bmatrix} = \begin{bmatrix} 0 & 1-\alpha \\ 1-\beta & 1-\alpha-\beta \end{bmatrix}.$$

As is easily verified, this matrix game has a single saddle point which coincides formally with (X,Y).

6°. Let the strategy X be arbitrary and let Y be the pure strategy $y=0$.

Let

$$H(x,y) = \sum_{n=1}^{\infty} 2^{-n}(x^n - \mu_n)y^n \sin\frac{1}{y} + \exp(-y^{-2}).$$

As above, here it is easy to verify that the strategies X and Y are optimal, $v_\Gamma = 0$, and the optimal strategies are unique.

7°. Finally, it remains only to remark that, throughout the entire discussion we can interchange the roles of the players, as well as of the strategies 0 and 1. □

5.6 Rational games with singular solutions. A class of games still narrower than the class of analytic games consists of the games that have rational payoff functions. However, even these games can have solutions whose structure is quite complicated. Let us present an example of a rational game on the unit square in which the only optimal strategy of player 1 has a Cantor set*) as its distribution function.

It will be convenient to begin with a somewhat unconventional definition of a Cantor distribution function.

Let us consider a sequence of independent random variables

$$X_1, X_2, \ldots, \tag{5.16}$$

each of which takes the value 0 with probability $\frac{1}{2}$ and the value 2 with probability $\frac{1}{2}$. Each realization of this sequence, written after a zero, can be considered as the ternary expansion of a number $x \in [0,1]$. The sequence (5.16) is so constructed that the only numbers x that appear are those whose ternary expansions do not contain the digit 1. Let X denote the random variable that takes only the values x that correspond to realizations in the sequence (5.16). The distribution of X is the Cantor distribution function C.

The distribution function C can be constructed in the following way. Set

$$C(0) = 0, \quad C(1) = 1, \tag{5.17}$$

*) A.N Kolmogorov and S.V. Fomin, Elements of the theory of functions and functional analysis, Moscow, 1976, p.341.

and for each interval $[a, b]$ for which C is defined only at the endpoints a and b, we set

$$C(x) = \frac{1}{3}(C(a) + C(b)) \text{ for } \frac{2a+b}{2} \leqq x \leqq \frac{a+2b}{3}.$$

Formally, this definition is equivalent to defining C by the functional equations

$$C\left(\frac{x}{3}\right) = \frac{1}{2}C(x),$$
$$C(1-x) = 1 - C(x)$$

with the boundary conditions (5.17).

The Cantor function C is continuous. It is everywhere differentiable, except on an uncountable set of measure zero (in fact, this is the set of values of the independent variable whose ternary expansion contains no 1's). Wherever the derivative of C exists, it is zero. The function C is the classical example of a singular distribution function.

Let us consider a game on the unit square with the payoff function

$$H(x,y) = \left(y - \frac{1}{2}\right)\left(\frac{1+(x-1/2)(y-1/2)^2}{1+(x-1/2)(y-1/2)^4} - \frac{1}{1+(x/3-1/2)^2(y-1/2)^4}\right).$$

The function H is analytic near the origin. In addition, it is evident that

$$H(x,y) = -H(x, 1-y).$$

Therefore every strategy Y of player 2, for which

$$P(Y > y) = P(Y < 1 - y), \quad y \in \left[0, \frac{1}{2}\right], \tag{5.18}$$

is equalizing, and therefore optimal, and

$$H(x, Y) \equiv 0 = v_\Gamma. \tag{5.19}$$

To show that the strategy C is optimal, it is sufficient to show that, for every y,

$$H(C, y) = \int_0^1 H(x,y)dC(x) = 0. \tag{5.20}$$

A term H that contains $x - \frac{1}{2}$ evidently contributes 0 to the integral, and the preceding equation can be rewritten in the form

$$\int_0^1 \left(\left(1+\left(x-\frac{1}{2}\right)^2\left(y-\frac{1}{2}\right)^4\right)^{-1} - \left(1+\left(\frac{x}{3}-\frac{1}{2}\right)\left(y-\frac{1}{2}\right)^4\right)^{-1}\right) dC(x) = 0.$$

However, if we introduce the new variable $x/e = t$ and use the facts that, first, $C(x/3) = \frac{1}{2}C(x)$, and then that $C(x) = \frac{1}{2}$ on $[\frac{1}{3}, \frac{2}{3}]$, we obtain, if we set

$$\left(1+\left(\frac{x}{3}-\frac{1}{2}\right)^2\left(y-\frac{1}{2}\right)^4\right)^{-1} = h(x,y),$$

$$\int_0^1 h(x,y)dC(x) = \int_0^{1/3} h(3t,y)dC(3t) = 2\int_0^{1/3} h(3t,y)dC(t)$$

$$= \int_0^{1/3} h(3t,y)dC(t) + \int_{2/3}^1 h(3t,y)dC(t) = \int_0^1 h(3t,y)dC(t)$$

and (5.20) is established.

It remains for us to establish the uniqueness of the optimal strategy C. For this purpose, we notice that (5.18) is sufficient for C to be an optimal strategy. Consequently the set of strategies contains a completely mixed strategy. But then, by 5.16 of Chapter 1 (also see 1.18) every optimal strategy X must be equalizing, i.e., the equation

$$H(X,y) = \int_0^1 H(x,y)dX(x) = v_\Gamma = 0, \quad y \in [0,1],$$

is satisfied.

The representation of the fractions in the expression of $H(x,y)$ as power series in the corresponding differences yields

$$\int_0^1 \sum_{n=0}^\infty (-1)^n \left(\left(x-\frac{1}{2}\right)^{2n}\left(y-\frac{1}{2}\right)^{4n+1} + \left(x-\frac{1}{2}\right)^{2n+1}\left(y-\frac{1}{2}\right)^{4n+3}\right) dX(x)$$

$$-\int_0^1 \sum_{n=0}^\infty (-1)^n \left(\frac{x}{3}-\frac{1}{2}\right)^{2n}\left(y-\frac{1}{2}\right)^{4n+1} dX(x)=0,$$

or, since the series converge in the interval of interest,

$$\sum_{n=0}^\infty (-1)^n \left(y-\frac{1}{2}\right)^{4n+1} \int_0^1 \left(\left(x-\frac{1}{2}\right)^{2n} - \left(\frac{x}{3}-\frac{1}{2}\right)^{2n}\right) dX(x)$$

$$-\sum_{n=0}^\infty (-1)^n \left(y-\frac{1}{2}\right)^{4n+3} \int_0^1 \left(x-\frac{1}{2}\right)^{2n+1} dX(x) = 0.$$

If a series vanishes identically so do its coefficients, i.e.,

$$\int_0^1 \left(\left(x-\frac{1}{2}\right)^{2n} - \left(\frac{x}{3}-\frac{1}{2}\right)^{2n}\right) dX(x) = 0,$$

$$\text{and} \quad \int_0^1 \left(x-\frac{1}{2}\right)^{2n-1} dX(x) = 0.$$

From these equations, we can calculate the moments

$$\mu_n = \int_0^1 x^n \, dX(x)$$

of the distribution X, one after another. These moments are uniquely determined by the equations, and in turn uniquely determine the optimal strategy, which accordingly is unique. Since we already know that C is an optimal strategy, we must evidently have $X = C$.

5.7 Example of the solution of a rational game. As an example of a rational game whose solution can be completely carried out, we consider the game whose payoff function is

$$H(x, y) = \frac{(1+x)^k (1+y)^k (1-xy)}{(1+xy)^{k+1}}, \quad \text{where } k > 0.$$

As a working hypothesis, let us assume that in this game player 2 has an optimal strategy with infinite spectrum. Then, by the theorem of 5.2, every optimal strategy of player 1 must be equalizing;

$$\int_0^1 H(x, y) dX(x) = v_\Gamma. \tag{5.21}$$

If, in the formula for H, we expand $(1-y)^{-k}$ and $(1+xy)^{-k-1}$ by the binomial theorem and substitute the results into our equation, we obtain

$$\begin{aligned} \int_0^1 (1+x)^k \left(\sum_{n=1}^{\infty} (-1)^n \binom{n+k}{k} x^n y^n\right) (1-xy) dX(x) \\ = v_\Gamma \sum_{n=1}^{\infty} (-1)^n \binom{n+k+1}{n} y^n. \end{aligned} \tag{5.22}$$

If we set

$$a_n = \int_0^1 (1+x)^k x^n \, dX(x), \tag{5.23}$$

for $n = 0, 1, \ldots$, we can rewrite (5.22) in the form

$$\sum_{n=1}^{\infty} (-1)^n \binom{n+k}{n} (y^n a_n - y^{n+1} a_{n+1}) = v_\Gamma \sum_{n=1}^{\infty} (-1)^n \binom{n+k-1}{n} y^n. \tag{5.24}$$

If we now compare the coefficients of corresponding powers on both sides of (5.24), we obtain

$$a_n\left(\binom{n+k}{n}+\binom{n+k-1}{n}\right)=v_\Gamma\binom{n+k+1}{n}$$

or after simplification

$$a_n = v_\Gamma\, k/(2n+k). \tag{5.25}$$

If now, with a view to substituting in (5.23), we set

$$dX(x) = M\frac{x^{k/2}-1}{(1+x)^k}dx, \tag{5.26}$$

where M is a normalizing constant which is determined by the condition

$$\int_0^1 dX(x) = 1, \tag{5.27}$$

we obtain

$$\frac{kv}{2n+k} = M\int_0^1 x^{n+k/2-1}\,dx = \frac{2M}{2n+k},$$

or $v_\Gamma = 2M/k$.

Knowing v_Γ we can use (5.25) to find all the a_n which together with v_Γ satisfy (5.24) and hence (5.21). Consequently the mixed strategies (5.26) and (5.27) are equalizing.

But the payoff function H of this game is symmetric (which, of course, does not mean that the game itself is symmetric). Therefore the same distribution X, considered as a mixed strategy of player 2, also equalizes its strategies. A pair of equalizing strategies of the players in a two-person game form an equilibrium situation, as we noticed in Chapter 1, 5.10.

§6 Separable games

6.1 Separable games. A two-person zero-sum game

$$\Gamma = \langle x, y, H\rangle \tag{6.1}$$

is said to be separable if x and y are compact sets, and the payoff function H is separable, i.e., has a representation

$$H(x,y) = \sum_{i=1}^{n} u_i(x)v_i(y),$$

where u_i and v_i are continuous functions of their arguments. We shall consider only separable games on the unit square; that is, we take $x = y = [0,1]$. □

Of course, the form of the representation of a separable function is not unique. We shall ordinarily represent H in the form

$$H(x,y) = \sum_{i=1}^{m}\sum_{j=1}^{n} a_{ij}u_i(x)v_j(y). \tag{6.2}$$

6.2. We rather seldom encounter separable payoff functions in actual game-theoretical practice. We should also not be too optimistic about being able to use the approximability of general payoff functions by separable functions: although such approximations can be applied to the value of a game (which is a continuous functional of the payoff function), it is usually not possible to obtain approximate expressions for players' optimal strategies by using them. Nevertheless, separable games are interesting objects of investigation as a natural generalization of finite matrix games (evidently, separable games are matrix games when u_i and v_j are constants). More precisely, if, in matrix games, each mixed strategy of a player is itself determined by a finite number of parameters, it follows that in separable games the effect of their use is determined by a finite number of parameters. Thus the space of a player's mixed strategies in a separable game has a natural embedding in a finite-dimensional Euclidean space. We may expect that results obtained in the theory of separable games will be quite instructive as prototypes of analogous results for the wider class of games in which the players' strategies are determined by finite sets of parameters.

6.3 "Moments" of players' strategies. For a situation (X,Y) in a separable game Γ with payoff function H defined by (6.2), we have

$$H(X,Y) = \int_0^1\int_0^1 H(x,y)dX(x)dY(y)$$

$$= \int_0^1\int_0^1 \sum_{i=1}^{m}\sum_{j=1}^{n} a_{ij}u_i(x)v_j(y)dX(x)dY(y)$$

$$= \sum_{i=1}^{m}\sum_{j=1}^{n} a_{ij}\int_0^1 u_i(x)dX(x)\int_0^1 v_j(y)dY(y),$$

or, if we set

$$\int_0^1 u_i(x)dX(x) = u_i(X), \quad \int_0^1 v_j(y)dY(y) = v_j(Y),$$

we may write

$$H(X,Y) = \sum_{i=1}^{m}\sum_{j=1}^{n} a_{ij}u_i(X)v_j(Y). \tag{6.3}$$

Consequently the players' payoffs do not depend on the payoffs themselves for strategies X and Y, but only on their numerical descriptions $u_i(X)$ and $v_j(Y)$, of the nature of moments. Therefore it is natural enough to introduce the vectors

$$u(X) = (u_1(X),\dots,u_m(X)),$$

$$v(Y) = (v_1(Y),\dots,v_n(Y)).$$

Strategies X' and X'' for which $u(X') = u(X'')$ are said to be *equivalent.* The equivalence of mixed strategies Y' and Y'' is defined similarly. We may evidently identify equivalent strategies. We have already used a similar process of identification in Chapter 1, 6.10; and in 2.8.

Consequently we may suppose that the arguments of payoff functions are not the individual strategies of the players, but are equivalence classes in the sense just defined. Correspondingly, we shall use the notation $H(u,v)$ to mean $H(X,Y)$, where X and Y are such that $u(X) = u$ and $v(Y) = v$.

Formula (6.3) shows that $H(u,v)$ is a bilinear function.

We denote by U_Γ the set of all m-component vectors $u(X)$. The set of n-component vectors $v(Y)$ for the same game is denoted by V_Γ.

6.4 Theorem. *For every separable game Γ the sets U_Γ and V_Γ are nonempty, bounded, convex, and closed.*

Proof. The nonemptiness of these sets is evident; their boundedness follows from the boundedness both of the payoff functions H, and of the probability measures describing the strategies.

Now let $u', u'' \in U_\Gamma$ and $\lambda \in [0,1]$. Then there are also mixed strategies X' and X'' for which $u' = u(X')$ and $u'' = u(X'')$.

In addition, we have, for $i = 1, \ldots, m$,

$$u_i(\lambda X' + (1-\lambda)X'') = \int_0^1 u_i(x)d(\lambda X' + (1-\lambda)X'')$$

$$= \lambda \int_0^1 u_i(x)dX' + (1-\lambda)\int_0^1 u_i(x)dX'' = \lambda u_i(X') + (1-\lambda)u_i(X''),$$

and the convexity of U_Γ is established.

We can show in a similar way that this set is closed. Let us consider a sequence $u^{(1)}, u^{(2)}, \ldots$ of vectors converging to $u^{(0)}$ that belong to U_Γ.

For each of these vectors $u^{(k)}$ we find a mixed strategy $X^{(k)}$ such that $u^{(k)} = u(X^{(k)})$. From the sequence $X^{(1)}, X^{(2)}, \ldots$ of strategies, thought of as distribution functions, we select a subsequence that converges at every point of continuity of the limit function. (This kind of convergence is guaranteed by Helly's first theorem). We may suppose that the whole sequence $X^{(1)}, X^{(2)}, \ldots$ converges; let $X^{(0)}$ be its limit. For $i = 1, \ldots, m$, set

$$u_i(X^{(0)}) = \int_0^1 u_i(x)dX^{(0)}.$$

If we now apply Helly's second theorem, on taking the limit inside a Stieltjes integral, we obtain

$$u_i(X^{(0)}) = \lim_{k\to\infty} \int_0^1 u_i(x)dX^{(k)} = \lim_{k\to\infty} u_i(X^{(k)}) = \lim_{k\to\infty} u_i^{(k)} = u_i^{(0)}.$$

Therefore, to the limit vector there corresponds a strategy (generally speaking, not just one) such that this vector belongs to U_Γ, which is thus shown to be closed. The required properties of V_Γ are established in a similar way. □

6.5. Vectors of the form

$$u(x) = \big(u_1(x), \ldots, u_m(x)\big), \quad x \in [0, 1],$$

evidently also belong to U_Γ since they correspond to pure strategies of player 1. Since the functions u_i are continuous, these vectors, as points, form a continuous curve, which we denote by K_Γ.

Similarly, we define a curve L_Γ corresponding to vectors of the form

$$v(y) = \big(v_1(y), \ldots, v_n(y)\big), \quad y \in [0, 1].$$

6.6 Theorem. In the notations of 6.3 and 6.5,

$$U_\Gamma = \operatorname{conv} K_\Gamma,$$
$$V_\Gamma = \operatorname{conv} L_\Gamma.$$

We shall prove only the first equation.

Since, as we have already noticed, $K_\Gamma \subset U_\Gamma$, and U_Γ is convex by the theorem in 6.4, we must also have

$$\operatorname{conv} K_\Gamma \subset U_\Gamma. \tag{6.4}$$

Now let $u^0 \in U_\Gamma \setminus \operatorname{conv} K_\Gamma$.

Since $\operatorname{conv} K_\Gamma$ is closed and convex, we may strictly separate u^0 from $\operatorname{conv} K_\Gamma$ (and hence also from K_Γ) by a hyperplane, i.e., we can find $c_1, \ldots, c_m$ such that

$$\sum_{i=1}^m c_i u_i^0 - \sum_{i=1}^m c_i u_i(x) \geqq \delta > 0 \tag{6.5}$$

for all $x \in [0, 1]$. Let u^0 correspond to the mixed strategy X^0:

$$u_i^0 = \int_0^1 u_i(x)dX^0(x).$$

Integration of (6.5) gives us

$$\sum_{i=1}^{m} c_i u_i^0 \int_0^1 dX^0(x) - \sum_{i=1}^{m} c_i \int_0^1 u_i(x)dX^0(x) \geqq \delta \int_0^1 dX^0(x),$$

or

$$\sum_{i=1}^{m} c_i u_i^0 - \sum_{i=1}^{m} c_i u_i^0 \geqq \delta > 0,$$

which is impossible. Therefore $U_\Gamma \subset \operatorname{conv} K_\Gamma$, which, together with (6.4), gives us what was required. □

6.7 Optimal strategies with finite spectrum.

Theorem. *Every strategy of player* 1 *in a separable game* Γ *with payoff function* H *is equivalent to a convex combination of at most* m *pure strategies. There is a similar proposition for the mixed strategies of player* 2.

Proof. Let X be any mixed strategy of player 1, and u a corresponding vector in U_Γ. Since K_Γ is connected, every $u \in U_\Gamma$ can be represented in the form*)

$$u = \sum_{k=1}^{m} \lambda_k u^{(k)},$$

where

$$u^{(k)} \in K_\Gamma, \quad \lambda_k \geqq 0, \quad \sum_{k=1}^{m} \lambda_k = 1.$$

This means that, for every y,

$$H(X, y) = \sum_{k=1}^{m} \lambda_k H(x^{(k)}, y) = H(X^*, y),$$

where X^* is a mixed strategy of player 1 that assigns probability λ_k to each pure strategy $x^{(k)}$ $(k = 1, \dots, m)$. □

6.8 Corollary. In a separable game Γ, player 1 has an optimal strategy which is a mixture of at most m pure strategies, and player 2 has an optimal strategy which is a mixture of at most n pure strategies.

Proof. By definition, a separable game has a continuous payoff function. Therefore, on the basis of what was said in, for example, 2.2, there is an equilibrium situation. Consequently we may speak of optimal strategies X and Y involving the players in Γ. By what was said previously, X and Y are equivalent to strategies X^* and Y^* which are mixtures of at most m and n pure strategies, respectively. But strategies that are equivalent to optimal strategies are evidently also optimal.

---

*) T. Bonnesen and W. Fenchel, Theorie der konvexen Körper, Springer, Berlin, 1934.

6.9 An auxiliary game. For the study of a separable game and at the same time for obtaining its solution, it is convenient, if we rely on what was said in 6.3, to consider the auxiliary game $\Delta_\Gamma = \langle U, V, H\rangle$. As we shall explain, here U and V are closed, bounded, convex subsets of the corresponding spaces $\mathbf{R}^m$ and $\mathbf{R}^n$, and the function H is, according to (6.3), bilinear:

$$H(u,v) = \sum_{i=1}^{m}\sum_{j=1}^{n} a_{ij}\, u_i\, v_i. \tag{6.6}$$

The connection between Γ and Δ_Γ is described by the following essentially trivial proposition.

Theorem. *For a situation (X^*, Y^*) to be a saddle point in the game Γ it is necessary and sufficient that the situation $\big(u(X^*), v(Y^*)\big)$ is a saddle point in Δ_Γ.*

For the proof, it is enough to note that, for every X and Y,

$$\begin{aligned} H(X,Y^*) &= H\big(u(X), v(Y^*)\big) \leqq H(X^*,Y^*) \\ &= H\big(u(X^*), v(Y^*)\big) \leqq H(X^*,Y) = H\big(u(X^*), v(Y)\big). \quad \square \end{aligned}$$

6.10 Scheme of finding solutions. What we have said leads to a process for finding saddle points in a separable game. The actual solution of Δ_Γ can be carried out by a version of the fixed point method, as follows.

We construct the curves K_Γ and L_Γ; then their convex hulls $\operatorname{conv} K_\Gamma$ and $\operatorname{conv} L_\Gamma$ (these will be the sets U and V); then we represent H as $U \times V \to \mathbf{R}$ and solve the auxiliary game.

Choose an arbitrary $u_0 \in U$ and construct the set $V(u_0)$ of the points $v \in V$ for which $H(u_0, v)$ attains its minimum. It is clear that if $v \in V(u_0)$, the situation (u_0, v) will be acceptable to player 2. Similarly, choose an arbitrary $v_0 \in V$ and construct the set $U(v_0)$ of points $u \in U$ for which $H(u, v_0)$ attains its maximum. If $u \in U(v_0)$ the situation (u, v_0) will be acceptable to player 1.

If $u \in U(v)$ and $v \in V(u)$, the pair (u, v), which is a situation in Δ_Γ, is said to be *fixed*. We see from what was said above that a fixed situation is an equilibrium situation.

We can systematize the search for a fixed situation.

For this purpose, let us write the payoff function (6.6) in the form

$$H(u,v) = \sum_{j=1}^{n}\left(\sum_{i=1}^{m} a_{ij}\, u_i\right) v_j.$$

The hyperplanes

$$\sum_{i=1}^{m} a_{ij}\, u_i = 0$$

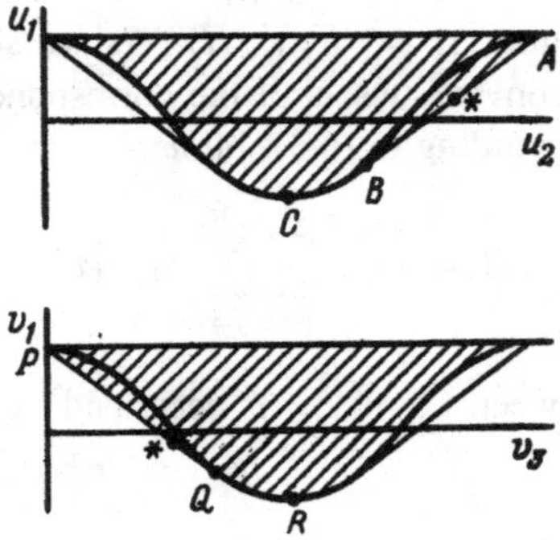

Fig. 3.3

divide the set U into a number of convex domains, inside each of which all these linear forms have a constant sign. If u_0 belongs to one of these domains, then in order to minimize $H(u_0, v)$ we must choose the largest possible values v_j with negative coefficients and the smallest values with positive coefficients. We form a transformation $V(u)$ in this way. The transformation $U(v)$ is formed similarly from the representation

$$H(u, v) = \sum_{i=1}^{m} \left(\sum_{j=1}^{n} a_{ij} v_j \right) u_i .$$

6.11. What we have just done can be illustrated by the following example.

Example. In a game Γ as in (6.1), let

$$x = y = [0, 1], \quad H(x, y) = \cos 2\pi x \cos 2\pi y + x + y.$$

Let us set

$$u_1 = \cos 2\pi x, \quad v_1 = \cos 2\pi x y,$$

$$u_2 = x, \quad v_2 = 1,$$

$$u_3 = 1, \quad v_3 = y.$$

The curves K_Γ and L_Γ in this case have the equations

$$u_1 = \cos 2\pi u_2 , \quad v_1 = \cos 2\pi v_3 . \tag{6.7}$$

Graphs of these curves are shown in Figure 3.3. (Here the convex hulls of K_Γ and L_Γ, i.e., the domains U and V, are shaded.)

Here the auxiliary game is $\Delta_\Gamma = \langle U, V, \tilde{H} \rangle$, where $\tilde{H}(u, v) = u_1 v_1 + u_2 + v_3$. Since u_3 and v_2 are constants, we shall write $u = (u_1 , u_2)$ and $v = (v_1 , v_3)$.

We need to find $u^* = (u_1^* , u_2^*)$ and $v^* = (v_1^* , v_3^*)$ such that the double inequality.

$$\tilde{H}((u_1 , u_2), (v_1^* , v_3^*)) \leqq \tilde{H}((u_1^* , u_2^*), (v_1^* , v_3^*)) \leqq \tilde{H}((u_1^* , u_2^*), (v_1 , v_3))$$

is satisfied for all $u = (u_1 , u_2) \in U$ und $v = (v_1 , v_3) \in V$.

We can argue as follows.

If $u_2' < u_2''$ we evidently have

$$\tilde{H}((u_1, u_2'), (v_1, v_3)) < \tilde{H}((u_1, u_2''), (v_1, v_3)),$$

whatever the values of u_1, v_1 and v_3 are, and in particular,

$$\tilde{H}((u_1^*, u_2'), (v_1^*, v_3^*)) < \tilde{H}((u_1^*, u_2'), (v_1^*, v_3^*)).$$

But this means that u_2' cannot be in $V(u^*)$. Therefore the point corresponding to u^* must lie on the line ABC. For a similar reason, the point corresponding to v^* must lie on PQR.

We shall need to know the coordinates of the point $Q = (v_1^Q, v_2^Q)$. On one hand, we have $1 - kv_3^Q = v_1^Q$, where k is the slope of the tangent PQ; and on the other hand,

$$k = -\left.\frac{dv}{dv_3}\right|_{v_3 = v_3^Q},$$

or, if we take account of (6.7),

$$\begin{aligned} 1 - kv_3^Q &= \cos 2\pi v_3^Q, \\ k &= 2\pi \sin 2\pi v_3^Q, \end{aligned} \tag{6.8}$$

from which

$$1 - 2\pi v_3^Q \sin 2\pi v_3^Q = \cos 2\pi v_3^Q,$$

or

$$v_3^Q = \frac{1 - \cos 2\pi v_3^Q}{2\pi \sin 2\pi v_3^Q} = \frac{1}{2\pi} \tan \pi v_3^Q.$$

If we solve this equation and use (6.8), we obtain $k_1 = 4.55$ and $v_3^Q = 0.37$. By symmetry, we have $u_2^B = 1 - 0.37 = 0.63$.

We first investigate the possible positions of u^* on AB and v^* on PQ. We have

$$\begin{aligned} v_1 &= 1 - kv_3, \\ u_1 &= 1 - k(1 - u_2) = 1 - k + ku_2, \end{aligned} \tag{6.9}$$

so that

$$\tilde{H}(u, v) = (1 - kv_3)(1 - k + ku_2) = 1 - k + (k+1)u_2 + (k^2 - k + 1)v_3 - k^2 u_2 v_3.$$

Hence we can obtain the following two representations of $\tilde{H}$ on $[AB] \times [PQ]$:

$$\tilde{H}(u, v) = (1 - k + (k^2 - k + 1)v_3) + (k + 1 - k^2 v_3)u_2, \tag{6.10}$$

$$\tilde{H}(u, v) = (1 - k + (k + 1)u_2) + (k^2 - k + 1 - k^2 u_2)v_3. \tag{6.11}$$

Suppose first that

$$k + 1 - k^2 v_3^* > 0. \tag{6.12}$$

Then $\tilde{H}(u, v^*)$ is an increasing function of u_2 and attains its maximum for $u_2^* = u_2^A = 1$. But then, in turn, the coefficient of v_3 is negative, and $\tilde{H}(u^*, v_3)$ is a decreasing function of v_3 and attains its minimum for $v_3^* = v_3^Q = 0.37$; in this case, as is easily calculated, $k + 1 + k^2 v_3^* < 0$. Consequently we have obtained a contradiction, and (6.12) cannot be satisfied at the saddle point (u^*, v^*).

Now let

$$k + 1 - k^2 v_3^* < 0. \tag{6.13}$$

Then $\tilde{H}(u, v^*)$ decreases and attains its minimum for $u_2^* = u_2^B = 0.63$. According to (6.11), the coefficient of v_3 is positive, and $H(u^*, v)$ attains its minimum for $v_3^* = v_3^P = 0$, which contradicts (6.13). Therefore (6.12) also cannot be satisfied at the saddle point (u^*, v^*).

There remains the possibility that $k + 1 - k^2 v_3^* = 0$, whence $v_3^* = (k+1)/k^2$. A similar analysis leaves us with only the possibility $k^2 - k + 1 - k^2 u_2^* = 0$ for u_2^*, so that $u_2^* = (k^2 - k + 1)/k^2$.

We see from (6.10) and (6.11) that $\tilde{H}(u, v^*)$ is actually independent of u, and $\tilde{H}(u^*, v)$ is independent of v, so that (u^*, v^*) is trivially a saddle point.

Together with (6.9), what we have found gives us $u_1^* = 1/k$, $v_1^* = -1/k$. Replacing k by its value, we obtain

$$u^* = (0.22\,; 0.83\,; 1.0), \quad v^* = (-0.22\,; 1.0\,; 0.27),$$

so that the value of the game is

$$v_\Gamma = u_1^* v_1^* + u_2^* + v_3^* = 0.62.$$

The point u^* is not on K, and therefore there is no pure strategy for player 1. However, it can be represented (in just one way) as a convex combination of points of K, namely u^A and u^B. It is easily calculated that

$$u^* = 0.54u^A + 0.46u^B,$$

i.e., player 1 should choose the pure strategy 1 with probability 0.54, and the pure strategy 0.63 with probability 0.46.

We find similarly that

$$v^* = 0.27v^P + 0.73v^R,$$

i.e., player 2 should play the pure strategy 0 with probability 0.27, and the pure strategy 0.37 with probability 0.73.

Let us investigate the possibility of the existence of other saddle points. To do this, we choose a value u^* and vary v (we emphasize that we are, as before, interested only in values v that correspond to a point on PQR). We see from (6.9) that as v varies along the interval PQ and its extension (in particular, up to Q'), H has the same sign. When v passes from a point of QQ' to a point of QR with the same ordinate, H increases strictly, as we have already seen. Therefore, there are no points on QR corresponding to optimal strategies of player 2 in the game Δ_Γ.

Similarly, we can see that there are no points on BC corresponding to optimal strategies of player 1.

Therefore the saddle point (u^*, v^*) that we found is the only one.

§7 Convex games

7.1 Definition and general method of solution. As we said above (see 3.3), a two-person zero-sum game

$$\Gamma = \langle \boldsymbol{x}, \boldsymbol{y}, H \rangle \tag{7.1}$$

is said to be convex if the set $\boldsymbol{y}$ is a convex subset of a real linear topological space and the partial payoff functions $H_x = H(x, \cdot) : \boldsymbol{y} \to \mathbf{R}$ are convex for every $x \in \boldsymbol{x}$.

According to 3.5, a convex game has a value if the set $\boldsymbol{y}$ is compact and the functions H_x are continuous. In that case, player 2 has a pure optimal strategy, and player 1 has, for every $\varepsilon > 0$, an ε-optimal strategy which is a mixture of finitely many pure strategies. If, in addition, $\boldsymbol{x}$ is a compact space and $\boldsymbol{y} \subset \mathbf{R}^n$, it follows from part 1 of the theorem in 3.13 that player 1 has an optimal strategy which is a mixture of at most $n + 1$ pure strategies. In the present section we discuss some solutions of games of this kind.

7.2. A general method for solving the convex games defined in (7.1) is as follows. First we must verify that the game is indeed convex. Second, we use the theorem of 1.14 and write the equation

$$v_\Gamma = \min_{y \in \mathcal{y}} \max_{x \in \mathcal{x}} H(x, y), \tag{7.2}$$

from which we obtain the value v_Γ of Γ and the set $\mathcal{T}(\Gamma)$ of the optimal strategies of player 2 as the set of those $y^* \in \mathcal{y}$ on which the minimum in (7.2) is attained. Third, we notice that each pure strategy x of player 1 that appears in the spectra of any of that player's optimal strategies is *essential*, i.e., that the equation

$$H(x, y^*) = v_\Gamma \tag{7.3}$$

is satisfied. Finally, in the fourth place, a mixture X^* of the essential strategies of player 1 will be optimal if it satisfies the equation

$$H(X^*, y^*) = \min_{y \in \mathcal{y}} H(X^*, y). \tag{7.4}$$

7.3. Finding the minimax in (7.2) is not in itself a game-theoretic problem, although it may present real difficulties. This is all the more the case for solving (7.3); consequently we shall not discuss general solutions of these problems, but restrict ourselves to the problem of finding solutions of (7.4) that have a specifically game-theoretic character.

The following auxiliary proposition will be useful.

Lemma. *If $X \in \mathbf{X}$ is a mixed strategy of player* 1 *in a convex game as in* (7.1), *the function*

$$H_X = \int_{\mathcal{x}} H_x \, dX(x) : \mathcal{y} \to \mathbf{R}$$

is convex.

Proof. Let $y', y'' \in \mathcal{y}$ and $\lambda \in [0, 1]$. Then

$$\begin{aligned} H(X, \lambda y' + (1-\lambda)y'') &= \int_{\mathcal{x}} H(x, \lambda y' + (1-\lambda)y'') dX(x) \\ &\leq \int_{\mathcal{x}} (\lambda H(x, y') + (1-\lambda)H(x, y'')) dX(x) \\ &= \lambda H(X, y') + (1-\lambda)H(X, y''). \qquad \square \end{aligned}$$

It follows from this lemma that any local minimum of the function $H(X, \cdot)$: $\mathcal{y} \to \mathbf{R}$ is actually its smallest value.

7.4 Optimality conditions for a mixed strategy of player 1. Let y^* be a pure optimal strategy of player 2 in the convex game Γ. It is known (Carathéodory's theorem*)) that y^* can be represented as a convex combination of at most $n+1$

---

*) See, for example, R.T. Rockafellar, Convex analysis, Princeton, 1970.

extreme points of $\mathcal{Y}$:

$$y^* = \sum_{r=0}^{n} \lambda_r y_r^0, \tag{7.5}$$

$$\lambda_r \geqq 0, \quad (r = 0, 1, \ldots, n), \quad \sum_{r=0}^{m} \lambda_r = 1. \tag{7.6}$$

Here if y^* is an interior point of $\mathcal{Y}$, all $\lambda_r > 0$; if y^* is on the boundary of $\mathcal{Y}$ and is an interior point of the m-dimensional intersection of $\mathcal{Y}$ with a supporting hyperplane of $\mathcal{Y}$, then $m+1$ of the numbers λ_r will be positive (for definiteness, let these be $\lambda_0, \ldots, \lambda_m$); finally, if y^* is an extreme point of $\mathcal{Y}$, there is just one positive λ_r. We shall for uniformity suppose that $\lambda_0, \ldots, \lambda_m > 0$ and $\lambda_{m+1}, \ldots, \lambda_n = 0$ with $1 \leqq m \leqq n$. By varying λ_r, we may evidently obtain any point y in a neighborhood of y^* that is in $\mathcal{Y}$.

Let us suppose that the function $H_x : \mathcal{Y} \to \mathbf{R}$ is differentiable with respect to y at $y = y^*$, in the following sense : the partial derivatives

$$\left. \frac{\partial H(x, y)}{\partial \lambda_r} \right|_{y=y^*}$$

exist for all r for which $\lambda_r > 0$, and the derivatives on the right exist for the other values of r. It is evident (from elementary analysis) that the differentiability of H in this sense is independent of the choice of the "representative" points $y_0^0, \ldots, y_n^0$. It is also clear that the differentiability in this sense of H_x implies the differentiability in the same sense of functions of the form H_X, where X is any mixed strategy of player 1 with a finite spectrum.

Finally, it is known that for $y = y^*$ to be a local minimum of the differentiable function H_X it is necessary, and if the function is convex, also sufficient, that

$$\left. \frac{\partial H(X, y)}{\partial \lambda_r} \right|_{y=y^*} - \lambda = 0 \text{ for all } r = 1, \ldots, m \tag{7.7}$$

(here λ is a Lagrange multiplier) and

$$\left. \frac{\partial H(X, y)}{\partial \lambda_r} \right|_{y=y^*} - \lambda \geqq 0 \text{ for } r = m+1, \ldots, n. \tag{7.8}$$

7.5 Finding optimal mixed strategies of player 1. Finding mixed strategies X^* of player 1 that satisfy (7.4) can be carried out as follows.

After solving equation (7.3), we enumerate the strategies of player 1, then choose any $n+1$ of them, which we denote by

$$x_0^0, \ldots, x_n^0$$

and find weights $\xi_0^0, \ldots, \xi_n^0$ such that the mixture X of these strategies with these weights satisfies (7.7) and (7.8), which then assume the form

$$\sum_{k=0}^{n} \xi_k \left.\frac{\partial H(x_k^0, y)}{\partial \lambda_r}\right|_{y=y^*} - \lambda = 0 \text{ for } r = 1, \ldots, m, \tag{7.9}$$

$$\sum_{k=0}^{n} \xi_k \left.\frac{\partial H(x_k^0, y)}{\partial \lambda_r}\right|_{y=y^*} - \lambda \geqq 0 \text{ for } r = m+1, \ldots, n. \tag{7.10}$$

We must also adjoin to this system the normalization condition

$$\sum_{k=0}^{n} \xi_k = 1. \tag{7.11}$$

If, for a given set of essential strategies, the system (7.9)–(7.11) has no solutions, this set cannot contain the spectrum of any optimal mixed strategies of player 1. However, from the automatic existence (by 3.13) of optimal strategies of the required form, it follows that the system (7.9)–(7.11) is solvable for some set of essential strategies. In particular cases, the solution can be obtained in a rather transparent form. We shall now discuss some such cases.

7.6. Let us make a comment. An optimal mixed strategy of player 1 is to be looked for as a mixture of some set of his essential strategies. Hence the problem necessarily leads to the study of a game in which the set of (mixed) strategies of player 1, with the assigned linear structure, is convex, and the payoff function under these mixed strategies is linear, i.e., convex. Consequently, convex games either are themselves "convexo-convex" (i.e., both sets of the players' strategies are convex, and the payoff function is convex in each argument), or their discussion can be reduced to the consideration of such games.

In these games, every strategy $x \in \boldsymbol{x}$ (pure or mixed) can be represented as a convex combination of a finite number of extreme points of $\boldsymbol{x}$:

$$x = \sum_i \xi_i x_i^0, \quad \xi_i \geqq 0, \quad \sum_i \xi_i = 1.$$

By the convexity of the function $H_y : \boldsymbol{x} \to \mathbf{R}$, we must have

$$H(X, y) = H\left(\sum_i \xi_i x_i^0, y\right) \leqq \sum_i \xi_i H(x_i^0, y),$$

so that

$$\sup_{X \in \boldsymbol{X}} H(X, y) \leqq \sup_{X \in \operatorname{extr} \boldsymbol{X}} H(x, y)$$

(where, as usual, extr$\boldsymbol{X}$ denotes the set of extreme points of $\boldsymbol{X}$), and the optimal strategies of player 1 can be constructed as convex mixtures of extreme points of $\boldsymbol{x}$. Of course, in case the payoff function is not strictly convex, some optimal strategies of player 1 may be overlooked.

7.7 Games of escape. There are various interesting interpretations of *games of escape* (also known as *one-step games of pursuit*), which are formulated in the following way.

In a two-person zero-sum game as in (7.1), let the set $\mathfrak{x}$ be a compact subset of n-dimensional Euclidean space $\mathbf{R}^n$ with the metric ρ, let $\mathfrak{y}$ be the convex closure of the subset, and let $H(x, y) = \rho(x, y)$.

Then in this game Γ player 1 chooses a point $x \in \mathfrak{x}$, trying to get as far as possible from the point $y \in \mathfrak{y}$, which is chosen with the opposite aim by player 2.

The game is convex because of the convexity of the square of the metric ρ in a Euclidean space. Indeed we have, for all x, y', y'' and $\lambda \in [0, 1]$,

$$\rho^2(x, \lambda y' + (1-\lambda)y'') = |x - (\lambda y' + (1-\lambda)y'')|^2$$
$$= |\lambda(x - y') + (1-\lambda)(x - y'')|^2$$
$$= \lambda^2|x - y'|^2 + (1-\lambda)^2|x - y''|^2 + 2\lambda(1-\lambda)(x - y')(x - y''),$$

and since, by the Bunyakovsky-Schwarz inequality,

$$(x - y')(x - y'') \leqq |x - y'|\,|x - y''|,$$

we must have

$$\rho^2(x, \lambda y' + (1-\lambda)y'')$$
$$\leqq \lambda^2|x - y'|^2 + (1-\lambda)^2|x - y''|^2 + 2\lambda(1-\lambda)|x - y'|\,|x - y''|$$
$$= (\lambda|x - y'| + (1-\lambda)|x - y''|)^2 = (\lambda\rho(x, y') + (1-\lambda)\rho(x, y''))^2.$$

The desired convexity follows immediately from this.

We notice that, in view of the symmetry of the metric, in the convex case, if the space $\mathfrak{x}$ is convex, Γ will also be convexo-convex.

Here the relation (7.2) can be interpreted as the determination of a point $y^* \in \mathfrak{y}$ such that the smallest ball with center at y^*, containing $\mathfrak{x}$, will have minimal radius (equal to v_Γ). We denote this ball by $S_{y^*,v}$. Equation (7.3) then describes the points of $\mathfrak{x}$ that lie on the boundary of $S_{y^*,v}$; and the equation (7.4), a necessary condition for (7.7) and (7.8) (or, equivalently, (7.9) and (7.10)), characterizes, as in the general case, the optimal strategy X^* of player 1.

In accordance with what we said at the beginning of 7.3, we shall suppose that both the value v_Γ of the game and the optimal strategy y^* of player 2 have already been found, and consider how to find the optimal strategy X^* of player 1.

In this case we have

$$H(x_k, y) = \rho(x_k, y) = \left(\sum_{j=1}^{n}(x_k^{(j)} - y^{(j)})^2\right)^{1/2},$$

or, by (7.5),

$$H(x_k, y) = \left(\sum_{j=1}^{n} \left(x_k^{(j)} - \sum_i \lambda_i y_i^{(j)} \right)^2 \right)^{1/2}$$

close to y^*. From this, after differentiating, rearranging as necessary, and remembering that the x_k satisfy (7.3) (or, what amounts to the same thing, (7.4)), we obtain

$$\left.\frac{\partial H(x_k, y)}{\partial \lambda_i}\right|_{y=y^*} = \frac{1}{v} \sum_{j=1}^{n} \left(\sum_i \lambda_i y_i^{(j)} - x_k^{(j)} \right) y_i^{(j)}.$$

Since x_k, y^*, and the y_i are vectors in $\mathbf{R}^n$, we may rewrite the right-hand side in terms of scalar products:

$$\left.\frac{\partial H(x_k, y)}{\partial \lambda_i}\right|_{y=y^*} = \frac{1}{v}(y^* - x_k) y_i^T, \tag{7.12}$$

so that (7.3) now assumes the form

$$\frac{1}{v} \left(y^* - \sum_{k=0}^{n} \xi_k x_k \right) y_i^T = \lambda, \quad i = 0, 1, \ldots, n. \tag{7.13}$$

The points y_i with $\lambda_i > 0$ are rather arbitrary points in the neighborhood of y^*. We require only that they are points of general position in a hyperplane that supports y at y^*, and that the simplex spanned by them contains y^*. Confining ourselves, in order to simplify the analysis, to the case when y^* is an interior point of y, we select them so that after indexing the x_k appropriately we can set

$$y_k = y^* + \alpha(x_k - y^*), \tag{7.14}$$

i.e.

$$x_k = y^* + (y_k - y^*)/\alpha.$$

Then (7.10) can be rewritten in the rather transparent form

$$\sum_{k=0}^{n} (\xi_k y_k - y^*) y_i^T = \lambda v, \quad i = 0, 1, \ldots, n, \tag{7.15}$$

from which we can find both λ and the weights ξ_k.

7.8. Games of pursuit are distinguished by great from more combinatorial variety. Let us consider the following example.

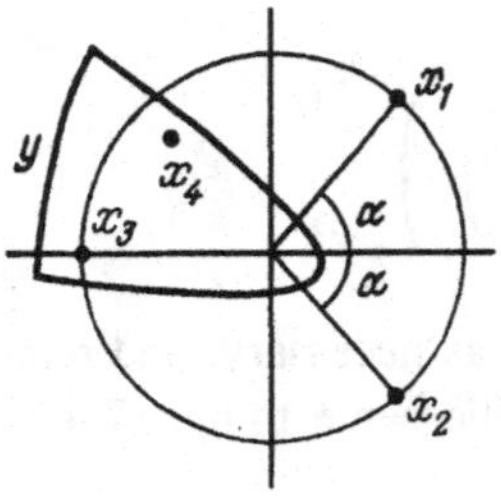

Fig. 3.4

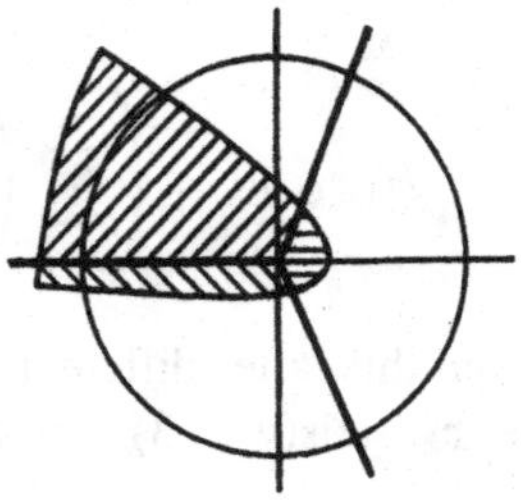

Fig. 3.5

Example. Let $n = 2$ and $\boldsymbol{x} = \{x_1, x_2, x_3, x_4\}$, where

$$x_1 \doteq (x_1^1, x_1^2) = (\cos\alpha, \sin\alpha), \quad |\alpha| \leqq \pi/2,$$

$$x_2 = (x_2^1, x_2^2) = (\cos\alpha, -\sin\alpha),$$

$$x_3 = (x_3^1, x_3^2) = (-1, 0),$$

$$x_4 = (x_4^1, x_4^2) = (-0.6, 0.6),$$

and $\boldsymbol{y}$ is any closed convex set that has the origin as an interior point (Figure 3.4). We notice that the three points x_1, x_2, x_3 lie on the unit circle.

Here the only solution y^* of (7.2) is the point (0,0), the center of the unit disk. In fact, the domain $\boldsymbol{y}$ that consists of the points for which the maximum $\max_{x \in \boldsymbol{x}} \rho(x, y)$ is attained at x_1, x_2 or x_3 is the intersection of $\boldsymbol{y}$ with the sectors sketched in Figure 3.5. Then the minimum of these maxima is attained at the origin. Therefore $y^* = (0, 0)$ and $v_\Gamma = 1$. The point x_4 lies inside the unit disk, and the maximum distance is not attained there.

The solution of equation (7.3) consists of the three points x_1, x_2, x_3.

To find the weights ξ_k we use the system (7.15), noting that in the present case $y^* = 0$, $v_\Gamma = 1$, so that we may take $y_i = \varepsilon x_i$ $(i = 1, 2, 3)$, where ε is a positive number so small that $\boldsymbol{y}$ contains an ε-neighborhood of the origin. Hence in the present case we may write (7.15) in the form

$$\begin{aligned} &(\xi_1 x_1 + \xi_2 x_2 + \xi_3 x_3) x_i^T = \lambda, \quad i = 1, 2, 3, \\ &\xi_1 + \xi_2 + \xi_3 = 1, \quad \xi_1, \xi_2, \xi_3 \geqq 0. \end{aligned} \tag{7.16}$$

However, in our case

$$|x_i|^2 = 1, \quad x_1 x_2 = \cos 2\alpha, \quad x_1 x_3 = x_2 x_3 = \cos\alpha.$$

Setting $\cos\alpha = t$, we have $\cos 2\alpha = 2t^2 - 1$, and (7.16) takes the form

$$\begin{aligned} \xi_1 + (2t^2 - 1)\xi_2 - t\xi_3 &= \lambda, \\ (2t^2 - 1)\xi_1 + \xi_2 - t\xi_3 &= \lambda, \\ -t\xi_1 - t\xi_2 + \xi_3 &= \lambda, \\ \xi_1 + \xi_2 + \xi_3 &= 1. \end{aligned}$$

The only solution of this system is

$$\xi_1 = \xi_2 = \frac{1}{2(t+1)}, \qquad \xi_3 = \frac{t}{t+1}.$$

We turn our attention to two special cases. If $\alpha = 0$, i.e., $t = 1$, the points x_1 and x_2 coincide, and their joint probability

$$2\frac{1}{2(t+1)} = \frac{1}{2}$$

is equal to the probability of x_3; this corresponds to the natural idea of the symmetry of this special case.

If $\alpha = \pm\pi/2$, i.e., $t = 0$, then x_1 and x_2 are diametrically opposite, and form the spectrum of the solution of the problem; however, the point x_3, although it remains essential, does not appear in the spectrum of the optimal strategy.

7.9 Blotto games. The games known as Blotto games form an important class of convex games; they have the following structure.

Definition. A *Blotto game* is a two-person zero-sum game (7.1) in which

$$\mathscr{X} = \left\{ x = (x_1, \ldots, x_n) : x_i \geqq 0, \sum_i x_i = A \right\}, \tag{7.17}$$

$$\mathscr{Y} = \left\{ y = (y_1, \ldots, y_n) : y_i \geqq 0, \sum_i y_i = D \right\} \tag{7.18}$$

and

$$H(x, y) = \sum_i F_i(x_i, y_i), \tag{7.19}$$

where all the functions F are convex and continuous in each variable. □

Consequently the sets $\mathscr{X}$ and $\mathscr{Y}$ are $(n-1)$-dimensional simplexes. The vertices of $\mathscr{X}$ with $x_i = A$ will be denoted by A_i, and similarly those of $\mathscr{Y}$ with $y_i = D$ will be denoted by D_i.

The convexity of Γ follows immediately from the hypothesis that the terms in (7.19) are convex.

Blotto games can be given the following interesting interpretation. Let player 1, "Colonel Blotto", have forces of strength A, which he may use for attacking n positions $1, 2, \ldots, n$ that are defended by the forces D of player 2, his opponent.

It is usual to suppose that Blotto's forces A are not less powerful than his opponent's forces D.

We also suppose that the opposing forces of both enemies cannot be redeployed during the engagement (let us think of attack and defence of mountain passes by squadrons of infantry), so that the payoff to Blotto at each point $i = 1, \ldots, n$ is the function F_i of his forces x_i and the forces y_i of his opponent that are applied at the same point.

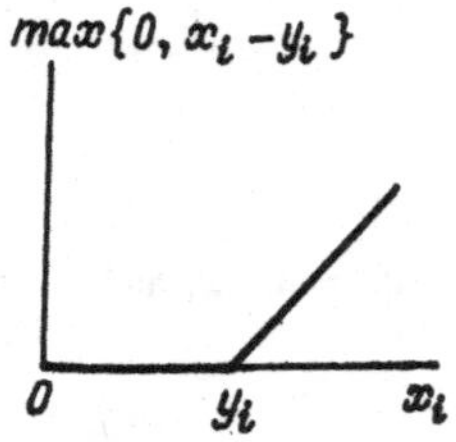

Fig. 3.6

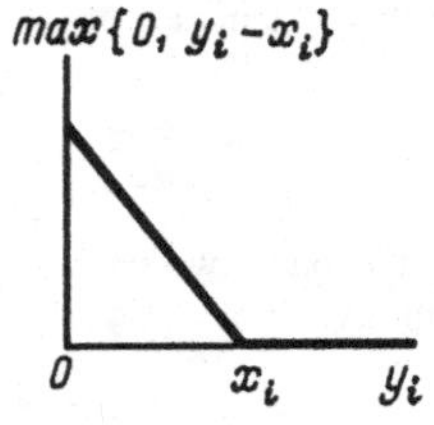

Fig. 3.7

7.10. We shall consider the case of the Blotto game when the direct result of the engagement at point i is proportional to the difference of the forces applied by the players at that point if the attacker has superior forces there, and zero otherwise. The proportionality coefficient $k_i > 0$ describes the generic effect of victory at point i. Therefore

$$H(x, y) = \sum_{i=1}^{n} k_i \max\{0, x_i - y_i\}. \tag{7.20}$$

Since the function $\max\{0, x_i - y_i\}$ is convex in y_i (Figure 3.6) the payoff function (7.17) is also convex, and hence so is Γ. We notice at the same time that the function $\max\{0, y_i - x_i\}$ is also convex in x_i (Figure 3.7) so that Γ is "convexo-convex".

7.11. As we noticed in 7.6, player 1 in Γ has an optimal strategy X^*, which is a mixture of the vertices $A_1, \ldots, A_n$ of the simplex x. Let, as usual, $X^*(A_i) = \xi_i$ $(i = 1, \ldots, n)$. For definiteness, let us take $\operatorname{supp} X^* = \{A_1, \ldots, A_r\}$, i.e., $\xi_i > 0$ for $i = 1, \ldots, r$ and $\xi_i = 0$ for $i > r$. Then, for each strategy y of player 2, we shall have $H(X^*, y) \geqq v_\Gamma$ and, in particular

$$H(X^*, D_j) \geqq v_\Gamma, \qquad j = 1, \ldots, r, \tag{7.21}$$

i.e.,

$$\sum_{i=1}^{r} \xi_i H(A_i, D_j) = \sum_{i=1}^{r} \xi_i k_i (A - D\delta(i, j))$$

$$= A \sum_{i=1}^{r} \xi_i k_i - D\xi_j k_j \geqq v_r \text{ for } j = 1, \ldots, r.$$

At the same time, we will have, for $j > r$,

$$H(X^*, D_j) = A \sum_{i=1}^{r} \xi_i k_i > v_\Gamma.$$

Now let $y^* = (y_1^*, \ldots, y_n^*)$ be a pure optimal strategy of player 2. If we suppose that $\Delta = \sum_{j>r} y_j > 0$, we may form a new distribution of strategies,

$y^0 = (y_1^0, \ldots, y_n^0)$, by setting

$$y_i^0 = \begin{cases} y_1 + \Delta & \text{if } i = 1, \\ y_i & \text{if } i = 2, \ldots, r, \\ 0, & \text{if } i > r. \end{cases}$$

Then

$$\begin{aligned} v_\Gamma = H(X^*, y^*) &= A \sum_{i \leqq r} \xi_i k_i - \sum_{i \leqq r} \xi_i k_i y_i \\ &> A \sum_{i \leqq r} \xi_i k_i - \sum_{i \leqq r} \xi_i k_i y_i^0 = H(X^*, y^0), \end{aligned}$$

which contradicts the optimality of y^*. Consequently $y_j^* = 0$ for $j > r$ (in other words, attacking with zero probability, i.e., not attacking, positions that do not need to be defended).

Let us suppose that all the left-hand sides in (7.21) are equal:

$$H(X^*, D_j) = A \sum_{i=1}^{r} \xi_i k_i - D \xi_j k_j = v_r^0 \geqq v_\Gamma, \quad j = 1, \ldots, r. \tag{7.22}$$

After subtracting the first equation from each of the others, we find that the probabilities ξ_j are inversely proportional to the k_j, i.e.,

$$\xi_j = \lambda_r / k_j. \tag{7.23}$$

Summing these equations over $j = 1, \ldots, r$, we obtain

$$\lambda_r = \left(\sum_{j=1}^{r} \frac{1}{k_j} \right)^{-1}, \tag{7.24}$$

and therefore

$$v_r^0 = A r \lambda_r - D \lambda_r = \lambda_r (rA - D). \tag{7.25}$$

The number v_r^0 depends both on r and on the set $(k_1, \ldots, k_r)$. We note, however, that if we replace one of these numbers by a larger one, the value of λ_r (and hence of v_r^0) is increased. Consequently we shall assume that $1, \ldots, r$ are the most significant indices, and that in general

$$k_1 \geqq k_2 \geqq \ldots \geqq k_r. \tag{7.26}$$

Choose r so that $v_r^0 = \max_{1 \leqq s \leqq n} v_s^0$. Let us show that $v_r^0 = v_\Gamma$ and that (7.23) is actually satisfied.

7.12. To these ends, we shall apply the theorem of 1.12 to show that

$$H(X, y^*) \leqq v_r^0 \leqq H(X^*, y) \tag{7.27}$$

for all $X \in \boldsymbol{X}$ and $y \in \boldsymbol{y}$.

For this purpose, we first show that if we choose X^* for each $y \in \boldsymbol{y}$ in accordance with (7.24), we shall have

$$\begin{aligned} H(X^*, y) &= \sum_{i=1}^{r} \xi_i H(A_i, y_i) = \sum_{i=1}^{r} \xi_i k_i (A_i - y_i) \\ &= \lambda_r \left(rA - \sum_{i=1}^{r} y_i \right) \geqq \lambda_r (rA - D) = v_r^0. \end{aligned} \tag{7.28}$$

It is harder to obtain the other side of the inequality. It follows from the maximal property of v_r^0 that $v_{r+1}^0 \leqq v_r^0$, i.e., that, by (7.25),

$$\frac{(r+1)A - D}{\lambda_r} \leqq \frac{rA - D}{\lambda_{r+1}},$$

or

$$((r+1)A - D)\frac{1}{\lambda_r} \leqq (rA - D)\left(\frac{1}{\lambda_r} + \frac{1}{k_{r+1}}\right),$$

from which

$$\frac{A}{\lambda_r} \leqq \frac{rA - D}{k_{r+1}},$$

or

$$k_{r+1} A \leqq \lambda_r (rA - D) = v_r^0. \tag{7.29}$$

Hence, by the way, it follows that $v_r^0 > 0$.

In just the same way, we obtain $k_r A \geqq v_r$ from $v_{r-1}^0 \leqq v_r^0$, and consequently (by (7.26))

$$k_j A \geqq v_r^0, \qquad j = 1, \ldots, r. \tag{7.30}$$

We now construct a strategy-distribution

$$y^* = (y_1^*, \ldots, y_r^*, 0, \ldots, 0)$$

such that we will have

$$k_j (A - y_j^*) = v_r^0, \qquad j = 1, \ldots, r. \tag{7.31}$$

For this purpose we should evidently take

$$y_j^* = A - \frac{v_r^0}{k_j}, \qquad j = 1, \ldots, r. \tag{7.32}$$

Moreover,

$$D = \sum_{j=1}^{r} y_j^* = rA - \frac{v_r^0}{\lambda_r}, \tag{7.33}$$

and it follows from (7.30) and (7.31) that $y_j^* > 0$. Consequently y^* is indeed a strategy of player 2.

However, (7.31) also means that $H(A_j, y^*) = v_r^0$, and hence for every $X \in \boldsymbol{X}$ we will have

$$H(X, y^*) = \sum_{i=1}^{n} \xi_i H(A_i, y^*) = \sum_{i=1}^{n} \xi_i k_i (A - y_i^*) \leqq \sum_{i=1}^{n} \xi_i v_r^0 = v_r^0 \qquad (7.34)$$

(since by the "convexo-convexity" of Γ it is enough to consider only the mixed strategies X whose spectra consist of the extreme strategies $A_1, \ldots, A_n$).

Hence it follows from (7.28) and (7.34) that (7.27) is actually satisfied for all $X \in \boldsymbol{X}$ and $y \in \boldsymbol{y}$. Then, by 1.12, this means that $v_r^0 = v_\Gamma$, $X^* \in \mathcal{S}(\Gamma)$, and $y^* \in \mathcal{T}(\Gamma)$.

7.13 Let us show that the pure optimal strategy y^* of player 2, constructed according to (7.31), is unique (among the player's pure strategies). In fact, let $\bar{y} = (\bar{y}_1, \ldots, \bar{y}_n)$ be an optimal pure strategy of player 2. Then

$$H(A_i, \bar{y}) = k_i (A - \bar{y}_i) \leqq v_\Gamma, \qquad i = 1, \ldots, r,$$

whence

$$\bar{y}_i \geqq A - \frac{v_\Gamma}{k_i}, \qquad i = 1, \ldots, r. \qquad (7.35)$$

But, according to (7.25)

$$\sum_{i=1}^{r} \bar{y}_i \geqq \sum_{i=1}^{r} \left(A - \frac{v_i}{k_i} \right) = rA - \lambda_r \lambda_r^{-1} (rA - D) = D,$$

and since the sum cannot exceed D, it is equal to D, and there is equality in (7.35) for all $i = 1, \ldots, r$. Therefore $\bar{y} = y^*$.

We also notice that the uniqueness of the optimal strategy of player 2 does not imply the uniqueness of the number r of positions attacked by player 1. It may, in fact, turn out that $v_r^0 = v_{r+1}^0 = 0$. But then, if we repeat the discussion that led to (7.29), we find from (7.25) that $k_{r+1} A = v_r^0$, after which (7.32) gives us $y_{r+1}^0 = 0$.

7.14. A somewhat paradoxical feature of the solution that we have obtained is that the more important positions are attacked with greater probability than the less important ones, whereas the least important positions are not attacked at all. At the same time the most important positions are defended "more than proportionally" to their importance.

Example. Let $n = 3$, $A = 18$, $D = 15$, $k_1 = 6$, $k_2 = 3$, $k_3 = 2$. Then, by (7.24), $\lambda_1 = 6$, λ_2, $\lambda_3 = 1$; by (7.25), $v_1^0 = 18$, $v_2^0 = 42$, $v_3^0 = 39$, and hence $r = 2$ and $v_\Gamma = 42$, $\xi_1 = 1/3$, $\xi_2 = 2/3$, and $\xi_3 = 0$. Finally, by (7.31), $y_1^* = 11$, $y_2^* = 4$, and $y_3^* = 0$.

7.15 Optimal allocation of limited resources. In conclusion, we consider the following techno-economic model.

Let an enterprise be capable of carrying out n kinds of activities. The vector $x = (x_1, \ldots, x_n)$ is a collection of activities, and carrying out the collection of activities means that activity i is carried out to the extent x_i. For carrying out the activities in the enterprise, there is available a quantity, which we assume equal to 1, of homogeneous resources. For carrying out activity i, let there be available an amount y_i of resources, so that the set of all distributions will be y in (7.18) with $D = 1$.

We denote by $F_i(x_i, y_i)$ the strain on the enterprise from carrying out the ith activity at the level x_i with amount y_i of resources. This measure can indicate, for example, the length of time for carrying out an amount x_i of activity of type i by the use of resources y_i. It is natural to consider the function $F_i(x_i, \cdot)$, with any given x_i, as being continuous, nonnegative, decreasing strictly monotonically, and convex. We also suppose that $F_i(\cdot, y_i)$ is continuous and increases strictly monotonically.

It is natural to think of the strain $H(x, y)$ from carrying out the enterprise, with the collection x of activities and the amount y of resources, as the greatest strain required for carrying out each activity in this collection, i.e., we suppose that

$$H(x, y) = \max_{1 \leqq i \leqq n} F_i(x_i, y_i). \tag{7.36}$$

The function H_x, as the convex envelope of a family of convex functions of y_i, is also convex. Hence the game Γ is also convex, and the preceding discussion is applicable to it. In particular, equation (7.2) is applicable. In the present case, it assumes the form

$$v_\Gamma = \min_{y \in \boldsymbol{y}} \sup_{x \in \boldsymbol{x}} \max_{1 \leqq i \leqq n} F_i(x_i, y_i),$$

or, if we use the possibility of permuting the extrema with the same name and the monotonic character of F_i and x_i,

$$v_\Gamma = \min_{y \in \boldsymbol{y}} \max_{1 \leqq i \leqq n} F_i(\sup_{x \in \boldsymbol{x}} x_i, y_i).$$

Since the set $\boldsymbol{x}$ is compact, the supremum of the argument of F_i is attained. Setting $\sup_{x \in \boldsymbol{x}} x_i = x_i^0$, we obtain

$$v_\Gamma = \min_{y \in \boldsymbol{y}} \max_{1 \leqq i \leqq n} F_i(x_i^0, y_i) = \max_{1 \leqq i \leqq n} F_i(x_i^0, y_i^*). \tag{7.37}$$

We denote by $x^{(i)}$ the strategy of player 1 for which the supremum of x_i is attained. We then look for the optimal strategy X^* of player 1 as a mixture of the pure strategies $x^{(i)}$ with probabilities ξ_i.

7.16. In order to immediately remove the combinatorial diversity of cases, we introduce the following hypothesis. We denote by $\mathcal{Y}^k$ the set of strategy-distributions of player 2 for which $y_k = 0$, and assume that

$$F_k(x_k^0, 0) > \min_{y \in \mathcal{Y}^k} \max_{i \neq k} F_i(x_i^0, y_i), \quad k = 1, \dots, n. \tag{7.38}$$

Informally speaking this assumption is perfectly natural: if an activity remains completely without guaranteed resources, then it becomes a bottleneck, at least for the case when all the other activities are optimally provided with resources.

7.17 Lemma. *Under the hypotheses* (7.38), *for each optimal strategy* $y^* = (y_1^*, \dots, y_n^*)$ *of player 2 we must have* $y_k^* > 0$ $(k = 1, \dots, n)$.

Proof. Suppose that some $y_k^* = 0$. At the same time, we have

$$\begin{aligned} v_\Gamma &= \min_{y \in \mathcal{Y}} \max_{1 \leqq i \leqq n} F_i(x_i^0, y_i) = \max_{1 \leqq i \leqq n} F_i(x_i^0, y_i^*) \\ &= \max\{F_k(x_k, y_k^*), \max_{i \neq k} F_i(x_i^0, y_i^*)\} \\ &= \max\{F_k(x_k^0, 0), \max_{i \neq k} F_i(x_i^0, y_i^*)\}. \end{aligned} \tag{7.39}$$

Let us now take the distribution $y^k \in \mathcal{Y}^k$ for which the minimum

$$\min_{y \in \mathcal{Y}^k} \max_{i \neq k} F_i(x_i^0, y_i)$$

is attained, and then, remembering (7.38), a value of ε for which

$$0 < \varepsilon < F_k(x_k^0, 0) - \max_{i \neq k} F_i(x_i^0, y_i^*), \tag{7.40}$$

and a positive δ so small that, when $y^\delta \in \mathcal{Y}$ with $|y_i^\delta - y_i^k| < \delta$ and $y_k^\delta > 0$,

$$|F_i(x_i^0, y_i^\delta) - F_i(x_i^0, y_i^k)| < \varepsilon \tag{7.41}$$

for $i = 1, \dots, n$ (this is possible because of the continuity of the finite number of functions F_i). Then, by (7.41) and (7.40),

$$\max_{i \neq k} F_i(x_i^0, y_i^\delta) < F_k(x_k^0, 0), \tag{7.42}$$

and, since the F_k are strictly monotonic in y and $y_k^\delta > 0$,

$$F_k(x_k^0, y_k^\delta) < F_k(x_k^0, 0). \tag{7.43}$$

As a result of (7.39), (7.42), and (7.43), it follows that

$$\max\{F_k(x_k^0, y_k^\delta), \max_{i \neq k} F_i(x_i^0, y_i^\delta)\} < F_k(x_k^0, 0) \leqq v_\Gamma,$$

and y^δ, as against X^*, leads to a smaller loss than the supposedly optimal strategy y^*. This contradiction establishes the lemma. □

7.18 Lemma. *Under the assumption* (7.38) *every optimal strategy* $X^* = (\xi_1, \ldots, \xi_n)$ *of player* 1 *is completely mixed, i.e.* $\xi_i > 0$ $(i = 1, \ldots, n)$.

Proof. For an optimal strategy y^* of player 2, we have

$$v_\Gamma = H(X^*, y^*) = \sum_{i=1}^{n} \xi_i H(x_i^0, y^*) = \sum_{i=1}^{n} \xi_i F_i(x_i^0, y_i^*).$$

Let us suppose that $\xi_k = 0$. By what we had above, $y_k^* > 0$. We define a new strategy y^0 for player 2 by setting

$$y_i^0 = \begin{cases} 0 & \text{if } i = k, \\ y_i^* + y_k^*/(n-1) & \text{if } i \neq k. \end{cases}$$

Then

$$H(X^*, y^0) = \sum_{i=1}^{n} \xi_i F_i(x_i^0, y_i^0).$$

For all nonzero coefficients ξ_i, we have, since F is strictly monotonic,

$$F_i(x_i^0, y_i^0) < F_i(x_i^0, y_i^*),$$

so that

$$H(X^*, y^0) < H(X^*, y^*),$$

and this contradicts the optimality of y^*. □

7.19. What was established in 7.18 lets us find an optimal strategy y^* of player 2. It follows from the inequality $\xi_k^* > 0$ that for an optimal strategy X^* of player 1,

$$H(X^*, y^*) = F_i(x_i^0, y_i^*) = v_\Gamma. \tag{7.44}$$

If we denote by φ_i the monotonic function inverse to $F_i(x_i^0, \cdot) : y \to \mathbf{R}$, we find from (7.44) that

$$y_i^* = \varphi_i(v_\Gamma), \tag{7.45}$$

or, by summation,

$$1 = \sum_{i=1}^{n} y_i^* = \sum_{i=1}^{n} \varphi_i(v_\Gamma). \tag{7.46}$$

Let us denote the sum of the φ_i's by φ. Then φ is also monotonic, and therefore has an inverse. It follows from (7.46) that $v_\Gamma = \varphi^{-1}(1)$, and if we subsitute this into (7.45) we obtain $y_i^* = \varphi_i(\varphi^{-1}(1))$.

Consequently the optimal strategy y^* of player 2 turns out to be unique.

Formula (7.45) has a natural interpretation in applied economics (and also in pure technology). The number v_Γ is a characteristic of the problem itself, i.e., is not connected with any particular kind of activity i; on the contrary, the function φ_i corresponds to the result achieved in activity i as a consequence of the resources expended on it; it is therefore the "resource-capacity" of an activity of this kind. Thus (7.45) shows that assigning resources to activity i must provide the same result, namely v_Γ, in accordance with its resource-capacity.

As we can easily see, special cases of this proposition are such technological principles as uniform strength, equireliability, etc.

7.20. Finally, we turn to the determination of the optimal strategy X^* of player 1. For this purpose, we form, in accordance with 7.5, the equations of the form (7.9) (the equations of the form (7.10) drop out, by the positivity of the components of the optimal strategies y^*), which in the present case appear in the form

$$\xi_k \left.\frac{\partial F_k(x_k^0, y_k)}{\partial y_k}\right|_{y=y^*} - \lambda = 0. \tag{7.47}$$

Hence we obtain, by a standard argument,

$$\lambda = \left(\sum_{k=1}^{n} \left(\left.\frac{\partial F_k(x_k^0, y_k)}{\partial y_k}\right|_{y=y^*} \right)^{-1} \right)^{-1},$$

and substitution into (7.47) provides what was required.

7.21 Remark. We can extend the preceding discussion to the case when

$$\lim_{y_k \to 0} F_k(x_k, y_k) = +\infty \tag{7.48}$$

for certain values of k among the $1, \dots, n$, and consequently the set y is a simplex without the $(n-1)$-dimensional face y^k on which $y_k = 0$. The fact that in this case we have to deal with a game in which the payoff function is unbounded has no influence either on the analysis of the game or on its solution.

For the values of k for which (7.48) holds, we choose an arbitrary $\varepsilon_k > 0$ and, setting $\bar{y}_i = (1 - \varepsilon_k)/(n-1)$ for $i \neq k$, we take

$$f_k > \max_{i \neq k} F_i(x_I^0, \bar{y}_i). \tag{7.49}$$

In accordance with (7.48), we find a $\tilde{y}_k > 0$ such that

$$F_k(x_k^0, \tilde{y}_k) > f_k. \tag{7.50}$$

Now we consider the $x \times \tilde{y}$-subgame $\tilde{\Gamma}$ of Γ in which

$$\tilde{y} = \{y : y = (y_1, \dots, y_n),\ y_k \geqq \tilde{y}_k, \sum_{i=1}^{n} y_i = 1\},$$

defining, by analogy with y_k, the set $\tilde{y}^k$ as the simplex y without the face $y_k = \tilde{y}_k$.

Condition (7.38) from 7.16 is automatically satisfied for this subgame: on the basis of (7.49) and (7.50) we will have

$$F_k(x_k^0, \tilde{y}_k) > \max_{i \neq k} F_i(x_i^0, \bar{y}_i) = \max_{i \neq k} F_i\left(x_i^0, \frac{1 - \varepsilon_k}{n-1}\right)$$

$$> \max_{i \neq k} F_i\left(x_i^0, \frac{1}{n-1}\right) \geqq \min_{\tilde{y}^k \in \tilde{y}^k} \max_{i \neq k} F_i(x_i^0, \tilde{y}^k).$$

Consequently, for the optimality of the strategy $\tilde{y}^*$ of player 2 in $\tilde{\Gamma}$ we must have $\tilde{y}_i^* > 0$ for $i = 1, \dots, n$. But since this is an interior point of $\tilde{y}$, and therefore of y, at which a local minimum (for y) is attained, by the convexity of the function H of y it must be a global minimum for H on the entire set y, i.e., an optimal strategy for player 2 in Γ.

From the maximin point of view of player 1, the games Γ and $\tilde{\Gamma}$ are, in general, identical.

7.22 Example. Let us consider the game Γ for which

$$H(x, y) = \max_{1 \leqq i \leqq n} F_i(x_i, y_i) = \max_{1 \leqq i \leqq n} \frac{x_i^r}{y_i^s},$$

where r and s are any positive numbers. Evidently the function F_i so defined satisfies the hypotheses listed in 7.15. In our case (see 7.19), we shall have

$$\varphi_i(t) = t^{-1/s}(x_i^0)^{r/s},$$

so that

$$u = \varphi(t) = t^{-1/s} \sum_{i=1}^{n} (x_i^0)^{r/s},$$

from which we have

$$t = \varphi^{-1}(u) = u^{-s} \left(\sum_{i=1}^{n} (x_i^0)^{r/s} \right)^{s}.$$

Therefore

$$v = \varphi^{-1}(1) = \left(\sum_{i=1}^{n} (x_i^0)^{r/s} \right)^{s} \tag{7.51}$$

and

$$y_i^* = \varphi_i(v) = (x_i^0)^{r/s} \left(\sum_{i=1}^{n} (x_i^0)^{r/s} \right)^{-1}. \tag{7.52}$$

Equation (7.47) becomes

$$\xi_k(-s) \frac{x_i^{0r}}{y_i^{*(s+1)}} - \lambda = 0,$$

from which, after natural transformations, we obtain

$$\xi_i = (x_i^0)^{-r/s} \left(\sum_{i=1}^{n} (x_i^0)^{-r/s} \right)^{-1}. \tag{7.53}$$

§8 Games with a simple payoff function

8.1 Simple two-person zero-sum games. In this and the next section we shall discuss two classes of two-person zero-sum games for which the analysis can form the basis of a rather systematic (although, of course, far from exhaustive) theory.

Definition. A two-person zero-sum game

$$\Gamma = \langle \boldsymbol{x}, \boldsymbol{y}, H \rangle \tag{8.1}$$

is said to be *simple* if the payoff function H attains only two values on the set $\boldsymbol{x} \times \boldsymbol{y}$ of situations. □

By passing, if necessary, to an affinely equivalent game, we may suppose, without loss of generality, that the two values assumed by the payoff function of a simple game are 0 and 1. The significance of the class of simple games becomes evident if we realize that we can reduce to such games all conflicts that obey the "threshold" principle: "all or nothing".

Let us set

$$\boldsymbol{z} = \{(x, y) : H(x, y) = 1\}.$$

It is clear that a simple game Γ with given sets $\boldsymbol{x}$ and $\boldsymbol{y}$ of strategies is completely determined by the set $\boldsymbol{z} \subset \boldsymbol{x} \times \boldsymbol{y}$, and it will be convenient to set $\Gamma = \langle \boldsymbol{x}, \boldsymbol{y}, \boldsymbol{z} \rangle$. Consequently a simple game can be represented as a multivalued function from $\boldsymbol{x}$ to $\boldsymbol{y}$ (or inversely), and if $\boldsymbol{x} = \boldsymbol{y}$, as a binary relation on $\boldsymbol{x}$. There are also natural geometrical representations, which we shall use extensively.

As usual, for each $x \in \boldsymbol{x}$ the set $\{y : (x, y) \in \boldsymbol{z}\}$ is called an *x-section* of $\boldsymbol{z}$ and is denoted by $\boldsymbol{y}(x)$, and the set $\{x : (x, y) \in \boldsymbol{z}\}$ is a *y-section* of $\boldsymbol{z}$ and denoted by $\boldsymbol{x}(y)$.

Definition. Let k be a positive integer. A family $\mathfrak{A}$ of subsets of a set A is called a *k-fold covering of A* if each element of A appears in at least k sets from $\mathfrak{A}$. A family $\mathfrak{A}$ is called a *k-fold packing* in A if each element of A appears in at most k sets from $\mathfrak{A}$.

According to this definition, every k-fold covering of multiplicity k is also an l-fold covering for every $l < k$, and a k-fold packing is also an l-fold packing for every $l > k$. We may, without danger of misunderstanding, take the multiplicity of a covering to be its largest multiplicity, and the multiplicity of a packing to be its smallest multiplicity.

A 1-fold covering is called simply a *covering* (or a *simple covering*), and an 1-fold packing is simply a *packing* (or a *simple packing*).

If $\boldsymbol{x}^* \subset \boldsymbol{x}$ and the family of x-sections of $\boldsymbol{z}$ corresponding to strategies in $\boldsymbol{x}^*$ forms a k-fold covering of $\boldsymbol{y}$, we say that $\boldsymbol{x}^*$ *generates a k-fold covering of* $\boldsymbol{y}$. Similarly if $\boldsymbol{y}^* \subset \boldsymbol{y}$ and the family of sections corresponding to strategies in $\boldsymbol{y}^*$ forms a k-fold covering of $\boldsymbol{x}$, we say that $\boldsymbol{y}^*$ *generates a k-fold packing* of $\boldsymbol{x}$. □

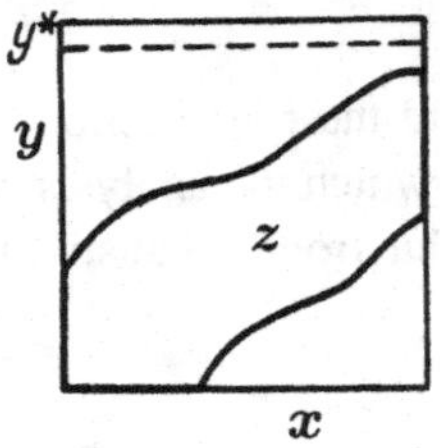

Fig. 3.8

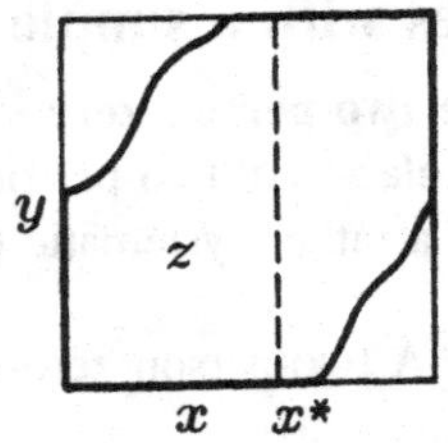

Fig. 3.9

We can estimate the upper and lower values of Γ by using the sets of strategies generated by multiple coverings of $\boldsymbol{y}$ and multiple packings of $\boldsymbol{x}$. These sets are also connected with the spectra of the optimal strategies of the players in the game.

8.2 The simplest cases. We begin by proving a few general propositions.

Theorem. *If, for the simple game in* (8.1), *the whole set $\boldsymbol{x}$ does not generate a covering of $\boldsymbol{y}$, then $v_\Gamma = 0, \mathcal{T}(\Gamma)$ consists of all the pure strategies that are not in $\bigcup_{x \in \boldsymbol{x}} \boldsymbol{y}(x)$, and all of their mixtures, and $\mathcal{S}(\Gamma)$ consists of all strategies of player* 1.

If the player 1 has strategies each of them generate a covering of $\boldsymbol{y}$, then $v_\Gamma = 1$, $\mathcal{S}(\Gamma)$ consists of all such strategies and all of their mixtures, and $\mathcal{T}(\Gamma)$ consists of all strategies of player 2.

Proof. From

$$y \in \boldsymbol{y} \setminus \bigcup_{x \in \boldsymbol{x}} \boldsymbol{y}(x)$$

it follows that we must have $(x, y) \notin \boldsymbol{z}$ for every $x \in \boldsymbol{x}$, i.e.,

$$\sup_x H(x, y) = 0. \tag{8.2}$$

Therefore, if Y is a mixture of such strategies, we have $\sup_x H(x, Y) = \inf_Y \sup_x H(x, Y) = 0$, from which we obtain $\bar{v}_\Gamma = 0$. It follows, because the payoffs in Γ are nonnegative, that $\underline{v}_\Gamma \geqq 0$. Therefore v_Γ exists and is zero. In addition, it follows from (8.2) that, for every $x \in \boldsymbol{x}$, $\inf_y H(x, y) = 1$, so that every strategy of player 1 is optimal.

The second part of the theorem can be established by a parallel argument. □

When Γ is a game on the unit square, this theorem eliminates any interest in the cases when the set $\boldsymbol{z}$ has the form sketched in Figures 3.8 or 3.9. In the first case, the strategies of the form y^* do not appear in any covering generated by $\boldsymbol{x}$; in the second, each strategy of the form x^* generates a covering of $\boldsymbol{y}$.

8.3 Theorem. *For a simple game Γ as in (8.1), let there exist in $\mathfrak{y}$ an infinite subset $\mathfrak{y}^*$ that generates a packing in $\mathfrak{x}$. Then $v_\Gamma = 0$, and every strategy of player 1 is optimal.*

If, in addition, $\mathfrak{x}$ contains a countable subset $\mathfrak{x}^*$ that generates a covering of $\mathfrak{y}$, then player 2 has no optimal strategies (but has ε-optimal strategies for every $\varepsilon > 0$).

Proof. Let $\mathfrak{y}^*$ generate a packing in $\mathfrak{x}$ and let $\{y_1, \dots, y_k, \dots\} \subset \mathfrak{y}^*$. Choose an arbitrary $Y \in \mathsf{Y}$ with $\operatorname{supp} Y = \{y_1, y_2, \dots\}$. It is clear that for every $x \in \mathfrak{x}$ we must have

$$H(x, Y) = \begin{cases} Y(y_k) & \text{if } x \in \mathfrak{x}(y_k), \\ 0 & \text{in the opposite case.} \end{cases} \tag{8.3}$$

Let us select k^* so that

$$\max_k Y(y_k) = Y(y_{k^*}),$$

and then choose $x_Y \in \mathfrak{x}(y_{k^*})$. Then it follows from (8.3) that for every $x \in \mathfrak{x}$,

$$H(x, Y) \leqq Y(y_{k^*}) = H(x_Y, Y),$$

and therefore

$$\sup_x H(x, Y) = H(x_Y, Y).$$

For a given packing, we can find a mixed strategy Y for which $\max_k Y(y_k)$ is arbitrarily small. Consequently, by (8.3), we must have

$$\bar{v}_\Gamma = \inf_Y \sup_x H(x, Y) = 0.$$

Since the values of the payoff function of Γ are nonnegative, we see that $\underline{v}_\Gamma = v_\Gamma = 0$.

Furthermore, it follows from

$$\underline{v}_\Gamma = \sup_X \inf_y H(X, y) = 0$$

that $\inf_y H(X, y) = 0$ for every strategy X, which must turn out to be optimal.

Now let the countable set $\{x_1, x_2, \dots\}$ generate a covering of $\mathfrak{y}$. Select any mixed strategy Y of player 2. There is evidently an x_n for which $Y(\mathfrak{y}(x_n)) = H(x_n, Y) > 0$. Therefore

$$\sup_x H(x, Y) > 0 = v_\Gamma,$$

i.e., the strategy Y is not optimal. However, the existence of ε-optimal strategies for player 2 for every $\varepsilon > 0$ follows, by the theorem of 1.6, from the existence of a value for Γ. □

8.4 The Fundamental theorem.

Theorem. *For a simple game Γ as in* (8.1), *if $\boldsymbol{x}$ contains an r-element subset that generates a k-fold covering of $\boldsymbol{y}$, then $\underline{v}_\Gamma \geqq k/r$.*

If $\boldsymbol{y}$ contains an s-element subset that generates a k-fold packing in $\boldsymbol{x}$, then $\bar{v}_\Gamma \leqq k/s$.

Proof. Let $\{x_1, \ldots, x_r\}$ generate a k-fold covering of $\boldsymbol{y}$. Let us take the strategy $X^* \in X$ for which $X^*(x_i) = 1/r$ for each $i = 1, \ldots, r$. For every $y \in \boldsymbol{y}$ we then have

$$H(X^*, y) = \frac{1}{r} \sum_{i=1}^{r} H(x_i, y). \tag{8.4}$$

The strategy y belongs to at least k sets $\boldsymbol{y}(x_i)$ from among those making up the k-fold covering of y. Therefore, we must have $H(x_i, y) = 1$ for k corresponding values of i. The remaining terms in the sum in (8.4) are nonnegative. Therefore

$$H(X^*, y) \geqq k/r, \tag{8.5}$$

from which we immediately find that

$$\underline{v}_\Gamma = \sup_X \inf_y H(X, y) \geqq k/r.$$

Now let $\{y_1, \ldots, y_s\}$ generate a k-fold packing in $\boldsymbol{x}$. Take the strategy $Y^* \in Y$ for which $Y^*(y_j) = 1/s$ for every $j = 1, \ldots, s$. For every $x \in \boldsymbol{x}$ we have

$$H(x, Y^*) = \frac{1}{s} \sum_{j=1}^{s} H(x, y_j). \tag{8.6}$$

The strategy x can belong to at most k of the sets $\boldsymbol{x}(y_j)$ that form the k-fold packing, so that in the sum in (8.6) at most k terms can be different from zero. Therefore

$$H(x, Y^*) \leqq k/s, \tag{8.7}$$

from which it follows that

$$\bar{v}_\Gamma = \inf_Y \sup_x H(x, Y) \leqq k/s. \qquad \square$$

8.5 Corollary. *If there simultaneously exist in $\boldsymbol{x}$ an n-element subset $\{x_1, \ldots, x_n\}$ that generates a k-fold covering of $\boldsymbol{y}$, and in $\boldsymbol{y}$ an n-element subset $\{y_1, \ldots, y_n\}$ that generates a k-fold packing in $\boldsymbol{x}$, then the game Γ has the value k/n, and the strategies X^* and Y^*, for which $X^*(x_i) = 1/n$ for every $i = 1, \ldots, n$ and $Y^*(y_j) = 1/n$ for every $j = 1, \ldots, n$, are optimal.*

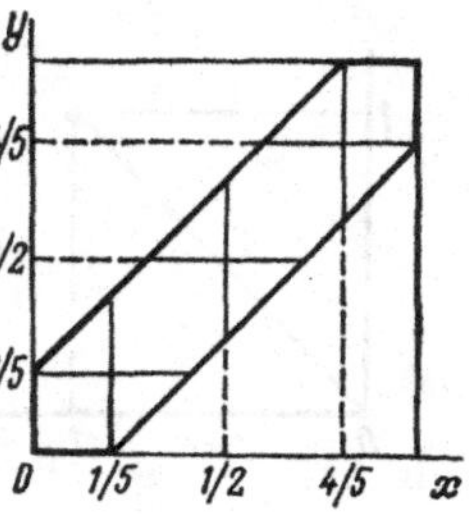

Fig. 3.10

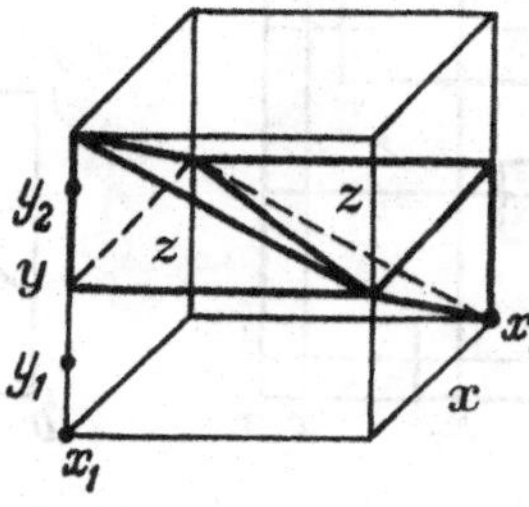

Fig. 3.11

Proof. That $\underline{v}_\Gamma = \bar{v}_\Gamma = v_\Gamma = k/n$ follows from the theorem of 8.4. In addition, if we set $r = n$ in (8.5) and go over to the mixed strategies, we obtain

$$H(X^*, Y) \geqq k/n \text{ for all } Y \in \boldsymbol{Y}$$

and setting $s = n$ in (8.7) and going over to the mixed strategies,

$$H(X, Y^*) \leqq k/n \text{ for all } X \in \boldsymbol{X},$$

from which the conclusion follows at once (see, for example, 1.11). □

8.6 Examples. Let us consider some examples.

Example 1. In a game Γ on the unit square, let $\boldsymbol{z}$ be a strip located along the diagonal of the situation square, as indicated in Figure 3.10.

It is clear that here the set of pure strategies $x_1 = 1/5$, $x_2 = 1/2$ and $x_3 = 4/5$ generate a covering of $\boldsymbol{y}$, and the set of strategies $y_1 = 0$, $y_2 = 1/2$ and $y_3 = 1$ generate a packing in $\boldsymbol{x}$. Hence $v = 1/3$ here, and one of the pairs of the players' optimal strategies,

$$X^* = \begin{bmatrix} 1/5 & 1/2 & 4/5 \\ 1/3 & 1/3 & 1/3 \end{bmatrix} \quad \text{and} \quad Y^* = \begin{bmatrix} 0 & 1/2 & 1 \\ 1/3 & 1/3 & 1/3 \end{bmatrix}$$

can be written down immediately.

Example 2. Let $\boldsymbol{x}$ be the unit square, $\boldsymbol{y} = [0, 1]$, and $\boldsymbol{z}$ the closed body consisting of two tetrahedra, as sketched in Figure 3.11.

Here a covering of $\boldsymbol{y}$ is generated by the pair x_1, x_2 of strategies; and a packing in $\boldsymbol{x}$, for example, by the pair of strategies $y_1 \in [0, 1/2)$ and $y_2 \in (1/2, 1]$. The value of the game is 1/2.

Example 3. Let $\boldsymbol{x} = \boldsymbol{y} = [0, 1]$, and let $\boldsymbol{z}$ have the form drawn in Figure 3.12.

Here the set of pure strategies $x_1 = 0$, $x_2 = 1/2$, and $x_3 = 1$ generates a 2-fold covering, and the set of strategies $y_1 = 0$, $y_2 = 1/2$ and $y_3 = 1$ generates a 2-fold packing in $\boldsymbol{x}$. Therefore, in this case $v_\Gamma = 2/3$.

Example 4. Let Γ be the simple game on the unit square in which the set $\boldsymbol{z}$ has the form sketched in Figure 3.13. It is evident that here the projections of the vertical links of the rather stretched out zigzag form a covering of $\boldsymbol{y}$, and the projections of the horizontal zigzag form a packing in $\boldsymbol{x}$ (in fact, a zigzag stretched enough so that the horizontal links have disjoint projections). The game falls under the theorem of 8.3.

The construction of the zigzag used in this example is typical for this whole class of games, and leads the way to many generalizations.

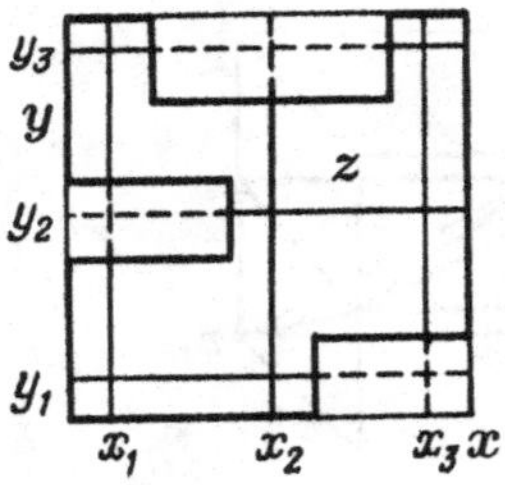

Fig. 3.12

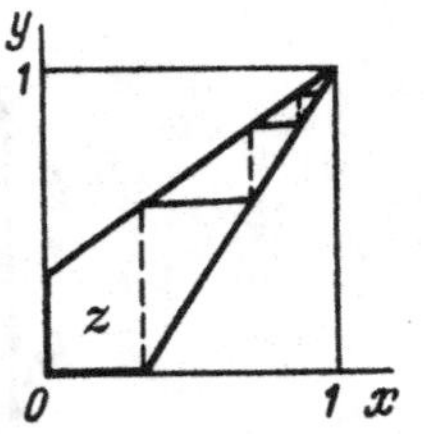

Fig. 3.13

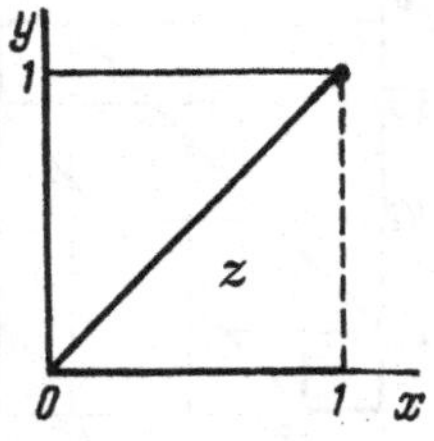

Fig. 3.14

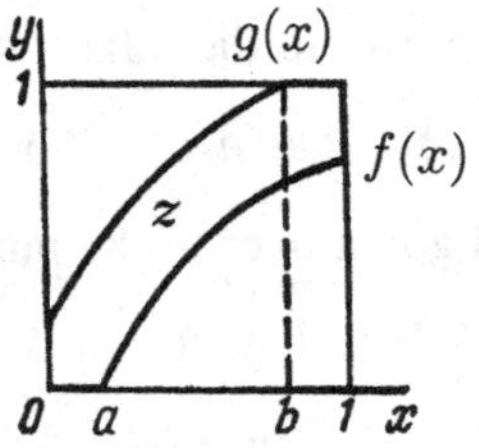

Fig. 3.15

Example 5. Let us consider the game on the unit square in 4.3; it is evidently simple.

Here (Figure 3.14) no finite set of cross sections generates a covering of $\boldsymbol{y}$. This means that in this case there is no basis for obtaining a positive lower bound for $\underline{v}_\Gamma$. On the other hand, the strategies $y_1 = c < 1$ and $y_2 = 1$ of player 2 generate a packing in $\boldsymbol{x}$, so that $\bar{v}_\Gamma \leqq 1/2$. Since it is also the case that no packing in $\boldsymbol{x}$ with three sets can be constructed, we must also have $\bar{v}_\Gamma \geqq 1/2$. Therefore $\bar{v}_\Gamma = 1/2$.

8.7 The case of simple coverings and packings. Let us consider a class of simple games on the unit square, of the following form.

Let

$$f, g : [0,1] \to [0,1];$$

these functions are assumed to be continuous for every $x \in [0,1]$, and to satisfy

$$f(x) \leqq g(x),\ f(x) = 0 \text{ on } [0,a],\ a \geqq 0;$$

$$g(x) = 1 \text{ on } [b,1],\ b < 1;$$

in addition, f is strictly increasing on $[a,1]$; and g, on $[0,b]$ (Figure 3.15).

We define a simple game Γ as in (8.1) by setting $\boldsymbol{x} = \boldsymbol{y} = [0,1]$ and

$$\boldsymbol{z} = \{(x,y) : x \in \boldsymbol{x},\ y \in \boldsymbol{y},\ f(x) \leqq y \leqq g(x)\}. \tag{8.8}$$

In what follows, we shall use the notation introduced in the present subsection in our discussions of the game Γ.

8.8. We first single out the case when the values of the functions f and g are equal at at least one point.

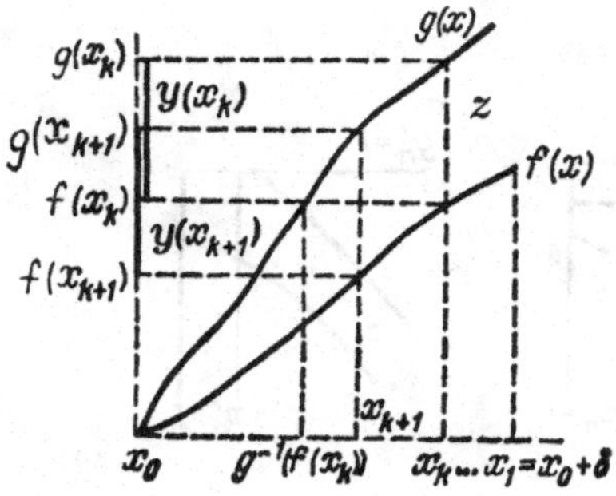

Fig. 3.16

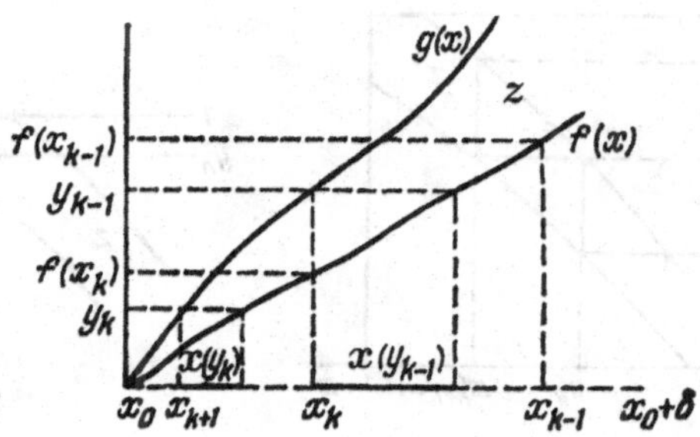

Fig. 3.17

Theorem. *If there is a point $x_0 \in [0,1]$ such that $f(x_0) = g(x_0)$, then $v_\Gamma = 0$, every strategy of player 1 is optimal, and player 2 has ε-optimal strategies for every $\varepsilon > 0$, but has no optimal strategies.*

Proof. In accordance with the theorem of 8.3, it is sufficient to show that y contains an infinite subset that generates a packing in x, and that x contains an infinite subset that generates a covering of y.

If $f(x) = g(x)$ for an infinite set of values of x, it is evident that the corresponding values of the functions are all different, and form a subset of x that generates a packing in y. The same x-values must appear in any subset of x that generates a covering of y. Consequently we have the hypotheses of the case discussed in the theorem of 8.3.

We therefore suppose, for the rest of this discussion, that the equation $f(x) = g(x)$ is satisfied for only a finite number of values of x. These must form a set of isolated points. Let x_0 be one of these points.

Then there is a $\delta > 0$ such that when $x_0 < 1$ we have $f(x) < g(x)$ for $x \in (x_0, x_0 + \delta]$, and when $x_0 > 0$ this inequality is satisfied for every $x \in [x_0 - \delta, x_0)$. For definiteness, let us consider the first case.

We choose any $x_1 \in (x_0, x_0 + \delta]$ and construct inductively the points $y_k \in (f(x_0), f(x_k))$, $x_{k+1} = g^{-1}(y_k)$, $k = 1, 2, \ldots$ (Figure 3.16). Evidently we have $f^{-1}(y_{k+1}) < x_{k+1} = g^{-1}(y_k)$, so that

$$x(y_k) \cap x(y_{k+1}) = [g^{-1}(y_k), f^{-1}(y_k)] \cap [g^{-1}(y_{k+1}), f^{-1}(y_{k+1})] = \emptyset,$$

and the sets $x(y_1), x(y_2), \ldots$ form an infinite packing in $(x_0, x_0 + \delta]$ and consequently in x. We also choose

$$x_1 = x_0 + \delta, \quad x_{k+1} \in (g^{-1}(f(x_k)), x_k), \quad k = 1, 2, \ldots$$

(Figure 3.17). Here $g(x_{k+1}) > f(x_k)$, so that

$$y(x_k) \cap y(x_{k+1}) = [f(x_k), g(x_k)] \cap [f(x_{k+1}), g(x_{k+1})] \neq \emptyset,$$

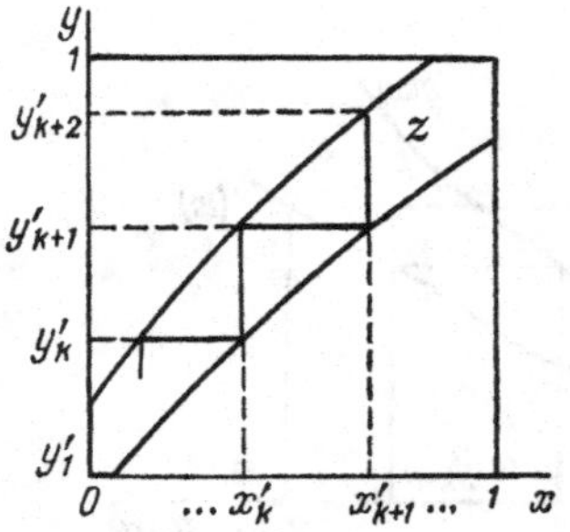

Fig. 3.18

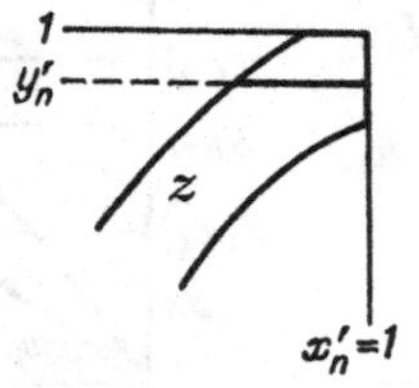

Fig. 3.19

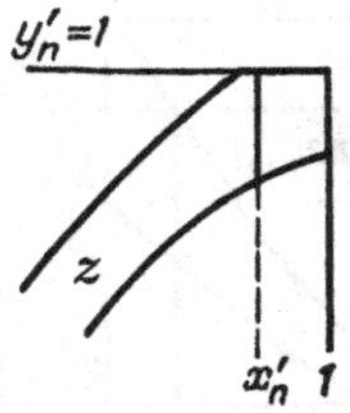

Fig. 3.20

and the sets $\boldsymbol{y}(x_1)$, $\boldsymbol{y}(x_2), \ldots$ form a countable covering of the interval $(0, g(x_0, \delta)]$. There are finitely many such intervals (not more than twice the number of common values of f and g); consequently they can all be covered by a countable number of sets of the form $\boldsymbol{y}(x_k)$. They still contain a finite number of points of coincidence of f and g, as well as a finite number of intervals of the form

$$(g(x_0 + \delta), f(x_0' - \delta)), \tag{8.9}$$

where x_0 and x_0' are consecutive points of equality of f and g (some of these intervals might be empty). Since, for all $x \in (x_0 + \delta, x_0' - \delta)$, the values of the difference $g(x) - f(x)$ differ from zero by a positive ε, each interval of the form (8.9) must cover a finite number of points of sets of the form $\boldsymbol{y}(x)$. □

8.9. The theorem just proved lets us assume to begin with that $f(x) < g(x)$ for all $x \in [0, 1]$ and, in particular, that $g(0) > 0$ and $f(1) < 1$. In this situation, the strip that forms the set $\boldsymbol{z}$ has vertical width at least equal to

$$\delta_v = \min_x \big(g(x) - f(x)\big) > 0,$$

and horizontal width at least

$$\delta_h = \min_y \big(f^{-1}(y) - g^{-1}(y)\big) > 0.$$

Let us describe, for the class of games under consideration, a process for constructing a class of finite subsets $\boldsymbol{x}^* \subset \boldsymbol{x}$ that generate a covering of $\boldsymbol{y}$, and an equal number of subsets $\boldsymbol{y}^* \subset \boldsymbol{y}$ that generate a packing in $\boldsymbol{x}$. By induction, we obtain

$$y_1' = 0, \tag{8.10}$$

$$x_k' = f^{-1}(y_k'), \quad k = 1, 2, \ldots, \tag{8.11}$$

$$y_{k+1}' = g(x_k'), \qquad k = 1, 2, \ldots \tag{8.12}$$

Geometrically, this process corresponds to constructing the zigzag sketched in Figure 3.18.

The length of each vertical link of this zigzag is at least δ_v, and that of each horizontal link is at least δ_h. The total length of horizontal links of the zigzag (and equally of the vertical links) does not exceed unity. Consequently the number of links in the zigzag is finite (it cannot exceed $2/\max\{\delta_v, \delta_h\}$), and after a finite number of steps the zigzag will bump into either the right-hand side of the situation square, at a point $y'_n \geqq f(1)$ or its upper side at a point $x'_n \geqq b$. In the first case (Figure 3.19) we take $x'_n = 1$ to complete the zigzag. In the second case (Figure 3.20) we set $y'_n = 1$ to complete the process.

8.10. It is easy to see that the set $\{x'_1, \dots, x'_n\}$ of strategies generates a covering of y. Here each intersection $y(x'_i) \cap y(x'_{i+1})$, $i = 1, \dots, n-2$, consists of a single point, and $y(x'_{n-1}) \cap y(x'_n)$ cannot be empty, but might contain an interval of finite length. If this actually occurs, it is evident that after a sufficiently small (mutually coordinated) translation to the left of $x'_1, \dots, x'_{n-1}$, together with x'_n, we shall, as before, generate a covering of y.

The set $\{y'_1, \dots, y'_n\}$ of strategies does not generate a packing in x: the cross-sections $x(y'_i) = [x'_{i-1}, x'_i]$ and $x(y'_{i+1}) = [x'_i, x'_{i+1}]$ intersect, at least at one point. However, since the zigzag bumps into the right-hand side of the square, the total length of its vertical links is less than unity, so that some connected family $(y'_2, \dots, y'_n\}$ of strategies will generate a packing in x. If, however, the zigzag bumps into the upper side of the square, the sum of the lengths of the horizontal links of the zigzag will be less than unity, and we again have the possibility of stretching the elements of the covering. In order to have a more precise description of the possibilities of mixing and stretching, we present the following constructions. For this purpose we set (literally "on the other hand")

$$y''_n = 1, \tag{8.13}$$

$$x''_k = g^{-1}(y''_k), \quad k = n,\ n-1, \dots, \tag{8.14}$$

$$y''_{k-1} = f(x''_k), \qquad k = n,\ n-1, \dots \tag{8.15}$$

Geometrically, this corresponds to constructing a zigzag in the direction opposite to the previous one. Since f and g are monotonic, it follows that this zigzag also has n links in each direction, and

$$0 = y'_n < y''_1 \leqq y'_2 < y''_2 \leqq \dots \leqq y'_n < y''_n = 1, \tag{8.16}$$

$$0 \leq x''_1 \leqq x'_1 < x''_2 \leqq x'_2 < \dots < x''_n \leqq x'_n \leqq 1 \tag{8.17}$$

(Figure 3.21).

8.11. We now construct a series of strategies that generates coverings of y and packings in x. The construction is based on the considerations that are at the base of the proof of the theorem of 8.7. Let us begin with the strategies of player 2.

We select, arbitrarily,

$$y_1 \in [y'_1, y''_1) \tag{8.18}$$

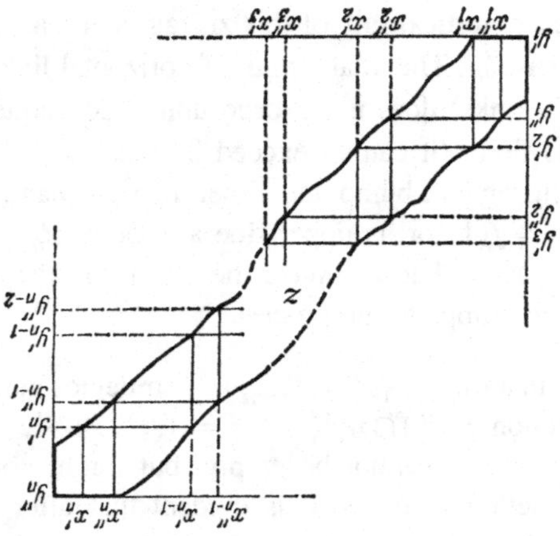

Fig. 3.21

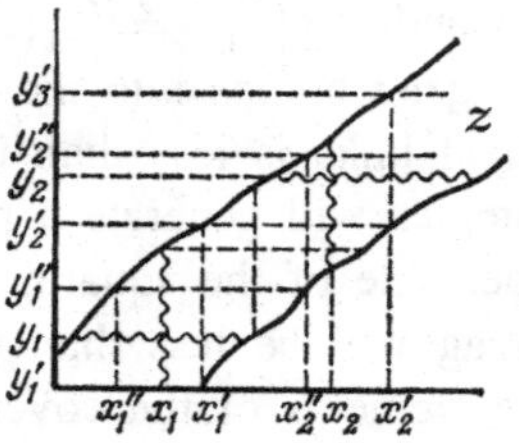

Fig. 3.22

and, by induction on k (equally arbitrarily)

$$y_{k+1} \in \big(g(f^{-1}(y_k)), y''_{k+1}\big) \text{ for } k = 1, \dots, n-1. \tag{8.19}$$

The process for constructing the set $\{y_1, \dots, y_n\}$ is rather intuitive geometrically: it is sketched in Figure 3.22.

It is clear that the cross-section $x(y_1)$ is the interval

$$[0, f^{-1}(y_1)]; \tag{8.20}$$

each cross-section $x(y_k)$ for $k = 2, \dots, n-1$ is an interval

$$[g^{-1}(y_k), f^{-1}(y_k)] \tag{8.21}$$

and, finally, the cross-section $x(y_n)$ is

$$[g^{-1}(y_n), 1]. \tag{8.22}$$

It follows from (8.18) that $y_{k+1} > g\big(f^{-1}(y_k)\big)$, i.e.,

$$g^{-1}(y_{k+1}) > f^{-1}(y_k),$$

so that the intervals (8.20)–(8.22) are pairwise disjoint, and form a packing in x.

On the other hand, we choose, arbitrarily,

$$x_1 \in [x_1'', x_1'], \tag{8.23}$$

$$x_{k+1} \in [x_{k+1}'', f^{-1}(g(x_k))] \text{ for } k = 1, \ldots, n-1 \tag{8.24}$$

(the construction of the set $\{x_1, \ldots, x_n\}$ resembles the construction of $\{y_1, \ldots, y_n\}$, except that here the cross-sections of the Π-shaped polygons are connected to each other). The cross-section $y(x_1)$ is the interval

$$[0, g(x_1)]; \tag{8.25}$$

the cross-sections $y(x_k)$ for $k = 2, \ldots, n-1$ are the intervals

$$[f(x_k), g(x_k)]; \tag{8.26}$$

and the cross-section $y(x_n)$ is the interval

$$[f(x_n), 1]. \tag{8.27}$$

It follows from (8.24) that $f(x_{k+1}) < g(x_k)$, i.e., the cross-sections in (8.24)–(8.26) intersect, and therefore the set $\{x_1, \ldots, x_n\}$ of strategies generates a covering of y.

Thus we have obtained the hypotheses of the corollary in 8.4. Therefore we have $v_\Gamma = 1/n$ and we have available a rather extensive assortment of equilibrium situations in Γ. We note that all the pure strategies of player 1 that appear with positive probabilities (equal to $1/n$) in the description of the optimal mixed strategies are to be taken one by one in horizontal zones

$$[x_k'', x_k'], \quad k = 1, \ldots, n, \tag{8.28}$$

and the strategies of player 2, with similar properties, are taken one by one in the vertical zones

$$[y_1', y_1''), (y_k', y_k'') \ (k = 2, \ldots, n-1), \ (y_n', y_n'']. \tag{8.29}$$

Since all the strategies $x = x_k$ and $y = y_k$ appear in some optimal strategy of a player, we must have

$$H(y, Y^*) = H(X^*, y) = v_\Gamma = 1/n \tag{8.30}$$

for all optimal X^* and Y^*.

As we shall see in what follows, the spectra of all the mixed strategies of player 1 in Γ are contained in the union of the zones (8.28); and the spectra of the optimal strategies of player 2, in the union of the zones (8.29).

8.12 A complete description of the sets of the players' optimal strategies. We turn to a complete description of the sets of optimal strategies of each player.

Lemma. *If X^* is an optimal strategy of player 1, and for some k*

$$x'_k < p < q < x''_{k+1}, \tag{8.31}$$

then $X^*([p,q]) = 0$.

If Y^* is an optimal strategy of player 2, and for some k

$$y''_k < p < q < y'_{k+1},$$

then

$$Y^*([p,q]) = 0.$$

Proof. Since f is a monotonic function, it follows from (8.31) that

$$f(x'_k) < f(p) < g(p) < g(x''_{k+1}),$$

and by (8.10) and (8.13),

$$y'_k < f(p) < g(q) < y''_{k+1}. \tag{8.32}$$

Using the arbitrariness in the choice of the strategies $y_1, \ldots, y_n$ in (8.28) and (8.19), and the continuity and monotonicity of f and g, we may choose these strategies so that $y_1, \ldots, y_k$ are arbitrarily close (from above) to the corresponding $y'_1, \ldots, y'_k$, and $y_{k+1}, \ldots, y_n$ are arbitrarily close (from below) to the corresponding $y''_{k+1}, \ldots, y''_n$. Then (8.32) yields

$$y_k < f(p) < g(q) < y_{k+1}.$$

Therefore we have, for $x \in [p,q]$,

$$H(x, y_k) = 0 \text{ for } k = 1, \ldots, n. \tag{8.33}$$

Now select a strategy $Y^* \in \boldsymbol{Y}$ for which $Y^*(y_k) = 1/n$ for $k = 1, \ldots, n$. It follows from (8.33) that

$$H(x, Y^*) = 0 \text{ for } x \in [p,q],$$

and consequently, for every mixed strategy $X \in \boldsymbol{X}$,

$$\int_p^{q+0} H(x, Y^*)dX(x) = 0.$$

Taking X to be an optimal strategy X^*, we therefore have

$$v_\Gamma = \frac{1}{n} = H(X^*, Y^*) = \int_0^1 H(x, Y^*)dX^*(x)$$

$$= \int_0^p H(x, Y^*)dX^*(x) + \int_q^{1+0} H(x, Y^*)dX^*(x).$$

But since Y^* is optimal, we must have $H(x, Y^*) \leqq 1/n$ for every $x \in \mathfrak{x}$. Taking account of this, we obtain

$$\frac{1}{n} \leqq \frac{1}{n}\left(\int_0^p dX^*(x) + \int_q^{1+0} dX^*(x)\right) = \frac{1}{n}(1 - X^*([p, q])),$$

and the conclusion follows.

The second part of the lemma is proved similarly. □

The conclusion of the lemma restricts the spectra of the optimal strategies of player 1 in Γ to the zones (8.28), and the spectra of player 2 to the closures of the zones (8.29). The second restriction will be made somewhat more precise later.

8.13. According to 1.11, for a strategy Y^* of player 2 in the game Γ to be optimal it is necessary and sufficient (since this is a two-person zero-sum game) that $H(x, Y^*) \leqq v_\Gamma$ for every $x \in \mathfrak{x}$, i.e., under our hypotheses, by (8.36) and 8.10,

$$Y^*[f(x), g(x)] \leqq 1/n \tag{8.34}$$

for all $x \in [0, 1]$. If a strategy $x \in \mathfrak{x}$ belongs to one of the zones $[x_k'', x_k']$ in (8.28), then $f(x) \leqq y_k' < y_k'' \leqq g(x)$. In this case the strategy x can appear, with positive probability, in the optimal mixed strategies. Consequently, by what we said in 8.11,

$$Y^*[f(x), g(x)] = Y^*[y_k', y_k''] = 1/n. \tag{8.35}$$

Therefore, under the conditions of an optimal strategy of player 2, the probability of choosing a pure strategy is $1/n$ for each zone of the form (8.29).

It can be established similarly that under the conditions of any optimal strategy of player 1, the probability of choosing a pure strategy is also $1/n$ for each zone of the form (8.28).

Now suppose that a pure strategy x in (8.34) does not belong to any zone (8.28). Here there may be several different cases.

First suppose that $x < x_1''$. In this case

$$H(x, Y^*) = Y^*([0, g(x)] \leqq Y^*([y_1', y_1'']) = 1/n,$$

i.e., (8.34) is a corollary of the right-hand equation in (8.35) and no other conditions are imposed on the optimal strategy. For the same reason, we can disregard the case when $x > x_n'$.

Now suppose that $x \in (x_k', x_{k+1}'')$. Then

$$f(x) \leqq y_k'' < y_k' < g(x).$$

This means that

$$Y^*([f(x), g(x)]) = Y^*([f(x), y_k'']) + Y^*((y_k'', y_{k+1}')) + Y^*([y_{k+1}', g(x)]).$$

By what was established in 8.11, the second term on the right reduces to zero, so that by using (8.32) we obtain

$$Y^*([f(x), y_k'']) + Y^*([y_{k+1}', g(x)]) \leqq 1/n. \tag{8.36}$$

However, on the other hand,

$$Y^*([y_k', y_k'']) = Y^*([y_k', f(x_k))) + Y^*([f(x), y_k'']) = 1/n.$$

Consequently (8.36) can be rewritten as

$$1/n - Y^*([y_k', f(x))) + Y^*([y_{k+1}', g(x)]) \leqq 1/n,$$

whence

$$Y^*([y_{k+1}', g(x)]) \leqq Y^*([y_k', f(x))). \tag{8.37}$$

We emphasize that the system of equations that we have obtained for all x that are not in the zones (8.28), together with equation (8.35) for all x, are not only necessary, but also sufficient for the validity of the inequalities (8.34) for all $x \in [0, 1]$, i.e., for the strategy Y^* to be optimal.

If we set $x = x_{k+1}''$ in (8.37), then by (8.15) and (8.14) we have

$$f(x_{k+1}'') = y_k'', \quad g(x_{k+1}'') = y_{k+1}'',$$

so that (8.37) acquires the form

$$Y^*([y_{k+1}, y_{k+1}'']) \leqq Y^*([y_k', y_k'']).$$

But by (8.35), the left-hand side here is equal to $1/n$. Therefore

$$1/n \leqq Y^*([y_k', y_k'']) \leqq Y^*([y_k', y_k'']) = 1/n.$$

This means that, under the conditions of any optimal strategy Y^* of player 2, the pure strategy y_k'' cannot be chosen with positive probability. The case $k = n$ is an exception. In other words, the distribution function of Y^* cannot have jumps at the points $y_1'', \dots, y_{n-1}''$.

Furthermore we can rewrite (8.37) as

$$Y^*((g(x), y_{k+1}'']) \geqq Y^*([f(x), y_k'']),$$

from which we can deduce by a similar argument that the left-hand end-points of the zones (8.29) cannot be jumps of the distribution function of Y^*, except for the case $k = 1$, i.e. the points $y_2', \dots, y_n'$.

A similar, but somewhat modified, situation is encountered if we consider an arbitrary optimal strategy X^* of player 1. From the inequality $H(X^*, y) \geqq 1/n$, which holds for every optimal strategy X^* and every strategy $y \in \mathcal{Y}$, it follows, first, that

$$X^*([x_k'', x_k']) = 1/n, \tag{8.38}$$

and, second, that

$$X^*([x_k'', g^{-1}(y))) \leqq X^*([x_{k+1}'', f^{-1}(x)]). \tag{8.39}$$

Since, in contrast to (8.37), the measure of the interval without its right-hand endpoint now appears on the left, the exclusion of jumps of the distribution function does not arise here.

8.14. Relations of the form (8.35) and (8.37)–(8.39) describe all the connections between distributions on consecutive intervals of the form (8.28) or (8.29) under the conditions of all optimal strategies of players 1 and 2. From the probabilistic point of view the aggregate of these connections is quite transparent. This means that, under the conditions of any optimal strategy of player 1, in distributions on the interval (8.28), as k increases, the concentration of probability increases toward the *left-hand* end of the interval; but in the probability distributions on the intervals (8.29), under the conditions of an optimal strategy Y^* of player 2, it is shifted, as k increases, toward the *right-hand* end of the interval. In the strategies of player 1, this shifting tendency is milder than in the strategies of player 2. The principal reason for this difference is that the set of situations that are favorable for player 1 is closed, whereas the set of situations that are favorable for player 2 is (relatively) open.

8.15. Since the functions f and g are continuous and strictly increasing, so are the functions f^{-1} and g^{-1}. Therefore passing from any $y \in [y'_k, y''_k]$ to $g(f^{-1}(y)) \in [y'_{k+1}, y''_{k+1}]$ establishes a single-valued monotonic mapping from the first of these intervals to the second.

Let ψ_k be single-valued continuous monotonic mappings from every interval $[y'_k, y''_k]$ onto $[0, 1]$. Let us suppose that these mappings are coordinated, in the sense that for $k = 1, \dots, n-1$,

$$\psi_k(y) = \psi_k(g(f^{-1}(y))),$$

for every $y \in [y'_k, y''_k]$, and arbitrary in other respects.

Every optimal strategy Y^* of player 2 induces, in a natural way, a conditional distribution on $[y'_k, y''_k]$ with some distribution function $\tilde{G}_k$:

$$\tilde{G}_k(y) = nY^*([y'_k, y)).$$

The mapping ψ_k carries this conditional distribution to a probability distribution on $[0, 1]$ with distribution function G_k:

$$G_k(z) = \tilde{G}(\psi_k^{-1}(z)) \text{ for } z \in [0, 1].$$

The inequality (8.39) can be rewritten in this case as

$$Y^*([y'_{k+1}, g(f^{-1}(y))]) \leqq Y^*([y'_k, y)) \text{ for } y \in [y'_k, y''_k].$$

If we apply the mappings ψ_{k+1} and ψ_k, we obtain

$$G_{k+1}(z+0) \leqq G_k(z) \text{ for } z \in [0, 1]. \tag{8.40}$$

It is clear that the validity of this inequality does not depend on the choice of the specific mapping ψ_k (these mappings must, of course, be chosen consistently).

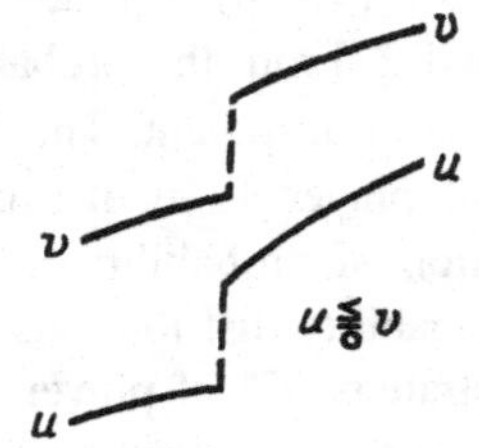

Fig. 3.23

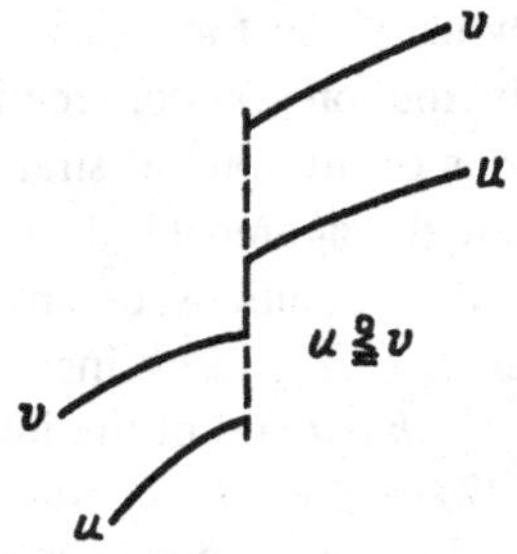

Fig. 3.24

The relation expressed by (8.40) between the distribution functions G_{k+1} and G_k has a kind of ordering. We shall indicate this by writing $G_{k+1} \underset{\circ}{\lesseqgtr} G_k$ (Figure 3.23).

This makes it possible to characterize the set of optimal strategies of player 2 in the game under consideration as the set of probability distributions for which the probability measure of each interval $[y'_k, y''_k]$ is $1/n$, and the graphs of the functions induced by these distributions form a sequence that is nondecreasing in the sense of the relation $\underset{\circ}{\lesseqgtr}$:

$$G_1 \underset{\circ}{\gtreqless} G_2 \underset{\circ}{\gtreqless} \dots \underset{\circ}{\gtreqless} G_n.$$

Turning to the optimal strategies of player 1, we may also introduce coordinated mappings φ_k of the intervals $[x''_k, x'_k]$ onto $[0, 1]$, for which

$$\varphi_k(x) = \varphi_{k+1}(f(g^{-1}(x))).$$

If we put the optimal strategies X^* into correspondence with conditional distributions on the intervals $[x''_k, x'_k]$, with distribution functions $\tilde{F}_k$:

$$\tilde{F}_k(x) = nY^*([x''_k, x)),$$

we may, by using the mappings φ_k as before, pass to probability distributions on $[0, 1]$ with the distribution functions $F_k : \tilde{F}(\varphi^{-1}(z))$ (Figure 3.24). Here inequality (8.39) assumes the form $F_k(z) \leqq F_{k+1}(z+0)$, which we write as $F_k \overset{\circ}{\leqq} F_{k+1}$. Therefore the graphs of the distribution functions F_k must be nondecreasing in the sense of $\overset{\circ}{\leqq}$:

$$F_1 \overset{\circ}{\leqq} F_2 \overset{\circ}{\leqq} \dots \overset{\circ}{\leqq} F_n.$$

8.16. We illustrate what has just been said by the example in which

$$f(x) = (1 - \delta)x, \quad g(x) = (1 - \delta)x + \delta$$

(the corresponding set z is a parallelogram, as shown in Figure 3.25).

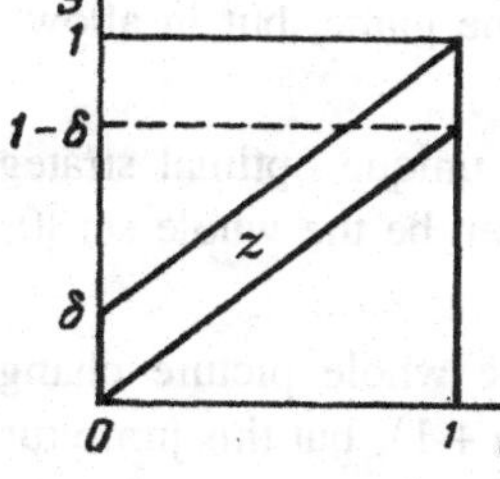

Fig. 3.25

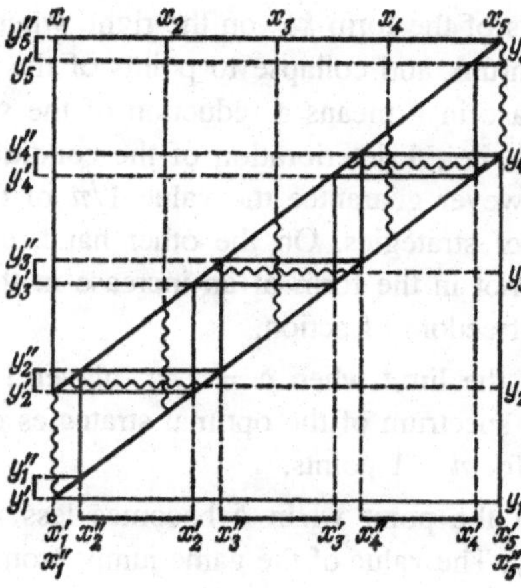

Fig. 3.26

The construction of zigzags corresponding to the points whose coordinates are described by (8.10)–(8.12) and (8.13)–(8.15) lead to the following values (Figure 3.26):

$$\begin{array}{llll} \downarrow\; y_1' = 0, & \downarrow\; x_1' = 0, & y_1'' = 1-(n-1)\delta, & x_1'' = 1, \\ \; y_2' = \delta, & \; x_2' = \delta', & y_2'' = 1-(n-2)\delta, & x_2'' = 1-(n-2)\delta', \\ \cdots & & & \\ y_{n-1}' = (n-2)\delta, & x_{n-1}' = (n-2)\delta', & \uparrow\; y_{n-1}'' = 1-\delta, & \uparrow\; x_{n-1}'' = 1-\delta', \\ y_n' = (n-1)\delta, & x_n = 1, & \; y_n'' = 1, & \; x_n'' = 1 \end{array}$$

Here $\delta' = \delta(1-\delta)$, $n = 1+[1/\delta]$, and the order of calculating the values y', x', y'' and x'' is shown by the arrows. We see that the value of the game is $1/n$, and that the zones where the spectra of the optimal strategies of player 1 are concentrated are the points 0 and 1, and $n-2$ intervals of the form

$$[1-(n-k)\delta', k\delta'], \text{ where } k = 1, \ldots, n-2, \tag{8.41}$$

each of length $1-(n-1)\delta'$. The zones of concentration of the spectra of the optimal strategies of player 2 are the intervals

$$\begin{aligned} &[0, 1-(n-1)\delta), \\ &(k\delta, 1-(n-k-1)\delta), \quad k = 1, \ldots, n-2, \\ &((n-1)\delta, 1], \end{aligned} \tag{8.42}$$

each of length $1-(n-1)\delta$.

If $1/(n-1) > \delta > 1/n$, then after a small decrease of δ, the value of $[1 \setminus \delta]$ and the value of the game itself are unchanged, the zones (8.42) are extended, and the set $[0, 1]$ is, step by step, almost filled, except for small neighborhoods of points of the form $k\delta'$ on the right, whereas the zones (8.41) are, on the other hand, shrunk and collapse to points of the form $k\delta$. This is rather natural, since a decrease in δ means a reduction of the set of situations that are favorable for player 1, i.e., a deterioration of the conditions of the game for that player, who can, however guarantee the value $1/n$ of the game, but only by a more careful choice of strategies. On the other hand, conditions are improved for player 2, though not in the form of an increase in the value of the game, but in allowing greater freedom of action.

In the limit when $\delta = 1/n$, player 1 will have a unique optimal strategy, and the spectrum of the optimal strategies of player 2 can be the whole set $[0, 1]$ except for $n - 1$ points.

At the point when δ becomes less than $1/n$, the whole picture changes abruptly. The value of the game jumps from $1/n$ to $1/(n+1)$, but this jump turns out to have relatively little effect on the optimal behavior of player 1, the spectra of whose optimal strategies still fill a significant part of the interval $[0, 1]$; player 2, on the other hand, has to play a very careful game with much shorter zones. With further decrease of δ the description of the dynamics repeats.

It is clear that what we have said in this subsection remains valid, with appropriate modifications, for the general games that we have considered, starting with 8.8.

§9 Games of timing

9.1. A large number of mathematical problems of analytic-computational type (for example, those connected with the solution of differential or integral equations) are solved by assuming a priori some form of the required solution. The choice of the form of the solution depends on the basis either of logical analysis of the problem, or even on purely intuitive ideas. In the course of the brief development of game theory, it has not, up to now, developed any inherent patterns for its intuitive reasoning. The basic steps on the road to the formation of this intuition must come from special theories about one or another kind of game-theoretic problems, particularly those that have transparent informal interpretations.

One of these classes of two-person zero-sum games consists of what are known as games of timing.

Definition. A *game of timing* is a two-person zero-sum game on the unit square with a payoff function H of the following form:

$$H(x, y) = \begin{cases} L(x, y) & \text{if } x < y, \\ \Phi(x) & \text{if } x = y, \\ M(x, y) & \text{if } x > y, \end{cases} \tag{9.1}$$

where, for every $x \in [0, 1]$,

$$\min\{L(x, x), M(x, x)\} \leqq \Phi(x) \leqq \max\{L(x, x), M(x, x)\}, \tag{9.2}$$

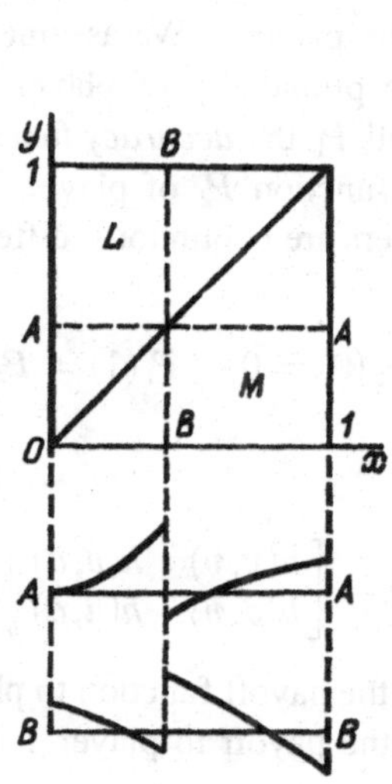

Fig. 3.27

and the functions L and M (which were called *semikernels* in 3.30) are continuous in both variables in the corresponding triangles, and, in addition,

$$\begin{aligned} &\text{increase strictly in } x \text{ for every } y, \\ &\text{decrease strictly in } y \text{ for every } x \end{aligned} \tag{9.3}$$

(Figure 3.27). □

In the course of this section we shall somewhat modify this system of conditions. In particular, we shall essentially weaken (9.2).

The term "games of timing" is explained by the following natural interpretation. Let each player be allowed, at any instant, to perform a single instantaneous action. The player who has acted first can alter the opponent's state and consequently the nature and the success of the opponent's action. Therefore, against the background of the general continuous dependence of the payoff function on time, there is a discontinuity if the values of its arguments coincide, since however small the difference may be in the times when the players perform their actions, it may be extremely important which player anticipates the other. The hypothesis that the payoff function is monotonic is included because in the course of time either player may acquire new possibilities (for example, by receiving new information).

9.2. We can hardly expect to discover general methods of finding explicit solutions of any wide class of games of timing. Moreover, in solving the majority of games of this kind it is necessary to apply special artificial methods, as we will see from the examples presented in this section.

These examples are based on a game known as a *single-shot duel*, which is as follows.

Let players 1 and 2 approach each other with constant speed, intending to fire their shots at instants x and y (which, under these conditions, we may take to be the distances between the players). We assume that a hit prevents a player from firing. Let $P_1(x)$ be the probability of player 1's disabling player 2 by a shot fired at distance x; we call P_1 the *accuracy function* of player 1. In the same way, we define the accuracy function P_2 of player 2. We shall suppose that the accuracy functions of the players are continuous, differentiable, strictly increasing, and satisfy the conditions

$$P_1(0) = P_2(0) = 0, \quad P_1(1) = P_2(1) = 1, \tag{9.4}$$

and are otherwise arbitrary.

Finally,

$$h = \begin{bmatrix} h(v,v) & h(v,d) \\ h(d,v) & h(d,d) \end{bmatrix}$$

is a matrix in which $h(v,v)$ is the payoff function to player 1 if he hits his opponent and is hit himself; $h(v,d)$ is the payoff to player 1 if he hits player 2, but is not hit himself; and so on.

It is usual to distinguish between "silent" duels, in which a player does not know when his opponent fires, unless he is hit himself; and "noisy" duels, in which a player always knows whether his opponent has fired. Strictly speaking, a noisy duel in the literal sense is not a game of timing or even a game in normal form (see section 22 of the Introduction), since in the play of such a game the state of a player's information may be changed (or not changed).

Formally, however, a noisy duel can quite naturally be interpreted as a game of timing. In fact, as a strategy $z \in [0,1]$, a player may consider his choosing to fire at the instant z if his opponent has not yet fired. If he has survived, it is expedient for the player to abandon his original plan of firing at time z, and fire at time 1, hitting his opponent with probability 1.

In a game of duel type we may rename the pure strategies of one of the players (say, for definiteness, of player 1), and denote by x the time of firing (or, what is essentially the same thing, the distance at which the shot is fired) at which the probability that player 1 will hit player 2 is x. Since we assumed that the accuracy function $P_1(x)$ is continuous, this renaming is a strategic equivalence transformation (see Chapter 1, 2.19 and Chapter 3, 1.13) and therefore does not change the character of the game. Hence when we are discussing specific examples of duel type, we may take $P_1(x) = x$ if this seems convenient.

9.3. By the theorem of 3.30, a game Γ on the unit square with payoff function H that satisfies (9.1) and (9.2) has an equilibrium situation, in general in the mixed strategies.

In analyzing games of timing, we shall systematically use the property of complementary slackness (see Theorem 4° in 1.18), namely that, at points of the spectrum of the optimal strategy of a player against the other player in a two-person zero-sum game, the payoff function attains the value of the game.

9.4 Lemma. *Let X and Y be respectively the optimal strategies of players 1 and 2 in the game Γ. Then the spectra of X and Y can be different only at the points 0 and/or 1.*

Proof. The set of points of the spectrum of X is closed. Consequently the set of points that are not in the spectrum of X is open (i.e., is the union of an at most countable family of intervals), possibly supplemented with the points 0 and/or 1. Let (x', x'') be one of the open subintervals of $(0, 1)$ that does not intersect the spectrum of X.

Let us consider the value of the payoff function $H(X, y)$ at $y \in (x', x'')$:

$$H(X,y) = \int_0^1 H(x,y)dX(x) = \int_0^{x'} L(x,y)dX(x) + \int_{x''}^1 M(x,y)dX(x).$$

It is clear that $H(X, y)$ is a continuous function of y in this interval. Now choose y' and $y'' \in (x', x'')$ with $y' < y''$ and consider the difference

$$H(X,y') - H(X,y'') =$$

$$\int_0^{x} (L(x,y') - L(x,y''))dX(x) + \int_{x''}^1 (M(x,y') - M(x,y''))dX(x).$$

Since L and M are strictly decreasing functions of their second argument, these integrals are nonnegative and at least one of them is positive. Therefore $H(X, y') > H(X, y'')$, and since $H(X, y'') \geqq v_\Gamma$, we must also have $H(X, y') > v_\Gamma$, which, together with the continuity of $H(X, y)$ at y', means, on the basis of 1.18 (see also 5.15, Chapter 1), that y' cannot be a point of the spectrum of the optimal strategy Y.

By the symmetry of these arguments, points other than 0 and 1 that are not in the spectrum of Y also cannot be points of spectrum of X. □

9.5 Lemma. *In a game Γ of timing, the spectrum of any optimal strategy of either player consists (possibly) of the point 0 and (also possibly) of an interval $[a, 1]$, where $a > 0$.*

Proof. Let X be an optimal strategy of player 1 in Γ. It is enough to establish that no points that are not in the spectrum of X can lie to the right of any point of the spectrum of X, excluding 0. Assuming the contrary, consider a point x'' that is not a point of the spectrum of X, and the set $\bar{x}$ of all points of the spectrum that lie to the left of it. Evidently $\bar{x}$ is a closed set. Let $\bar{\bar{x}} = \max \bar{x}$. Evidently $\bar{\bar{x}} < x''$. Let us take a point $x' \in (\bar{\bar{x}}, x'')$.

Now choose any sufficiently small $\varepsilon > 0$ and estimate the difference

$$H(X,\bar{x}) - H(X,\bar{x}-\varepsilon) = \int_0^1 \big(H(x,\bar{x}) - H(x,\bar{x}-\varepsilon)\big)\,dX(x)$$

$$= \int_0^{\bar{x}-\varepsilon} \big(L(x,\bar{x}) - L(x,\bar{x}-\varepsilon)\big)\,dX(x) + \int_{\bar{x}-\varepsilon}^{\bar{x}} \big(L(x,\bar{x}) - M(x,\bar{x}-\varepsilon)\big)\,dX(x)$$

$$+ \int_{\bar{x}}^1 \big(M(x,\bar{x}) - M(x,\bar{x}-\varepsilon)\big)\,dX(x).$$

Since L and M are continuous, the first and third integrals in this sum tend to zero with ε. The second integral tends to zero only if the distribution function F_X does not have a jump at $\bar{x}$. Consequently if F_X is continuous at $\bar{x}$, the function $H(X,y)$ is continuous at the same point. In that case, by complementary slackness (see Theorem 4° in 1.16) we must have

$$H(X,\bar{x}) = v_\Gamma. \tag{9.5}$$

At the same time

$$H(X,\bar{x}) = \int_0^1 H(x,\bar{x})dX(x) = \int_0^{x'} H(x,\bar{x})dX(x) + \int_{x''}^1 H(x,\bar{x})dX(x)$$

$$< \int_0^{x'} H(x,x')dX(x) + \int_{x''}^1 H(x,x')dX(x) = H(X,x') = v_\Gamma,$$

which evidently contradicts (9.5).

Therefore, to the right of points of the spectrum of X that are different from zero and that are points of continuity of F_x, there can be only points of the spectrum of X.

As a result, the spectrum of X can consist of a set of (evidently at most countably many) isolated points that form its "left flank", together with an interval $[a, 1]$ that may (or may not) reduce to the single point 1. It remains only to show that there can be at most one isolated left-point of the spectrum, namely 0.

Let x_0 be a nonzero isolated point of the spectrum of X. By the preceding lemma, it must also be a point of the spectrum of Y. If x_0 is not an isolated point of the spectrum of Y, then every neighborhood of it must contain other points of the spectrum of Y which, by what has already been established, must also be points of the spectrum of X, which do not exist. Consequently x_0 is also an isolated point of the spectrum of Y. This in turn means that the distribution function F_Y has a jump at x_0. But then, by complementary slackness, (9.5) must be satisfied. Repeating the arguments that follow that formula, we again arrive at a contradiction. □

In what follows, we shall always assume for definiteness that $a > 0$. The analysis of the case $a = 0$ differs only in some unimportant details, and we shall not give it special consideration.

9.6 Lemma. *For some $x_0 \in [0, 1]$, let*

$$L(x_0, x_0) \leqq \Phi(x_0) \leqq M(x_0, x_0). \tag{9.6}$$

Then an optimal strategy of players in Γ cannot have points of its spectrum to the left of x_0.

Proof. Consider an $\bar{x} \times \bar{y}$-subgame $\bar{\Gamma}$ of Γ, where $\bar{x} = \bar{y} = [x_0, 1]$. This game has a solution $(\bar{X}, \bar{Y})$ for which

$$H(x, \bar{Y}) \leqq H(\bar{X}, \bar{Y}) \leqq H(\bar{X}, y) \text{ for all } x \in \bar{x} \text{ and } y \in \bar{y}.$$

Now select any $x < x_0$. Then, since L is monotonic, we can use the left side of (9.4) to obtain

$$H(x, \bar{Y}) = \int_{x_0}^{1} L(x, y) d\bar{Y}(y) < \int_{x_0}^{1} L(x_0, y) d\bar{Y}(y) < L(x_0, x_0) d\bar{Y}(x_0)$$

$$+ \int_{x_0+0}^{1} L(x_0, y) d\bar{Y}(y) \leqq \Phi(x_0) d\bar{Y}(x_0) + \int_{x_0+0}^{1} L(x_0, y) d\bar{Y}(y)$$

$$= H(x_0, \bar{Y}) \leqq H(\bar{X}, \bar{Y}).$$

Similarly (using the right-hand side of (9.6)) we find that

$$H(\bar{X}, \bar{Y}) \leqq H(\bar{X}, y) \text{ for all } y \in [0, 1].$$

Therefore the mixed strategies $\bar{X}$ and $\bar{Y}$, which are optimal in $\bar{\Gamma}$, become, after having been extended by zero to mixed strategies in Γ, optimal in the latter game.

If, finally, X is any optimal strategy of player 1 in Γ, the set of points of its spectrum must, by the lemma of 9.2, coincide with the spectrum of Y, i.e., it is a subset of $[x_0, 1]$. This is also valid for points of the spectrum of every strategy Y of player 2. □

By the lemma, we may confine our attention to the case when the double equation in (9.6) can be satisfied only when $x_0 = 0$, i.e., when

$$L(x, x) > M(x, x) \text{ for } x \in (0, 1]. \tag{9.7}$$

9.7 Lemma. *In the game Γ the only points that can be discontinuities of the distribution function of the optimal strategies of the players are the points 0 and 1.*

Proof. Select an optimal strategy X of player 1, and suppose that some $x_0 \in (0,1)$ is a discontinuity of F_X.

By the assumption made at the end of the preceding subsection, we have $L(x_0,x_0) > M(x_0,x_0)$. This means that either $L(x_0,x_0) > \Phi(x_0)$ or $\Phi(x_0) > M(x_0,x_0)$. Suppose that the first inequality is satisfied.

By the lemma in 9.5, the entire interval $[x_0,1]$ must belong to the spectrum of X. Construct on $(x_0,1)$ a sequence $x_1, x_2, \ldots$ of discontinuities of F_X that converges to x_0. At these points the function $H(X,y)$ of y is continuous, and therefore, by complementary slackness, we must have $H(X,x_n) = v(\Gamma)$.

On the other hand, we have

$$H(X,x_0) = \int_0^{x_0-0} L(x,x_0)dX(x) + \Phi(x_0)dX(x_0)$$
$$+ \int_{x_0+0}^{x_n} M(x,x_0)dX(x) + \int_{x_n}^{1} M(x,x_0)dX(x)$$

and

$$H(X,x_n) = \int_0^{x_0-0} L(x,x_n)dX(x) + L(x_0,x_n)dX(x_0)$$
$$+ \int_{x_0+0}^{x_n} M(x,x_n)dX(x) + \int_{x_n}^{1} M(x,x_n)dX(x),$$

so that

$$H(X,x_0) - H(X,x_n)$$
$$= \int_{x_n}^{x_0-0} (L(x,x_0) - L(x,x_n))dX(x) + (L(x_0,x_n) - \Phi(x_0))dX(x_0)$$
$$- \int_{x_0}^{x_n} (L(x,x_n) - M(x,x_0))dX(x) + \int_{x_n}^{1} (M(x,x_0) - M(x,x_n))dX(x).$$

In the last equation, the first and last integrals on the right tend to zero with increasing n because L and M are uniformly continuous. The second integral tends to zero since

$$\lim_{n\to\infty} F_X(x_n) = F_X(x_0+0).$$

Finally, the term that is outside of the integrals has the limit

$$(L(x_0,x_0) - \Phi(x_0))dX(x_0),$$

which is not zero. This, however, is in contradiction with the relation

$$H(X, x_0) = v_\Gamma.$$

Now let

$$L(x_0, x_0) = \Phi(x_0) > M(x_0, x_0).$$

If there is a nonzero point of the spectrum to the left of x_0, then if we take a sequence, converging to x_0 from the left, of points $x_1, x_2, \ldots$ of continuity of F_X, we obtain, by reasoning as before,

$$0 = \lim((X, x_0) - H(X, x_n)) = (\Phi(x_0) - M(x_0, x_0))dX(x_0) > 0,$$

i.e., a contradiction.

If, however, there are no points of the spectrum to the left of x_0, let us take $y < x_0$ and write

$$H(X, x_0) = L(0, x_0)dX(0) + \Phi(x_0)dX(x_0) + \int_{x_0+0}^{1} M(x, x_0)dX(x),$$

$$H(X, y) = L(0, y)dX(0) + M(x_0, y)dX(x_0) + \int_{x_0+0}^{1} M(x, y)dX(x),$$

from which

$$\begin{aligned} H(X, x_0) &- H(X, y) \\ &= (L(0, x_0) - L(0, y))\,dX(0) + (\Phi(x_0) - M(x_0, y))\,dX(x_0) \\ &+ \int_{x_0+0}^{1} (M(x, x_0) - M(x, y))\,dX(x). \end{aligned}$$

Since L and M are uniformly continuous, it follows that as y approaches x_0 the first and third terms become arbitrarily small, and the second is close to $(\Phi(x_0) - M(x_0, x_0))dX(x_0)$.

If, moreover, $dX(x_0) > 0$, this number is but again positive by hypothesis. But this means that for y close to x we have, on the left, $H(X, x_0) = v_\Gamma > H(X, y)$, i.e., player 2 has a pure strategy y which does not let player 1 obtain the value of the game, even by the choice of the best strategy. Therefore $dX(x_0) = 0$. □

9.8. It turns out that the players' optimal strategies that are found in games of the class considered here are actually optimal for games of a more extensive class that is obtained by replacing (9.2) by the less restrictive hypothesis

$$\min\{L(0,0), M(0,0)\} \leqq \Phi(0) \leqq \max\{L(0,0), M(0,0)\}, \tag{9.8}$$

$$M(1,1) \leqq \Phi(1) \leqq L(1,1). \tag{9.9}$$

In fact, let $\tilde{\Gamma}$ be a game of the more general type with $\tilde{\Phi}$ in place of Φ. Let us now denote the payoff function by $\tilde{H}$. From $\tilde{\Gamma}$ we construct a game Γ by replacing the function $\tilde{\Phi}$ by a function Φ for which

$$\Phi(0) = \tilde{\Phi}(0), \quad \Phi(1) = \tilde{\Phi}(1),$$

where the values $\Phi(x)$ for $x \in (0,1)$ satisfy (9.2) and are otherwise arbitrary.

Let X and Y be equilibrium situations in Γ. This means that

$$H(x,Y) \leqq H(X,Y) \leqq H(X,y) \text{ for all } x \text{ and } y \in [0,1]. \tag{9.10}$$

By what was proved in the preceding subsection, the functions F_X and F_Y have no jumps on $(0,1)$. Therefore it follows immediately that

$$\tilde{H}(x,Y) = H(x,Y),\ \tilde{H}(X,Y) = H(X,Y),\ \tilde{H}(X,y) = H(X,y),$$

whence (9.10) can be rewritten in the form

$$\tilde{H}(x,Y) \leqq \tilde{H}(X,Y) \leqq \tilde{H}(X,y) \text{ for all } x,y \in [0,1],$$

i.e., (X,Y) is an equilibrium in $\tilde{\Gamma}$.

In particular, it follows from this that there are equilibrium situations (in the mixed strategies) for games of the form (9.1) that satisfy only (9.8) and (9.9), but not (9.2). Thus, by what we said in 9.6, hypothesis (9.9) can be replaced by the more general condition

$$\min\{L(1,1), M(1,1)\} \leqq \Phi(1) \leqq \max\{L(1,1), M(1,1)\}.$$

9.9. Differentiability of semikernels. Let us now impose further restrictions on the semikernels L and M, by supposing that the partial derivatives

$$\frac{\partial L}{\partial x}, \frac{\partial L}{\partial y}, \frac{\partial^2 L}{\partial x \partial y}, \frac{\partial M}{\partial x}, \frac{\partial M}{\partial y}, \frac{\partial^2 M}{\partial x \partial y} \tag{9.11}$$

exist, and are integrable in an appropriate sense.

Lemma. *Under the hypothesis that the derivatives* (9.11) *exist and are integrable, the distribution functions of each optimal strategy of players* 1 *and* 2 *are differentiable except perhaps at* 0 *and* 1.

Proof. Let X be an optimal strategy of player 1, and let $a = \min \operatorname{supp} X$. Choose $y \in [a,1)$ and write

$$\begin{aligned} v_\Gamma = H(X,y) &= L(0,y)dX(0) + \int_a^y L(x,y)dX(x) \\ &+ \int_y^{1-0} M(x,y)dX(x) + M(1,y)dX(1), \end{aligned} \tag{9.12}$$

or, after integrating by parts,

$$v_\Gamma = L(0,y)dX(0) + L(y,y)F_X(y) - L(a,y)F_X(a)$$
$$+ \int_a^y \frac{\partial L(x,y)}{\partial x} F_X(x)dx + M(1,y)F_X(1-0) - M(y,y)F_X(y)$$
$$+ \int_y^1 \frac{\partial M(x,y)}{\partial x} F_X(x)dx + M(1,y)dX(1).$$

Letting $dX(0) = \alpha$, $dX(1) = \beta$, and noticing that $F_X(a) = \alpha$ and $F_X(1-0) = 1-\beta$, and that, by (9.7), $L(y,y) \neq M(y,y)$, we may write

$$F_X(y) = (L(y,y) - M(y,y))^{-1} \Bigg(v_\Gamma - \alpha L(0,y) - \alpha L(a,y)$$
$$- M(1,y)(1-\beta) + \int_a^y \frac{\partial L(x,y)}{\partial x} F_X(x)dx + \int_y^1 \frac{\partial M(x,y)}{\partial x} F_X(x)dx \Bigg).$$

By our hypotheses, the function of y on the right is differentiable. Consequently, F_X is differentiable for $y \in (a,1)$. If $y \in (0,a)$, it is trivial that $\frac{\partial}{\partial y} F_X(y) = 0$.

In a similar way, the distribution function F_Y of the optimal strategy of player 2 is differentiable on $(0,1)$. □

9.10. Let us, as usual, denote the density of the optimal strategy X by f_X and rewrite (9.12) in the form

$$v_\Gamma = \alpha L(0,y) + \int_a^y L(x,y)f_X(x)dx + \int_y^1 M(x,y)f_X(x)dx + \beta M(1,y). \quad (9.13)$$

Differentiation with respect to y yields

$$0 = \alpha \frac{\partial L(0,y)}{\partial y} + L(y,y)f_X(y) + \int_a^y \frac{\partial L(x,y)}{\partial y} f_X(x)dx - M(y,y)f_X(y)$$
$$+ \int_y^1 \frac{\partial M(x,y)}{\partial y} f_X(x)dx + \beta \frac{\partial M(1,y)}{\partial y},$$

which, after the obvious transformations, assumes the form

$$
\begin{aligned}
f_X(y) = (-L(y,y) + M(y,y))^{-1} \Bigg(& \alpha \frac{\partial}{\partial y} L(0,y) + \beta \frac{\partial}{\partial y} M(1,y) \\
& + \int_a^y \frac{\partial L(x,y)}{\partial y} f_X(x)dx + \int_y^1 \frac{\partial M(x,y)}{\partial y} f_X(x)dx \Bigg) .
\end{aligned}
\tag{9.14}
$$

We can, for an optimal strategy Y of player 2 with jumps γ at 0 and δ at 1, and for any pure strategy $x \in [a, 1)$ of player 1, write (as for the cases (9.13) and (9.14))

$$
\begin{aligned}
v_\Gamma = H(x, Y) = {} & \gamma M(x,0) + \delta L(x,1) \\
& + \int_a^x M(x,y) f_Y(y)dy + \int_x^1 L(x,y) f_Y(y)dy,
\end{aligned}
$$

and obtain, after differentiation and appropriate transformations,

$$
\begin{aligned}
f_Y(x) = (L(x,x) - M(x,x))^{-1} \Bigg(& \gamma \frac{\partial}{\partial x} M(x,0) + \delta \frac{\partial}{\partial x} L(x,1) \\
& + \int_a^x \frac{\partial M(x,y)}{\partial x} f_Y(y)dy + \int_x^1 \frac{\partial L(x,y)}{\partial y} f_Y(y)dy \Bigg) .
\end{aligned}
\tag{9.15}
$$

9.11. We can summarize as follows.

Under the preceding hypotheses, the distribution functions for the players' optimal strategies in a game of timing can only consist of jumps at the point 0 and 1, together with a density on $[a, 1)$ that satisfies the integral equations (9.14) or (9.15).

The functions f_X and f_Y are defined and nonnegative on $[a, 1)$. Together with the still unknown, but nonnegative, parameters α, β, γ, δ, and a, they satisfy the probabilistic normalizations

$$
\alpha + \int_a^1 f_X(x)dx + \beta = 1, \tag{9.16}
$$

$$
\gamma + \int_a^1 f_Y(x)dx + \delta = 1. \tag{9.17}
$$

In addition, the strategy a appears in the spectrum of each of the strategies X and Y; also the distribution functions F_X and F_Y are continuous at the point a, by

9.7, so that the functions $H(X,y)$ and $H(x,Y)$ are also continuous at a. Thus, by complementary slackness, we must have

$$H(X,a) = v_\Gamma \text{ and } H(a,Y) = v_\Gamma. \tag{9.18}$$

It is sometimes convenient to write these conditions as the single equation $H(X,a) = H(a,Y)$.

The discussion that we have just carried out for the strategy a is equally applicable to any strategy $z \in [a,1)$, and we have, for each such strategy,

$$H(X,z) = H(z,Y) = v_\Gamma. \tag{9.19}$$

It is sometimes convenient to apply a corollary of these equations, namely

$$\frac{\partial H(X,y)}{\partial y} \equiv 0 \text{ and } \frac{\partial H(x,Y)}{\partial x} \equiv 0. \tag{9.20}$$

The situation at the point 1 is less trivial. If F_Y has a jump at 1, then, again by complementary slackness, we must have

$$H(X,1) = v_\Gamma. \tag{9.21}$$

However, if F_Y is continuous at 1, this equation is a corollary of the continuity of $H(X,\cdot)$. A similar discussion applies to the equation

$$H(1,Y) = v_\Gamma. \tag{9.22}$$

Further, once more by complementary slackness, we have the following alternatives:

$$\alpha = 0 \text{ or } H(0,Y) = v_\Gamma, \tag{9.23}$$

$$\gamma = 0 \text{ or } H(X,0) = v_\Gamma. \tag{9.24}$$

If, finally, $a = 0$, and one of the distribution functions, say F_X, has a jump at the point a, then (again, by the theorem of 1.18) only the second of the conditions (9.18) remains and the first condition is replaced by the condition $a = 0$. In the same way, if $a = 0$ and F_Y has a jump at $a = 0$, the second condition is replaced by $a = 0$.

Any of the conditions (9.16)–(9.24) can be used to determine unknown parameters of the optimal strategies in the game.

9.12. The solution of Γ is substantially simplified if the game is symmetric, i.e. if

$$L(x,y) = -M(y,x) \text{ and } \Phi(x) = 0 \text{ for all } x \text{ and } y. \tag{9.25}$$

In this case, equations (9.14) and (9.15) coincide, and can be written in the single form

$$f_Z(z) = \frac{1}{-2L(z,z)}\left(\alpha\frac{\partial L(0,z)}{\partial z} - \beta\frac{\partial L(z,1)}{\partial z} + \int_a^z \frac{\partial L(x,z)}{\partial z} f_Z(x)dx - \int_z^1 \frac{\partial L(z,x)}{\partial z} f_Z(x)dx\right). \tag{9.26}$$

Naturally, in this case one can take $\alpha = \gamma$ and $\beta = \delta$. In addition, the solution of the problem is essentially facilitated by the fact that in the symmetric case v_Γ must be 0, so that the value of the game is no longer unknown and does not require any of the equations in the system.

9.13. The integral equations (9.14) and (9.15) can sometimes be replaced by systems of differential equations. In some special cases this provides a simple and rapid solution.

We have one of these cases when L and M are separable:

$$L(x,y) = \sum_{i=1}^{l} u_i(x)v_i(y),$$

$$M(x,y) = \sum_{j=1}^{m} s_j(x)t_j(y).$$

Then the integral equations (9.14) and (9.15) assume the form

$$f_X(y) = \frac{1}{M(y,y) - L(y,y)}\left(\sum_{i=1}^{l} v_i'(y)\left(\alpha u_i(0) + \int_a^y u_i(x)f_X(x)dx\right)\right.$$
$$\left. + \sum_{j=1}^{m} t_j'(y)\left(\beta s_j(1) + \int_y^1 s_j(x)f_X(x)dx\right)\right), \tag{9.27}$$

$$f_Y(x) = \frac{1}{L(x,x) - M(x,x)}\left(\sum_{j=1}^{m} s_j'(x)\left(\gamma t_j(0) + \int_a^x t_j(y)f_Y(y)dy\right)\right.$$
$$\left. + \sum_{i=1}^{l} u_i'(x)\left(\delta v_i(1) + \int_x^1 v_i(y)f_Y(y)dy\right)\right). \tag{9.28}$$

Now set

$$\int_z^1 u_i(x)f_Y(x)dx = U_i(z), \quad \int_z^1 v_i(y)f_Y(y)dy = V_i(z),$$

$$\int_z^1 s_j(x)f_X(x)dx = S_i(z), \quad \int_z^1 t_j(y)f_Y(y)dy = T_j(z),$$

from which, by differentation,

$$-u_i(z)f_X(z) = U_i'(z), \quad -v_i(z)f_Y(z) = V_i'(z),$$
$$-s_j(z)f_X(z) = S_j'(z), \quad -t_j(z)f_Y(z) = T_j'(z).$$

In addition, it is clear that

$$U_i(1) = 0 \text{ for } i = 1,\dots,l; \quad S_j(1) = 0 \text{ for } j = 1,\dots,m, \tag{9.29}$$

$$V_i(1) = 0 \text{ for } i = 1,\dots,l; \quad T_j(1) = 0 \text{ for } j = 1,\dots,m. \tag{9.30}$$

With this notation, (9.27) and (9.28) can be written as follows:

$$f_X(y) = \frac{1}{M(y,y) - L(y,y)}\left(\sum_{i=1}^{l} v_i'(y)\left(\alpha u_i(0) + U_i(a) - U_i(y)\right)\right.$$
$$\left. + \sum_{j=1}^{m} t_j'(y)\left(\beta s_j(1) + S_j(y)\right)\right),$$

$$f_Y(x) = \frac{1}{L(x,x) - M(y,y)}\left(\sum_{j=1}^{m} s_j'(x)\left(\gamma t_j(0) + T_j(a) - T_j(x)\right)\right.$$
$$\left. + \sum_{i=1}^{l} u_j'(x)\left(\delta v_i(1) + V_i(x)\right)\right).$$

If we multiply the first equation by $u_i(y)$ and $s_j(y)$, and the second by $t_j(x)$ and $v_j(x)$, we obtain

$$-U_i'(y) = \frac{u_i(y)}{M(y,y) - L(y,y)}\left(\sum_{i=1}^{l} v_i'(y)\left(\alpha u_i(0) + U_i(a) - U_i(y)\right)\right.$$
$$\left. + \sum_{j=1}^{m} t_j'(y)\left(\beta s_j(1) + S_j(y)\right)\right) \quad \text{for } i = 1,\dots,l; \tag{9.31}$$

$$-S_j'(y) = \frac{s_j(y)}{M(y,y) - L(y,y)}\left(\sum_{i=1}^{l} v_i'(y)\left(\alpha u_i(0) + U_i(a) - U_i(y)\right)\right.$$
$$\left. + \sum_{j=1}^{m} t_j'(y)\left(\beta s_j(1) + S_j(y)\right)\right) \quad \text{for } j = 1,\dots,m; \tag{9.32}$$

$$-T_j'(x) = \frac{t_j(x)}{L(x,x) - M(x,x)}\left(\sum_{j=1}^{m} s_j'(x)\left(\gamma t_j(0) + T_j(a) - T_j(x)\right)\right.$$
$$\left. + \sum_{i=1}^{l} u_j'(x)\left(\delta v_i(1) + V_i(x)\right)\right) \quad \text{for } j = 1,\dots,m; \tag{9.33}$$

$$-V_i'(x) = \frac{v_i(x)}{L(x,x) - M(x,x)} \left(\sum_{j=1}^{m} s_j'(x) \left(\gamma t_j(0) + T_j(a) - T_j(x) \right) \right.$$
$$\left. + \sum_{i=1}^{l} u_i'(x) \left(\delta v_i(1) + V_i(x) \right) \right) \quad \text{for } i = 1, \ldots, l. \tag{9.34}$$

Equations (9.31) and (9.32) together form a system of linear differential equations for the functions $U_i(a) - U_i(y)$ and $S_j(y)$.

Equations (9.29) are not initial conditions for the system (9.31), (9.32). However, since at this point we know that the system has a solution under these conditions, we can proceed rather pragmatically. Here it is enough to determine at least one of the unknown functions. For example, having found $S_j(y)$, we can obtain

$$f_X(y) = -\frac{S_j'(y)}{s_j(y)}. \tag{9.35}$$

We can say the same for solutions of the system (9.33), (9.34), under the conditions (9.30).

Finding the parameters α, β, γ, δ, and a requires some sort of special reasoning.

9.14 A general example. Let us consider a noisy duel as an example. Let the accuracy function of player 1, in accordance with what we said at the end of 9.2, be x, let the accuracy function of player 2 have the simplest form, $P_2(y) = y$, and let

$$h = \begin{bmatrix} 1 & 1 \\ 0 & 0 \end{bmatrix}. \tag{9.36}$$

This matrix corresponds to the state of affairs when player 1 attempts only to defeat his opponent, irrespective of his own survival. In this case, as we can easily calculate directly, the payoff function is

$$H(x,y) = \begin{cases} x, & \text{if } x \leqq y, \\ x(1-y), & \text{if } x > y. \end{cases} \tag{9.37}$$

Strictly speaking, this game does not exactly fall into our class of games of timing: the function $L(x,y)$, i.e. x, is not a strictly decreasing function of y. Nevertheless, it is only natural to use the theory discussed above as an example of a heuristic solution of a game. Of course, the situation we have obtained needs to be checked to verify that it is an equilibrium.

Here we may take

$$u(x) = x, \quad v(y) = 1,$$
$$s(x) = x, \quad t(y) = 1 - y.$$

In our case we have

$$M(y,y) - L(y,y) = s(y)t(y) - u(y)v(y) = y(1-y) - y = -y^2,$$

and substitution into formulas (9.31)–(9.34) yields

$$-U'(y) = -\frac{y}{y^2}(-1)(\beta + S(y)), \tag{9.38}$$

$$-S'(y) = -\frac{y}{y^2}(-1)(\beta + S(y)), \tag{9.39}$$

$$-T'(x) = \frac{1-x}{x^2}(\gamma + T(a) - T(x) + \delta + V(x)), \tag{9.40}$$

$$-V'(x) = \frac{1}{x^2}(\gamma + T(a) - T(x) + \delta + V(x)). \tag{9.41}$$

It follows from (9.39) that

$$-S'(y) = \frac{1}{y}(\beta + S(y)),$$

or, after solving this differential equation,

$$S(y) + \beta = \frac{c_1}{y}. \tag{9.42}$$

Setting $y = 1$ here, we find from (9.39) that $c_1 = \beta$, and if we differentiate and use (9.42), that

$$S'(y) = -\frac{\beta}{y^2},$$

i.e., by (9.35),

$$f_X(x) = -\frac{S'(y)}{s(y)} = \frac{\beta}{y^3}. \tag{9.43}$$

In addition, we can rewrite (9.41) as

$$-x^2V'(x) = \gamma + T(a) - T(x) + \delta + V(x). \tag{9.44}$$

The right-hand side is a differentiable function of x. Consequently V is twice differentiable, and we have

$$-2xV'(x) - x^2V''(x) = -T'(x) + V'(x),$$

or, after using (9.40) and (9.41),

$$-2xV'(x) - x^2V''(x) = xV'(x),$$

from which we naturally obtain

$$V'(x) = \frac{c_2}{x^2}, \tag{9.45}$$

so that

$$f_Y(x) = -\frac{V'(x)}{v(x)} = -\frac{c_2}{x^3}. \tag{9.46}$$

It remains to determine the parameters c_2, β, γ, δ, and a.

We begin with the analysis of alternative (9.23). Let us suppose that $\alpha > 0$. Then we must have $H(0, Y) = v_\Gamma$. But from condition (9.37) in this game we see that $H(0, Y) = 0$, so that we must have $v_\Gamma = 0$. However, on the other hand, if we take the strategy of player 1 to be a uniform distribution X on $[0, 1]$, we evidently obtain

$$H(\bar{X}, y) = \int_0^y L(x, y)dx + \int_y^1 M(x, y)dx = \int_0^y x dx + \int_y^1 x(1-y)dx$$

$$= \frac{y^2}{2} + (1-y)\frac{1}{2}(1-y^2) = \frac{1}{2}(1-y-y^2).$$

By a simple calculation,

$$\min_{y \in [0,1]} H(\bar{X}, y) = H\left(\bar{X}, \frac{\sqrt{3}}{3}\right) = \frac{1}{2}\left(1 - \frac{\sqrt{3}}{3} + \frac{\sqrt{3}}{9}\right) > 0,$$

so that

$$v_\Gamma \geqq \max_X \min_y H(X, y) \geqq \min_y H(\bar{X}, y) > 0,$$

and we have a contradiction. Consequently $\alpha = 0$.

In particular, it follows from what has been established that the normalization condition (9.16), after taking account of (9.43), assumes the form

$$\beta \int_a^1 \frac{dx}{x^3} + \beta = \frac{\beta}{2}\left(\frac{1}{a^2} + 1\right) = 1. \tag{9.47}$$

It follows easily from this that $\beta > 0$.

If we now assume that $\delta > 0$, then, by what we said in 9.11, we must have $H(X, 1) = v_\Gamma$, and hence $H(X, 1) = H(X, a)$, which in the present case means that

$$\beta \int_a^1 \frac{dx}{x^2} + \beta = \beta(1-a) \int_a^1 \frac{dx}{x^2} + \beta(1-a).$$

But since $\beta > 0$, this is possible only if $a = 0$, and then the integral on the left of (9.47) diverges, so that (9.47) is impossible. Therefore $\delta = 0$.

If we take account of this, the equation $H(a, Y) = v_\Gamma$ yields

$$\gamma a - \frac{c_2 a}{2}\left(\frac{1}{a^2} - 1\right) = v_\Gamma,$$

and then by the normalization condition for Y,

$$\gamma - \frac{c_2}{2}\left(\frac{1}{a^2} - 1\right) = 1. \tag{9.48}$$

By comparing these two equations, we find that $v_\Gamma = a$. Consequently the equation $H(X, a) = v_\Gamma$ can be written as

$$\frac{\beta(1-a)}{a} = a. \tag{9.49}$$

Comparing the value of β, obtained from this, with the value from (9.47), we obtain, after simplifying,

$$a^2 + 2a - 1 = 0, \tag{9.50}$$

from which (taking the positive root) $a = \sqrt{2} - 1$. Therefore, by (9.49), $\beta = a^2/(1-a) = (2-\sqrt{2})/2$. Since a is a root of the equation (9.50), we can rewrite (9.48) as

$$\gamma - \frac{c_2}{a} = 1, \tag{9.51}$$

and since $\beta > 0$ we have $H(1, Y) = v_\Gamma$, which now assumes the form $\gamma - c_2 = v_\Gamma = a$. From this and (9.51) we obtain $\gamma = 0$ and $c_2 = -a = 1 - \sqrt{2}$.

Therefore we finally obtain

$$F_X(x) = \begin{cases} 0, & \text{if } x \leqq \sqrt{2}-1, \\ \frac{2-\sqrt{2}}{2} \int\limits_{\sqrt{2}-1}^{x} \frac{dx}{x^3} = \frac{2-\sqrt{2}}{4}\left((1+\sqrt{2})^2 - \frac{1}{x^2}\right), & \text{if } \sqrt{2}-1 \leqq x \leqq 1, \\ 1, & \text{if } x = 1+0, \end{cases}$$

$$F_Y(y) = \begin{cases} 0, & \text{if } y \leqq \sqrt{2}-1, \\ \frac{\sqrt{2}-1}{2}\left((1+\sqrt{2})^2 - \frac{1}{y^2}\right), & \text{if } \sqrt{2}-1 \leqq y \leqq 1+0. \end{cases}$$

Direct verification shows that

$$H(x, Y) \leqq a \leqq H(X, y) \text{ for all } x \text{ and } y \in [0, 1],$$

from which it follows that X and Y actually form an equilibrium in the game, and that a is its value. Therefore $v_\Gamma = \sqrt{2} - 1 = 0.414$.

9.15 Special classes of games of timing. It often happens in the solution of individual special classes of separable games of timing (including those of duel type) that it is sometimes more convenient not to apply the general theory presented in 9.13, but to apply some special technique.

As an example, let us consider the class of duels with arbitrary accuracy functions P_1 and P_2 and the matrix

$$h = \begin{bmatrix} 0 & 1 \\ -1 & 0 \end{bmatrix} \tag{9.52}$$

(i.e., player 1 considers hitting his antagonist and his own survival as equally successful). In the interest of retaining symmetry of notation in the general discussion, we shall omit the reduction $P_1(x) = x$ that was shown at the end of 9.2 to be possible.

As is easily calculated, in this case we are dealing with a game in which the payoff function is

$$H(x,y) = \begin{cases} P_1(x) - P_2(y) + P_1(x)P_2(y) & \text{for } x < y, \\ P_1(x) - P_2(x) & \text{for } x = y, \\ P_1(x) - P_2(y) - P_1(x)P_2(y) & \text{for } x > y. \end{cases} \tag{9.53}$$

On the basis of what we said above, the solution of this game Γ is the pair (X, Y) of strategies consisting of the densities f_X and f_Y on an interval $[a, 1]$ and (possibly) jumps at 0 and 1, of amounts α and β for X, and γ and δ for Y.

In this case the integral equation (9.14), after some obvious transformations (including those that used (9.16)), can be written in the form

$$f_X(y) = \frac{P_2'(y)}{2P_1(y)P_2(y)}\left(1 - \int_a^y P_1(x)f_X(x)dx + \int_y^1 P_1(x)f_X(x)dx\right). \tag{9.54}$$

With the abbreviation $P_1(x)f_X(x) = h(x)$, we obtain the equation

$$2h(y) = \frac{P_2'(y)}{P_2(y)}\left(1 - \int_a^y h(x)dx + \int_y^1 h(x)dx\right), \tag{9.55}$$

or

$$-2h(y)\left(1 - \int_a^y h(x)dx + \int_y^1 h(x)dx\right)^{-1} = -\frac{P_2'(y)}{P_2(y)}.$$

On the left (as well as on the right), the numerator is the derivative of the denominator. Consequently, if we integrate and then exponentiate, we obtain

$$1 - \int_a^y h(x)dx + \int_y^1 h(x)dx = \frac{c_1}{P_2(y)},$$

where c_1 is a constant of integration. Differentiation and renaming a variable yield

$$2h(x) = 2P_1(x)f_X(x) = \frac{c_1 P_2'(x)}{(P_2(x))^2}, \tag{9.56}$$

from which

$$f_X(x) = \frac{c_1 P_2'(x)}{2P_1(x)(P_2(x))^2}. \tag{9.57}$$

Similarly, in this case (9.15) accuires the form

$$f_Y(x) = \frac{P_1'(x)}{2P_2(x)P_1(x)}\left(1 - \int_a^x P_1(y)f_Y(y)dy + \int_x^1 P_1(y)f_Y(y)dy\right), \tag{9.58}$$

from which

$$f_Y(y) = \frac{c_2 P_1'(y)}{2P_2(y)(P_1(y))^2}. \tag{9.59}$$

For a complete determination of the optimal strategies X and Y, it remains to calculate the pure strategy a, the constants c_1 and c_2, the amounts of the jumps α and β of F_X at 0 and 1, and the jumps γ and δ of F_Y at the same points. Besides this, we have to determine the value v_Γ of the game.

9.16. We use the following considerations to determine these eight parameters.

We first suppose that neither of the spectra of X and Y contains points other than 0 and 1. In that case we consider a $\{0,1\} \times \{0,1\}$-subgame of Γ. This a 2×2-matrix game, whose matrix (by (9.53)) is

$$\begin{pmatrix} 0 & -1 \\ 1 & 0 \end{pmatrix}.$$

The situation $(1,1)$ (lower right-hand corner) is a saddle point. Therefore (for example, by the theorem on the independence of irrelevant alternatives in 2.18 of Chapter 1) this situation is also a saddle point in the whole game Γ. But then we must have $H(x,1) \leqq H(1,1)$, i.e., by (9.53), $2P_1(x) - 1 \leqq 0$, which is impossible since $P_1(x) > \frac{1}{2}$.

This contradiction shows that X and Y must have points of their spectra in the open interval $(0,1)$. This means, in the first place, that $a < 1$; and, in the second place, that $c_1 \neq 0$ and $c_2 \neq 0$.

Since $a < 1$, it follows that $(a,1) \neq \emptyset$. But every pure strategy $y \in (a,1)$ of player 2 is a point of that player's optimal strategy Y. Therefore, by 1.18,

$$H(X,y) = v_\Gamma \text{ for every } y \in (a,1). \tag{9.60}$$

For the same reason we have

$$H(x,Y) = v_\Gamma \text{ for every } x \in (a,1). \tag{9.61}$$

The strategy $y = 1$ is a point of the spectrum of the optimal strategy Y, but $H(X,y)$ may have a discontinuity at $y = 1 - 0$. Therefore, by what we said in 1.18, there is an alternative:

$$\delta = 0 \text{ or } H(X,1) = v_\Gamma. \tag{9.62}$$

Similarly we have the alternative:

$$\beta = 0 \text{ or } H(1,Y) = v_\Gamma. \tag{9.63}$$

Finally, we have the normalizing equations (9.16) and (9.17).

9.17. Let us analyze the relations obtained so far.

Equation (9.60) can be written in the form

$$\alpha H(0,y)+\int_a^1 H(x,y)f_X(x)dx+\beta H(1,y)=v_\Gamma,$$

or, if we use (9.53),

$$-\alpha P_2(y)+\int_a^y (P_1(x)-P_2(y)+P_1(x)P_2(y))f_X(x)dx$$

$$+\int_y^1 (P_1(x)-P_2(y)-P_1(x)P_2(y))f_X(x)dx+\beta(1-2P_2(y))=v_\Gamma.$$

If we group some of the terms and replace the density f_X in the others by the formula in (9.57), we obtain

$$-P_2(y)\left(\alpha+\int_a^1 f_X(x)dx+\beta\right)+\frac{c_1}{2}(1+P_2(y))\int_a^y\frac{P_2'(x)dx}{(P_2(x))^2}$$

$$+\frac{c_1}{2}(1-P_2(y))\int_y^1\frac{P_2'(x)dx}{(P_2(x))^2}+\beta(1-P_2(y))=v_\Gamma.$$

If we use the normalizing condition (9.16) and calculate the integrals, we obtain

$$-P_2(y)+\frac{c_1}{2}(1+P_2(y))\left(\frac{1}{P_2(a)}-\frac{1}{P_2(y)}\right)$$

$$+\frac{c_1}{2}(1-P_2(y))\left(\frac{1}{P_2(y)}-1\right)+\beta(1-P_2(y))=v_\Gamma.$$

or, collecting terms,

$$P_2(y)\left(-1+\frac{c_1}{2P_2(a)}+\frac{c_1}{2}-\beta\right)$$

$$+\frac{1}{P_2(y)}\left(-\frac{c_1}{2}+\frac{c_1}{2}\right)+\left(\frac{c_1}{2P_2(a)}-\frac{3c_1}{2}+\beta\right)=v_\Gamma.$$

Since this equation is an identity for $y\in(a,1)$, the coefficient of $P_2(y)$ must vanish:

$$1+\beta=\frac{c_1}{2}\left(1+\frac{1}{P_2(a)}\right) \tag{9.64}$$

(the coefficient of $1/P_2(y)$ is identically zero); in addition,

$$\beta + \frac{c_1}{2}\left(\frac{1}{P_2(a)} - 3\right) = v_\Gamma. \tag{9.65}$$

A parallel argument, applied to (9.61), yields

$$1 + \delta = \frac{c_2}{2}\left(1 + \frac{1}{P_1(a)}\right) \tag{9.66}$$

and

$$\delta + \frac{c_2}{2}\left(\frac{1}{P_1(a)} - 3\right) = -v_\Gamma. \tag{9.67}$$

9.18 Lemma. $\alpha = \gamma = 0$.

Proof. We have

$$H(X,0) = \alpha H(0,0) + \int_a^1 H(x,0) f_X(x)dx + \beta H(1,0)$$

$$= \int_a^1 P_1(x) f_X(x)dx + \beta = \frac{c_1}{2}\left(\frac{1}{P_1(a)} - 1\right) + \beta.$$

Together with (9.65) and the fact that $c_1 = 0$, this yields $H(X,0) > v_\Gamma$, so that $\alpha = 0$.

The equation $\gamma = 0$ is obtained similarly by calculating $H(0,Y)$ and using (9.67). □

9.19 Lemma. $\beta\delta = 0$.

Proof. Suppose that $\beta > 0$ and $\delta > 0$. Then both (9.62) and (9.63) are satisfied.

The first of these can be rewritten in the form

$$H(X,1) = \int_a^1 H(x,1) f_X(x)dx + \beta H(1,1) = v_\Gamma,$$

or, by (9.53),

$$\int_a^1 (2P_1(x) - 1) f_X(x)dx = -\int_a^1 f_X(x)dx + 2\frac{c_1}{2}\int_a^1 \frac{P_2'(x)dx}{(P_2(x))^2} = v_\Gamma,$$

whence

$$-1 + \beta + c_1\left(\frac{1}{P_2(a)} - 1\right) = v_\Gamma, \tag{9.68}$$

and symmetrically

$$-1+\delta+c_2\left(\frac{1}{P_1(a)}-1\right)=-v_\Gamma. \tag{9.69}$$

Subtracting (9.68) from (9.65) gives us

$$\frac{c_1}{2}\left(1+\frac{1}{P_2(a)}\right)=1, \tag{9.70}$$

which, together with (9.64), leads to $\beta=0$.

Similarly, subtracting (9.69) from (9.67) yields

$$\frac{c_2}{2}\left(1+\frac{1}{P_1(a)}\right)=1, \tag{9.71}$$

so that we obtain $\delta=0$ from (9.66). □

9.20. From what we have said, the normalizing conditions (9.16) and (9.17) become the form

$$\frac{c_1}{2}\int\limits_a^1\frac{P_2'(x)dx}{P_1(x)(P_2(x))^2}+\beta=1, \tag{9.72}$$

$$\frac{c_2}{2}\int\limits_a^1\frac{P_1'(y)dy}{P_2(y)(P_1(y))^2}+\delta=1, \tag{9.73}$$

and therefore at least one of the numbers β and δ (although for the time being, we do not know which) vanishes.

First suppose that $\beta=0$. Then because $c_1>0$, (9.72) and (9.64) yield

$$\int\limits_a^1\frac{P_2'(x)dx}{P_1(x)(P_2(x))^2}=1+\frac{1}{P_2(x)}. \tag{9.74}$$

This relation can be considered as if it were the characteristic equation for a game Γ with the payoff function (9.53). A second characteristic equation,

$$\int\limits_a^1\frac{P_1'(y)dy}{P_2(y)(P_1(y))^2}=1+\frac{1}{P_1(a)} \tag{9.75}$$

is obtained by using the alternative hypothesis that $\delta=0$. We recall that by 9.2 we can let $P_1(x)=x$ in (9.74) and (9.75). However, this does not usually lead to any simplification of the general discussion.

Taking a to be variable, we find that the ratio of the left-hand side of (9.74) to the right-hand side becomes infinite as $a\to 0$ (by L'Hospital's rule). Consequently, when a is sufficiently small the left-hand side is strictly larger than the right. At the same time, when $a \to 1-0$ the left-hand side of (9.74) tends to zero, whereas the right-hand side tends to 2. Consequently (9.74) can be solved for a. Since, in addition, the absolute value of the derivative of the left side is always greater than that of the right, the root of (9.74) is unique.

If we substitute the value of the root a of (9.74) into (9.72) (or into (9.64)), with $\beta = 0$, we obtain the value of c_1; then we find v_Γ from (9.65). Dropping the parameter δ from (9.66) and (9.67), we obtain

$$1 - v_\Gamma = c_2 \left(\frac{1}{P_1(a)} - 1 \right), \tag{9.76}$$

which gives us c_2. Finally, substituting c_2 and a into (9.73) gives us δ. If it turns out that $\delta \geqq 0$, we obtain the solution. However, if δ turns out to be negative, this means that the assumption that $\beta = 0$ leads to a contradiction. We must then make the alternative assumption $\delta = 0$ and turn to a similar discussion based on (9.75), which then necessarily leads to a conclusion, since there are no other possibilities.

9.21 Example: power accuracy function. Let $P_1(x) = x$ and $P_2(x) = x^k$. By (9.57) and (9.59) we will have

$$f_X(x) = \frac{kc_1}{2} x^{-(k+2)}, \tag{9.77}$$

$$f_Y(y) = \frac{c_2}{2} y^{-(k+2)}. \tag{9.78}$$

Here the characteristic equation (9.74) takes the form

$$\int_a^1 \frac{kx^{k-1}\,dx}{x^{2k+1}} = 1 + \frac{1}{a^k},$$

i.e.,

$$\frac{k}{k+1}\left(\frac{1}{a^{k+1}} - 1\right) = 1 + \frac{1}{a^k},$$

or

$$(2k+1)a^{k+1} + (k+1)a - k = 0. \tag{9.79}$$

Notice that the second characteristic equation (9.75) in this case has the form

$$(k+2)a^{k+1} + (k+1)a^k - 1 = 0.$$

9.22 Two special cases. Symmetric game. Let $k = 1$. Then the characteristic equation (9.77) becomes

$$3a^2 + 2a - 1 = 0,$$

and this equation has the single root $a = \frac{1}{3}$ on the interval $[0, 1]$, so that $P_1(a) = P_2 = a = \frac{1}{3}$.

In addition, we have successively

from (9.64) (setting $\beta = 0$) $c_1 = \frac{1}{2}$,

from (9.65) $v_\Gamma = 0$ (which is natural, by the symmetry of Γ),

from (9.75) $c_2 = \frac{1}{2}$,

from (9.73) $\delta = 0$.

Here the optimal strategies are the same for both players. Each of them consists only of the density $1/(4t^3)$ on the interval $[\frac{1}{3}, 1]$.

9.23 An example of an asymmetric game. Let $k = 2$. Then the characteristic equation (9.79) has the form

$$5a^3 + 3a - 2 = 0.$$

This equation has the root $a = 0.48109$. We also obtain:

from (9.64) (setting $\beta_1 = 0$)

$$c_1 = 2\left(1 + \frac{1}{a^2}\right)^{-1} = 0.375895,$$

from (9.65)

$$v_\Gamma = \frac{c_1}{2}\left(\frac{1}{a^2} - 3\right) = 0.248208,$$

from (9.75)

$$c_2 = (1 - v_\Gamma)\left(\frac{1}{a} - 1\right)^{-1} = 0.0697001,$$

and finally, from (9.73), which now takes the form

$$\frac{c_2}{2}\int_a^1 \frac{dx}{x^4} + \delta = 1,$$

we have

$$\delta = 1 - \frac{c_2}{6}\left(\frac{1}{a^3} - 1\right) = 0.072883.$$

Therefore, on the basis of (9.77) and (9.78), the optimal strategies X and Y of players 1 and 2 in this game have, on the interval $[0.481, 1)$, the respective densities

$$f_X(x) = 0.376x^{-4} \text{ and } f_Y(y) = 0.035y^{-4},$$

and in addition Y has the jump 0.073 at the point 1.

Notes and references for Chapter 3

§1. Two general surveys of work concerned with two-person zero-sum games were given by E.B. Yanovskaya [10,21]. The concept of a value of a two-person zero-sum game as the common value of the minimaxes of payoff functions and the realization of the optimality principle was introduced by J. von Neumann [1], and that of the ε-optimality of strategies, by A. Wald [1,2]. The elementary connections between ε-saddle points of a game and minimaxes of its payoff functions were described by N.N. Vorob'ev in [20]; and the case $\varepsilon = 0$, by him in [2]. The theorem of **1.15** on the equality of values for different mixed strategies in two-person zero-sum games was found by E.B. Yanovskaya [21].

General two-person zero-sum games with opposite reflexive preference relations of the players (instead of payoff functions), including two-person zero-sum games with vector payoff functions, was discussed by V.V. Podinovskiĭ [3].

A necessary condition for the realizability of a maximin was introduced by Yu.B. Germeĭer [1].

The book by V.F. Dem'yanov and V.N. Malozemov [1] is devoted to the problem of calculating a minimax. Yu.V. Germeĭer and V.V. Fedorov [1,2] investigated the possibility of calculating a maximin by the method of penalty functions.

Generalizations of concepts of a maximin in various directions were constructed by D. Blackwell [1], A.M. Rubinov [1], and also V.V. Podinovskiĭ and V.D. Nogin [1]. The connection between the maximin problem and convex analysis was discussed by R.T. Rockafellar [1].

The property of having a saddle point in the pure strategies is sometimes transferred from a family of subgames to a whole game. For this see the papers of L.S. Shapley [2], S. Karlin [3], V.P. Gradusov [1] and J. Kindler [1].

§2. The axiomatic foundation of the maximin principle in **2.2–2.4**, based on the concept of the players' possibilities, was given by N.N. Vorob'ev. The axiomatic foundation of the maximin principle described in **2.5–2.10**, based on the concept of preference, follows the paper [1] of J. Milnor; and the axiomatic foundation of the value of a two-person zero-sum game, as a result of reasonable behavior of the players (**2.11, 2.12**), on the paper [1] of E.Ĭ. Vilkas. A different version of the axiomatics that produces the value of a two-person zero-sum game was found by S.H. Tijs [1]. A discussion of the fundamental nature of the maximin principle is contained in the papers of Milnor [1] and G. Owen [1].

A sharpening of the maximin principle for two-person zero-sum games, guaranteeing the uniqueness of the optimal situation, is given in the papers of R.C. Buck [1] and S. Huyberechts [1].

An extension of the axiomatic approach to the definition of sets of equilibrium situations for arbitrary noncooperative games for two players is given in the paper by M.J.M. Jansen and S.H. Tijs [1]; and for noncooperative games for n players, by E.Ĭ. Vilkas [8].

§3. The $\mathcal{S}$-topology on spaces of strategies in two-person zero-sum games was suggested by Teh-Tjoe Tie [1], and the $\mathcal{J}$-topology, by A. Wald [2]. The existence of saddle points in the mixed strategies in games with continuous payoff functions on the unit square was established J. Ville [1].

The systematic discussion of convex two-person zero-sum games was initiated by H.F. Bohnenblust, S. Karlin and L.S. Shapley [1] (and also by H. Kneser [1]). They established the propositions of **3.4** and **3.5** for the case $\bar{y} \subset \mathbf{R}^n$ (the proof given here of the theorem in **3.5** was given by N.N. Vorob'ev, Jr. [1]). The theorem in **3.6** was proved by J.E.L. Peck and A.L. Dulmage [1]. The theorems in **3.7** and **3.9–12** are actually contained in H.F. Bohnenblust, S. Karlin and L.S. Shapley [1], and the proofs given here were found by N.N. Vorob'ev, Jr. [1,2]. Teh-Tjoe Tie [1] proved the existence of a value, under some restrictions, for the $\mathcal{S}$-topology of games, and T. Parthasarathy [1] put these conditions into a more transparent form. Ky Fan [1] introduced concavo-convex shaped functions of two variables with the aim of proving the theorem of **3.15**. His application to concavo-convex shaped games leads to the theorem of **3.16** (Parthasarathy [2]). The constructions, arguments, and theorems in **3.18–3.24** were found by Khoang Tuĭ [1], and their corollary in **3.26**, by Wu Wen-tzün [1]. Another foundation for ideas connected with the general minimax theorems, not including (but not included by) Khoang Tuĭ's theorem, was established by Bui Công Cuong [1,2]. M. Sion [1] introduced quasi-concavo-convex games and proved the theorem of **3.28**. N.J. Young [1,2] and E.B. Yanovskaya [20], using V.I. Ptak's [1] lemma, obtained a further very general minimax theorem.

The theorem of subsection 3.30 was found by E.B. Yanovskaya [4]. For another theorem on minimaxes for games on the unit square with discontinuous payoff functions, see E.B. Yanovskaya [11,13,14] (the last paper also considers the case of games not of the two-person zero-sum type), and Parthasarathy [3].

§4. The example in **4.2** of a game that does not have a value in the mixed strategies was suggested by A. Wald [1], but the author has not succeeded in establishing the origin of the very "fashionable" example in **4.3**. For further examples see M. Sion and P. Wolfe [1] and E.B. Yanovskaya [8].

The solution of games with countably additive strategies was suggested for consideration by Karlin [1]; the discussion and general theorem in **4.6–4.12** were given by E.B. Yanovskaya [8]. In [18] she also extended this result to general noncooperative games.

§5. The theorem that the set of games with unique solutions is everywhere dense in the class of games on the unit square was found by S. Karlin [3]. An essential sharpening was proposed by G.N. Dyubin [1,2]. He showed that the "majority" (namely, a set of second Baire category) of the games on the unit square with a continuum of payoffs consists of games for which the only optimal strategies of the players have continuous, but singular, distribution functions. The rest of the results of this section were found by O. Gross [1] and by I.L. Glicksberg and O. Gross [1]. See also S. Karlin's monograph [3]. An interesting discussion of the mathematical methods for studying such games was found by V.K. Domanskiĭ [1]. An analytically delicate investigation of "bell-shaped" games on the unit square was carried out by S. Karlin [5] and appeared in his book [3].

§6. Separable games on the unit square were first discussed by O. Helmer [1] and M. Dresher [1]. Detailed discussions of separable games are given in the paper by M. Dresher, S. Karlin and L.S. Shapley [1] and also in S. Karlin's monograph [3]. See also D. Gale and O. Gross [1].

§7. A general technique for solving convex games was found by M. Dresher and S. Karlin [1], and is included in S. Karlin's book [3]. An analysis of Blotto games is given by D.W. Blackett [1]. The exposition in **7.15–7.21** of a method of solving convex games, connected with the optimal allocation of limited resources under undetermined requirements, was worked out by N.N. Vorob'ev in [22] (and originally for the purpose of solving undetermined (including statistically undetermined) problems of structural mechanics in [7]). A general theory of game-theoretic problems connected with the distribution of limited resources was developed in J. Danskin's book [2]. Games on the unit square with modified convexity properties were analyzed by S. Karlin [3,4].

§8. Determining values and pairs of optimal strategies for the examples in **8.16** were studied in M. Dresher's book [2]. The other results in this section were obtained by N.N. Vorob'ev [7,15]. For further development of this problem see M.M. Lutsenko [1,2]. The difficulty of solving more general games on the unit square with discontinuous payoff functions can be seen from the examples of games analyzed by G.B. Brown [1] and N.A. Nikitina [1]. A still more difficult problem is the investigation of two-person zero-sum games on products of compact sets other than line segments, and with discontinuous payoff functions. This is attested by the investigations of A.I. Sobolev [1] and G. Owen [2] and the discussion by E. Borel [1] of games on the product of two triangles.

§9. The first publications on this topic are the papers of M. Shiffman [1] and S. Karlin [2] (see S. Karlin's monograph [3]). The account here reproduces in an elementary form (i.e., without the use of intergal operators) the basic results for games of duel type. See also S. Karlin's monograph [3], which presents the parameters of optimal strategies for some examples of games of the kind described in **9.14–9.22**. A different class of duels was studied by R. Restrepo [1], A.S. Mikhaĭlova [1,2], and others. A detailed study of problems of this type, and a corresponding bibliography, are given in S. Karlin's monograph [3]. Approximation properties of optimal strategies for games of this class are discussed by W. Vogel [1]. The question of the uniqueness of optimal strategies is discussed by E.B. Yanovskaya [3]. For duels with continuous firing see Karlin's book [3], and E.B. Yanovskaya [5,7].

It may seem paradoxical that the solution of two-person zero-sum games of duel type has a structure similar to solutions of two-person zero-sum duels, as can be seen from the papers of D.P. Sudzhyute [1–6].

Chapter 4
Matrix games

§1 Basic concepts and propositions

1.1 Matrix games. In 4, Chapter 1, a matrix game was defined as a finite two-person zero-sum game, i.e., as a game $\Gamma = \langle x, y, H\rangle$ in which the sets x and y of the players' strategies are finite. Unless the contrary is stated, we shall also always suppose that $x = \{1, \dots, m\}$ and $y = \{1, \dots, n\}$.

The strategies of the first player in a matrix game will be thought of as the horizontal rows in a table; and the strategies of the second player, as the vertical rows. If the entries are filled by the values of the payoff function of player 1, we obtain a matrix called the *payoff matrix* of the game, or (for short) the *matrix of the game*. A matrix game together with the matrix A of payoffs will be denoted by $\Gamma(A)$ or Γ_A.

Of course, all propositions about arbitrary two-person zero-sum games are still valid for matrix games. We shall reproduce some of these here, with simplified proofs, so that this chapter will be logically independent of the preceding chapters.

1.2 Maximin principle and mixed strategies. The basic optimality principle in matrix games (just as in general two-person zero-sum games) is the maximin principle (see § 2, Chapter 3), which consists of finding equilibrium situations in the sense of Nash (see 2.11, Chapter 1), i.e., in the present case, saddle points of the payoff function. This principle is that, in a game Γ_A with matrix $A = \|a_{ij}\|_{i=1,\dots,m; j=1,\dots,n}$ we are to choose a row i^* and column j^* of the matrix such that

$$a_{ij^*} \leqq a_{i^*j^*} \leqq a_{i^*j}$$

for every $i = 1, \dots, m$ and $j = 1, \dots, n$.

Since many matrices do not have saddle points, it turns out to be necessary to construct mixed extensions of matrix games (cf. 5.2 of Chapter 1 and 1.14 of Chapter 3).

Definition. A *mixed extension* of a matrix game Γ_A is an infinite two-person zero-sum game

$$\tilde{\Gamma}(A) = \langle X, Y, H\rangle$$

where X and Y are the sets of probability measures on the sets $x = \{1, \dots, m\}$ and $y = \{1, \dots, n\}$, and

$$H(X,Y) = \sum_{i=1}^{m}\sum_{j=1}^{n} X(i) a_{ij} Y(j) = XAY^T, \tag{1.1}$$

for all $X \in \boldsymbol{X}$ and $Y \in \boldsymbol{Y}$, where we have a scalar product on the right, and ^{T} denotes transposition. □

In what follows we shall usually take $X(i) = \xi_i$ and $Y(j) = \eta_j$. The sets $\boldsymbol{X}$ and $\boldsymbol{Y}$ are conveniently interpreted as sets of vectors:

$$\boldsymbol{X} = \left\{ (\xi_1, \ldots, \xi_m) : \xi_i \geqq 0, i = 1, \ldots, m, \sum_{i=1}^{m} \xi_i = 1 \right\},$$

$$\boldsymbol{Y} = \left\{ (\eta_1, \ldots, \eta_n) : \eta_j \geqq 0, j = 1, \ldots, n, \sum_{j=1}^{n} n_j = 1 \right\}.$$

In this interpretation, $\boldsymbol{X}$ and $\boldsymbol{Y}$ are considered as simplexes of the corresponding dimensions, whose points (the mixed strategies) are given by their barycentric coordinates.

A situation (X^*, Y^*) is a saddle point in the mixed strategies if

$$XAY^{*T} \leqq X^* AY^{*T} \leqq X^* AY^T \text{ for every } X \in \boldsymbol{X} \text{ and } Y \in \boldsymbol{Y}. \tag{1.2}$$

As usual, $A_i.$ is the ith row of A, and $A._j$ is its jth column. It is easy to verify that a necessary and sufficient condition for a situation (X^*, Y^*) to be a saddle point in Γ_A is that

$$A_i.Y^{*T} \leqq X^* AY^{*T} \leqq X^* A._j \text{ for } i = 1, \ldots, m \text{ and } j = 1, \ldots, n \tag{1.3}$$

(for general noncooperative games, this was done in 5.13, Chapter 1; also see Theorem 2°, 1.18, Chapter 3).

1.3 Realizability of the maximin principle. In accordance with what was said in 1.9, Chapter 3, in order to prove the existence in a matrix game Γ_A of saddle points in the mixed strategies it is necessary and sufficient to prove the existence and equality of the matrix extrema

$$\max_X \inf_Y XAY^T \text{ and } \min_Y \sup_X XAY^T.$$

Since XAY^T is a linear function of Y for each given X, we must have

$$XAY^T \geqq \min_j XA._j.$$

Since, in addition, XAY^T depends continuously on Y, this function attains a maximum on the compact set $\boldsymbol{Y}$, and we have

$$\min_Y XAY^T \geqq \min_j XA._j.$$

Since the opposite inequality is always valid, the two minima are equal.

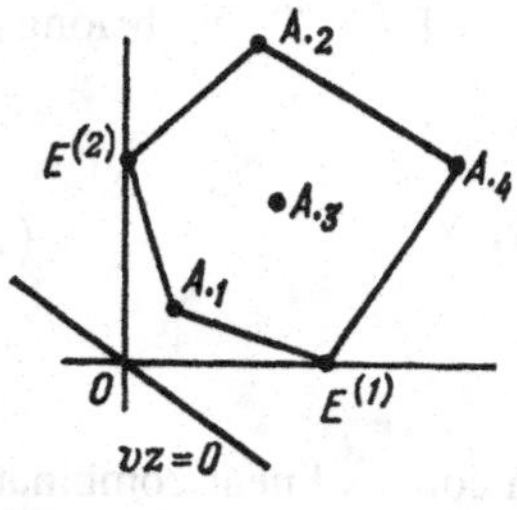

Fig. 4.1

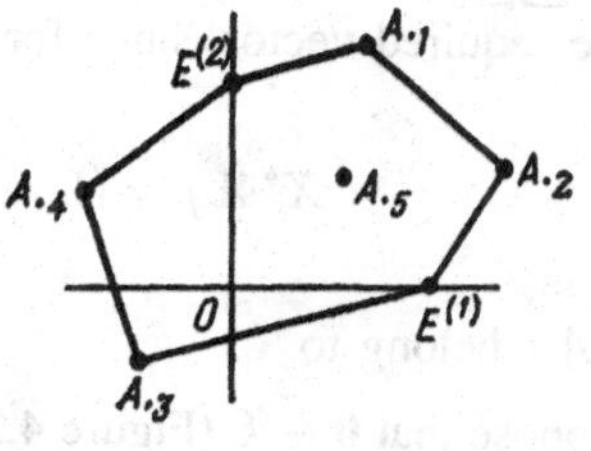

Fig. 4.2

It is clear, moreover, that $\min_j XA_{\cdot j}$ is a piecewise linear function, since it is continuous and therefore attains a maximum on the compact set $\boldsymbol{X}$:

$$\max_X \min_j XA_{\cdot j},$$

which (by what was said above) must equal $\max_X \min_Y XAY^T$.

A similar argument establishes the existence and coincidence of the extrema

$$\min_X \max_j XA_{\cdot j} \text{ and } \min_X \max_Y XAY^T.$$

Consequently, to prove the existence of a saddle point in a matrix game Γ_A it is sufficient to establish the equality

$$\max_X \min_j XA_{\cdot j} = \min_Y \max_i A_{i\cdot}Y^T. \tag{1.4}$$

The common value of these extrema is called the *value* of Γ_A (cf. 1.10, Chapter 3) and is denoted by v_A or $v(A)$.

The proposition that concerns us follows from the theorems in 3.10 and 6.6, Chapter 1, as well as from the corresponding theorems of Chapter 3. Here we give two additional proofs that depend on quite different considerations.

1.4 Lemma (on two alternatives). *For an $m \times n$-matrix game A, there is either a vector $X^* = (\xi_1^*, \ldots, \xi_m^*) \in \boldsymbol{X}$ such that $X^* A_{\cdot j} \geqq 0$ for all $j = 1, \ldots, n$, or a vector $Y^* = (\eta_1^*, \ldots, \eta_n^*) \in \boldsymbol{Y}$ such that $A_{i\cdot}Y^{*T} \leqq 0$ for all $i = 1, \ldots, m$.*

Proof. Let $\mathscr{C}$ denote the convex hull of the vectors $A_{\cdot j}$ $(j = 1, \ldots, n)$ and of the basis vectors $E^{(i)}$ $(i = 1, \ldots, n)$, for which the ith component is equal to 1 and the other components are 0 (Figure 4.1). The set $\mathscr{C}$ is evidently a polytope, and therefore closed, bounded, and nonempty, and has a finite number of vertices.

First suppose that $\mathbf{0} \notin \mathscr{C}$. Then there is a hyperplane that separates $\mathbf{0}$ from $\mathscr{C}$, in other words, there exists a hyperplane $UZ = \mathbf{0}$ (i.e. passing through 0) for which $UZ > 0$ for all $Z \in \mathscr{C}$. In particular, since all the $E^{(i)} \in \mathscr{C}$,

$$UE^{(i)} = u_i > 0 \text{ for } i = 1, \ldots, m,$$

and therefore $\sum_{i=1}^{m} u_i > 0$. Therefore the vector $X^* = U/\sum_{i=1}^{m} u_i$ belongs to $\boldsymbol{X}$. This is the required vector, since for $j = 1, \dots, n$

$$X^* A_{.j} = (UA_{.j})/\sum_{i=1}^{m} u_i > 0, \tag{1.5}$$

since all the $A.j$ belong to $\mathscr{C}$.

Now suppose that $\mathbf{0} \in \mathscr{C}$ (Figure 4.2). Then $\mathbf{0}$ is a convex linear combination of the vertices of $\mathscr{C}$:

$$\mathbf{0} = \sum_{j=1}^{n} \alpha_j A_{.j} + \sum_{i=1}^{m} \varepsilon_i E^{(i)}, \tag{1.6}$$

where $\alpha_j, \varepsilon_i \geqq 0$ $(i = 1, \dots, m\,; j = 1, \dots, n)$ and

$$\sum_{j=1}^{n} \alpha_j + \sum_{i=1}^{m} \varepsilon_i = 1. \tag{1.7}$$

Equation (1.6) means that for $i = 1, \dots, m$

$$\sum_{j=1}^{n} \alpha_j a_{ij} + \varepsilon_i = 0, \tag{1.8}$$

and since $\varepsilon_i \geqq 0$, we must have

$$\sum_{j=1}^{n} \alpha_j a_{ij} \leqq 0. \tag{1.9}$$

If we had $\sum_{j=1}^{n} \alpha_j = 0$, then since all the terms are nonnegative they would all have to be zero. But then by (1.8) all the ε_i must be zero, which evidently contradicts (1.7). Consequently $\sum_{j=1}^{n} \alpha_j > 0$, and we may set

$$\eta_j^* = \alpha_j \left(\sum_{j=1}^{n} \alpha_j \right)^{-1}.$$

Evidently $Y^* = (\eta_1^*, \dots, \eta_n^*) \in \boldsymbol{Y}$ and here, by (1.8),

$$A_{i.} Y^{*T} = \sum_{j=1}^{n} \alpha_j a_{ij} \left(\sum_{j=1}^{n} \alpha_j \right)^{-1} \leqq 0 \tag{1.10}$$

for every $i = 1, \dots, m$. Hence Y^* is the required vector. □

1.5 Theorem. *Every matrix game Γ_A has an equilibrium point.*

Proof. By what was said above, it is sufficient for our purposes to establish the minimax equation (1.4).

Let us suppose that, in the lemma proved above, the first alternative occurs, i.e., there is a vector $X \in \mathbf{X}$ for which $XA_j \geqq 0$ for $j = 1, \ldots, n$. Then evidently $\min_j XA_{.j} > 0$ and therefore $\max_X \min_j XA_{.j} > 0$. If we had the second alternative, a similar discussion would lead to $\min_Y \max_i A_{.j} Y^T \leqq 0$. Consequently the double inequality

$$\max_X \min_j XA_{.j} < 0 < \min_Y \max_i A_{i.} Y^T$$

is impossible.

Now consider the matrix $A(t) = \| a_{ij} - t \|$. If we apply the preceding discussion to this matrix, we find that is impossible to have

$$\max_X \min_j XA_{.j} - t < 0 < \min_Y \max_i A_{i.} Y^T - t,$$

i.e.,

$$\max_X \min_j XA_{.j} < t < \min_Y \max_i A_{i.} Y^T$$

for every real t. But this is possible only when

$$\max_X \min_j XA_{.j} \leqq \min_Y \max_i A_{i.} Y^T.$$

Since the opposite inequality is evidently always valid (see 1.5, Chapter 3), we must have (1.4). □

1.6 Another proof of the existence theorem. The proof just presented leads to the desired result, although in a somewhat roundabout way, since in the proof of the lemma on the two alternatives we had to use a theorem on the separation of convex sets. We now present a different proof of the lemma on the two alternatives, one that is, to be sure, somewhat formal, but does not depend on properties of convex sets. It consists of two parts.

1.7 Theorem. *If S is a subspace of $\mathbf{R}^n$, and $S^\perp$ is its orthogonal complement, then either S or $S^\perp$ contains a vector $Y \geq 0$ (i.e., does not merely intersect the nonnegative orthant in the zero vector).*

Proof. Let the set $A \subset \mathbf{R}^n$ consist of all the points of the positive orthant that are not interior points of the unit ball (the case corresponding to $n = 2$ is drawn in Figure 4.3). Let $P : R \to S$ be the projection of the vectors of $\mathbf{R}^n$ on S, and let the projection $Y = (\eta_1, \ldots, \eta_n)$ of some vector $Z \in A$ have the minimum length among the projections of the vectors of A on S. Supposing that $\eta_i < 0$, we consider the nonnegative vector

$$W = (0, \ldots, 0, -\eta_i, 0, \ldots, 0),$$

for which evidently $Z + W = A$. Then

$$|Y + W|^2 = |Y|^2 - \eta_i^2 < |Y|^2,$$

and therefore (since the projection operator is an idempotent and does not increase the length of a projected vector)

$$|P(Z + W)| = |Y + PW| = |PY + PW| = |P(Y + W)| \leqq |Y + W| < |Y|.$$

But since $Z + W \in A$, this contradicts the definition of Y. Consequently $Y \geqq 0$. If $Y \neq 0$, the theorem is proved. If $Y = 0$, then $Z \in S^\perp$ and evidently $Z \geq 0$. □

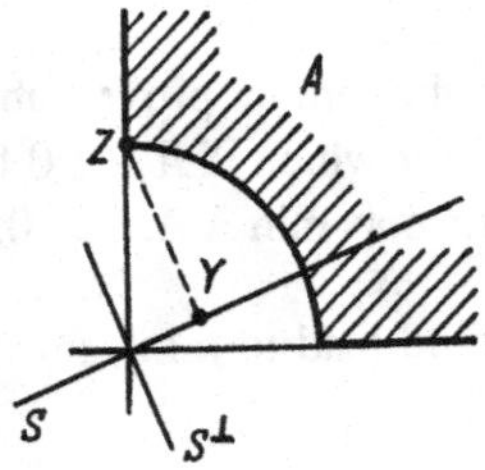

Fig. 4.3

1.8 Lemma on the two alternatives. *Let A be any $m \times n$-matrix. Then either $\mathbf{R}^m$ contains a vector $X \geqq 0$ such that $XA \geqq 0$ or $\mathbf{R}^n$ contains a vector $Y \geqq 0$ such that $AY^T \leqq 0$.*

Proof. The set of all vectors of the form (X, XA) for $X \in \mathbf{R}^m$, as well as the set of all vectors of the form $(-AY^T, Y^T)$ for $Y \in \mathbf{R}^n$, is the Euclidean space $\mathbf{R}^{m+n}$. Moreover, the subspace $\mathbf{R}^m$ of all X and the subspace $\mathbf{R}^n$ of all Y are orthogonal complements. In this situation the preceding theorem provides exactly what is required. □

1.9 Properties of the optimal strategies of the players and of the value of a game. The common value of the minimaxes in (1.4) is the value of the matrix game Γ_A, which is denoted by $v(A)$ (or v_A), and the players' strategies for which the outside extrema are attained in (1.4) are the players' optimal strategies. The set of optimal strategies of player 1 in Γ_A will be denoted by $\mathcal{S}(A)$; and the set of optimal strategies of player 2, by $\mathcal{T}(A)$ (cf. 1.7, Chapter 3). If (X^*, Y^*) is a pair of these optimal strategies, they satisfy inequality (1.2). Such pairs form saddle points of payoff functions of Γ_A, i.e., are equilibrium situations. The set of all equilibrium situations of Γ_A is denoted by $\mathcal{C}(A)$ (cf. 1.8, Chapter 3). Evidently $\mathcal{C}(A) = \mathcal{S}(A) \times \mathcal{T}(A)$. It is also clear (cf. 1.8, Chapter 3) that the payoff function is constant on $\mathcal{C}(A)$ and assumes the value v_A.

Matrix games Γ_B and Γ_A of the same dimension are called *affinely equivalent* (cf. 1.13, Chapter 1) if there are $k > 0$ and $a \in \mathbf{R}$ such that $B = kA + aE$ (where the matrix E has the same format as A and B and all its entries are unity). It can be verified directly (see 1.16, Chapter 1) that in this case

$$\mathcal{S}(B) = \mathcal{S}(A), \quad \mathcal{T}(B) = \mathcal{T}(A), \text{ and } v_B = kv_A + a.$$

It follows from (1.4) that

$$\max_i \min_j a_{ij} \leqq v_A \leqq \min_j \max_i a_{ij}.$$

Because the outer extrema in (1.4) are assumed, it follows that if player 1 has a pure optimal strategy, then

$$v_A = \max_i \min_j a_{ij},$$

and if player 2 has a pure optimal strategy then

$$v_A = \min_j \max_i a_{ij}.$$

1.10. Let us quote, in connection with matrix games, the following useful chain of propositions. In what follows, we shall, as before, deal with the game Γ_A. For pairs of symmetric propositions, we shall, as a rule, prove only the parts that concern the first player.

If v is any number, $X \in \boldsymbol{X}$, and

$$v \leqq XA_{\cdot j} \text{ for } j = 1, \ldots, n, \tag{1.11}$$

then $v \leqq v_A$; similarly, if $Y \in \boldsymbol{Y}$ and

$$A_{i\cdot}Y^T \leqq v \text{ for } i = 1, \ldots, m,$$

then $v \geqq v_A$.

In fact, if in (1.11) we turn to the optimal strategy Y^* of player 2, and then to the optimal strategy X^* of player 1, we obtain

$$v \leqq XAY^{*T} \leqq X^*AY^{*T} \leqq v_A.$$

1.11. In connection with matrix games we repeat the criteria for the optimality of strategies and for the value of the game in 1.12, Chapter 3. If $X \in \boldsymbol{X}$ and $Y \in \boldsymbol{Y}$, and v is a number, and we have the inequality

$$A_{i\cdot}Y^T \leqq v \leqq XA_{\cdot j} \text{ for } i = 1, \ldots, m \text{ and } j = 1, \ldots, n, \tag{1.12}$$

then v is the value of Γ_A and $(X, Y) \in \mathscr{C}(A)$.

In fact, if we turn, on both sides of (1.12), to the mixed strategies X and Y, we obtain $XAY^T \leqq v \leqq XAY^T$, i.e. $v = XAY^T$, and by substituting in (1.12) we obtain

$$A_{i\cdot}Y^T \leqq XAY^T < XA_{\cdot i} \text{ for } i = 1, \ldots, m, \text{ and } j = 1, \ldots, n,$$

i.e., (X, Y) is a saddle point.

1.12. If, for $X^0 \in \boldsymbol{X}$,

$$v_A \leqq X^0 A_{\cdot j} \text{ for } j = 1, \ldots, n, \tag{1.13}$$

then X^0 is an optimal strategy of player 1; and if, for $Y^0 \in \boldsymbol{Y}$,

$$A_{i\cdot}Y^{0T} \leq v_A \text{ for } i = 1, \ldots, m,$$

then Y^0 is an optimal strategy of player 2.

In fact, from (1.13) we have

$$\max_X \min_j XA_{\cdot j} = v_A \leqq \min_j X^0 A_{\cdot j} \leqq \max_X \min_j XA_{\cdot j}.$$

Here all the terms must be equal; in particular, the maximum on the right must be attained for X^0, and this strategy is optimal. Hence we immediately find that

$$v_A = \min_j XA_{\cdot j}$$

is a necessary and sufficient condition for the optimality of the strategy X, and

$$v_A = \max_i A_{i\cdot}Y^T$$

is a necessary and sufficient condition for the optimality of the strategy Y.

1.13 Sets of all the optimal strategies of the players. It follows from what we have said that $\mathcal{S}(A)$ is the set of vectors $X \in \boldsymbol{X}$ that satisfy (1.13). Hence it is a polytope.

By symmetry, $\mathcal{T}(A)$ is also a polytope.

Consequently, for matrix games we have reproduced the result of 5.2, Chapter 1, in a somewhat stronger form.

1.14 Complementary slackness. For matrix games, the results of 5.15, Chapter 1 and of 4°, 1.18, Chapter 3 now appear in the following form.

Theorem. *If X^* is an optimal strategy of player* 1 *and we have, for some j_0,*

$$X^* A_{\cdot j_0} > v_A, \tag{1.14}$$

then we must have $\eta^*_{j_0} = 0$ in every optimal strategy $Y^* = (\eta^*_1, \dots, \eta^*_n)$ of player 2.

Symmetrically, if Y^* is any optimal strategy of player 2, *and $A_{i_0\cdot} Y^{*T} < v_A$ for some i_0, then we must have $\xi^*_{i_0} = 0$ in every optimal strategy $X^* = (\xi^*_1, \dots, \xi^*_m)$ of player* 1.

Proof. We have

$$X^* A_{\cdot j} \geqq v_A \text{ for } j = 1, \dots, n.$$

If we multiply these inequalities termwise by the corresponding η_j, we obtain

$$X^* A_{\cdot j} \eta^*_j \geqq v_A \eta^*_j \tag{1.15}$$

which yields, after summing on j,

$$X^* A Y^{*T} \geqq v_A \sum_{j=1}^{n} \eta^*_j = v_A,$$

where the left-hand side is evidently actually equal to v_A, so that in fact this inequality is an equality. But the termwise addition of inequalities in the same sense yields an equality only in the case when in fact it was equalities that were added, i.e., if, in (1.15), all the inequalities are equalities, and, in particular, $X^* A_{\cdot j_0} \eta^*_{j_0} = v_A \eta^*_{j_0}$. This, however, is consistent with (1.14) only if $\eta^*_j = 0$. □

1.15 Completely mixed strategies. Let an optimal strategy X of player 1 in the matrix game Γ_A be completely mixed, i.e., all its components ξ_i are strictly positive (see the definition in 1.3, Chapter 2). Then the theorem of the preceding subsection shows that, for every optimal strategy Y of player 2,

$$A_{i\cdot} Y^T = v_A \text{ for } i = 1, \dots, m. \tag{1.16}$$

Similarly, if an optimal strategy Y of player 2 is completely mixed, then, for every optimal strategy X of player 1,

$$X A_{\cdot j} = v_A \text{ for } j = 1, \dots, n. \tag{1.17}$$

We now consider a class of matrix games for which the fact just established leads to an exhaustive analysis of the games.

1.16 Definition. A matrix game Γ_A is said to be *completely mixed* if all the optimal strategies of the players are completely mixed. □

Let the game Γ_A be completely mixed, and let its payoff matrix A be nonsingular (it will be shown in § 3 that the nonsingularity of A can be deduced from the property that the game is completely mixed), and therefore square. We are going to rewrite the systems (1.16) and (1.17) in matrix form, adjoining the condition of norming of the strategies*)

$$AY^T = J_n^T v_A, \tag{1.18}$$

$$J_n Y^T = 1, \tag{1.19}$$

$$XA = v_A J_n, \tag{1.20}$$

$$XJ_n^T = 1. \tag{1.21}$$

Since A is nonsingular, we can obtain

$$Y^T = A^{-1} J_n^T v_A \tag{1.22}$$

from (1.18).

Multiplying on the left by J_n and using (1.19) gives us

$$1 = J_n A^{-1} J^T v_A,$$

i.e.,

$$v_A = (J_n A^{-1} J_n^T)^{-1}, \tag{1.23}$$

and after substitution in (1.22) we obtain

$$Y^T = \frac{A^{-1} J_n^T}{J_n A^{-1} J_n^T}. \tag{1.24}$$

Similarly, from (1.20) and (1.21) we obtain

$$X = \frac{J_n A^{-1}}{J_n A^{-1} J_n^T}. \tag{1.25}$$

We see that for matrix games of the type that we consider there is only a single saddle point (X, Y), where the optimal strategies X and Y are given explicitly by (1.24) and (1.25), and the value is obtained from (1.23).

1.17 Inversion of complementary slackness (Farkas's lemma). Furthermore we will find it useful to have a proposition that is in a sense a converse of what we proved in 1.14. As preparation, we present the following lemma, often called Farkas's lemma, which is not directly related to game theory.

---

*) For each positive integer n, J_n is the n-dimensional vector all of whose components are 1.

Lemma. *Let the vectors $Z^{(0)}, Z^{(1)}, \dots, Z^{(n)} \in \mathbf{R}^n$ have the property that, for arbitrary $W \in \mathbf{R}^m$, it follows from $WZ^{(i.)^t} \geqq 0$ for $i = 1, \dots, n$ that $WZ^{(0)^t} \geqq 0$.*
Then $Z^{(0)}$ belongs to the convex cone $\mathscr{C}$ generated by the vectors $Z^{(1)}, \dots, Z^{(n)}$.

For the proof, we suppose that $Z^{(0)} \notin \mathscr{C}$. This means that we can find a vector $\bar{W}$ and a number c such that

$$\bar{W}U^T < c \text{ for all } U \in \mathscr{C}, \tag{1.26}$$

but $\bar{W}Z^{0T} > c$. Since $0 \in \mathscr{C}$, it follows that $c > 0$. If now there is a $\bar{Z} \in \mathscr{C}$ such that $W\bar{Z}^T > 0$, then if we multiply $\bar{Z}$ by a sufficiently small scalar $t > 0$ we can make $\bar{W}(t\bar{Z})^T > c$. This, however, contradicts (1.26). Therefore $\bar{W}Z^T \leqq 0$ and therefore $(-\bar{W})Z^T \geqq 0$ for all $z \in \mathscr{C}$ and, in particular, $(-\bar{W})Z^{(i)T} \geqq 0$ for $i = 1, \dots, n$. But then, by hypothesis, $(-W)Z^{(0)T} \geqq 0$, so that $\bar{W}Z^{(0)T} \leqq 0$, and this contradicts the fact that $\bar{W}Z^{(0)T} > c > 0$. □

1.18 Theorem. *If we have $\eta_{j_0} = 0$ for every optimal strategy $Y^* = (\eta_1^*, \dots, \eta_m^*)$ of player 2 in the game Γ_A then there is an optimal strategy $X^* = (\xi_1^*, \dots, \xi_m^*)$ of player 1 for which $X^* A_{\cdot j_0} > v_A$.*

There is a parallel proposition for the components of an optimal strategy of player 1.

The structure of the proof is reminiscent of the proof of the lemma in 1.4. If we turn, if necessary, to an affinely equivalent game (see 1.9), we may suppose, without loss of generality, that $v_A = 0$.

We consider the cone $\mathscr{C}$ generated by the basis vectors $E^{(i)}$ $(i = 1, \dots, m)$ and all the columns $A_{\cdot j}$ of A, omitting the column $A_{\cdot j_0}$.

We suppose first that $-A_{\cdot j_0} \in \mathscr{C}$. Then we may write

$$-A_{\cdot j_0} = \sum_{j \neq j_0} \alpha_j A_{\cdot j} + \sum_{i=1}^{m} \varepsilon_j E^{(i)},$$

where $\alpha_j \geqq 0$ $(j \neq j_0)$ and $\varepsilon_i \geqq 0$ $(i = 1, \dots, m)$. The last sum is the nonnegative vector $(\varepsilon_1, \dots, \varepsilon_m)$. Therefore

$$\sum_{j \neq j_0} \alpha_j A_{\cdot j} + A_{\cdot j_0} \leqq 0. \tag{1.27}$$

If we set

$$\alpha_j \left(1 + \sum_{j \neq j_0} \alpha_j\right)^{-1} = \eta_j \text{ for } j \neq j_0 \text{ and}$$

$$\left(1 + \sum_{j \neq j_0} \alpha_j\right)^{-1} = \eta_{j_0},$$

$(\eta_1, \dots, \eta_m) = Y$, and we find from (1.27) that

$$\sum_{j \neq j_0} A_{\cdot j}\eta_j + A_{\cdot j_0}\eta_{j_0} \leqq 0,$$

i.e., $AY^T \leqq 0$ or, in other words,

$$A_{i\cdot}Y^T \leqq 0 \text{ for } i = 1, \ldots, m.$$

Therefore Y is an optimal strategy of player 2 in Γ_A, with $\eta_{j_0} > 0$; and this contradicts the hypothesis of the theorem.

Consequently we must have $-A_{\cdot j_0} \notin \mathscr{C}$. Then, by the lemma in 1.17, there must be a vector $W = (\omega_1, \ldots, \omega_n)$ for which

$$WE^{(i)T} \geqq 0 \text{ for } i = 1, \ldots, m, \tag{1.28}$$

$$WA_{\cdot j} \geqq 0 \text{ for } j = 1, \ldots, n, \tag{1.29}$$

and moreover

$$WA_{\cdot j_0} > 0. \tag{1.30}$$

It follows from (1.28) that $W \geqq 0$; and from (1.30), that $W \neq 0$. Consequently $\sum_{i=1}^{m} \omega_1 > 0$. Let us set

$$\xi_i^* = \omega_i \left(\sum_{i=1}^{m} \omega_i \right)^{-1} \quad \text{for } i = 1, \ldots, m \text{ and } X^* = (\xi_1^*, \ldots, \xi_m^*).$$

Then if we divide (1.29) and (1.30) term by term by $\omega_1 + \ldots + \omega_m$ we obtain $X^* A_{\cdot j} \geqq 0$ for $j = 1, \ldots, n$, and also $X^* A_{\cdot j_0} > 0$. The first of these relations means that X^* is an optimal strategy of player 1 in Γ_A; the second, that it is the one required. □

1.19. The combination of the propositions in 1.14 and 1.18 gives us the equivalence of the following two characteristics of the pure strategy j of player 2 in the matrix game Γ_A:

a) there is an optimal strategy X^* of player 1 for which $X^* A_{\cdot j} > v_A$;

a′) for every optimal strategy $Y^* = (\eta_1^*, \ldots, \eta_n^*)$ of player 2 we must have $\eta_j^* = 0$.

Hence the equivalence of the following conditions follows by contraposition:

b) for every optimal strategy X^* of player 1, we must have $X^* A_{\cdot j} = v_A$;

b′) there is an optimal strategy $Y^* = (\eta_1^*, \ldots, \eta_n^*)$ of player 2 for which $\eta_j^* > 0$.

A pure strategy j of player 2 for which the conditions a) or a′) are satisfied is said to be *essential*; and a pure strategy for which the conditions b) or b′) are satisfied is said to be *superfluous*.

Essential and superfluous strategies of player 1 are defined similarly.

A subgame of Γ_A in which all the pure strategies of the players are essential strategies in Γ_A is called an *essential subgame* of Γ_A.

1.20 Symmetry of matrix games. The value v_A of a matrix game Γ_A, and the sets $\mathcal{S}(A)$ and $\mathcal{T}(A)$ of optimal strategies of the players in Γ_A are functions of the game. We are going to establish some properties of these functions. We begin with such combinatorial properties as symmetry.

In 1.19 of Chapter 1 we defined the symmetry of a two-person zero-sum game. In connection with matrix games, this assumes the following form.

Definition. Two matrix games A and B are said to be *(mirror) symmetric* if $B = -A^T$. A single matrix game is said to be *symmetric* if it is (mirror) symmetric to itself. □

Evidently, if a matrix game has an $m \times n$ matrix, the mirror-symmetric game has an $n \times m$ matrix, and the payoff matrix of a symmetric game is square.

1.21 The sets of the players' optimal strategies in a matrix game, as well as its value, are skew-symmetric functions of its payoff matrix. The precise meaning of this statement is formulated in the following theorem.

Theorem.

1) $\mathcal{S}(-A^T) = \mathcal{T}(A)$;

2) $\mathcal{T}(-A^T) = \mathcal{S}(A)$;

3) $v(-A^T) = -v(A)$.

Proof. Take $X^* \in \mathcal{S}(A)$ and $Y^* \in \mathcal{T}(A)$. That is to say,

$$A_i.Y^{*T} \leqq X^* AY^{*T} \leqq X^* A_{\cdot j} \text{ for } i = 1, \dots, m;\ j = 1, \dots, n.$$

If we transpose and change signs, we obtain

$$Y^*(-A^T)_{\cdot i} \geqq Y^*(-A^T)X^{*T} \geqq (-A^T)_j.X^{*T}$$

for $j = 1, \dots, n$; $i = 1, \dots, m$, i.e. $Y^* \in \mathcal{S}(-A^T)$ and $X^* \in \mathcal{T}(-A^T)$. Moreover,

$$v(A) = X^* AY^{*T} = -Y^*(-A^T)X^{*T} = -v(-A^T). \square$$

Corollary. *If the matrix game Γ_A is symmetric, then $v_A = 0$ and $\mathcal{S}(A) = \mathcal{T}(A)$.*

1.22. These properties of symmetric games facilitate their analysis and solution. Therefore it is interesting to have methods for symmetrizing games, i.e., for the construction, for any matrix game $\Gamma(A)$, of a symmetric matrix game $\Gamma(A^s)$ such that every solution of $\Gamma(A)$ can be obtained from some solution of $\Gamma(A^s)$, and every solution of $\Gamma(A^s)$ determines a solution of $\Gamma(A)$. We are not going to go into the semantics of the expressions "can be obtained" and "determines", since we do not envisage the construction of a general theory of symmetrization, and shall confine ourselves to a single instance. Other examples of symmetrization will be given in § 3.

Let us take any $m \times n$-game Γ_A with payoff matrix A and construct from it the game $\Gamma(A^s)$ with the payoff matrix $A^s = \|a^s_{kl}\|$, where the values of the index k are pairs of the form (i_1, j_2) with $i_1 \in \boldsymbol{x}$ and $j_2 \in \boldsymbol{y}$; the values of l are also pairs, but of the form (i_2, j_1) with $i_2 \in \boldsymbol{x}$ and $j_1 \in \boldsymbol{y}$, and

$$a^s_{kl} = a^s_{(i_1,j_2)(i_2,j_1)} = a_{i_1,j_1} - a_{i_2,j_2}. \tag{1.31}$$

The $mn \times mn$-game $\Gamma(A^s)$ can usefully be thought of as the simultaneous participation of the players in two copies of the game $\Gamma(A)$, where in one copy each of them assumes the role of player 1, and in the other, that of player 2.

1.23 Theorem. *The construction of* $\Gamma(A^s)$ *from* $\Gamma(A)$ *is a symmetrization in the sense described above.*

Proof. That $\Gamma(A^s)$ is symmetric follows immediately from its definition.

We take $X^* \in \mathcal{S}(A)$ and $Y^* \in \mathcal{T}(A)$ and set

$$Z^*(i,j) = X^*(i)Y^*(j).$$

The function Z^*, considered as a measure on $\boldsymbol{x} \times \boldsymbol{y}$, is a mixed strategy for each of the players in $\Gamma(A^s)$. Let us show that Z^* is an optimal strategy for each of them.

In fact, by the choice of X^* and Y^* we have

$$XAY^{*T} \leqq X^*AY^{*T} \leqq X^*AY^T \text{ for all } X \in \boldsymbol{X} \text{ and } Y \in \boldsymbol{Y},$$

from which we have

$$XAY^{*T} - X^*AY^T \leqq 0 \leqq X^*AY^T - XAY^{*T}$$

or, by (1.31),

$$ZA^sZ^{*T} \leqq 0 \leqq Z^*A^sZ^T.$$

But this means that $Z^* \in \mathcal{S}(A^s) = \mathcal{T}(A^s)$.

Conversely, let Z be any mixed strategy of one of the players in the game $\Gamma(A^s)$. This is a measure on the product $\boldsymbol{x} \times \boldsymbol{y}$. Let us consider its projections on the sets $\boldsymbol{x}$ and $\boldsymbol{y}$:

$$Z_{\boldsymbol{x}}(i) = \sum_j Z(i,j) \text{ and } Z_{\boldsymbol{y}}(j) = \sum_i Z(i,j)$$

and show that if $Z = Z^*$ is an optimal strategy of a player in $\Gamma(A^s)$ then its projections $Z^*_{\boldsymbol{x}}$ and $Z^*_{\boldsymbol{y}}$ are optimal strategies of players 1 and 2 in $\Gamma(A)$.

In fact, if Z^* is an optimal strategy, then because $v(A^s) = 0$ we must have

$$ZA^sZ^{*T} \leqq 0 \leqq Z^*A^sZ^T$$

for every measure Z on $\boldsymbol{x} \times \boldsymbol{y}$. Then we have successively

$$\sum_{(i_1,j_2)} \sum_{(i_2,j_1)} Z(i_1,j_2)a^s_{(i_1,j_2)(i_2,j_1)}Z^*(i_2,j_1) \leqq 0,$$

$$\sum_{(i_1,j_2)} \sum_{(i_2,j_1)} Z(i_1,j_2)((a_{i_1j_1} - a_{i_2j_2})Z^*(i_2,j_1) \leqq 0,$$

$$\sum_{i_1,j_1} Z_{\boldsymbol{x}}(i_1)a_{i_1j_1}Z^*_{\boldsymbol{y}}(j_1) - \sum_{i_2,j_2} Z^*_{\boldsymbol{x}}(i_2)a_{i_2j_2}Z_{\boldsymbol{y}}(j_2) \leqq 0,$$

or, finally

$$Z_x A Z_y^{*T} \leqq Z_x^* A Z_y^T$$

for all $Z_y \in \boldsymbol{X}$ and $Z_x \in \boldsymbol{Y}$. But by 1.11 this also means that $Z_x^* \in \mathcal{S}(A)$ and $Z_y^* \in \mathcal{T}(A)$. □

1.24 Analytic properties of the maximin principle. We are going to consider the set of the $m \times n$-matrix games for given m and n as an mn-dimensional space, with element-wise convergence. It turns out that the value of a game is a continuous function of its matrix, and that the sets of optimal strategies of each player are upper semicontinuous functions of the matrix.

Theorem. *Let $A^{(1)}, A^{(2)}, \ldots$ be a sequence of $m \times n$-matrices converging to the matrix A. Then:*

1) $\lim_{k\to\infty} v_{A^{(k)}} = v_A$;

2) *if $\boldsymbol{D}_1$ is an open subset of $\boldsymbol{X}$, containing $\mathcal{S}(A)$, then from some k onward we have $\mathcal{S}(A^{(k)}) \subset \boldsymbol{D}_1$;*

3) *If $\boldsymbol{D}_2$ is an open subset of $\boldsymbol{Y}$, containing $\mathcal{T}(A)$, then from some k onward we have $\mathcal{T}(A^{(k)}) \subset \boldsymbol{D}_2$.*

Proof. Proposition 1) follows directly from the definition of a saddle point.

Proposition 2) will be proved by contradiction. Suppose that for some open set $\boldsymbol{D}_1 \subset \boldsymbol{X}$ there are arbitrarily large values of k for which $X^{(k)} \in \mathcal{S}(A)^{(k)} \setminus \boldsymbol{D}_1 \subset \boldsymbol{X} \setminus \boldsymbol{D}_1$. By the compactness of $\boldsymbol{X} \setminus \boldsymbol{D}_1$ there is a subsequence of these points $X^{(k)}$ that has a limit $X^{(0)} \in \boldsymbol{X} \setminus \boldsymbol{D}_1$.

However, for each k we must have

$$X^{(k)} A_{\cdot j}^{(k)} \geqq v(A^{(k)}) \text{ for } j = 1, \ldots, n.$$

If we take the limit over the subsequence, we obtain

$$X^{(0)} A_{\cdot j}^{(k)} \geqq v(A) \text{ for } j = 1, \ldots, n,$$

i.e., $X^{(0)} \in \mathcal{S}(A) \subset \boldsymbol{D}_1$, and this is a contradiction.

Proposition 3) is obtained from 2) by a change of notation. □

1.25 Corollary. *If $A^{(k)} \to A$, $X^{(k)*} \in \mathcal{S}(A^{(k)})$, and $X^{(0)}$ is one of the limit points of the sequence $X^{(1)*}, X^{(2)*}, \ldots$, then $X^{(0)} \in \mathcal{S}(A)$.*

In fact, if this were not the case, the set $\mathcal{S}(A)$ and the point $X^{(0)}$ would have disjoint neighborhoods, whereas, according to the theorem, $X^{(0)}$ must be a limit point of each neighborhood of $\mathcal{S}(A)$.

There is a similar proposition for the sets of optimal strategies of player 2. □

1.26 Differentiation of the values of a game. Let A and H be two matrices of the same order. In this case we may suppose that the sets of pure and of mixed strategies of the first and second players in the games Γ_A and Γ_H coincide. Since $\mathcal{S}(A) \subset \boldsymbol{X}(= \boldsymbol{X}(H))$ and $\mathcal{T}(A) \subset \boldsymbol{Y}(= \boldsymbol{Y}(H))$, we can consider a $\mathcal{S}(A) \times \mathcal{T}(A)$-subgame of the mixed extension $\langle \boldsymbol{X}, \boldsymbol{Y}, H\rangle$ of Γ_H. Let us denote this subgame by $\Gamma(A/H)$.

The sets $\mathcal{S}(A)$ and $\mathcal{T}(A)$ are closed convex polytopes, and the function H is bilinear on their product and therefore may be considered as convex on $Y \in \mathcal{T}(A)$ and concave on $X \in \mathcal{S}(A)$. Therefore by Theorem 3.10, Chapter 1 (see also the corollary in 3.5, Chapter 3), the game $\Gamma(A/H)$ has a value $v(\Gamma(A/H)) = v(A/H)$ (see also, in the same connection, the discussion that will be given in 3.12).

What we have just said can be used to discuss the differentiation of the values of a matrix game, in the sense of the following definition.

Definition. If A and H are two matrices of the same order, the *right-hand derivative* of $v(A)$ in the H "direction" is the limit

$$\lim_{\alpha\to+0} \frac{1}{\alpha}(v(A+\alpha H) - v(A)),$$

and the *left-hand derivative* is the analogous limit as $\alpha \to -0$. □

1.27. It turns out that, in the sense just defined, the value function $v(A)$ has both right- and left-hand derivatives at every "point" (matrix A).

Theorem. *For any two matrices A and H of the same order,*

$$1)\quad \lim_{\alpha\to+0} \frac{1}{\alpha}(v(A+\alpha H) - v(A)) = v(A/H); \tag{1.32}$$

$$2)\quad \lim_{\alpha\to-0} \frac{1}{\alpha}(v(A+\alpha H) - v(A)) = v(-A^T/H^T). \tag{1.33}$$

Proof. Let $(X^*, Y^*) \in \mathcal{S}(A)$ and $(X^{(\alpha)*}, Y^{(\alpha)*}) \in \mathcal{C}(A+\alpha H)$. Evidently

$$\begin{aligned} v(A+\alpha H) &\leqq X^{(\alpha)*}(A+\alpha H)Y^{*T} \\ &= X^{(\alpha)*}AY^{*T} + \alpha X^{(\alpha)*}HY^{*T} \leqq v_A + \alpha X^{(\alpha)*}HY^{*T} \end{aligned} \tag{1.34}$$

and similarly

$$v(A+\alpha H) \geqq v(A) + \alpha X^* HY^{(\alpha)*T}. \tag{1.35}$$

Therefore we must have, when $\alpha > 0$,

$$X^* HY^{(\alpha)*T} \leqq \frac{1}{\alpha}(v(A+\alpha H) - v(A)) \leqq X^{(\alpha)*}HY^{*T}, \tag{1.36}$$

and consequently, on the right,

$$\frac{1}{\alpha}(v(A+\alpha H) - v(A)) \leqq \max_{X\in\mathcal{S}(A+\alpha H)} XHY^{*T}.$$

Since $A + \alpha H \to A$ as $\alpha \to 0$, according to 1.24 all the limit points of a sequence of values $\mathcal{S}(A + \alpha H)$ must belong to $\mathcal{S}(A)$. Therefore

$$\lim_{\alpha \to +0} \frac{1}{\alpha}(v(A + \alpha H) - v(A)) \leqq \max_{x \in \mathcal{S}(A)} XHY^{*T},$$

and since this is true for every strategy $Y \in \mathcal{T}(A)$,

$$\lim_{\alpha \to +0} \frac{1}{\alpha}(v(A + \alpha H) - v(A)) \leqq \min_{Y \in \mathcal{T}(A)} \max_{x \in \mathcal{S}(A)} XHY^{T}. \tag{1.37}$$

Operating similarly on the left-hand side of (1.36), we have

$$\max_{x \in \mathcal{S}(A)} \min_{Y \in \mathcal{T}(A)} XHY^{T} \leqq \lim_{a \to -0} \frac{1}{\alpha}(v(A + \alpha H) - v(A)). \tag{1.38}$$

However, the left-hand side of (1.38) and the right-hand side of (1.37) are respectively $\underline{v}(A/H)$ and $\overline{v}(A/H)$. However, as we observed, the game $\Gamma(A/H)$ has a value, so that $\underline{v}(A/H) = \overline{v}(A/H) = v(A/H)$. Together with (1.37) and (1.38), this yields (1.32).

Now let $\alpha < 0$. Taking X^*, Y^*, $X^{(\alpha)*}$ and $Y^{(\alpha)*}$ as before, we obtain (1.34) and (1.35). However, since $\alpha < 0$, instead of (1.36) we have

$$X^{(\alpha)*} HY^{*T} \leqq \frac{1}{\alpha}(v(A + \alpha H) - v(A)) \leqq X^* HY^{(\alpha)*T}.$$

Here we obtain (1.33) by maximizing on Y^* and $Y^{(\alpha)*}$ and minimizing on X^* and $X^{(\alpha)*}$, and carrying out a discussion parallel to that used in part 1). □

1.28 Example. Let

$$A = \begin{bmatrix} 1 & 1 \\ 0 & 1 \end{bmatrix}, \quad H = \begin{bmatrix} 0 & 1 \\ 0 & 0 \end{bmatrix}.$$

(i.e., we have in mind differentiating the value of the game with a variable matrix, obtained from A by varying its upper left-hand element.)

In this case (cf. 1.2)

$$X = \{\xi = X(1) : 0 \leqq \xi \leqq 1\}, \quad Y = \{\eta = Y(1) : 0 \leqq \eta \leqq 1\}. \tag{1.39}$$

The following analysis of the simplest matrix games presents no difficulty. For form's sake, we refer to the next section, which is completely independent of the example that we are studying here.

In the notation of (1.39), we have

$$\mathcal{S}(A) = 1, \quad \mathcal{T}(A) = [0, 1].$$

Therefore $\Gamma(A/H)$ is a 1×2-game with payoff matrix $(0, 1)$. Evidently, in this case $\mathcal{T}(A/H) = 1$ and $v(A/H) = 0$.

To calculate the left derivative, we write

$$-A^T = \begin{bmatrix} -1 & 0 \\ -1 & -1 \end{bmatrix}, \quad H^T = \begin{bmatrix} 0 & 0 \\ 1 & 0 \end{bmatrix}.$$

By 1.21, we shall have

$$\mathcal{S}(-A^T) = [0,1], \quad \mathcal{T}(-A^T) = 1.$$

Therefore $\Gamma(-A^T/H^T)$ is a 2×1-game with payoff matrix $\begin{bmatrix} 0 \\ 1 \end{bmatrix}$. Evidently, in this case $\mathcal{S}(-A^T/H^T) = 0$ and $v(-A^T/H^T) = 1$.

Among other things, in this case the value of the derivative in which we are interested can also be calculated directly. Consider the game $\Gamma_{A(\alpha)}$ with

$$A(\alpha) = \begin{bmatrix} 1 & 1+\alpha \\ 0 & 1 \end{bmatrix}.$$

Evidently when $\alpha < 0$ we must have $v(\alpha) = 1/(1-\alpha)$, from which $(d/d\alpha)v(\alpha) = 1/(1-\alpha)^2$, which is 1 when $\alpha = 0$. On the other hand, when $\alpha > 0$ we must have $v(\alpha) \equiv 1$, from which $(d/d\alpha)v(\alpha) \equiv 0$.

§2 Solution of matrix games of small format

The content of this section is quite elementary. In it we present and justify methods for finding solutions of matrix games in the simplest cases. It can be used by anyone, even by someone who does not really know what a saddle point is.

2.1 Analysis of a 2×2-matrix game. The smallest possible number of strategies for a player is 2. Consequently 2×2-matrix games are the simplest possible. Nevertheless, the class of 2×2-matrix games is a four-parameter family, and its analysis is not trivial, although quite elementary. Although passing to affinely equivalent games (see 1.9) makes it possible to reduce the number of parameters in 2×2-games from four to two, the amount of simplification achieved in this way appears to be problematical. Consequently we shall not use this reduction.

Let us consider a game Γ_A with the payoff matrix

$$A = \begin{bmatrix} a_{11} & a_{12} \\ a_{21} & a_{22} \end{bmatrix}. \tag{2.1}$$

To each such game we can apply the methods previously used for finding a solution (including the method of solving 2×2-bimatrix games in 2.1, Chapter 2), as well as the methods that will be discussed in the following sections. We can, however, analyze the game Γ_A directly.

Let us first look for saddle points of Γ_A in the pure strategies. Since the set of these is rectangular (see 1.9), there may be a) four, b) two, c) one, or d) no saddle points. Let us examine these possibilities.

Case a) occurs when all four elements of A are equal and equal to the value of the game, and *every situation is an equilibrium situation.*

Case b) occurs when an equilibrium situation in the pure strategies consists of either a row or a column of the matrix. Let us suppose that it consists of a row (say, for definiteness, the first row). This will occur when

$$a_{11} = a_{12}, \quad a_{21} \leqq a_{11}, \quad a_{22} \leqq a_{12},$$

and at least one of the two inequalities is an equation. *Here player* 1 *has a unique optimal strategy, but all strategies are optimal for player* 2. Symmetrically, *if two equilibrium situations in the pure strategies form a column of the matrix* (2.1), *player* 2 *has a unique optimal strategy, but all strategies are optimal for player* 1.

We now turn to case c). For definiteness, let the only equilibrium situation in the pure strategies lie at the upper left-hand corner of the matrix (2.1). This means that the first pure strategy is optimal for each player. In this case we must have $v_A = a_{11}$ and

$$a_{21} \leqq a_{11} \leqq a_{12}. \tag{2.2}$$

c_1) If both inequalities here are equations, then, as is easily verified, we would be subject to the conditions of case a) or b).

c_2) Suppose that

$$a_{21} < a_{11} \leqq a_{12}. \tag{2.3}$$

In this case the first pure strategy of player 1 *is that player's unique optimal strategy*. In fact, let us suppose that we also have $X = (\xi, 1-\xi) \in \mathcal{S}(A)$ with $\xi < 1$. Then, by the optimality of the strategy $A_{\cdot 1}$, we must have $XA_{\cdot 1} = v_A = a_{11}$, i.e. $\xi a_{11} + (1-\xi)a_{21} = a_{11}$; but this contradicts (2.3).

Let us find the optimal strategies of player 2. For this purpose we note that in this case we must have $a_{22} > a_{12}$ (since if $a_{22} < a_{12}$ we have the conditions of case b); and if $a_{22} = a_{12}$, we have conditions like those of case c_1)). Let

$$Y = (\eta, 1-\eta) \in \mathcal{T}(A). \tag{2.4}$$

That $Y \in \mathcal{T}(A)$ follows from (2.2). Take $\eta < 1$. According to (2.4) we must have $A_2.Y^T \leqq v_A = a_{11}$, i.e.

$$a_{21}\eta + a_{22}(1-\eta) \leqq a_{11} = a_{12},$$

or

$$\eta \leqq \frac{a_{22} - a_{12}}{a_{22} - a_{21}} = \eta_0.$$

Therefore *in this case* $\mathcal{T}(A)$ *is an interval* $[\eta_0, 1]$.

c_3) If $a_{21} = a_{11} < a_{12}$, a discussion like that for case c_2) shows that here $a_{22} > a_{21}$, *player* 2 *has a unique optimal (pure) strategy* $A_{\cdot 1}$, and $\mathcal{S}(A) = [\xi_0, 1]$, where

$$\xi_0 = \frac{a_{22} - a_{21}}{a_{22} - a_{12}}.$$

c_4) If $a_{21} < a_{11} < a_{12}$, a modification of the discussion in cases c_2) and c_3) leads to the conclusion that *here the players have no optimal mixed strategies.*

Finally, *in case d) there are only mixed optimal strategies*. Let (X, Y) be a pair of them with $X = (\xi, 1-\xi)$, $Y = (\eta, 1-\eta)$, and $0 < \xi, \eta < 1$. Then by 1.14 we must have

$$XA_{\cdot 1} = XA_{\cdot 2} = v_A,$$

i.e.

$$\xi a_{11} + (1-\xi)a_{21} = \xi a_{12} + (1-\xi)a_{22},$$

whence

$$\xi = \frac{a_{22} - a_{21}}{a_{11} - a_{12} - a_{21} + a_{22}},$$

so that

$$X = (a_{22} - a_{21}, a_{11} - a_{12})(a_{11} - a_{12} - a_{21} + a_{22})^{-1} \tag{2.5}$$

and

$$v_A = \frac{a_{11}a_{22} - a_{21}a_{12}}{a_{11} - a_{21} - a_{12} + a_{22}} = \frac{\begin{vmatrix} a_{11} & a_{12} \\ a_{21} & a_{22} \end{vmatrix}}{a_{11} - a_{12} - a_{21} + a_{22}}. \tag{2.6}$$

Similarly we find that

$$Y = (a_{22} - a_{12}, a_{11} - a_{21})(a_{11} - a_{12} - a_{21} + a_{22})^{-1}. \tag{2.7}$$

Let us notice that in this case Γ_A has exactly one equilibrium situation (namely, in the mixed strategies).

It is easy to see that formulas (2.5), (2.6), and (2.7) are special cases of (4.14), (4.16) and (4.15), which will be introduced in due course.

2.2 Graphical (grapho-analytical) methods. If we undertake any detailed analysis of games of order greater than 2×2, it turns out to be so complicated as to be quite overwhelming. Consequently, in all further instances we shall restrict our attention to describing how to find equilibrium situations in special cases that may be representative.

Intuitive geometric representations turn out to be very useful.

Let Γ_A be any $m \times n$-game. As we said in 1.3, the set $\mathcal{S}(A)$ consists of the vectors X for which $\max_X \min_i XA_{\cdot j}$ is attained. We suppose that the simplex $\mathbf{X}$ of mixed strategies of player 1 lies in a "horizontal" coordinate hyperplane, and the payoffs H are in a perpendicular "vertical" hyperplane. Then the equations

$$XA_{\cdot j} = H, \quad j = 1, \ldots, n, \tag{2.8}$$

will be the equations of hyperplanes passing through $\mathbf{X}$. The lower envelope of these hyperplanes, with the equation

$$\min_j XA_{\cdot j} = H \tag{2.9}$$

forms (if we use terminology borrowed from three-space) a convex-upward polyhedron resembling a tent or an obelisk. As is easily seen, the points in this polyhedron represent the possibilities for player 1 (see 2.2, Chapter 3). The projection on the simplexes in the base of the polyhedron of its highest point (or edge or side, respectively) of the polyhedron give us $\mathcal{S}(A)$, and its altitude gives the value v_A of the game.

A similar discussion of the lower envelope of the hyperplanes $A_i.Y^T = H$ over the simplex of mixed strategies of player 2 provides a description of the set of possibilities for player 2 and a description of the set $\mathcal{T}(A)$.

This intuitive description can be useful in the analysis of games of any size, and for $2 \times n$- and $m \times 2$-games, with arbitrary m and n (and, to some extent, for $3 \times n$- and $m \times 3$-games), it provides complete solutions. It is natural to call this method of solving games *grapho-analytical*, since it allows one to establish graphically the qualitative properties of their solutions (for example, their spectra), knowledge of which can lead to explicit forms and analytic expressions for the solutions.

2.3. Let us consider a game Γ_A with a $2 \times n$-payoff matrix

$$A = \begin{bmatrix} a_{11} & a_{12} & \dots & a_{1n} \\ a_{21} & a_{22} & \dots & a_{2n} \end{bmatrix}. \tag{2.10}$$

If we set $X = (\xi, 1 - \xi)$, equation (2.8) assumes the form

$$\xi a_{1j} + (1 - \xi)a_{2j} = H, \tag{2.11}$$

and (2.9) becomes

$$\min_j(\xi a_{1j} + (1 - \xi)a_{2j}) = H. \tag{2.12}$$

There are now two possibilities.

First, it may be that the graph of (2.12) has a single highest point M.

If this point has the abscissa 0 or 1 (Figure 4.4), player 1 has a pure optimal strategy and there is a line (2.11) passing through M and not decreasing in the neighborhood of M. To this line there evidently correspond pure optimal strategies of player 2. All mixtures of these strategies are also optimal.

In addition, there may also be lines (2.11) that pass through M and decrease. These do not correspond to optimal strategies of player 2. However, some of their "mixtures" with strictly decreasing lines, mixtures that do not decrease, also belong to $\mathcal{T}(A)$. Finally, we pass to the convex hull of all the strategies.

If, however, the abscissa ξ_M of M is strictly between zero and one (Figure 4.5), there are at least one line (2.11) passing through M with positive slope and at least one with negative slope. Let the corresponding strategies in (2.11) have indices j' and j''. Solving the corresponding equations yields

$$\xi_M = \frac{a_{2j''} - a_{2j'}}{a_{1j'} - a_{1j''} - a_{2j'} + a_{2j''}},$$

and

$$\eta_M = \frac{a_{1j'}a_{2j''} - a_{1j''}a_{2j'}}{a_{1j'} - a_{1j''} - a_{2j'} + a_{2j''}}.$$

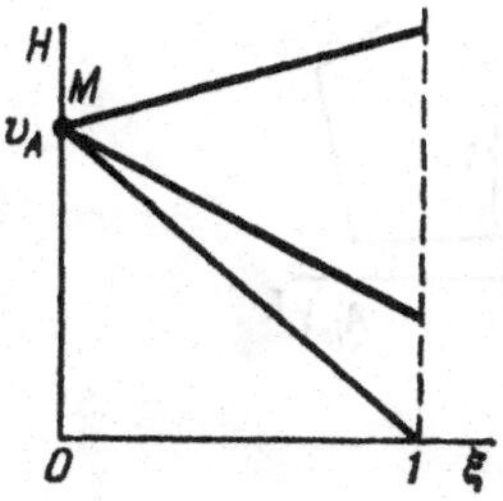

Fig. 4.4

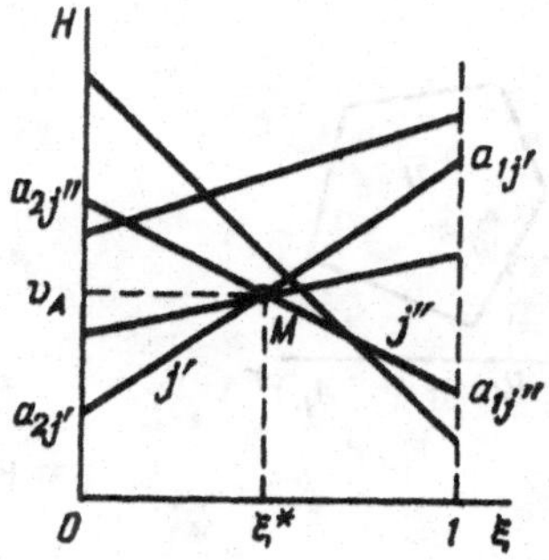

Fig. 4.5

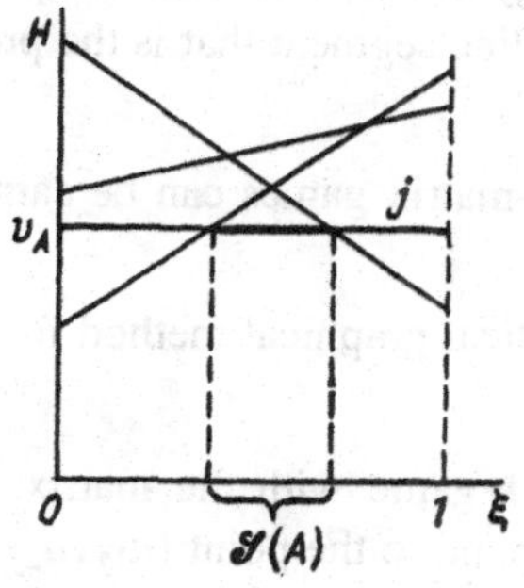

Fig. 4.6

Since the lines $XA_{\cdot j}' = H$ and $XA_{\cdot j}'' = H$ intersect at a point c with ordinate v_A and have slopes of opposite sign, it follows that there are points $\eta_{j',j''} \in (0,1)$ such that a combination of these lines with weights $\eta_{j',j''}$ and $1 - \eta_{j',j''}$ is a horizontal line passing through this point. In particular, the ordinates of the points of intersection of this horizontal line with the lines $\xi = 0$ and $\xi = 1$ will be equal to v_A (and therefore equal to each other); i.e., we will have the equation

$$a_{1j'}\eta_{j',j''} + a_{1j''}(1 - \eta_{j',j''}) = a_{2j'}\eta_{j',j''} + a_{2j''}(1 - \eta_{j',j''}).$$

Solving this equation, we obtain

$$\eta_{j',j''} = \frac{a_{2j''} - a_{1j''}}{a_{1j'} - a_{1j''} - a_{2j'} + a_{2j''}}.$$

The mixed strategy of player 2, composed of the strategies $A_{\cdot j}'$ and $A_{\cdot j}''$ with the weights that we have found, will give player 1 the amount v_A independently of what that player does. Consequently this mixed strategy of player 2 is leveling (see 5°, 1.16, Chapter 3) and is therefore an optimal strategy.

If we combine in this way each strategy-line of positive slope with each strategy-line of negative slope, we obtain a set $Y_1, \ldots, Y_k$ of optimal strategies of player 2. Evidently $\mathcal{T}(A)$ is their convex hull.

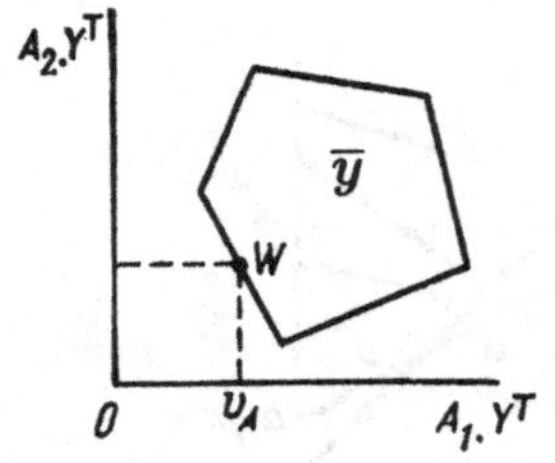

Fig. 4.7

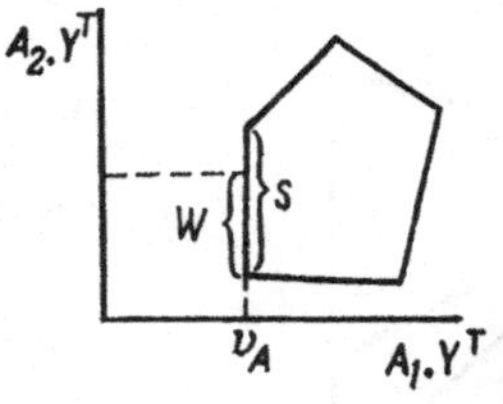

Fig. 4.8

A second possibility is that the graph of (2.12) has a horizontal top part (Figure 4.6). In this case there is a strategy $A_{.j}$ of player 2 for which $a_{1j} = a_{2j}$. As is easily verified, this strategy of player 2 is the unique optimal strategy for this player, and the set $\mathcal{S}(A)$ is the line segment that is the projection of the horizontal top.

The discussion of $m \times 2$-matrix games can be carried out in a similar way.

2.4. There is another very practical graphical method for solving $2 \times n$ (or $m \times 2$) matrix games.

Let us consider the $2 \times n$-game with the matrix A in (2.10) and let each strategy $A_{.j}$ of player 2 correspond to the point (a_{1j}, a_{2j}) of the plane. We denote the set of these points, as the set of all strategies of player 2, by $\bar{y}$. If we extend this correspondence to the mixed strategies, connecting the mixed strategy Y with the point $(A_1.Y^T, A_2.Y^T)$, we obtain a polygon, the convex hull of $\bar{y}$, which we denote by $\bar{Y}$.

Let us now construct a square with a corner at 0 and sides parallel to the coordinate axes, touching $\bar{Y}$ (Figure 4.7). It is clear that the largest coordinate of a point W lying on both the square and the polygon is equal to v_A. Let the set of points W lie on the side S of $\bar{Y}$, and let Y^* (Figure 4.8) consist of all representations of points W as convex combinations of points of $\bar{y}$ (evidently these can only be points on the side S; in general there can be more than two). Evidently Y^* is $\mathcal{T}(A)$. The argument in the preceding subsection shows how to describe the optimal strategies of player 1 in the various cases that may arise here.

2.5 Example. Let us consider the game with the 2×4-matrix

$$A = \begin{bmatrix} 2 & 2 & 2 & 3 \\ 1 & 3 & 4 & 1 \end{bmatrix}.$$

In this case the polygon $\bar{Y}$ in 2.4 is a triangle, and the square touches $\bar{Y}$ along the interval $w'w''$ (Figure 4.9). It is clear that $v_A = 2$. The representations of the points w' and w'' as convex

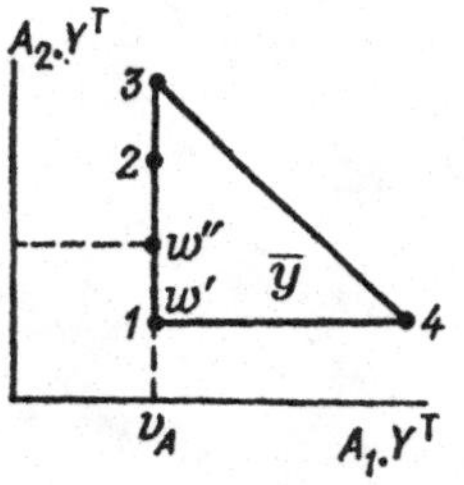

Fig. 4.9

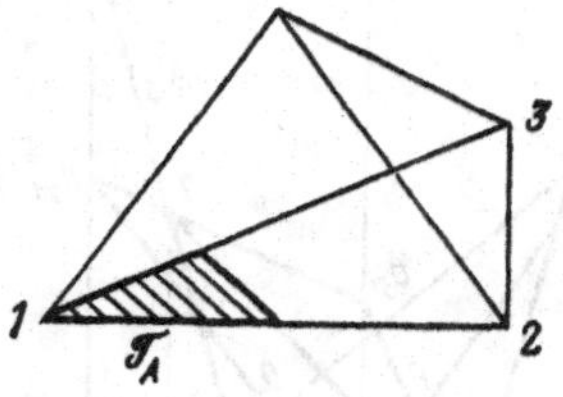

Fig. 4.10

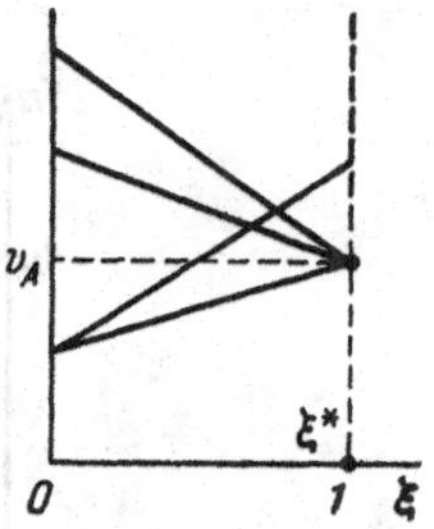

Fig. 4.11

combinations of the vertices *1*, *2*, and *3* are

$$
\begin{aligned}
w' &= 1,\\
w'' &= \frac{1}{2}1 + \frac{1}{2}2,\\
w'' &= \frac{2}{3}1 + \frac{1}{3}3.
\end{aligned}
\tag{2.13}
$$

The set of mixed strategies of player 2 in this game can be represented as a tetrahedron (Figure 4.10). The shaded triangle in the diagram is the convex hull of the points whose barycentric coordinates are the coefficients of the right-hand sides in (2.13). These points also constitute $\mathcal{T}(A)$.

To determine $\mathcal{S}(A)$ we reproduce the construction described in 2.2 (Figure 4.11). We see from the figure that $\mathcal{S}(A)$ consists only of the pure strategy of player 1.

2.6. Although the graphical solution of matrix games described in 2.2 is also available for $3 \times n$- and $m \times 3$-games, in practice carrying out the corresponding constructions requires detailed drawings, constructed according to the rules of descriptive geometry. We confine ourselves to the example of constructing the solution of the 3×3-matrix game with the payoff matrix

$$A = \begin{bmatrix} 0 & 20 & 24 \\ 18 & 8 & 24 \\ 15 & 10 & 0 \end{bmatrix}.$$

From the purely informational point of view, it would be enough to construct a frontal ("façade") projection. However, for full visualization we have added a vertical projection (plan). The figure completely corresponds to what was said in 2.2 and hardly needs any additional comments. The analytic part of the solution is as follows.

We see from the figure that $\mathcal{S}(A)$ corresponds to the highest points of the interval RQ. In every case, one of the projections of the two points R or Q falls in $\mathcal{S}(A)$.

The point Q lies in the plane $\xi_3 = 0$ on the intersection of the lines with the equations

$$\frac{H_Q - a_{11}}{a_{31} - a_{11}} = \frac{\xi_1(Q) - 1}{-1}, \quad \frac{H_Q - a_{12}}{a_{32} - a_{12}} = \frac{\xi_1(Q) - 1}{-1},$$

from which $H_Q = 12$, and the point P is in the plane $\xi_2 = 0$ at the intersection of the lines with the equations

$$\frac{H_P - a_{11}}{a_{31} - a_{11}} = \frac{\xi_1(P) - 1}{-1}, \quad \frac{H_P - a_{12}}{a_{32} - a_{12}} = \frac{\xi_1(P) - 1}{-1},$$

from which again $H_P = 12$. Therefore the interval PQ is horizontal. Consequently the part RQ is also horizontal, so that $\mathcal{S}(A)$ is all of $R'Q'$, and $v_A = 12$.

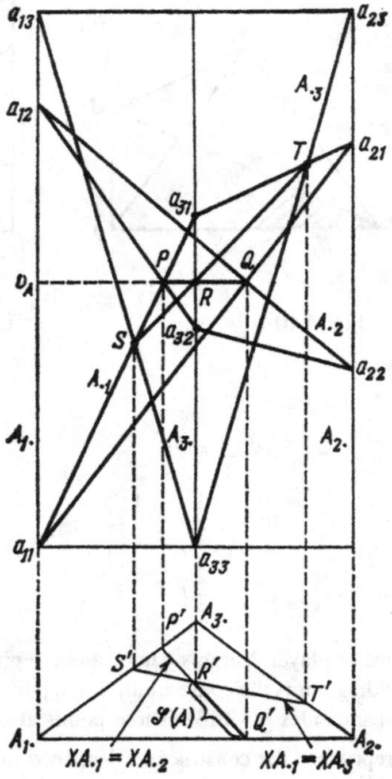

Fig. 4.12

From (2.14) it follows that $\xi_1(Q) = 1/3$, so that $Q' = (1/3, 2/3, 0)$.

To calculate the barycentric coordinates of R it is simplest to use the facts that $XA_{.j} = v_A$ $(j = 1, 2, 3)$ and $XJ_3^T = 1$, i.e.

$$18\xi_2 + 15\xi_3 = 12,$$

$$20\xi_1 + 8\xi_2 + 10\xi_3 = 12,$$

$$24\xi_1 + 24\xi_2 = 12,$$

$$\xi_1 + \xi_2 + \xi_3 = 1,$$

from which $R' = (1/4, 1/4, 1/2)$.

Finally, by the theorem on complementary slackness (see 1.14) $\mathcal{T}(A)$ is the set of nonnegative solutions of the system of equations $A_{i.}Y^T = v_A$ $(i = 1, 2, 3)$ and $J_3Y^T = 1$, i.e. of the system

$$20\eta_2 + 24\eta_3 = 12,$$

$$18\eta_1 + 8\eta_2 + 24\eta_3 = 12,$$

$$15\eta_1 + 10\eta_2 = 12,$$

$$\eta_1 + \eta_2 + \eta_3 = 1.$$

Hence it follows that $\mathcal{T}(A)$ consists of the single mixed strategy $(2/5, 3/5, 0)$.

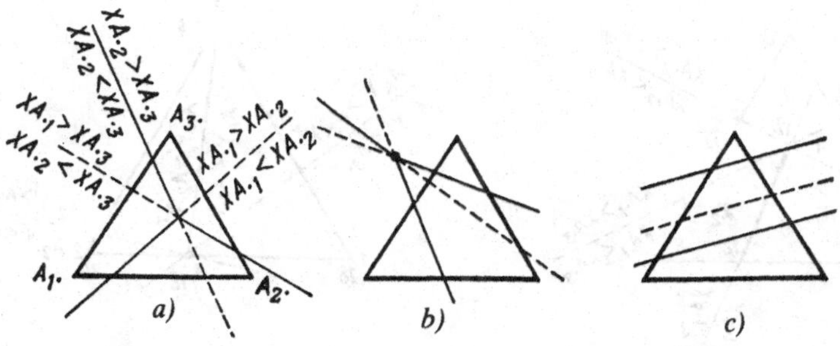

Fig. 4.13

2.7. For the solution of 3×3-games (and with some artifices also for $3 \times n$-games, where $n > 3$) one can apply a different method of graphical analysis.

The simplex of mixed strategies of player 1 in a 3×3-game Γ_A is a triangle, and all of the constructions in 2.6 are realized in three-space. We shall consider them in plan, i.e. look at them from above. In the present case we have

$$\max_X \min_j XA_{\cdot j} = \max_X \min\{XA_{\cdot 1}, XA_{\cdot 2}, XA_{\cdot 3}\}.$$

Let $\boldsymbol{X}_j$ $(j = 1, 2, 3)$ denote the set of $X \in \boldsymbol{X}$, for which an interior minimum is attained at $XA_{\cdot j}$. Then we can write

$$\max_X \min_j XA_{\cdot j} = \max\{\max_{X \in \boldsymbol{X}_1} XA_{\cdot 1}, \max_{X \in \boldsymbol{X}_2} XA_{\cdot 2}, \max_{X \in \boldsymbol{X}_3} XA_{\cdot 3}\}.$$

Therefore we need to find the sets $\boldsymbol{X}_1$, $\boldsymbol{X}_2$, and $\boldsymbol{X}_3$, maximize the corresponding linear forms on these sets, and compare the maxima of these forms.

Evidently the sets $\boldsymbol{X}_j$ of strategies are polygons whose sides are either on the lines whose equations are

$$XA_{\cdot j} = XA_{\cdot j'}, \quad j' \neq j,$$

or on the sides of the simplex $\boldsymbol{X}$ (Figure 4.13, a, b, c). We see that the largest possible number of sides of any polygon $\boldsymbol{X}_j$ is five. But the set of points of the polygon at which a linear form is maximized is either a vertex or a pair of vertices with the side joining them, or possibly the whole polygon. Therefore it is enough to evaluate the linear forms at the vertices of the polygons (as is easily done; we have to calculate at most six values) and compare them with each other. The largest of these values will be v_A, and the convex hull of the points at which it is attained is $\mathcal{S}(A)$.

The problem of finding $\mathcal{T}(A)$ is similar. Its solution is easier because we already know v_A.

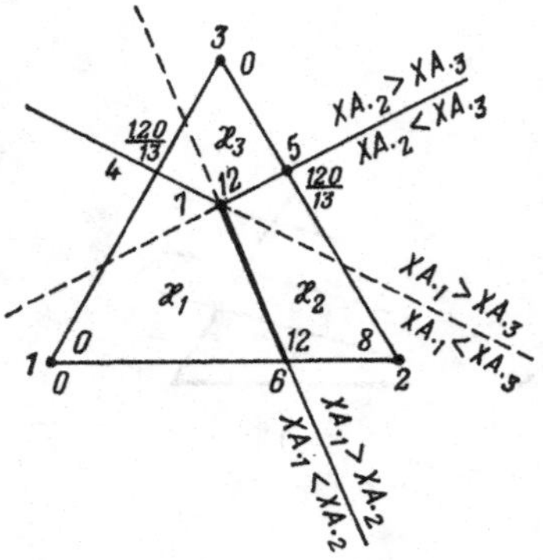

Fig. 4.14

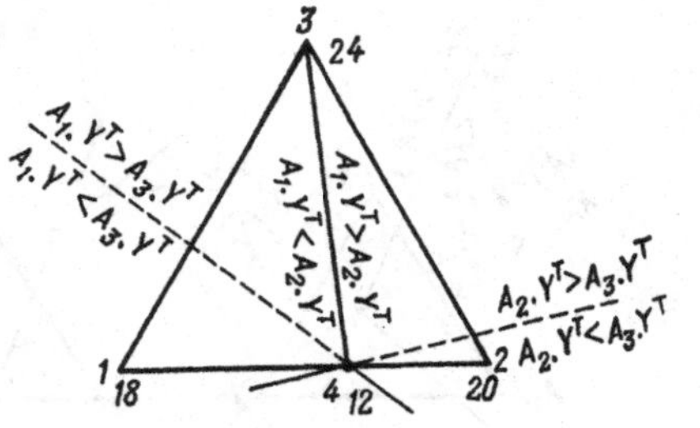

Fig. 4.15

2.8 Example. To compare the calculations, we consider the game with the same payoff matrix A as in 2.6.

The process of constructing the sets X_1, X_2, and X_3 is illustrated in Figure 4.14. The possible extreme strategies correspond to the points labeled 1–7. The values of the linear forms calculated at these points are indicated on the figure. The largest one is 12. It occurs at points 6 and 7, and is the value of the game. The line segment joining points 6 and 7 is $\mathcal{S}(A)$.

The set $\mathcal{T}(A)$ can be described on the basis of the set $\mathcal{S}(A)$ and the number v_A. Figure 4.15 gives a graphical basis for calculating it independently.

§3 Matrix games and linear programming

3.1 Fundamental concepts of linear programming. We present some fundamental terminology of linear programming.

Definition. The standard (maximization) $m \times n$-*problem of linear programming* is the problem of finding an m-vector X that satisfies the inequalities

$$XA \leqq B, \tag{3.1}$$

$$X \geqq 0, \tag{3.2}$$

and maximizes the product

$$XC^T \to \max, \tag{3.3}$$

where A is an $m \times n$-matrix, and B and C are an n-vector and an m-vector, respectively.

We denote this problem by $P^+(A, B, C)$. A vector X that satisfies (3.1) and (3.2) is called an *admissible solution* of the problem. The set of these solutions is denoted by $A^+(A, B)$. Vectors in $A^+(A, B)$ that satisfy (3.3) are called *optimal solutions* (or simply *solutions*) of the problem $P^+(A, B, C)$; we denote this set by $S^+(A, B, C)$. □

The set $S^+(A, B, C)$ might be empty. This can happen, in the first place, if the problem does not have any admissible solutions: $A^+(A, B) = \emptyset$; and in the second place if the linear form XC^T does not have a maximum on $A^+(A, B)$, i.e., if the set $A^+(A, B)$ is unbounded in the corresponding direction and XC^T takes arbitrarily large values on it.

Definition. The *problem dual to* $\boldsymbol{P}^+(A, B, C)$ is the problem $\boldsymbol{P}^-(A^T, C, B)$, which is to find an n-vector Y that satisfies

$$AY^T \geqq C^T, \tag{3.4}$$

$$Y \geqq 0, \tag{3.5}$$

and minimizes the product BY^T:

$$BY^T \to \min. \qquad \square$$

In a way similar to that used above, we define the set $\boldsymbol{A}^-(A^T, C)$ of admissible solutions and the set $S^-(A^T, C, B)$ of optimal solutions.

In a similar way, we define the *minimization* problem $\boldsymbol{P}^-(A, B, C)$ in which the inequalities corresponding to (3.1) have the opposite sense, and the linear form XC^T is to be minimized; and also the dual maximization problem $\boldsymbol{P}^+(A^T, B, C)$.

In this system of definitions of problems dual to $\boldsymbol{P}^-(A^T, C, B)$ we encounter the problem $\boldsymbol{P}^+(A^{TT}, B, C)$, i.e. the problem $\boldsymbol{P}^+(A, B, C)$, that we introduced originally. Therefore we may speak of pairs of mutually dual problems.

3.2. Let us consider the pair of dual linear programming problems $P^+(A, B, C)$ described by (3.1)–(3.3), and $\boldsymbol{P}^-(A^T, B, C)$, described by (3.4)–(3.6).

In (3.1) we multiply on the right by Y^T, using (3.5), to obtain

$$XAY^T \leqq BY^T;$$

and similarly in (3.4) we multiply on the left by X, using (3.5), to obtain

$$XAY^T \geqq XC^T.$$

Therefore we must have, for all $X \in \boldsymbol{A}^+(A, B)$ and $Y \in \boldsymbol{A}^-(A^T, C)$,

$$XC^T \leqq BY^T. \tag{3.7}$$

Equality in these relations is evidently a sufficient condition for

$$X \in S^+(A, B, C), \qquad Y \in S^-(A^T, C, B).$$

Its necessity is less trivial; it will follow from the theorem in 3.4.

This proposition is sometimes called the duality theorem of linear programming.

If there is equality in (3.7), the common value of the left and right sides of (3.7) is called the *value* of the pair of dual problems $\boldsymbol{P}^+(A, B, C)$ and $\boldsymbol{P}^-(A^T, C, B)$. We denote it by $v(A, B, C)$.

3.3 Equivalence of the theories of matrix games and of linear programming. The solution of every matrix game can be reduced to the solution of a pair of dual linear programming problems.

Let us consider an arbitrary matrix game Γ_A with an $m \times n$-payoff matrix A, all of whose elements are positive (if this is not the case, we can bring about positivity by passing to an affinely equivalent game).

Let $\langle \boldsymbol{X}, \boldsymbol{Y}, A\rangle$ be a mixed extension of Γ_A and let $X \in \boldsymbol{X}$. Set $v_X = \min_j XA_{\cdot j}$. It follows from the positivity of A that $v_X > 0$.

We have

$$v_X \leqq XA_{\cdot j}, \qquad j = 1, \ldots, n. \tag{3.8}$$

It follows from 1.10 that $v_X \leqq v_A$, so that by 1.12 the equation

$$v_X \to \max_X v_X = v_A \tag{3.9}$$

is a necessary and sufficient condition for $X \in \mathcal{S}(A)$.

Let us consider the vector $\tilde{X} = X/v_A$. It follows from (3.8) that

$$\tilde{X}A \geqq J_n. \tag{3.10}$$

Since X is normalized and nonnegative, it follows that

$$\tilde{X}J_m^T = 1/v_X, \tag{3.11}$$

$$\tilde{X} \geq \mathbf{0}. \tag{3.12}$$

It follows from (3.9) and (3.7) that $X \in \mathcal{S}(A)$ is equivalent to

$$\tilde{X}J_m^T \to \min. \tag{3.13}$$

The relations (3.10), (3.12) and (3.13) define a linear programming problem $P^-(A, J_n, J_m)$, and

$$v(A, J_n, J_m) = 1/v_A, \tag{3.14}$$

$$\frac{S^-(A, J_n, J_m)}{v(A, J_n, J_m)} = \mathcal{S}(A). \tag{3.15}$$

Similarly, if we introduce

$$v_Y = \max_i A_{i\cdot} Y^T$$

for $Y \in \mathbf{Y}$, and set $\tilde{Y} = Y/v_Y$, we obtain a problem $P^+(A^T, J_m, J_n)$ for which

$$\frac{S^+(A^T, J_m, J_n)}{v(A, J_n, J_m)} = \mathcal{T}(A). \tag{3.16}$$

3.4. Let us select a pair of dual problems $P^+(A, B, C)$ and $P^-(A^T, C, B)$ with an $m \times n$-matrix A, and consider the $(m+n+1) \times (m+n+1)$-matrix, written in block form,

$$M = \begin{bmatrix} \mathbf{0} & -A & C^T \\ A^T & \mathbf{0} & -B^T \\ -C & B & \mathbf{0} \end{bmatrix}.$$

The matrix game Γ_M with payoff matrix M is symmetric, and therefore, by 1.23,

$$v_M = 0, \tag{3.17}$$

$$\mathcal{S}(M) = \mathcal{T}(M). \tag{3.18}$$

Let $Z = (U, V, t)$ be a strategy of a player in this game, where U is an m-vector, V is an n-vector, and t is a scalar. It is clear that

$$Z \geq 0, \tag{3.19}$$

$$ZJ_{m+n+1}^T = 1. \tag{3.20}$$

Solving the problems $P^+(A, B, C)$ and $P^-(A^T, C, B)$ is equivalent to finding an optimal strategy in the game Γ_M. The following theorem gives a precise description of this equivalence.

Theorem. *Let there be given a pair $P^+(A, B, C)$ and $P^-(A^T, C, B)$ of linear programming problems, and the corresponding matrix M. Then:*

1) *if there is an optimal strategy $Z = (U, V, t)$ in which $t > 0$, then*

$$X = U/t \in S^+(A, B, C), \tag{3.21}$$

$$Y = V/t \in S^-(A^T, C, B); \tag{3.22}$$

2) *if $t = 0$ for every optimal strategy $Z = (U, V, t)$, then*

$$S^+(A, B, C) = S^-(A^T, C, B) = \emptyset;$$

3) *if*

$$\begin{aligned} &X \in S^+(A, B, C), \\ &Y \in S^-(A, B, C), \\ &t = (XJ_m^T + J_n Y^T + 1)^{-1}, \end{aligned} \tag{3.23}$$

then

$$Z = t(X, Y, 1) \in \mathcal{S}(M) = \mathcal{T}(M). \tag{3.24}$$

Proof. 1) Let $Z = (U, V, t)$ and $t > 0$. By (3.17) the condition that Z is optimal can be written as

$$ZM \geqq \mathbf{0}^{m+n+1}. \tag{3.25}$$

We may write (3.19), (3.20), and (3.21) in a "block" form:

$$(U, V, t) \geqq \mathbf{0}^{m+n+1},$$

$$(U, V, t)J_{m+n+1}^T = 1.$$

$$(U, V, t) \begin{bmatrix} \mathbf{0} & -A & C^T \\ A^T & \mathbf{0} & -B^T \\ C & B & \mathbf{0} \end{bmatrix} \geqq \mathbf{0}^{n+m+1},$$

or

$$U \geq \mathbf{0}, V \geq \mathbf{0}, t \geqq \mathbf{0}, \tag{3.26}$$

$$UJ_m^T + VJ_n^T + t = 1, \tag{3.27}$$

$$AV^T - tC^T \geqq \mathbf{0}, \tag{3.28}$$

$$-UA + tB \geqq \mathbf{0}, \tag{3.29}$$

$$UC^T - BV^T \geqq 0. \tag{3.30}$$

It follows from the condition $t > 0$ that we may consider $X = U/t$ and $Y = V/t$ instead of U and V. Dividing each of (3.29), (3.28) and (3.26) by t, we obtain the statements of the problems $P^+(A, B, C)$ and $P^-(A^T, C, B)$. If we apply the duality theorem to these, we obtain (3.7): $XC^T \leqq BY^T$. On the other hand, dividing (3.30) by t yields $XC^T \geqq BY^T$, i.e., there is equality in (3.7), so that (3.21) and (3.22) are satisfied.

2) Now let $t = 0$ *for every* strategy $Z = (U, V, t) \in \mathcal{S}(M)$. Then we have the hypotheses of the theorem of 1.18, so that *for some* optimal strategy $Z = (U, V, t)$ we have strict inequality in (3.30): $UC^T - BV^T > 0$, and the system (3.28)–(3.30) assumes the form

$$AV^T \geqq \mathbf{0}, \tag{3.31}$$

$$UA \leqq \mathbf{0}, \tag{3.32}$$

$$UC^T > BV^T. \tag{3.33}$$

In the last inequality it is evident that either the larger number is positive or the smaller is negative.

If

$$UC^T > 0 \tag{3.34}$$

we choose $X \in S^+(A, B, C)$ and consider vectors of the form $X + \alpha U$ with $\alpha > 0$. By (3.1) and (3.32) we have

$$(X + \alpha U)A = XA + \alpha UA \geqq B,$$

and by (3.2) and (3.26),

$$X + \alpha U \geqq \mathbf{0}^m.$$

Therefore $X + \alpha U \in A^+(A, B)$ for every $\alpha > 0$. But

$$(X + \alpha U)C^T = XC^T + \alpha UC^T$$

increases unboundedly with α, by (3.34). Therefore $S^+(A, B, C) = \emptyset$. Consequently there cannot be a Y for which (3.7) will hold for all $X \in A^+(A, B)$. Consequently $A^-(A^T, C) = \emptyset$ and hence $S^-(A^T, C, B) = \emptyset$.

Similarly, the assumption that $BV^T < 0$ leads, on the basis of (3.4), (3.5), (3.26) and (3.31), to $S^-(A^T, C, B) = \emptyset$ and $A^+(A, B) = \emptyset$ (and therefore to $S^+(A, B, C) = \emptyset$).

3) Finally, let $X \in S^+(A, B, C)$ and $Y \in S^-(A^T, B, C)$. Take t and Z as in (3.23) and (3.24). Evidently we now have $Z \geqq 0$ and $ZJ^T_{m+n+1} = 1$, so that, in the game with payoff matrix M, the vector Z can be interpreted as a strategy (of any player).

In addition

$$ZM = (tU, tV, t)\begin{bmatrix} 0 & -A & C^T \\ A^T & 0 & -B^T \\ -C & B & 0 \end{bmatrix}$$

$$= t(YA^T - C, -XA + B, XC^T - YB^T).$$

Since X and Y were assumed to be optimal, there is equality in (3.4), (3.2), and (3.7). Therefore all the components of the vector ZM are nonnegative; consequently $Z \in \mathcal{S}(M)$. □

This theorem makes it possible to reduce the determination of (all the) saddle points of matrix games to the determination of (all the) optimal solutions of corresponding dual pairs of linear programming problems. Consequently, to find solutions of matrix games we can use any methods of solving linear programming problems (in particular, the simplex well-known method).

3.5 Application to the symmetrization of matrix games. By using what we have said above, we can amplify the content of 1.23–1.25 by another method of symmetrizing matrix games.

By 3.3, the description of the solution of the $m \times n$-game Γ_A with $v_A = 0$ (although the last restriction, as has repeatedly been noticed, is not essential) reduces to the description of the solution of a pair of dual problems $P^-(A, J_n, J_m)$ and $P^+(A^T, J_m, J_n)$, which, by 3.4, is equivalent to the solution of the $(m+n+1) \times (m+n+1)$-matrix game with payoff matrix

$$\begin{bmatrix} 0 & -A & J^T_m \\ A^T_m & 0 & -J^T_n \\ -J_m & J_n & 0 \end{bmatrix},$$

which is evidently skew-symmetric.

3.6 Hide-and-seek games. Establishing a correspondence between matrix games and pairs of dual linear programming problems suggests the question of discovering various informal interpretations of classes of matrix games to which there correspond informal interpretations of classes of (pairs of) linear programming problems.

One such class consists of games that can be interpreted as "hide-and-seek" games, in the following formulation.

Let

$$B = \|b_{ij}\|_{i,j=1,\dots,n}$$

be a square $n \times n$-matrix. We consider a "hiding game on B", in which player 2 is hidden on one of the n^2 entries of B, and player 1 chooses a line (row or column) of B. If player 2 is hidden at (i, j) and player 1 chooses either row i or column j, then player 1 receives the amount b_{ij} from player 2.

Consequently we have a $2n \times n^2$-matrix game in which the strategies of player 1 are pairs (α, k), where $\alpha = 1$ or 2 ($\alpha = 1$ corresponds to searching on rows, and $\alpha = 2$, to columns), $k = 1, \dots, n$; and the strategies of player 2 are evidently pairs (i, j) where $i, j = 1, \dots, n$. The elements $a_{(\alpha,k)(i,j)}$ of the payoff matrix A of this game are evidently as follows:

$$a_{(\alpha,k)(i,j)} = \begin{cases} b_{ij} & \text{if } \alpha = 1 \text{ and } k = i, \text{ or } \alpha = 2 \text{ and } k = j, \\ 0 & \text{otherwise.} \end{cases} \tag{3.35}$$

The linear programming problem $P^+(A^T, J_{2n}, J_{n^2})$ corresponding to this matrix game is

$$A\tilde{Y}^T \leqq J_{2n}^T, \tag{3.36}$$

$$J_{n^2}\tilde{Y}^T \to \max, \tag{3.37}$$

$$\tilde{Y} \geq 0,$$

where $\tilde{Y}$ is an n^2-vector. We denote its components by $\tilde{\eta}_{(i,j)}$.

In terms of coordinates, (3.36) has the form

$$\sum_{(i,j)} a_{(\alpha,k)(i,j)}\,\tilde{\eta}_{(i,j)} \leqq 1, \ \alpha = 1, 2\,; \ k = 1, \dots, n,$$

or, by (3.55),

$$\sum_j b_{ij}\,\tilde{\eta}_{(i,j)} \leqq 1 \text{ for } \alpha = 1, \tag{3.38}$$

$$\sum_i b_{ij}\,\tilde{\eta}_{(i,j)} \leqq 1 \text{ for } \alpha = 2. \tag{3.39}$$

The maximization (3.37) turns into

$$\sum_{(i,j)} \tilde{\eta}_{(i,j)} \to \max. \tag{3.40}$$

Let us introduce new variables, setting

$$b_{ij}\,\tilde{\eta}_{(i,j)} = \zeta_{ij}, \quad i, j = 1, \dots, n. \tag{3.41}$$

Then (3.38), (3.39) and (3.40) become

$$\sum_j \zeta_{ij} \leqq 1, \quad \sum_i \zeta_{ij} \leqq 1, \quad \zeta_{ij} \geqq 0, \tag{3.42}$$

$$\sum_{(i,j)} (\zeta_{ij}/b_{ij}) \to \max. \tag{3.43}$$

The maximization problem (3.43) with the constraints (3.42) is the well-known *assignment problem* with the effectiveness matrix $\|b_{ij}^{-1}\|_{i,j=1,\dots,n}$. There are integral-valued solutions of this problem, in which each ζ_{ij} is 0 or 1. The integral-valued solution of the problem actually consists of finding a permutation π that associates a row-index with a column-index in such a way that

$$\sum_i b_{i,\pi i}^{-1} \to \max, \tag{3.44}$$

where we maximize over all the permutations of n symbols.

Let the maximum in (3.44) be obtained for the permutation π_0.

Then

$$\zeta_{ij} = \begin{cases} 1 & \text{if } j = \pi_0 i, \\ 0 & \text{otherwise,} \end{cases}$$

or, if we return to the variables $\tilde{\eta}_{(i,j)}$, we have

$$\tilde{\eta}_{(i,j)} = \begin{cases} b_{ij}^{-1} & \text{if } j = \pi_0 i, \\ 0 & \text{otherwise.} \end{cases}$$

According to 3.4, to pass from the solution $\tilde{Y}$ of the linear programming problem (3.36), (3.37) to the optimal strategy Y of player 2 in the original matrix game, it is enough to normalize the vector solution; in the present case this yields

$$\eta_{(i,j)} = b_{ij}^{-1} \left(\sum_i b_{ij}^{-1} \right)^{-1} \quad \text{if } j = \pi_0 i, \text{ and 0 otherwise,} \tag{3.45}$$

and the normalizing factor will be the value v_A of the game.

It is evident that to each permutation π_0 that maximizes (3.44) there corresponds an optimal strategy Y_{π_0} of player 2, described by (3.45), in the original matrix game. Here it is clear that if, for two permutations π_0' and π_0'' that maximize (3.44), we have $b_{i\pi_0' i} = b_{i\pi_0'' i}$ for each i, $i = 1, \ldots, n$, then $Y_{\pi_0'} = Y_{\pi_0''}$. We also notice that all the probabilities determined by (3.45) are positive, i.e., that the spectrum V_{π_0} consists of n pure strategies of the form $(i, \pi_0 i)$, where $i = 1, 2, \ldots, n$. The convex hull of all such "integral-valued", "permuted" optimal strategies Y_π is the set of all optimal strategies of player 2.

It remains only to find the optimal strategies of player 1. We consider the linear programming problem

$$\tilde{X} A \geqq J_{n^2}^T, \tag{3.46}$$

$$\tilde{X} J_{2n}^T \to \min, \tag{3.47}$$

$$\tilde{X} \geqq \mathbf{0},$$

where $\tilde{X}$ is a $2n$-vector. We denote its first n components by $\tilde{\xi}_{1i}$ and the last n by $\tilde{\xi}_{2i}$.

By (3.35), we can rewrite (3.46) in the form

$$(\tilde{\xi}_{1i} + \tilde{\xi}_{2i}) b_{ij} \geqq 1, \quad i, j = 1, \ldots, n. \tag{3.48}$$

Recalling that $X = \tilde{X}/v_A$ is an optimal strategy of player 1, and setting $\xi_{ak} = \tilde{\xi}_{ak}/v_A$ we can rewrite (3.48) in the form

$$\xi_{1i} + \xi_{2i} \geqq b_{ij}^{-1} v_A, \quad i, j, = 1, \ldots, n. \tag{3.49}$$

For there to be equality in (3.49) for a specific pair (i, j), it is enough for it to be contained in the spectrum of an optimal strategy of player 2, or in the spectrum of a "permuted" of such a strategy. After finding all these strategies of player 2 and replacing the corresponding inequalities in (3.49) by equations, we obtain a system of relations that describe the set of optimal strategies of player 1.

3.7 Polyhedral games. The application of the transition to dual problems in linear programming allows us to solve games that are more general than matrix games.

Definition. A *polyhedral game* is a two-person zero-sum game

$$\Gamma = \langle \mathfrak{x}, \mathfrak{y}, H \rangle \tag{3.50}$$

in which $\mathfrak{x}$ and $\mathfrak{y}$ are closed convex bounded polytopes in finite-dimensional Euclidean spaces, and the payoff function H is bilinear. □

As an example of a polyhedral game we may take a mixed extension of any matrix game. The polytopes of the players' strategies in it are the corresponding simplexes.

Evidently polyhedral games are convexo-concave, and therefore by the corollary in 3.5 of Chapter 3, each player has a pure optimal strategy. The actual solution of a polyhedral game depends on the way the polytopes $\mathfrak{x}$ and $\mathfrak{y}$ are defined.

3.8. If the polytopes $\mathfrak{x}$ and $\mathfrak{y}$ of strategies in the polyhedral game (3.50) are given as the convex hulls of their vertices:

$$\mathfrak{x} = \operatorname{conv}[x^{(1)}, \ldots, x^{(m)}],$$
$$\mathfrak{y} = \operatorname{conv}[y^{(1)}, \ldots, y^{(n)}],$$

then each strategy $x \in \mathfrak{x}$ and $y \in \mathfrak{y}$ can be represented (in general in more than one way) as convex combinations of the vertices of the associated polytopes:

$$x = \sum_{i=1}^{m} \xi_i x^{(i)}, \quad \text{where} \sum_{i=1}^{m} \xi_i = 1 \quad \text{and } \xi_i \geqq 0,\ i = 1, \ldots, m, \tag{3.51}$$

$$y = \sum_{j=1}^{n} \eta_j y^{(j)}, \quad \text{where} \sum_{j=1}^{n} \eta_j = 1 \quad \text{and } \eta_j \geqq 0,\ j = 1, \ldots, n. \tag{3.52}$$

Since H is bilinear, it follows that

$$H(x, y) = \sum_{i=1}^{m} \sum_{j=1}^{n} H(x^{(i)}, y^{(j)}) \xi_i \eta_j,$$

so that, instead of the strategies x and y, we may consider the equivalent mixed strategies

$$X = (\xi_1, \ldots, \xi_m) \text{ and } Y = (\eta_1, \ldots, \eta_n).$$

Therefore Γ can be reduced to the $m \times n$-matrix game with payoff matrix $\|h_{ij}\|$, where

$$h_{ij} = H(x^{(i)}, y^{(j)}).$$

It is clear that (3.51) and (3.52) establish a correspondence between the solutions of this matrix game and the solutions of the original polyhedral game Γ.

3.9. However, if the polytopes $\boldsymbol{x}$ and $\boldsymbol{y}$ in (3.50) are not specified by their vertices, but in some other way, the construction of a matrix game corresponding to the given polyhedral game can be difficult.

For example, let $\boldsymbol{x}$ and $\boldsymbol{y}$ be given by their faces, i.e. by the systems of inequalities

$$xM \geqq P \text{ and } Ny^T \geqq Q. \tag{3.53}$$

Here M and N are $m \times p$- and $n \times q$-matrices (where p and q are collections of restrictions determined by $\boldsymbol{x}$ and $\boldsymbol{y}$; from these restrictions, we obtain the equations of hyperplanes on which the faces of $\boldsymbol{x}$ and $\boldsymbol{y}$ lie). Here $\boldsymbol{x}$ is described by a system of inequalities in one sense; and $\boldsymbol{y}$, in the opposite sense. This is done only for symmetry in the discussion but evidently causes no loss of generality. In addition, we may assume that $\boldsymbol{x}$ and $\boldsymbol{y}$ lie in the nonnegative orthants of the enveloping Euclidean spaces. (This also does not decrease the generality of our constructions.) We shall not include a corresponding condition of nonnegativity in (3.53).

We denote the coefficient matrix of the bilinear form H by A. Then

$$H(x, y) = xAy^T. \tag{3.54}$$

For a mixed extension of the $m \times n$-matrix game (which is polyhedral) the system (3.53) assumes the form

$$xJ_m^T \leqq 1, \qquad J_n y^T \geqq 1,$$
$$-xJ_m^T \leqq -1, \quad -J_n y^T \geqq -1.$$

(Here restrictions in the form of equations have been replaced by pairs of inequalities with opposite senses.)

3.10. Let us describe a method for solving a polyhedral game Γ for which the sets $\boldsymbol{x}$ and $\boldsymbol{y}$ are defined by (3.53) and the payoff matrix is given by (3.54). The method reduces to finding the maximum

$$\max_{x \in \boldsymbol{x}} \min_{y \in \boldsymbol{y}} xAy^T,$$

and the strategies $x \in \boldsymbol{x}$ for which the outer extremum is attained. Let us fix a point $x \in \boldsymbol{x}$. Finding $\min_{y \in \boldsymbol{y}} xAy^T$ corresponds to finding a solution, depending on x, of the linear programming problem

$$(xA)y^T \to \min,$$
$$Ny^T \geqq Q^T,$$
$$v \geqq \mathbf{0}.$$

Since $\boldsymbol{y}$ is nonempty and bounded, this is a solvable problem. Consequently (this follows, for example, from the theorem in 3.4) we can also solve the dual problem

$$vQ^T \to \max,$$
$$vN \leqq xA,$$
$$v \geqq \mathbf{0}.$$

This problem also depends on x: in particular, the polytopes of admissible solutions depend on x; we denote these by $v(x)$. By the duality theorem (3.2) we have

$$\min_{y \in \boldsymbol{y}} xAy^T = \max_{v \in v(x)} vQ^T.$$

Since this equation holds for every $x \in \boldsymbol{x}$, we must have

$$\max_{x \in \boldsymbol{x}} \min_{y \in \boldsymbol{y}} xAy^T = \max_{x \in \boldsymbol{x}} \max_{v \in v(x)} vQ^T. \tag{3.55}$$

The double maximum on the right is evidently attained on the optimal solutions (x, v) of the linear programming problem

$$\begin{aligned} (x, v)(\mathbf{0}, Q)^T &\to \max, \\ xM &\leqq P, \\ (x, v)\begin{pmatrix} -A \\ N \end{pmatrix} &\leqq \mathbf{0}, \\ (x, v) &\geqq \mathbf{0}. \end{aligned} \tag{3.56}$$

For the values of x that appear in these solutions (x, v), the maximum is also attained on the left of (3.55). These values of x form the set $\mathcal{S}(\Gamma)$.

Similarly, $\mathcal{T}(\Gamma)$ consists of the values $y \in \boldsymbol{y}$ that appear in the optimal solution (y, u) of the problem

$$\begin{aligned} (\mathbf{0}, P)(y, u)^T &\to \min, \\ Ny^T &\geqq Q, \\ (-A, M)(y, u)^T &\geqq \mathbf{0}, \\ (y, u) &\geqq \mathbf{0}. \end{aligned} \tag{3.57}$$

This discussion is essentially an example of the solution of a matrix game by a "nonmatrix" method, i.e., without constructing the payoff matrix.

3.11. The property of independence of irrelevant alternatives which, according to 2.18, Chapter 1, equilibrium situations have in every noncooperative game, can be sharpened for polyhedral games.

Theorem. *Let $\Gamma = \langle \boldsymbol{x}, \boldsymbol{y}, A \rangle$ be a polyhedral game, and let Γ' be a polyhedral $\boldsymbol{x}' \times \boldsymbol{y}'$-subgame. Then if*

$$\mathcal{S}(\Gamma) \cap \boldsymbol{x}' = \emptyset, \tag{3.58}$$

there is an optimal strategy $\bar{x}$ of player 1 in Γ' that lies on the boundary of $\boldsymbol{x}'$.

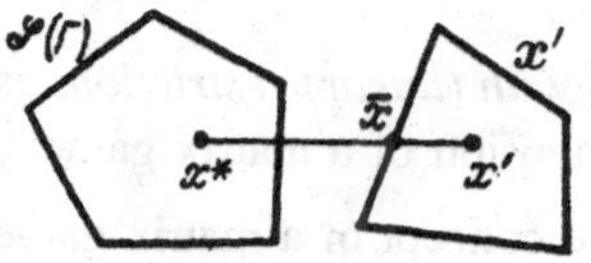

Fig. 4.16

Proof. Choose any $x^* \in \mathscr{S}(\Gamma)$. By hypothesis, $x^* \notin \mathbf{x}'$.

Let us suppose that there exists a

$$y^* \in \mathscr{T}(\Gamma) \cap \mathbf{y}' \neq \emptyset,$$

and select any $x' \in \mathscr{S}(\Gamma)$. We are interested in the case when x' is not on the boundary of $\mathbf{x}'$. Join x' and x^* by a line segment (Figure 4.16). Since $\mathbf{x}'$ is convex there is a point $\bar{x}$ (and only one) on this line segment that belongs to the boundary of $\mathbf{x}'$. We set

$$\bar{x} = \lambda x^* + (1 - \lambda)x'. \tag{3.59}$$

Taking account successively of the optimality of x^* in Γ, of the linearity of the function xAy^{*T} in x, and of the optimality of x' in Γ', we obtain the trio of inequalities

$$x^* Ay^{*T} \geqq \bar{x}Ay^{*T} \geqq x' Ay^{*T} \geqq xAy^{*T} \tag{3.60}$$

for every $x \in \mathbf{x}'$. In addition, by the optimality of y^* in Γ and in Γ', we have

$$\left.\begin{aligned} x^* Ay^{*T} &\leqq x^* Ay^T, \\ x' Ay^{*T} &\leqq x' Ay^T \end{aligned}\right\} \quad \text{for every } y \in \mathbf{y}'.$$

Consequently, by (3.59) we have

$$\bar{x}Ay^{*T} \leqq \bar{x}Ay^T \text{ for all } y \in \mathbf{y}',$$

and since it follows by (3.60) that $xAy^{*T} \leqq \bar{x}Ay^{*T}$, the pair $(\bar{x}, y^*)$ is a saddle point in the game Γ'.

Suppose now that $\mathscr{T}(\Gamma) \cap \mathbf{y}' = \emptyset$. As above, take $y^* \in \mathscr{T}(\Gamma)$, $y \in \mathscr{T}(\Gamma')$, and find a point $\bar{y}$, belonging to the boundary of $\mathbf{y}'$, on the line segment joining y^* and y'. We are interested in the case when $y' \neq \bar{y}$. If we now use the linearity of the scalar product, the optimality of the strategies x' and y' in Γ', and the fact that $x' \neq \bar{x}$ and $y' \neq \bar{y}$, we obtain

$$x^* Ay'^T \leqq \bar{x}Ay'^T \leqq x' Ay'^T \leqq x' A\bar{y}^T \leqq x' Ay^{*T} \leqq x^* Ay^{*T} \leqq x^* Ay'^T, \tag{3.61}$$

i.e., all the terms of this chain of inequalities are actually equal.

From this and the inclusion $(x', y') \in \mathscr{C}(\mathscr{T}')$, we obtain, for every $x \in \mathbf{x}$,

$$xAy'^T \leqq x' Ay'^T \leqq \bar{x}Ay'^T.$$

Therefore, from $(\bar{x}, y') \notin \mathscr{C}(\Gamma')$ there follows the existence of $y' \in \mathbf{y}'$ such that

$$\bar{x}Ay'^T > \bar{x}Ay''^T. \tag{3.62}$$

But it follows from $(x', y') \in \mathscr{C}(\Gamma')$ that

$$x' Ay'^T \leqq x' Ay''^T. \tag{3.63}$$

It follows from this, (3.59), (3.62), (3.63), and the equality of all the terms in (3.61), that

$$x^* Ay^{*T} = x^* Ay'^T > x^* Ay''^T,$$

which contradicts $(x^*, y^*) \in \mathscr{C}(\Gamma)$. Therefore (3.62) is not satisfied for any $y'' \in \mathbf{y}'$, so that $(\bar{x}, y') \in \mathscr{C}(\Gamma')$. □

We emphasize that the equality in (3.43) was obtained under the assumption that x' and y' are not on the boundaries of $\mathbf{x}'$ and $\mathbf{y}'$. Under this assumption, the games Γ and Γ' will have equal values.

3.12 Matrix games with restrictions.

Definition. A matrix game *with (linear) restrictions* is a polyhedral game which is a subgame of a mixed extension of a matrix game. □

As a matter of fact, the concept of a matrix game with restrictions is equivalent to the general concept of a polyhedral game, because every bounded subset z of the space $\mathbf{R}^p$ can be homothetically embedded in a p-dimensional simplex spanned by the unit vectors of $\mathbf{R}^{p+1}$.

One of the simplest examples of a matrix game with restrictions is the mixed extension of a matrix game Γ_A with an $m \times n$-payoff matrix, in which, however, player 1 can use pure strategies $1, \ldots, n$ with nonincreasing probabilities, i.e., the player's mixed strategies $X = (\xi_1, \ldots, \xi_m)$ satisfy $\xi_1 \geqq \ldots \geqq \xi_m$.

This example illustrates a player's possibly psychological inclination to arrange the pure strategies according to subjective preferences and to choose the more preferred strategies with greater probability, and all the more when repeating the game more frequently.

It is easily verified that this game leads, as described in 3.9, to an $m \times n$-matrix game Γ_B in which

$$B_r = \frac{1}{r} \sum_{i=1}^{r} A_{i\cdot}, \quad r = 1, \ldots, m.$$

3.13. From the point of view of matrix games with restrictions, it is natural to consider the problem of finding the derivatives of the value $v(A)$ of the matrix game Γ_A in the direction H (see 1.26). As follows from the theorem in 1.27, these derivatives are the values of the matrix games with restrictions,

$$\Gamma(A/H) = \langle \mathcal{S}(A), \mathcal{T}(A), H \rangle \text{ and } \Gamma(-A^T/H^T) = \langle \mathcal{T}(A), \mathcal{S}(A), H^T \rangle.$$

Here it is relevant to recall that according to 1.13 the sets $\mathcal{S}(A)$ and $\mathcal{T}(A)$ are convex polytopes lying in the corresponding simplexes.

Suppose that we already know the value v_A of Γ_A. Then, by 1.3, and the description in terms of inequalities in (3.53), we have

$$\mathcal{S}(A) = \{X : -XA \leqq -J_n v_A,\ -XJ_m^T \leqq -1,\ XJ_m^T \leqq 1\},$$
$$\mathcal{T}(A) = \{Y : -AY^T \geqq -J_m^T v_A,\ -J_n Y^T \geqq -1,\ J_n Y^T \geqq -1\}.$$

Then, for the linear programming problem described by (3.36) and (3.57), we must have

$$M = (-A, -J_m^T, J_m^T), \quad N = (-A^T, -J_n^T J_n^T)^T,$$
$$P = (-J_n v_A, -1, 1), \quad Q = (-J_m v_A, -1, 1).$$

The coefficient matrix of the bilinear form in (3.54) will be H in this case. On the basis of the pair of dual linear programming problems in (3.56) and (3.57), we can write

$$XA \geqq J_n v_A, \qquad AY^T \leqq J_m^T v_A,$$

$$XJ_m^T = 1, \qquad J_n Y^T = 1,$$

$$XH + V(A^T, J_n^T, -J_n^T) \geqq 0, \qquad HY^T + (A, J_m^T, -J_m^T)U^T \leqq 0,$$

$$X, V \geqq \mathbf{0}, \qquad Y, U \geqq \mathbf{0},$$

$$V(J_m v_A, -1, 1)^T \to \min, \qquad (J_n v_A, 1, -1)U^T \to \max.$$

The common value v_A of these problems will be the right derivative with respect to H. The value of the left derivative is calculated similarly.

§4 Description of all equilibrium situations in matrix and bimatrix games

4.1 Saddle points in matrix games. In 1.5, Chapter 2, we outlined the fundamental method for enumerating the equilibrium situations in bimatrix games. Let us describe the algorithm first for the matrix case, and then for bimatrix games.

As we established in 1.13 (also see 5.12, Chapter 1), the sets of the players' optimal strategies in each matrix game are closed convex (and nonempty) polytopes. As we know, such sets are the convex hulls of their extreme points (i.e., vertices). Consequently, in order to have a complete description of the sets of the players' optimal strategies, it is enough to describe the extreme points of these sets. (The set of all extreme points of a set A is denoted by extr$A$.)

A solution of this problem was given, for a special case, in 1.16. A complete analysis is contained in the following theorem.

The extreme points of the set of optimal strategies of a player in a matrix game are called, for short, the player's *extreme strategies* in the game. The theorem in the next two subsections is usually called the *Shapley-Snow theorem.*

4.2 Theorem. *Let Γ_A be an $m \times n$-matrix game, with $v_A \neq 0$ and moreover $X = (\xi_1, \dots, \xi_m) \in \mathcal{S}(A)$ and $Y = (\eta_1, \dots, \eta_n) \in \mathcal{T}(A)$.*

Then in order for $X \in$ extr $\mathcal{S}(A)$ and $Y \in$ extr $\mathcal{T}(A)$ it is necessary and sufficient that there is a nondegenerate submatrix B of A for which

$$X_B = \frac{J_r B^{-1}}{J_r B^{-1} J_r^T}, \tag{4.1}$$

$$Y_B = \frac{B^{-1} J_r^T}{J_r B^{-1} J_r^T}, \tag{4.2}$$

$$v_A = \frac{1}{J_r B^{-1} J_r^T}, \tag{4.3}$$

where r is the number of rows in B, and X_B and Y_B are subvectors of X and Y, consisting of the components of X and Y that correspond to the rows (or columns) of A that occur in the formation of B.

Proof. *Sufficiency.* Let there exist a nonsingular matrix B for which (4.1)–(4.14) are satisfied. We can calculate directly that

$$X_B J_r^T = \frac{J_r B^{-1} J_r^T}{J_r B^{-1} J_r^T} = 1,$$

i.e., the components of X_B sum to unity. Consequently the components of X that do not appear in X_B are all zero.

In addition, it follows from (4.1) and (4.3) that

$$X_B B = \frac{J_r B^{-1}}{J_r B^{-1} J_r^T} B = J_r \frac{1}{J_r B^{-1} J_r^T} = J_r v_A,$$

i.e.,

$$X_B B_{\cdot j} = v_A \text{ for every column } B_{\cdot j} \text{ of } B. \tag{4.4}$$

Now suppose that $X \notin \operatorname{extr} \mathcal{S}(A)$. This means that there are also different optimal strategies $X' = (\xi_1', \ldots, \xi_m')$ and $X'' = (\xi_1'', \ldots, \xi_m'')$, such that $X = (X' + X'')/2$.

The preceding equation means that

$$\xi_i = (\xi_i' + \xi_i'')/2 \tag{4.5}$$

for $i = 1, \ldots, n$. Since $\xi_i' \geqq 0$ and $\xi_i'' \geqq 0$, it follows from (4.5) that $\xi_i' = \xi_i'' = 0$ if $\xi_i = 0$. Therefore different vectors X' and X'' can be distinguished only by their subvectors X_B' and X_B'', i.e. $X_B' \neq X_B''$.

In addition, since X' and X'' are optimal strategies, for every column of B we have

$$X' A_{\cdot j} = X_B' B_{\cdot j} \geqq v_A, \quad X'' A_{\cdot j} = X_B'' B_{\cdot j} \geqq v_A.$$

However, on the other hand, if we use (4.4) we have

$$\frac{1}{2}(X_B' + X_B'') B_{\cdot j} = v_A.$$

Consequently, for all columns of B we must have $X_B' B_{\cdot j} = X_B'' B_{\cdot j} = v_A$; therefore $X_B' B = X_B'' B = J_r v_A$, and hence $(X_B' - X_B'')B = \mathbf{0}$.

However, this is impossible for different vectors X_B' and X_B'', since we assumed that B is nonsingular.

Necessity. We assume that, in addition to the other hypotheses of the theorem, $X \in \operatorname{extr} \mathcal{S}(A)$, $Y \in \operatorname{extr} \mathcal{T}(A)$. We shall construct B in the following way. We divide the set of rows of A into three classes.

The first class consists of the rows $A_{i\cdot}$ for which $\xi_i > 0$. It is clear that

$$A_{i\cdot} Y^T = v_A \tag{4.6}$$

for these values of i.

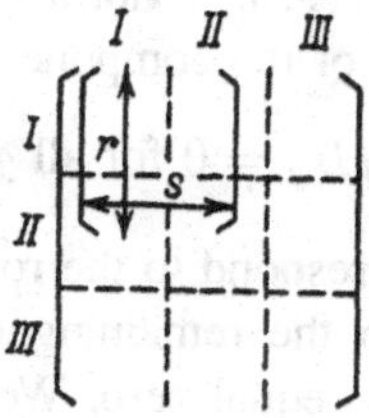

Fig. 4.17

The second class consists of the rows $A_{i\cdot}$ for which $\xi_i = 0$ and (4.6) is still satisfied.

Finally, the third class consists of the other rows of A, i.e., the rows $A_{i\cdot}$ for which $\xi_i = 0$ and $A_{i\cdot}Y^T < v_A$.

In addition, we shall need to consider vectors of the form Z in which the components correspond to the rows of A (these include, for example, the columns of A). In all these cases we shall denote by Z^1 a subvector of Z that consists of components of a row of the first class.

We now introduce a similar decomposition of the set of columns of A. Thus, the first class consists of the columns $A_{\cdot j}$ for which $\eta_j > 0$ (these satisfy $XA_{\cdot j} = v_A$); the second, of those for which $\eta_j = 0$ but $XA_{\cdot j} = v_A$; the third, of those for which $XA_{\cdot j} > v_A$ (for these, η_j must be 0). If the components of the vector V consist of the columns of A, we denote by V^1 the subvector of V consisting of the components that correspond to columns that belong to the first class.

We call a set of rows A *distinguished* if:

a) it contains all the rows of the first class;

b) it contains no rows of the third class;

c) from rows of the second class, it contains only the $A_{i\cdot}$ for which the subvectors $A^1_{i\cdot}$ are linearly independent of the analogous subvectors $A^1_{i\cdot}$ for rows, already included in the set.

We may collect the rows of the second class, in a fixed distinguished set, in any order; hence under a certain order of enumeration the distinguished set will involve some rows of the second class and under another order — the other rows.

We arrange the distinguished set of columns of A in a similar way. Let us now fix one of the distinguished sets of rows of A and one of the distinguished sets of columns. The matrix of the elements that are in the intersection of the rows and columns of the fixed sets will be denoted by B. Let it have r rows and s columns. The construction is shown schematically in Figure 4.17.

Let us show that the matrix B is the one required.

We first prove that the matrix we have constructed is nonsingular, i.e., that neither its rows nor its columns are linearly dependent.

Let us suppose that there is linear dependence for a row of B. This means that there is a vector $G_B = (\gamma_{i_1}, \ldots, \gamma_{i_r})$, not all of whose components are zero, such that $G_B B = 0$, or, in terms of the components,

$$G_B B_{\cdot j} = 0 \text{ for all } j. \tag{4.7}$$

The components of G_B correspond to the rows of B. We extend this vector by components corresponding to the remaining components of the rows of A, taking all the new components to equal zero. We denote the resulting vector by G. We have

$$GA_{\cdot j} = G_B B_{\cdot j} + \sum \gamma_i a_{ij} = G_B B_{\cdot j} = 0$$

(the summation is over the indices i that were not used in forming B).

Furthermore, $B_{i\cdot} Y_B^T = A_{i\cdot} Y^T = v_A$, so that $BY_B^T = v_A J_r^T$. Therefore

$$(G_B B) Y_B^T = G_B (BY_B^T) = G_B J_r^T v_A = 0,$$

and since $v_A \neq 0$ by hypothesis, we must have $G_B J_r^T = 0$.

Recalling that G differs from G_B only by having a number of zero components, we obtain

$$GJ_m^T = 0. \tag{4.8}$$

By hypothesis, some of the components of G are not zero. Let $\gamma_i \neq 0$. If row i belongs to the second class, saying that it is not equal to zero means that the row $B_{i\cdot}$ is linearly dependent on the others. Consequently this linear dependence must also occur for the corresponding subvectors $A_{i\cdot}^1$. But this contradicts the construction of the set of rows. Consequently the only components of G that are different from 0 are those corresponding to rows in the first class, i.e., those for which $\gamma_i \neq 0$ implies $\xi_i > 0$. In other words, the only nonzero components in G are those that appear in G^1.

Let us now consider vectors of the form $X_\alpha = X + \alpha G$. Since nonzero components of G correspond to positive components of X, the components of X_α are nonnegative when α is sufficiently small in absolute value.

In addition, by (4.8),

$$(X + \alpha G) J_m^T = XJ_m^T + \alpha GJ_m^T = XJ_m^T = 1. \tag{4.9}$$

Consequently, when α is sufficiently small in absolute value, the vectors of the form X_α are mixed strategies of player 1.

Let us show that these strategies X_α are optimal. For this purpose, we consider the scalar product $X_\alpha A_{\cdot j}$. If column j of A occurs in the formation of B, we have, taking account of (4.7),

$$X_\alpha A_{\cdot j} = (X + \alpha G) A_{\cdot j} = XA_{\cdot j} + \alpha GA_{\cdot j} = XA_{\cdot j} + \alpha G_B B_{\cdot j} = XA_{\cdot j} = v_A.$$

Now let j not occur in the formation of B. In this case the column $A_{\cdot j}$ must belong either to the second or third class.

We first consider the case when it belongs to the second class. Then for this j the vector $A^1_{\cdot j}$ depends linearly on the vectors $A_{\cdot k}$ that are subvectors of the columns belonging to the set described above. Notice that $G^1 A^1_{\cdot j} = 0$ by (4.7). Let

$$A^1_{\cdot j} = \sum_k \lambda_k A^1_{\cdot k}$$

describe the linear dependence. We then obtain

$$\begin{aligned}(X + \alpha G)A_{\cdot j} &= XA_{\cdot j} + \alpha G A_{\cdot j} = XA_{\cdot j} + \alpha G^1 A^1_{\cdot j} \\ &= XA_{\cdot j} + \alpha G^1 \sum_k \lambda_k A^1_{\cdot k} = XA_{\cdot j} + \alpha \sum_k \lambda_k (G^1 A^1_{\cdot k}) XA_{\cdot j} = v_A . \quad (4.10)\end{aligned}$$

Finally, let $A_{\cdot j}$ belong to the third class. In this case $XA_{\cdot j} > v_A$, so that, for sufficiently small α, we must have

$$(X + \alpha G)A_{\cdot j} = XA_{\cdot j} + \alpha G A_{\cdot j} > v_A . \tag{4.11}$$

Relations (4.9), (4.10), and (4.11) show that

$$X_\alpha A_{\cdot j} \geqq v_A \text{ for } j = 1, \dots, n.$$

A reference to 1.12 shows that X_α is optimal.

We now select a number α, so small in absolute value that both X_α and $X_{-\alpha}$ are optimal strategies of player 1. But we have $X = (X_\alpha + X_{-\alpha})/2$, so that X is not an extreme point of the set of optimal strategies of player 1. The resulting contradiction shows that the rows of B cannot be linearly dependent. It can be shown similarly that its columns cannot be linearly dependent. Consequently B is square and nonsingular.

In particular, we can therefore speak of a matrix B^{-1} inverse to B.

The rest of the discussion reproduces what was said about B in 1.16.

It follows from the construction of B that

$$X_B B = v_A J_r, \ X_B J_r^T = 1, \ BY_B^T = J_r^T v_A,$$

from which the validity of (4.1)–(4.3) follows. □

4.3. We present a more complete form of the result that we have obtained.

Theorem. *In the game Γ_A, let*

$$X \in \mathcal{S}(A), \quad Y \in \mathcal{T}(A). \tag{4.12}$$

*Then a necessary and sufficient condition for $X \in \operatorname{extr} \mathcal{S}(A)$ and $Y \in \operatorname{extr} \mathcal{T}(A)$ is that there is a square $r \times r$-submatrix B of A for which**)

$$J_r \operatorname{adj} B^{-1} J_r^T \neq 0, \tag{4.13}$$

*) As usual, for a square matrix M we denote its determinant by $\det M$; and by $\operatorname{adj} M$, the *adjoint* of M, which is obtained from M by replacing each element m_{ij} by the algebraic complement of m_{ji} in $\det M$. If M is nonsingular, $\operatorname{adj} M = M^{-1} \det M$.

$$X_B = \frac{J_r B^{-1}}{J_r \operatorname{adj} B^{-1} J_r^T}, \tag{4.14}$$

$$Y_B = \frac{B^{-1} J_r^T}{J_r \operatorname{adj} B^{-1} J_r^T}, \tag{4.15}$$

$$v_A = \frac{\det B}{J_r \operatorname{adj} B^{-1} J_r^T}. \tag{4.16}$$

For 2×2-games, special cases of (4.14), (4.15) and (4.16) were obtained in § 2 in the corresponding forms (2.5), (2.6), and (2.7).

Proof. If $v_A \neq 0$, we have the hypotheses of the preceding theorem, and by the relation $\operatorname{adj} B = B^{-1} \det B$ formulas (4.14)–(4.16) are equivalent to (4.1)–(4.3), while (4.13) follows from $J_r B^{-1} J_r^T \neq 0$, since $v_A \neq 0$ and $\det B \neq 0$.

If $v_A = 0$, we consider the game Γ' with the payoff matrix

$$A' = \|a_{ij} + 1\|_{i=1,\dots,m;\ j=1,\dots,n}.$$

The games A and A' are affinely equivalent. Therefore the two inclusions (4.12) are equivalent, $X \in \mathcal{S}(A')$ and $Y \in \mathcal{T}(A')$, and since $v_{A'} = 1 \neq 0$, we find ourselves in the hypotheses of the case already discussed: a necessary and sufficient condition for $X \in \operatorname{extr} \mathcal{S}(A')$ and $Y \in \operatorname{extr} \mathcal{T}(A')$ is that there exists a nonsingular matrix B' for which

$$X_{B'} = \frac{J_r \operatorname{adj} B'}{J_r \operatorname{adj} B' J_r^T}, \tag{4.17}$$

$$Y_{B'} = \frac{\operatorname{adj} B' J_r^T}{J_r \operatorname{adj} B' J_r^T}, \tag{4.18}$$

$$v_{A'} = \frac{\det B}{J_r \operatorname{adj} B' J_r^T}. \tag{4.21}$$

It remains only to transfer these formulas from the submatrix B' of the matrix A' to the corresponding submatrix B of A. Here we first notice that the passage from A to B consists of striking out rows and columns with the same indices as in deriving B from A. Therefore

$$X_{B'} = X_B, \quad Y_{B'} = Y_B, \tag{4.20}$$

and moreover

$$v_{A'} = v_A + 1. \tag{4.21}$$

Let us replace, in the determinants $\det B$ and $\det B'$, the jth column by the unit column J_r^T, and denote the resulting determinants by $\det(B \underset{\cdot j}{\|} J_r^T)$ and $\det(B' \underset{\cdot j}{\|} J_r^T)$.

If we subtract from each column of $\det(B' \underset{\cdot j}{\|} J_r^T)$ the ith column J_r^T, we evidently obtain the equal determinant $\det(B \underset{\cdot j}{\|} J_r^T)$. On the other hand, we can expand each determinant by elements of the jth column:

$$\sum_i B'_{ij} = \det(B' \underset{\cdot j}{\|} J_r^T) = \det(B \underset{\cdot j}{\|} J_r^T) = \sum_i B_{ij}, \tag{4.22}$$

where B'_{ij} and B_{ij} are the cofactors of the elements a'_{ij} and a_{ij} of the determinants.

The middle equation in (4.22) can be written in the form

$$\operatorname{adj} B' J_r^T = \operatorname{adj} B J_r^T. \tag{4.23}$$

From this, we immediately obtain

$$J_r \operatorname{adj} B' J_r^T = J_r \operatorname{adj} B J_r^T. \tag{4.24}$$

As for (4.23), but operating on rows instead of columns, we obtain

$$J_r \operatorname{adj} B' = J_r \operatorname{adj} B. \tag{4.25}$$

Finally, because each of the r columns of $\det B'$ is the sum of a column of $\det B$ and J_r^T, the determinant $\det B'$ decomposes into 2^r terms, of which $2^r - r - 1$ have equal (unit) columns, and therefore vanish. Therefore, by (4.22), we have

$$\det B' = \det B + \sum_j \det(B' \underset{\cdot j}{\|} J_r^T) = \det B + \sum_{i,j} B_{ij} = \det B + J_r \operatorname{adj} B J_r^T. \tag{4.26}$$

As a result, (4.20), (4.23), and (4.24) yield (4.14); (4.20), (4.24), and (4.25) yield (4.15); and (4.21), (4.24), and (4.26) yield (4.16). □

4.4. We notice, as a corollary of this theorem, the following useful proposition.

Theorem. *If $m < n$ in an $m \times n$-game, then among the strategies of player* 2 *there are those that are mixtures of at most m pure strategies.*

If $m > n$, then among the strategies of player 1 there are those that are mixtures of at most n pure strategies.

Proof. If $m < n$, we select an extreme strategy of player 2. The number of pure strategies that appear in its spectrum cannot exceed the order of a square submatrix of B, which in turn cannot be larger than m. If $m > n$ the argument is similar. □

4.5. Another corollary establishes a connection between the spectral properties of the optimal strategies of matrix games and the algebraic properties of their payoff matrices.

Theorem. *If the matrix game Γ_A is completely mixed, it has just one equilibrium situation, the matrix A is square, and (if its order is $r \times r$) it has $J_r \operatorname{adj} A J_r^T \neq 0$; if, in addition, $v_A \neq 0$, the matrix A is nonsingular.*

Proof. We select a pair of extreme strategies of the players in Γ_A, induced by an $r \times r$-submatrix of B as in the theorem in 4.3. On one hand, the spectra of these strategies are in the sets of rows (columns) of B; but on the other hand, they cover the set of rows (columns) of the whole matrix A. Therefore $B = A$, so that A is square, and has $J_r \operatorname{adj} A J_r^T = 0$. It then follows that each player in Γ_A has just one extreme strategy, since the bounded polytope, having only a single vertex, consists just of this vertex. □

4.6. The theorem of 4.2 (like that of 4.3) is really an algorithm that enumerates, at least in principle, all the equilibrium situations in any matrix game.

Let us enumerate all the square nonsingular submatrices of the game A: $B_1, \ldots, B_k$, and for each B_l (of order $r_l \times r_l$) let us find the vectors X_l and Y_l by formulas (4.14) and (4.15). If any components of these vectors are negative, the matrix B_l does not lead to the players' extreme strategies, and must be excluded from further consideration. If, however, all the components of X_l and Y_l are nonnegative, we have only to extend these vectors by zero components to vector strategies X_l^* and Y_l^*, and then check whether the situation (X_l^*, Y_l^*) is an equilibrium situation. For this purpose, we have only to check the inequalities

$$A_{i\cdot} Y_l^{*T} \leqq X_l^* A Y_l^{*T} \leqq X_l^* A_{\cdot j}$$

for $i = 1, \ldots, m$ and $j = 1, \ldots, n$. If it then turns out that $X_l^* \in \mathcal{S}(A)$ and $Y_l^* \in \mathcal{T}(A)$, by the theorem of 4.2 we have $X_l^* \in \operatorname{extr} \mathcal{S}(A)$ and $Y_l^* \in \operatorname{extr} \mathcal{T}(A)$; fixing on the resulting equilibrium situation, we consider the resulting submatrix. In the contrary case, B_l is discarded.

Therefore the determination of the players' extreme strategies in a matrix game reduces to finding its optimal strategies as convex mixtures of extreme strategies.

In practice, application of this algorithm is quite tedious, even for games of rather low format.

4.7 Bimatrix games. The idea underlying the theorem of 4.2 and at the base of the algorithm for enumerating the equilibrium situations in matrix games can also be adapted to solving the analogous problem for bimatrix games (see 1.3, Chapter 1). What was established in the later subsections can actually be modified to apply to the "constructivization" of the theorem in 1.5, Chapter 2. However, we shall not dwell on that theorem in the present discussion.

As we have noticed, in matrix games the players' optimal strategies form convex polytopes. This allowed us to reduce the description of these sets to the enumeration of their extreme points. In bimatrix games, as we have had occasion to notice (in the example of the "battle of the sexes" game: see 2.16, Chapter 1, and 2.4, Chapter 2), the set of a player's equilibrium strategies is not necessarily convex, and the method previously described is not directly applicable. Instead, here we can use a somewhat weaker property, which reminds us of convexity, was established in the theorem in 5.12 of Chapter 1, and was used for a similar reason in 1.5, Chapter 2.

4.8. Let us consider a bimatrix $m \times n$-game $\Gamma = \Gamma_{A,B}$, which will be the same until the end of this subsection.

Let X be a strategy of player 1 in Γ and

$$\mathscr{C}(X) = \{Y \in \boldsymbol{Y} : (X, Y) \in \mathscr{C}(\Gamma)\}.$$

As we noticed in 1.5, Chapter 2, the set $\mathscr{C}(X)$ is a closed convex polytope. This set is not empty if X is an equilibrium strategy of player 1 in Γ.

The same can be said for the set

$$\mathscr{C}(Y) = \{X \in \boldsymbol{X} : (X, Y) \in \mathscr{C}(\Gamma)\},$$

where Y is any strategy of player 2.

Let us set, for arbitrary finite sets $\bar{\boldsymbol{X}} \subset \boldsymbol{X}$ and $\bar{\boldsymbol{Y}} \in \boldsymbol{Y}$ of mixed strategies of the players,

$$\mathscr{C}(\bar{\boldsymbol{X}}) = \bigcap_{X \in \bar{\boldsymbol{X}}} \mathscr{C}(X) \text{ and } \mathscr{C}(\bar{\boldsymbol{Y}}) = \bigcap_{Y \in \bar{\boldsymbol{Y}}} \mathscr{C}(Y).$$

Evidently, $\mathscr{C}(\bar{\boldsymbol{X}})$ and $\mathscr{C}(\bar{\boldsymbol{Y}})$ are also closed convex polytopes. In particular, they might be empty.

Definition. A strategy Y of player 2 is called an *extreme equilibrium strategy* if there is a finite set $\bar{\boldsymbol{X}}$ of strategies of player 1 such that $Y \in \operatorname{extr}\operatorname{conv}\mathscr{C}(\bar{\boldsymbol{X}})$.

Extreme equilibrium strategies of player 1 are defined analogously. □

We now discuss an effective description of the set $\bar{Y}$ of extreme equilibrium strategies of player 2. Here we shall explain why this set is finite. In the same way, the set $\boldsymbol{X}$ of extreme equilibrium strategies of player 1 is finite.

4.9 Lemma. *If $\bar{Y}$ is an extreme equilibrium strategy of player 2, and*

$$\alpha = \max A_{i\cdot} Y^T, \tag{4.27}$$

then $(\bar{Y}, \bar{\alpha})$ is an extreme solution of the system

$$AY^T \leqq \alpha J_m^T, \tag{4.28}$$

$$Y \geqq 0, \tag{4.29}$$

$$J_n Y^T = 1 \tag{4.30}$$

Proof. By the definition of an extreme equilibrium strategy, there is a finite set $\bar{\boldsymbol{X}} = \{X_1, \ldots, X_p\}$ such that $\bar{Y}$ is an extreme solution of the system

$$A_{i\cdot} \leqq X_k AY^T, \quad i = 1, \ldots, m; \quad k = 1, \ldots, p, \tag{4.31}$$

$$X_k B_{\cdot j} \leqq X_k BY^T, \quad j = 1, \ldots, n; \quad k = 1, \ldots, p, \tag{4.32}$$

$$Y \geq 0, \tag{4.33}$$

$$J_n Y^T = 1. \tag{4.34}$$

It follows from (4.27) that, for $k = 1, \ldots, p$,

$$X_k A\bar{Y}^T = \bar{\alpha}, \tag{4.35}$$

and (4.28) is an immediate corollary of (4.31) and (4.35). Then $\bar{Y}$ automatically satisfies (4.29) and (4.30), by (4.33) and (4.34). Therefore $(\bar{Y}, \bar{\alpha})$ is a solution of (4.28)–(4.30).

Suppose now that $(\bar{Y}, \bar{\alpha})$ is not an extreme solution of the system (4.28)–(4.30). This means that there are different solutions (Y', α') and (Y'', α'') of this system, such that

$$(\bar{Y}, \bar{\alpha}) = ((Y', \alpha') + (Y'', \alpha''))/2.$$

We shall show that in this case Y' and Y'' are solutions of the system (4.31)–(4.34).

From (4.28), we have

$$A_{i\cdot} Y'^T \leqq \alpha', \quad A_{i\cdot} Y''^T \leqq a'', \quad i = 1, \ldots, m;$$

therefore

$$\alpha = \max_i A_{i\cdot} \bar{Y}^T = \max_i (A_{i\cdot} Y'^T + A_{i\cdot} Y''^T)/2$$

$$\leqq (\max_i A_{i\cdot} Y'^T + \max_i A_{i\cdot} Y''^T)/2 \leqq (\alpha' + \alpha'')/2.$$

Consequently

$$\alpha' = \max_i A_{i\cdot} Y'^T, \tag{4.36}$$

$$\alpha'' = \max_i A_{i\cdot} Y''^T. \tag{4.37}$$

From this, together with (4.35),

$$\bar{\alpha} = X_k A\bar{Y}^T = (X_k AY'^T + X_k AY''^T)/2 \leqq (\alpha' + \alpha'')/2 = \bar{\alpha}.$$

Therefore

$$\alpha' = X_k AY'^T,$$

$$\alpha'' = X_k AY''^T.$$

Comparing these equations with (4.36) and (4.37) yields

$$\max_i A_{i\cdot} Y'^T = X_k AY'^T,$$

$$\max_i A_{i\cdot} Y''^T = X_k AY''^T.$$

Therefore the vectors Y' and Y'' satisfy (4.31). In addition, it follows immediately from (4.29) and (4.30) that Y' and Y'' also satisfy (4.33) and (4.34). It remains to verify (4.32).

However, since $(X_k, \bar{Y})$ is an equilibrium situation,

$$X_k BY'^T \leqq X_k B\bar{Y}^T,$$
$$X_k BY''^T \leqq X_k B\bar{Y}^T$$

and since

$$X_k B\bar{Y}^T = (X_k BY'^T + X_k BY''^T)/2,$$

we have

$$X_k B\bar{Y}^T = X_k BY'^T = X_k BY''^T$$

and therefore

$$X_k B_{\cdot j} \leqq X_k BY'^T = X_k BY''^T, \quad j = 1, \ldots, m; \quad k = 1, \ldots, p.$$

Consequently Y' and Y'' are solutions of the system (4.31)–(4.35), and we have obtained a contradiction.

The proof of the lemma shows that we must look for extreme equilibrium strategies for (corresponding) subvectors of extreme solutions of systems of the form (4.28)–(4.30).

4.10 Lemma. *If* $(\bar{Y}, \bar{\alpha})$ *is an extreme solution of the system* (4.28)–(4.30), *there is an* $s \times s$*-submatrix* D *of* A *such that the matrix*

$$\bar{D} = \begin{bmatrix} D & -J_s^T \\ J_s & 0 \end{bmatrix} \tag{4.38}$$

is nonsingular.

Here

$$\bar{Y}_D = (\det \bar{D})^{-1} \left(\sum_{i=1}^{s} D_{i1}, \ldots, \sum_{i=1}^{s} D_{is} \right), \tag{4.39}$$

the other components of $\bar{Y}$ *are zero, and*

$$\bar{\alpha} = (\det D)/(\det \bar{D}). \tag{4.40}$$

Proof. Let J be the set of indices i for which the equation $A_{i\cdot} \bar{Y}^T = \bar{\alpha}$ is satisfied in (2.28). Since we are considering extreme solutions, J is not empty. Thus $A_{i\cdot} \bar{Y}^T < \bar{\alpha}$ for $i \notin J$. Let r be the number of elements of J.

We also denote by K the set of indices j for which $\bar{y}_j > 0$. Thus $\bar{y}_j = 0$ for $k \notin K$. Let K contain s elements.

Let us show that the system of equations

$$A_{JK} Y_K^T = \alpha J_r^T, \tag{4.41}$$
$$J_s Y_K^T = 1 \tag{4.42}$$

has the unique solution

$$Y_K = \bar{Y}_K, \quad \alpha = \bar{\alpha}.$$

In fact, let there be two solutions (Y', α') and (Y'', α'') of the system (4.41), (4.42). Then we set

$$\bar{Y}'_K = \bar{Y}_K + \varepsilon(Y'_K - Y''_K),$$
$$\bar{Y}''_K = \bar{Y}_K + \varepsilon(Y''_K - Y'_K),$$

supposing that the other components of $\bar{Y}'$ and $\bar{Y}''$ are zero, and that ε is positive and otherwise undefined for the present.

It is clear that since $\bar{Y}_K > \mathbf{0}^s$, for sufficiently small ε we have

$$\bar{Y}' \geq 0, \quad \bar{Y}'' \geq 0. \tag{4.43}$$

Moreover, since $\bar{Y}'$ and $\bar{Y}''$ satisfy (4.42), we must have

$$J_n \bar{Y}' = J_n \bar{Y}'' = 1. \tag{4.44}$$

Thus $\bar{Y}'$ and $\bar{Y}''$ are mixed strategies of player 2.

In addition, we have

$$A_{i\cdot} \bar{Y}'^T = \bar{\alpha} + \varepsilon(\alpha' - \alpha''),$$
$$A_{i\cdot} \bar{Y}''^T = \bar{\alpha} + \varepsilon(\alpha'' - \alpha')$$

for $i \in J$, and also

$$A_{i\cdot} \bar{Y}'^T = A_{i\cdot} \bar{Y}^T + \varepsilon A_{JK}(Y'^T_J - Y''^T_J),$$
$$A_{i\cdot} \bar{Y}''^T = A_{i\cdot} \bar{Y}^T + \varepsilon A_{JK}(Y''^T_J - Y'^T_J)$$

for $i \in J$. By (4.38) we can find δ so small that

$$(\bar{\alpha} + \delta) J_m^T \geqq A\bar{Y}'^T, \tag{4.45}$$

$$(\bar{\alpha} + \delta) J_m^T \geqq A\bar{Y}''^T \tag{4.46}$$

(of course, if ε is suitably chosen).

Relations (4.43)–(4.46) show that both $(\bar{Y}', \bar{\alpha} + \delta)$ and $(\bar{Y}'', \bar{\alpha} - \delta)$ are solutions of (4.28)–(4.30). Since

$$(\bar{Y}, \alpha) = ((\bar{Y}', \alpha + \delta), (\bar{Y}'', \alpha - \delta))/2,$$

and the solution $(\bar{Y}, \bar{\alpha})$ is, by hypothesis, extreme, δ must be 0. But then $\alpha' = \alpha''$ and $\bar{Y}' = \bar{Y}''$, so that $Y'_K = Y''_K$, and the uniqueness of the solution of the system (4.41)–(4.42) is established.

Hence there follows the linear independence of the columns of the $(r+1) \times (s+1)$ matrix

$$\begin{bmatrix} A_{JK} & -J_r^T \\ J_s & 0 \end{bmatrix}.$$

Therefore this matrix has $s+1$ linearly independent rows. Then we may choose the rows so that one of them is the last row of the matrix. The selected rows form the required matrix $\bar{D}$. This is the matrix of the system (4.41), (4.42). After all this, (4.39) and (4.40) are obtained by elementary calculations with determinants. □

It follows from the lemma that the extreme solution of the system (4.28)–(4.30) is gven by a matrix of the form $\bar{D}$.

4.11. If we combine what has been said in the preceding two subsections, we find that the candidates for extreme equilibrium strategies of player 2 can only be the strategies defined in the hypotheses of the lemma in 2.10. The candidates for extreme equilibrium strategies of player 1 are defined similarly. The set of the first of these is denoted by $\tilde{Y}$; and that of the second, by $\tilde{X}$. Both $\tilde{X}$ and $\tilde{Y}$ are finite.

We emphasize that the components of the vectors in $\tilde{X}$, and equally of those in $\tilde{Y}$, are expressed rationally in terms of the elements of the matrices A and B.

4.12 Theorem. *In a bimatrix game $\Gamma_{A,B}$ for every $\bar{X} \subset \tilde{X}$ the set $\mathscr{C}(\bar{X})$ (and for every $\bar{Y} \subset \tilde{Y}$, the set $\mathscr{C}(\bar{Y})$) has finitely many effectively determined extreme points.*

Proof. Evidently

$$\operatorname{extr}\mathscr{C}(\bar{X}) \subset \tilde{Y}, \quad \operatorname{extr}\mathscr{C}(\bar{Y}) \subset \tilde{X},$$

from which there follows the finiteness of the set of extreme points in which we are interested.

Let us now select any $\bar{X} \subset \tilde{X}$ and describe the set $\mathscr{C}(\bar{X})$; to do this, it is enough to give the set $\operatorname{extr}\mathscr{C}(\bar{X})$ and form its convex hull. But $\operatorname{extr}\mathscr{C}(\bar{X}) \subset \tilde{Y}$. Therefore, in the first place, we can determine which strategies in $\tilde{Y}$ belong to $\mathscr{C}(\bar{X})$, i.e., which vectors $Y \in \tilde{Y}$ satisfy the inequalities

$$A_{i\cdot}Y^T \leqq XAY^T, \quad i = 1, \ldots, m; \quad X \in \bar{X}, \tag{4.47}$$

$$XB_{\cdot j} \leqq XBY^T, \quad j = 1, \ldots, n; \quad Y \in \bar{Y}. \tag{4.48}$$

This enables us to make a direct search for strategies in $\tilde{Y}$. If no such strategies are found in $\tilde{Y}$, then $\mathscr{C}(\bar{X}) = \emptyset$. In the second place, for a precise determination of $\operatorname{extr}\mathscr{C}(\bar{X})$ we must exclude from the strategies in $\tilde{Y} \cap \mathscr{C}(\bar{X})$ those that are not extreme, i.e., those that are representable as convex combinations of the others.

Similarly, $\mathscr{C}(\bar{Y})$ is determined for every $\bar{Y} \subset \tilde{Y}$. □

4.13. The determination for $\bar{X} \subset \tilde{X}$ of $\mathscr{C}(\bar{X})$, and for each $\bar{Y} \subset \tilde{Y}$ of $\mathscr{C}(\bar{Y})$, can be rationalized to some extent. For example, the following procedure is appropriate.

First we define, as in the proof of the preceding theorem, the set $\operatorname{extr}\mathscr{C}(X)$ for all $X \in \tilde{X}$. If it happens for some X that $\operatorname{extr}\mathscr{C}(X) = \emptyset$, this X is not an extreme equilibrium strategy of player 1 and can be excluded from further consideration. Evidently the set

$$\bigcup_{X \in \tilde{X}} \operatorname{extr}\mathscr{C}(X)$$

so obtained is the set of all the extreme equilibrium strategies of player 2, and the other strategies in Y that do not belong to this set may be discarded.

Then we form, from the remaining elements of $\boldsymbol{X}$, pairs $\{X_i, X_j\}$, finding the corresponding $\operatorname{extr}\mathscr{C}(X_i, X_j)$, and excluding the pairs for which the resulting set is empty. Then, from the remaining pairs, we form triples (i.e., we consider only the triples for which none of the three pairs forming the triple has been discarded); having discarded the unwanted triples, then (if this is still possible) we form quadruples, and continue this process until it breaks off of itself.

4.14 Theorem.

$$\mathscr{C}(\Gamma) = \bigcup_{\bar{\boldsymbol{X}} \subset \tilde{\boldsymbol{X}}} \operatorname{conv} \bar{\boldsymbol{X}} \times \mathscr{C}(\bar{\boldsymbol{X}}) = \bigcup_{\bar{\boldsymbol{Y}} \subset \tilde{\boldsymbol{Y}}} \mathscr{C}(\bar{\boldsymbol{Y}}) \times \operatorname{conv} \bar{\boldsymbol{Y}}.$$

Proof. Let $(X, Y) \in \mathscr{C}(\Gamma)$. Then by definition (see 4.8) $\operatorname{extr}\mathscr{C}(Y) \subset \tilde{\boldsymbol{X}}$. Denote $\operatorname{extr}\mathscr{C}(Y)$ by $\bar{\boldsymbol{X}}$. Since $X \in \mathscr{C}(Y)$, and $\mathscr{C}(Y)$ is a convex set which is bounded and has finitely many extreme points, we must have, by Carathéodory's theorem, $X \in \operatorname{conv} \bar{\boldsymbol{X}}$. Furthermore, for every $X' \in \mathscr{C}(Y)$, and consequently for every $X' \in \bar{\boldsymbol{X}}$, the situation (X', Y) is an equilibrium situation.

Therefore, by definition

$$Y \in \bigcap_{X' \in \bar{\boldsymbol{X}}} \mathscr{C}(X') = \mathscr{C}(\bar{\boldsymbol{X}}).$$

Consequently we find that

$$(X, Y) \in \operatorname{conv} \bar{\boldsymbol{X}} \times \mathscr{C}(\bar{\boldsymbol{X}}), \tag{4.49}$$

from which there follows

$$\mathscr{C}(\Gamma) \subset \bigcup_{\bar{\boldsymbol{X}} \in \tilde{\boldsymbol{X}}} \operatorname{conv} \bar{\boldsymbol{X}} \times \mathscr{C}(\bar{\boldsymbol{X}}). \tag{4.50}$$

On the other hand,

$$(X, Y) \in \bigcup_{\bar{\boldsymbol{X}} \subset \tilde{\boldsymbol{X}}} \operatorname{conv} \bar{\boldsymbol{X}} \times \mathscr{C}(\bar{\boldsymbol{X}})$$

means that (4.49) holds for some $\bar{\boldsymbol{X}} \subset \tilde{\boldsymbol{X}}$. But $Y \in \mathscr{C}(\bar{\boldsymbol{X}})$ is equivalent to $\bar{\boldsymbol{X}} \subset \mathscr{C}(Y)$; but since the last set is convex we must also have $\bar{\boldsymbol{X}} \subset \mathscr{C}(Y)$. In particular, if $X \in \operatorname{conv} \bar{\boldsymbol{X}}$ we must also have $X \in \mathscr{C}(Y)$, i.e., $(X, Y) \in \mathscr{C}(\Gamma)$. Therefore

$$\bigcup_{\bar{\boldsymbol{X}} \subset \tilde{\boldsymbol{X}}} \operatorname{conv} \bar{\boldsymbol{X}} \times \mathscr{C}(\bar{\boldsymbol{X}}) \subset \mathscr{C}(\Gamma).$$

Together with (4.50), this gives us the first of the required equations. The second can be found by a similar discussion. □

§5 Solution of matrix games with matrix payoffs of a special form

5.1 Solution of a special kind of block-matrix games. Up to now, relatively few classes of multiparametric games are known to admit an analytic solution. In a certain sense, this can be understood from the extremely limited possibilities of reducing games to their subgames (as in the "factorization" of a game with a given matrix into games with some or others of its submatrices).

We can attempt, for example, to replace, in a matrix game, some of its block-subgames (that is, matrices) by their values (that is, by numbers) and to discuss the original game in terms of the resulting "quotient game". As a rule, this can be done only under rather restrictive hypotheses.

Let us be concerned with an $m \times n$-matrix A that is represented in block form:

$$A = \begin{bmatrix} A^{11} & \dots & A^{1q} \\ \dots & \dots & \dots \\ A^{p1} & \dots & A^{pq} \end{bmatrix}.$$

We shall suppose that each matrix $A^{kl}, k = 1, \dots, p$ and $l = 1, \dots, q$, has dimensions $m_k \times n_l$, and let $v(A^{kl}) = v_{kl}$. If row i is part of (A^{kl}), the indices i and k will sometimes be denoted by i_k or $k(i)$. The notations j_l and $l(j)$ will be used similarly. We consider the matrix

$$V = \begin{bmatrix} v_{11} & \dots & v_{1q} \\ \dots & \dots & \dots \\ v_{p1} & \dots & v_{pq} \end{bmatrix}.$$

A strategy in $X(A)$ is an m-vector with components ξ_i corresponding to row i of A; strategies in $X(A^{kl})$ are m_k-vectors with components $\xi^k_{i_k}$ cooresponding to row i_k of A^{kl}; finally, strategies in $X(V)$ are p-vectors with components ξ^V_k corresponding to row k of the matrix V.

If we are given any $X^V \in X(V)$ and $X^k \in X(A^{kl})$ $(k = 1, \dots, p)$, then if we set, for $i = 1, \dots, m$,

$$\xi_i = \xi^V_{k(i)} \xi^{k(i)}_i , \tag{5.1}$$

we obtain a vector X that evidently belongs to $X(A)$. Similarly, if we consider strategies $Y^V \in Y(V)$ and $Y^l \in Y(A^{kl})$ $(l = 1, 2, \dots, q)$ with components η^V_l and $\eta^l_{j_l}$, and set $\eta_j = \eta^V_{l(j)} \eta^{l(j)}_j$, we can construct the strategy $Y = (\eta_1, \dots, \eta_n) \in Y(A)$.

5.2. The possibility (although only incomplete) of obtaining a solution of Γ_A from solutions of Γ_V and all the games $\Gamma(A^{kl})$ is determined by the following theorem.

Theorem. 1) *If, in the notation of* 5.1,

$$\bigcap_{l=1}^{q} \mathcal{S}(A^{kl}) = \mathcal{S}^k \neq \emptyset \text{ for } k = 1,\ldots,p, \tag{5.2}$$

then $v_A \geqq v_V$.

2) *If*

$$\bigcap_{k=1}^{p} \mathcal{T}(A^{kl}) = T^l \neq \emptyset \text{ for } l = 1,\ldots,q, \tag{5.3}$$

then $v_A \leqq v_V$.

3) *If* (5.2) *and* (5.3) *are satisfied, then* $v_A = v_V$; *from* $X^V \in \mathcal{S}(V)$ *and* $X^k \in \mathcal{S}^k$ $(k = 1,\ldots,p)$ *we can construct a strategy* $X \in \mathcal{S}(A)$ *by using* (5.1); *and from* $Y^V \in \mathcal{T}(V)$ *and* $Y^l \in \mathcal{T}^l$ $(l = 1,\ldots,q)$ *we can construct a strategy* $Y \in \mathcal{T}(A)$.

Proof. 1) Take

$$X^V \in \mathcal{S}(V), \quad X^k \in \mathcal{S}^k, \quad k = 1,\ldots,p,$$

and from these, construct $X \in \boldsymbol{X}(A)$ by using (5.1). Since the strategies X^V and X^k are optimal, we have

$$\begin{aligned} v_V &= \max_{X_V} \min_l X^V V_{\cdot l} = \min_l X^V V_{\cdot l} = \min_l \sum_{k=1}^{p} \xi_k^V v_{kl} \\ &= \min_l \sum_{k=1}^{p} \xi_k^V \min_{j_l} X^k A^{kl}_{\cdot j_l} \leqq \min_l \min_{j_l} \sum_{k=1}^{p} \xi_k^V \sum_{i_k=1}^{m_k} \xi_{i_k}^k a_{i_k j_l}^{kl} \\ &= \min_j \min_{j_l} \sum_{k=1}^{p} \sum_{i_k=1}^{m_k} \xi_k^V \xi_{i_k}^k a_{i_k j_l}^{kl} = \min_j \sum_{i=1}^{m} \xi_i a_{ij} \\ &= \min_j XA_{\cdot j} \leqq \max_X \min_j X_{\cdot j} = v_A. \end{aligned} \tag{5.4}$$

2) Established similarly.

3) It follows from 1) and 2) that we now have $v_V = v_A$. Therefore there is equality in (5.4), from which, by 1.12, it follows that $X \in \mathcal{S}(A)$. The inclusion $Y \in \mathcal{T}(A)$ is proved similarly. □

5.3. Notice that the converse of part 3) of the theorem in 5.2 is not valid: not every strategy $X \in \mathcal{S}(A)$ can be constructed in the described way from strategies $X^V \in \mathcal{S}(V)$ and $X^k \in \mathcal{S}^k$ $(k = 1,\ldots,p)$. As an example, consider the game with the payoff matrix

$$A = (A^{11}, A^{12}) = \begin{bmatrix} 2 & 0 & 2 & 0 \\ 0 & 2 & 0 & 2 \end{bmatrix}.$$

Here $\mathcal{S}(A^{11}) = \mathcal{S}(A^{12})$ and consists of the single vector $(1/2, 1/2)$, and the hypothesis (5.3) is satisfied automatically since $q = 1$. Therefore the hypothesis of part 3) of the theorem in 5.2 is satisfied.

Furthermore, $\mathcal{T}(A^{11}) = \mathcal{T}(A^{12})$ and consists of the vector $(1/2, 1/2)$; and $V = (1, 1)$. Therefore,

$$\mathcal{T}(V) = \mathbf{Y}(V) = \{(\eta^V, 1 - \eta^V)\}_{\eta^V \in [0,1]}.$$

Consequently the construction in 5.1 yields

$$Y = \left(\frac{1}{2}\eta^V, \frac{1}{2}\eta^V, \frac{1}{2}(1 - \eta^V), \frac{1}{2}(1 - \eta^V)\right). \tag{5.5}$$

It is also clear that $(1/2, 0, 0, 1/2) \in \mathcal{T}(A)$, but this strategy cannot be represented in the form (5.5).

5.4. As an example of a specific application of the theorem we consider the 4×4-game with the payoff matrix

$$W = \begin{bmatrix} A & B \\ C & D \end{bmatrix} = \begin{bmatrix} a_{11} & a_{12} & b & b \\ a_{21} & a_{22} & b & b \\ c & c & d_{11} & d_{12} \\ c & c & d_{21} & d_{22} \end{bmatrix}.$$

Here every strategy in the games Γ_B and Γ_C is optimal. Therefore the hypotheses of part 3) of the preceding theorem are automatically satisfied. Consequently

$$v_W = v \begin{bmatrix} v_A & b \\ c & v_D \end{bmatrix}$$

and the optimal strategies in Γ_W can be found by solving the 2×2-games Γ_A, Γ_D, and Γ_V and applying the construction in 5.1.

5.5 Minkowski-Leontief matrices. As we saw in 1.16, finding optimal strategies in a completely mixed game reduces to the solution of a system of linear algebraic equations, which is an essentially simpler problem. Consequently, establishing the property that a game is completely mixed is an important step toward its solution. Here we present two classes of games that turn out to be completely mixed.

Let an $n \times n$-matrix $A = \|a_{ij}\|$ have the property that there is a $q > 0$ such that

$$a_{ij} \leqq q \text{ for } i \neq j, \tag{5.6}$$

$$\sum_{i=1}^{n} a_{ij} > nq. \tag{5.7}$$

Matrices whose elements satisfy these inequalities are usually called *Minkowski-Leontief matrices.*

Theorem. *A matrix game with a Minkowski-Leontief payoff matrix is completely mixed and has a positive value.*

Proof. Let A be the Minkowski-Leontief $n \times n$-matrix and let $X = (1/n, \ldots, 1/n)$. Then, by (5.7),

$$XA_{\cdot j} = \sum_{i=1}^{n} \frac{1}{n} a_{ij} = \frac{1}{n} \sum_{i=1}^{n} a_{ij} > q$$

for $j = 1, \ldots, n$. Therefore $v_A > q$.

In addition, let $X = (\xi_1, \ldots, \xi_n) \in \mathcal{S}(A)$ and $\xi_{i_0} = 0$. Then by (5.6) we have

$$XA_{\cdot i_0} = \sum_{i=1}^{n} \xi_i a_{ii_0} = \sum_{i \neq i_0} \xi_i a_{ii} \leqq \sum_{i=1}^{n} \xi_i q = q > v_A,$$

which contradicts the optimality of X. Consequently all the optimal strategies of player 1 must be completely mixed.

Now let $Y = (\eta_1, \ldots, \eta_n) \in \mathcal{T}(A)$. By the existence of the completely mixed strategy X of player 1, we must, by 1.15, have

$$A_{i\cdot} Y^T = v^A \text{ for } i = 1, \ldots, n.$$

Let us suppose that $\eta_{j_0} = 0$. Then we will have

$$A_{j_0 \cdot} Y^T = \sum_{i=1}^{n} a_{j_0 j} \eta_j = \sum_{j \neq j_0} a_{j_0 j} \eta_j \leqq q < v_A,$$

which is impossible. Consequently every optimal strategy of player 2 is completely mixed. □

5.6 Corollary. *Every Minkowski-Leontief matrix is nonsingular, and in a game with an $n \times n$ payoff matrix A of this kind the players have unique (completely mixed) optimal strategies X^* and Y^*:*

$$X^* = v_A J_n A^{-1},$$
$$Y^{*T} = A^{-1} J_n^T v_A;$$

furthermore

$$v_A = (J_n A^{-1} J_n^T)^{-1}.$$

The proof is immediate from the theorem in 4.5. □

5.7 A diagonal game. The diagonal matrix

$$A = \begin{bmatrix} a_1 & 0 & \ldots & 0 \\ 0 & a_2 & \ldots & 0 \\ \multicolumn{4}{c}{\ldots\ldots\ldots\ldots\ldots} \\ 0 & 0 & \ldots & a_n \end{bmatrix},$$

with $a_i > 0$, $i = 1, \ldots, n$, is a special Minkowski-Leontief matrix. Let us note that the $\Gamma(A)$-matrix game is not a diagonal noncooperative game in the sense of 1.8, Chapter 2.

For this matrix, hypotheses (5.6) and (5.7) are satisfied with $q = 0$.

We find immediately that this game is completely mixed and has a positive value, and that the unique optimal strategies of the players can be given explicitly.

In the present case we have

$$A^{-1} = \begin{bmatrix} a_1^{-1} & 0 & \dots & 0 \\ 0 & a_2^{-1} & \dots & 0 \\ \dots & \dots & \dots & \dots \\ 0 & 0 & \dots & a_n^{-1} \end{bmatrix},$$

so that

$$J_n A^{-1} J_n^t = \sum_{i=1}^{n} \frac{1}{a_i},$$

from which

$$v_A = \left(\sum_{i=1}^{n} \frac{1}{a_i} \right)^{-1},$$

$$X = \left(\sum_{i=1}^{n} \frac{1}{a_i} \right)^{-1} \left(\frac{1}{a_i}, \dots, \frac{1}{a_n} \right),$$

$$Y = \left(\sum_{i=1}^{n} \frac{1}{a_i} \right)^{-1} \left(\frac{1}{a_i}, \dots, \frac{1}{a_n} \right).$$

We make an additional remark: If a diagonal element $a_i \leqq 0$ in the diagonal matrix A, then $v_A = 0$, the ith pure strategy of player 2 is optimal, and all pure strategies of player 1 are optimal except perhaps the ith (if $a_i < 0$). It is not difficult to complete the analysis of a game of this kind.

5.8 A block-diagonal matrix game. A combination of the results of 5.2 and 5.7 leads to a method for solving block-diagonal matrix games, i.e. games with matrices of the form

$$A = \begin{bmatrix} A^{(1)} & \mathbf{0} & \dots & \mathbf{0} \\ \mathbf{0} & A^{(2)} & \dots & \mathbf{0} \\ \dots & \dots & \dots & \dots \\ \mathbf{0} & \mathbf{0} & \dots & A^{(p)} \end{bmatrix},$$

where $A^{(k)}$ is any (not necessarily square) matrix, and the boldface zeros denote zero submatrices of the corresponding dimensions. The matrix A, in its block representation, satisfies the hypotheses of part 3) of the theorem in 5.2. Therefore the solution of Γ_A reduces to the solution of games $\Gamma_{A^{(1)}}, \dots, \Gamma_{A^{(p)}}$ and of Γ_V, which in this case is diagonal. If we eliminate the uninteresting case when one of the numbers $v(A^{(k)})$ is not positive (cf. the remark at the end of 5.7), we have

$$v_A = \left(\sum_{k=1}^{p} \frac{1}{v(A^{(k)})} \right)^{-1},$$

$$X = \left(\sum_{k=1}^{p} \frac{1}{v(A^{(k)})} \right)^{-1} \left(\frac{X^{(1)}}{v(A^{(1)})}, \dots, \frac{X^{(p)}}{v(A^{(p)})} \right),$$

$$Y = \left(\sum_{k=1}^{p} \frac{1}{v(A^{(k)})} \right)^{-1} \left(\frac{Y^{(1)}}{v(A^{(1)})}, \dots, \frac{Y^{(p)}}{v(A^{(p)})} \right),$$

where $X^{(k)}$ and $Y^{(k)}$ are any optimal strategies of players 1 and 2 respectively in each of the games $\Gamma(A^{(k)})$.

5.9 Quasicyclic games. Another class of matrix games that, as we shall explain, turn out to be completely mixed, is obtained by the following construction.

Let $L_0, \dots, L_{n-1}$ be real intervals for which

$$\max L_{i-1} < \min L_i, \quad i = 1, \dots, n-1. \tag{5.8}$$

An $n \times n$-matrix $A = \|a_{ij}\|_{i,j=1,\dots,n}$ is said to be *quasicyclic* if $a_{ij} \in L_{j-i}$ and the difference $j - i$ is taken modulo n.

Let us suppose that the strategy $X = (\xi_1, \dots, \xi_n) \in \mathcal{S}(A)$ contains a zero component: $\xi_{i_0} = 0$. Since $a_{ij} \in L_{j-i}$, and $a_{i,j-1} \in L_{j-i-1}$ (here $j - 1$ is taken modulo n), it must follow from (5.8) that $a_{ij} > a_{i,j-1}$ when $j \neq i$. We have

$$XA_{\cdot i_0} = \sum_{i=1}^{n} \xi_i a_{ii_0} = \sum_{i \neq i_0} \xi_i a_{ii_0} > \sum_{i \neq i_0} \xi_i a_{i,i_0-1} = XA_{\cdot i_0-1} \geqq v_A.$$

Consequently, by 1.14, we must have $\eta_{i_0} = 0$ in every $Y = (\eta_1, \dots, \eta_n) \in \mathcal{T}(A)$.

However, on the other hand, if $i \neq k$ it follows from (5.8) that $a_{i+1,k} < a_{ik}$ (again, $i + 1$ is taken modulo n). Therefore,

$$A_{i_0+1\cdot} Y^T = \sum_{i=1}^{n} a_{i_0+1,j} \eta_j = \sum_{j \neq i_0} a_{i_0+1,j} \eta_j < \sum_{j \neq i_0} a_{i_0 j} \eta_j = A_{i_0\cdot} Y^T \leqq v_A.$$

It follows that $\xi_{i_0+1} = 0$. Repeating this argument, we arrive at $X = 0$, which is not true. Consequently $X > 0$. If we start from the assumption $\eta_{j_0} = 0$, we obtain $Y = 0$, so that actually $Y > 0$.

Therefore a game with a quasicyclic payoff matrix is completely mixed.

5.10 An "attack-defense" game. If a matrix game is not completely mixed, it is rarely possible to obtain a formulation of an algorithm that describes, in full detail, a method for solving the game and the results of the solution. The classes (unfortunately, very narrow!) of matrix games for which such algorithms can be constructed are all the more instructive.

We shall present an example of one such class of games that have an intuitive interpretation in terms of attack and defense.

Let player 1 intend to attack, with one unit of force, one of the objectives $C_1, \dots, C_n$, which have positive values $a_1, \dots, a_n$. Player 2, also having available a single unit of force, defends one of these objectives. We suppose that when an undefended objective C_i is attacked, it is certainly destroyed (player 1 wins a_i), and that a defended objective survives an attack with probability $p > 0$ (player 1 wins, on the average, $(1-p)a_i$). In a different, but equivalent, formulation of the problem we may suppose that an undefended objective is completely destroyed by an attack, but a defended one retains a fraction p of its usefulness.

Evidently, the problem of the choice by player 1 of an objective to attack, and the choice by player 2 of an objective to defend, lead to a matrix game A with payoff matrix

$$A = \begin{bmatrix} (1-p)a_1 & a_1 & \dots & a_1 \\ a_2 & (1-p)a_2 & \dots & a_2 \\ \dots & \dots & \dots & \dots \\ a_n & a_n & \dots & (1-p)a_n \end{bmatrix}.$$

5.11. Let us undertake an analysis of this game, Γ_A. We first consider the conditions under which each optimal strategy of player 2 in Γ_A will be pure. Evidently in these cases the pure optimal strategy of player 2 is unique.

We denote by a' the largest of the values $a_1, \dots, a_n$. Suppose for definiteness that $a' = a_1 = a_2 = \dots a_k$. Let a'' denote the value next after a' in magnitude. We shall suppose that $a'' = a_{k+1} = \dots = a_l$.

Let us suppose that the unique optimal strategy of player 2 is to defend objective C_j. We show that in this case $j = k = 1$.

We assume that there is an undefended objective of maximal value a'. This means that there is an integer f, different from j, and not exceeding k. An attack by player 1 on C_f yields the payoff a'. Therefore, under our hypotheses, the value v_A of the game under consideration must satisfy the inequality

$$v_A \geqq a'. \tag{5.9}$$

On the other hand, let us consider a mixed strategy Y of player 2 that consists of defending each of $C_1, \dots, C_k$ with probability $1/k$. Then an attack by player 1 on any of the objectives of maximal value yields

$$(1-p)a'/k + (k-1)a'/k = a'(1-p/k) < a'.$$

An attack by player 1 on any other objective yields at most $a'' < a'$.

Therefore the adoption by player 2 of the strategy Y decreases the guaranteed payoff to player 1 by an amount less than a'. The adoption by player 2 of that player's optimal strategy can produce only a further decrease in the guaranteed payoff to player 1. Consequently $v_A < a'$, which contradicts (5.9).

Therefore every objective of maximal value should be protected with positive probability. But since by hypothesis player 2 plays a pure strategy, this is possible only if the objective of maximal value is protected, and this is unique.

5.12. Now choose any $X = (\xi_1, \dots, \xi_n) \in \mathcal{S}(A)$. Since $(X, 1) \in \mathcal{C}(A)$ we have

$$a_{i1} \leqq XA_{\cdot 1} \leqq XA_{\cdot j}, \qquad i, j = 1, \dots, n.$$

Consequently nonzero components of X can correspond only to the maximal element of the column $A_{\cdot 1}$.

According to the relation between $(1-p)a'$ and a'' (only these numbers can be maximal in the column $A_{\cdot 1}$), we may, in general, distinguish three different cases.

1°. $(1-p)a' > a''$. In this case $(1-p)a'$ is the maximal element of column $A_{\cdot 1}$. Therefore the single optimal strategy of player 1 is pure, and consists of attacking the most valuable objective; $v_A = (1-p)a'$.

2°. $(1-p)a' = a''$. For the strategy X of player 1 to be optimal, it is necessary and sufficient under our hypotheses that

$$v_A = (1-p)a' = XA_{\cdot 1} \leqq XA_{\cdot j}, \qquad j = 1, \ldots, n.$$

For $1 < j \leqq l$, this gives us

$$(1-p)a' \leqq \xi_1 a' + \xi_j a''(1-p) + (1-\xi_1-\xi_j)a'', \tag{5.10}$$

and for $j > l$,

$$(1-p)a' \leqq \xi_1 a' + (1-\xi_1)a''.$$

The last inequality is automatically satisfied, since $p > 0$. The inequality (5.10) can be rewritten as

$$(1-p)a' \leqq \xi_1 a' + \xi_j(1-p)^2 a' + (1-\xi_1-\xi_j)(1-p)a',$$

or

$$1-p \leqq \xi_1 + \xi_j(1-p)^2 + (1-\xi_1-\xi_j)(1-p),$$

or

$$0 \leqq \xi_1 p - \xi_j p(1-p),$$

i.e.,

$$\xi_j \leqq \xi_1/(1-p). \tag{5.11}$$

Consequently, in this case, the optimal strategy of player 1 is any available strategy, pure or mixed, whose nonzero components correspond to maximal elements (not necessarily all of them), in the column chosen by player 2, that satisfy (5.11).

3°. $(1-p)a' < a''$. Here the nonzero components of X can only be $\xi_2, \ldots, \xi_l$. The further conditions for its optimality consist of

$$v_A = a'' = XA_{\cdot 1} \leqq XA_{\cdot j}.$$

For $1 < j \leqq l$, this gives us

$$a'' \leqq \xi_j a''(1-p) + (1-\xi_j)a'',$$

which is impossible when $\xi_j > 0$. The resulting contradiction shows that in this case some of the optimal strategies of player 2 must be mixed.

5.13. We now suppose that player 2 has mixed optimal strategies, that Y is one of these, and that $\operatorname{supp} Y = R$. Set $|R| = \rho$. Since Y is not a pure strategy, it follows that $\rho \geqq 2$.

Again let $X = (\xi_1, \ldots, \xi_n)$. Player 2, defending one of the objectives C_r $(r \in R)$ pays player 1

$$XA_{\cdot r} = \sum_{s \in R} a_s \xi_s - pa_r \xi_r = v_A.$$

Comparing any two equations of this system, we obtain

$$a_r \xi_r = \text{const}, \qquad r \in R. \tag{5.12}$$

Since all $a_r > 0$, this equation shows that either all ξ_r $(r \in R)$ are zero, or all are positive. We begin with the first possibility.

Under its hypotheses, only undefended objectives should be attacked. Since at least one objective is always attacked with positive probability, R contains at most $n - 1$ strategies. It is clear that the optimal procedure for player 1 is to attack the objective among those not belonging to R (i.e., undefended) that has the largest value. Let this be objective C_t.

Then the pair (t, Y) must, under our hypotheses, form an equilibrium situation. In particular, therefore,

$$A_{t\cdot} Y^T \leqq a_{tj}, \quad 1 \leqq j \leqq n. \tag{5.13}$$

But under our hypotheses C_t is undefended, and therefore

$$A_{t\cdot} Y^T = a_t. \tag{5.14}$$

Now let player 2 play, instead of Y, the tth pure strategy. The player's disadvantage is then evidently equal to $a_{tt} = (1-p)a_t$, which with (5.13) and (5.14) gives us a contradiction.

Thus the assumption that all the probabilities ξ_r $(r \in R)$ are zero turns out to be inconsistent, and we must suppose that all these probabilities are positive.

The content of our inference is quite natural: it means that under the hypotheses of an equilibrium situation only attackable objectives should be defended. Since by hypothesis R contains at least two pure strategies, we conclude that player 1, in each optimal strategy, attacks at least two objectives. This means that in this case a player cannot have pure optimal strategies.

5.14. We turn now directly to looking for equilibrium situations (X, Y) in Γ_A, where $X = (\xi_1, \ldots, \xi_n)$ and $Y = (\eta_1, \ldots, \eta_n)$.

A necessary and sufficient condition for $(X, Y) \in \mathscr{C}(\Gamma)$ is that

$$A_{i\cdot} Y^T \leqq XAY^T \leqq XA_{\cdot j}, \quad i, j = 1, \ldots, n. \tag{5.15}$$

We denote the spectrum supp X of the "most mixed" optimal strategy X by S. It was established above that $R \subset S$. We denote the number of strategies in R by ρ, and in S, by σ.

The relation (5.15), taking into account that $\eta_r > 0$ for $r \in R$ and that $\xi_s > 0$ for $s \in S$, can be rewritten as the following system:

$$A_{r\cdot}Y^T = v_A, \qquad r \in R, \tag{5.16}$$

$$A_{s\cdot}Y^T = v_A, \qquad s \in S \setminus R, \tag{5.17}$$

$$A_{i\cdot}Y^T \leqq v_A, \qquad t \notin S, \tag{5.18}$$

$$XA_{\cdot r} = v_A, \qquad r \in R, \tag{5.19}$$

$$XA_{\cdot s} \geqq v_A, \qquad s \in S \setminus R, \tag{5.20}$$

$$XA_{\cdot t} \geqq v_A, \qquad t \notin S. \tag{5.21}$$

Equation (5.16) means that

$$A_{r\cdot}Y^T = (1-p)a_r\eta_r + a_r(1-\eta_r) = a_r(1-p\eta_r) = v_A, \qquad r \in R. \tag{5.22}$$

Comparing one of these equations with each of the others (this is possible since $|R| \geqq 2$), we obtain

$$a_r(1-p\eta_n) = a_r(1-p\eta_{r'}),$$

or

$$\frac{a_r}{a_{r'}}(1-p\eta_r) = 1-p\eta_{r'};$$

summing over r and setting

$$\sum_{r' \in R} \frac{1}{a_{r'}} = \alpha_R,$$

we will have

$$a_r\alpha_R(1-p\eta_r) = \rho - p,$$

from which

$$\eta_r = \frac{1}{p}\left(1 - \frac{\rho - p}{a_r\alpha_R}\right). \tag{5.23}$$

Since $y_r > 0$, we must have, for all $r \in R$,

$$a_r > \frac{\rho - p}{\alpha_R} = v_A. \tag{5.24}$$

Furthermore, if we substitute the expression for η_r into (5.22), we obtain

$$\frac{\rho - p}{\alpha_R} = v_A. \tag{5.25}$$

Therefore, if we form Y from the component η_r defined by (5.23) and adjoin a set of zero components, we obtain by using (5.24) a strategy of player 2 that, by (5.25), satisfies (5.16). Then it follows from (5.24) that pure strategies in R

correspond to defending objectives whose values are strictly larger than the value of the game. However, we still do not know the value of the game, because we still have not found α_R and ρ.

Equation (5.17) means that

$$\sum_j a_s \eta_j = a_s \sum_j \eta_j = a_s = v_A. \tag{5.26}$$

We observe, in this connection, that $S \setminus R$ is nonempty only in the case when there are objectives in our game whose value is equal to the value of the game. Conversely, if $a_s = v_A$, we have $s \in S \setminus R$.

Suppose that $a_t > v_A$ for some $t \notin S$. Then

$$A_{t.} Y^T = a_t > v_A,$$

which contradicts the optimality of Y. Since we must have $t \in S \setminus R$ when $a_t = v_A$, we obtain

$$a_t < v_A \tag{5.27}$$

for $t \notin S$.

Therefore $A_{t.} Y^T < v_A$ for $t \in S$; and (5.18) follows.

As a result of what we have just said, (5.24) is not only necessary, but also sufficient, for $r \in R$.

We now consider the equations (5.19), which we can rewrite in the form

$$xA_{.r} = \sum_{i \in R \setminus r} \xi_i a_i + (1-p) a_r \xi_r + \sum_{i \in S \setminus r} x_i \xi_i = v_A.$$

Since the first sum contains $\rho - 1$ terms and by (5.12) they are all equal (we denote their common value by λ), and a_i is v_A for $i \in S \setminus R$, we obtain, if we denote $\sum_{i \in S \setminus R} \xi_i$ by $X(S \setminus R)$,

$$(\rho - p)\lambda + v_A X(S \setminus R) = v_A,$$

i.e.,

$$\lambda = \frac{v_A}{\rho - p}(1 - X(S \setminus R)) = \frac{1 - X(S \setminus R)}{\alpha_R}. \tag{5.28}$$

We notice that (5.28), together with (5.12), in turn gives us (5.19).

We now turn to the inequality (5.20). For $s \in S \setminus R$, it follows that

$$XA_{.s} = \sum_{i \in R} \xi_i a_i + \sum_{i \in S \setminus (R \cup s)} \xi_i a_i + (1-p)\xi_s a_s \geqq v_A,$$

or, in the notations we have introduced,

$$\rho\lambda + v_A X(S \setminus R) - p\xi_s a_s \geqq v_A. \tag{5.29}$$

But by (5.28) we must have

$$X(S \setminus R) = 1 - X(R) = 1 - \lambda\alpha_R. \tag{5.30}$$

Therefore, by (5.25), the inequality (5.29) can be rewritten as

$$\rho\lambda + \frac{\rho - p}{\alpha_R}(1 - \lambda\alpha_R) - p\frac{\rho - p}{\alpha_R}\xi_s \geqq \frac{\rho - p}{\alpha_R},$$

or, after simplifying, as

$$\xi_s \leqq \lambda\alpha_R/(\rho - p). \tag{5.31}$$

We notice that, conversely, we can deduce (5.20) from (5.31).

Finally, (5.21) is automatically satisfied under our hypotheses: for every $t \notin S$,

$$XA_{\cdot t} = \sum_{r \in R} \xi_r a_r + \sum_{s \in S \setminus R} \xi_s a_s > a_s \sum_{s \in S} \xi_s = a_s = v_A.$$

Thus v_A and the vectors X and Y are expressed in terms of ρ and α_R by means of (5.12), (5.23), (5.25), and (5.31).

5.15. To complete the solution of our game, it remains to determine the set R and the parameter λ.

For this purpose we write the values

$$a_1, a_2, \ldots, a_n$$

in nonincreasing order. By (5.24), (5.26) and (5.27), we may suppose that $\{1, \ldots, \rho\} = R$, and $\{\rho + 1, \ldots, \sigma\} = S \setminus R$ (where, of course, $a_{\rho+1} = \ldots = a_\sigma$).

Evidently (5.24) must be satisfied for $r = \rho$, but not for larger values of r. Hence in finding ρ we may use the following procedure. For $r = \rho$ we rewrite (5.24) in the form

$$\frac{a_\rho}{a_1} + \ldots + \frac{a_\rho}{a_\rho} > \rho - p,$$

or

$$\frac{a_1 - a_\rho}{a_\rho} + \ldots + \frac{a_\rho - a_\rho}{a_\rho} < p. \tag{5.32}$$

If ρ is increased by 1 on the left side of this inequality, then, first, each term is increased; and, second, another nonnegative term is added (actually, zero). Therefore the left side of (5.32) is a nondecreasing function of ρ.

Consequently either (5.32) is satisfied for all values of ρ, and then $\rho = n$; or there is a value of ρ for which (5.32) is satisfied, whereas it fails to be valid for $\rho + 1$. Evidently such a value of ρ is unique and is the one required. We consider these two cases separately.

First let $\rho = n$. In this case it is evident that $\sigma = 0$, and α_R can be found immediately as $a_1^{-1} + \ldots + a_n^{-1}$. Since we know ρ and α_R, we obtain $v_A(n - p)/\alpha_R$ from (5.25).

Since $(S \setminus R) = \emptyset$ here, we find from (5.30) that $\lambda = 1/\alpha_R$, whence by the definition of λ

$$\xi_i = 1/(a_i\alpha_R), \qquad i = 1, \ldots, n,$$

and from (5.23)

$$\eta_i = \frac{1}{p}\left(1 - \frac{n-p}{a_i \alpha_R}\right), \qquad i = 1, \dots, n.$$

Now let $\rho < n$. Here the value of α_R can be calculated at once, and σ is determined as the number of objects whose common value follows immediately after a_ρ. The determination of this value gives us the value of v_A, by (5.26).

Finally we find λ. For this purpose we rewrite (5.31), taking (5.30) into consideration:

$$(1 - X(S \setminus R))/(\rho - p) \geqq \xi_s. \tag{5.33}$$

Summing over $s \in S \setminus R$, we obtain

$$(\sigma \setminus \rho)\,(1 - X(S \setminus R))/(\rho - p) \geqq X(S \setminus R),$$

from which

$$X(S \setminus R) \leqq (\sigma - \rho)/(\sigma - p),$$

so that, using (5.30) again,

$$\lambda = (1 - X(S \setminus R))/\alpha_R \geqq (\rho - p)/\alpha_R(\sigma - p).$$

An upper bound for λ is found from the condition $X(R) = \lambda\alpha_R \leqq 1$. Therefore

$$(\rho - p)/\alpha_R(\sigma - p) \leqq \lambda \leqq 1/\alpha_R,$$

from which (5.30) and (5.33) at once give us

$$0 \leqq \xi_s \leqq (\sigma - \rho)/(\sigma - p).$$

Thus the value of the game is $(\rho - p)/\alpha_R$, where ρ and α_R are determined by (5.32), and the players' optimal strategies are subject to the conditions

$$\xi_r = \lambda/\alpha_R, \quad r \in R,$$

$$X(S \setminus R) \leqq (\sigma - \rho)/(\sigma - p), \quad 0 \leqq \xi_s, \quad s \in S \setminus R,$$

$$\xi_t = 0, \quad t \notin S,$$

$$y_r = \frac{1}{p}\left(1 - \frac{\rho - p}{\alpha_R}\right), \quad r \in R,$$

$$y_t = 0, \quad t \notin R.$$

The optimal strategy of player 2 is unique in all cases. Player 1 has a unique optimal strategy only when $\sigma = \rho$.

§6 Approximate methods for solving matrix games

6.1 The necessity of approximate methods. If the payoff matrix A of the matrix game Γ_A has any general form (i.e., depends on a considerable number of parameters and does not have any particular combinatorial structure), it is simply impossible to say anything about analytic expressions describing the sets $\mathcal{S}(A)$ and $\mathcal{T}(A)$ of optimal strategies of the players in Γ_A (or even about the individual optimal strategies), or about the value v_A of the game. Solving this problem algorithmically would have to be very complicated and of a ramified nature. In § 4 we described an algorithm based on a search of the possible forms of the extreme points of $\mathcal{S}(A)$ and $\mathcal{T}(A)$. However, applying it evidently requires an excessively large number of operations.

A more promising approach is to use algorithms of iterative type, where at each step finding an approximately optimal solution is needed to construct the next approximation, which is closer to the optimal solution. All the well known alternating gradient methods are based on this idea. As a matter of fact, they include the simplex method, which, by what was said in 3.4, can be used for finding the value of a matrix game and certain optimal strategies in it.

From among the iterative algorithms we naturally select those in which the required approximately optimal solutions are determined by the nature of their "nonoptimality", which realizes the feedback that decreases the nonoptimality at the next step. Here the steps may be taken to be discrete (units), or infinitesimal, i.e., the process of stepwise improvement is taken to be continuous. Evidently the basic method of analysis in the first case involves difference techniques, and in the second case, the apparatus of differential equations. We shall consider both of these versions.

6.2 The method of differential equations. We first notice that on the basis, for example, of what was said in 1.20–1.23 and 3.5, in searching for optimal strategies of the players we may confine ourselves to symmetric matrix games for which the payoff matrix $A = \|a_{ij}\|$ is skew symmetric, i.e., $a_{ij} = -a_{ji}$ for all i and j.

Let A be a skew-symmetric $n \times n$-matrix, and let $X = \{\xi_1, \ldots, \xi_n\}$ be an arbitrary n-vector. Let us set

$$u_k = A_k X^T, \qquad k = 1, \ldots, n,$$

$$\varphi_k = \varphi_k(X) = \max\{0, u_k\},$$

$$\Phi = \Phi(X) = \sum_{k=1}^{n} \varphi_k(X),$$

$$\Psi = \Psi(X) = \sum_{k=1}^{n} \varphi_k^2(X).$$

We shall assume that all the variables listed here depend on a real parameter t, and consider the system of differential equations

$$\frac{d\xi_k}{dt} = \varphi_k - \Phi\xi_k, \quad k = 1, \ldots, n, \tag{6.1}$$

with initial conditions

$$\xi_k(0) = \xi_k^0, \qquad k = 1, \ldots, n. \tag{6.2}$$

The system (6.1) of differential equations with the initial conditions (6.2) has a unique solution, which is continuous.*)

As a preliminary, we establish two fundamental properties of this solution.

6.3. If $\xi_k^0 \geqq 0$, then $\xi_k(t) \geqq 0$ for all $t > 0$.

In fact, suppose that $\xi_k(t') < 0$ for some $t' > 0$. This means that there is a $t'' \in [0, t')$ such that $\xi_k(t'') = 0$ and $\xi_k(t) < 0$ for all $t \in (t'', t']$.

Since $\xi_k(t') < \xi_k(t'')$, the interval (t'', t') must contain a point t''' at which $\xi_k'(t''') < 0$. But since $\varphi_k(t) \geqq 0$ and $\Phi(t) \geqq 0$ we find from (6.1) that

$$0 \geqq \xi_k'(t''') = \varphi_k(t''') - \Phi(t''')\xi_k(t''') \geqq 0,$$

which is impossible.

6.4. If $\sum_{k=1}^{n} \xi_k^0 = 1$, we have

$$\sum_{k=1}^{n} \xi_k(t) = 1 \text{ for all } t > 0.$$

In fact, suppose that, for some t', we had

$$\sum_{k=1}^{n} \xi_k(t') > 1.$$

Then there is a $t'' \in [0, t')$ such that $\sum_{k=1}^{n} \xi_k(t'') = 1$ and

$$\sum_{k=1}^{n} \xi_k(t) > 1 \text{ for all } t \in (t'', t']. \tag{6.3}$$

Since, for these values of t,

$$\sum_{k=1}^{n} \xi_k(t) > \sum_{k=1}^{n} \xi_k(t''),$$

the sum must be strictly increasing, at least at one point t''' of (t'', t'), i.e.

$$\frac{d}{dt} \sum_{k=1}^{n} \xi_k(t''') > 0. \tag{6.4}$$

*) See, for example, I.G. Petrovskiĭ, Ordinary differential equations, Prentice-Hall, 1966.

If we now add all the equations in (6.1) term by term, we obtain

$$\frac{d}{dt}\sum_{k=1}^{n}\xi_k(t) = \sum_{k=1}^{n}(\varphi_k(t) - \Phi(t)\xi_k(t))$$

$$= \Phi(t) - \Phi(t)\sum_{k=1}^{n}\xi_k(t) = \Phi(t)\left(1 - \sum_{k=1}^{n}\xi_k(t)\right).$$

Together with (6.3) and (6.4), this yields, for $t = t'''$,

$$0 < \frac{d}{dt}\sum_{k=1}^{n}\xi_k(t''') = \Phi(t''')\left(1 - \sum_{k=1}^{n}\xi_k(t''')\right) \leqq 0,$$

and we have a contradiction.

Similarly, we can eliminate the possibility that

$$\sum_{k=1}^{n}\xi_k(t) < 1.$$

Therefore, if the vector $X(0)$ is a strategy, this property is preserved for $X(t)$ for all $t > 0$.

6.5. We now turn to the objective of our discussion.

Formally, we have

$$\frac{d\varphi_k^2}{dt} = 2\varphi_k\frac{d\varphi_k}{dt}. \tag{6.5}$$

In the case when $\varphi_k = u_k > 0$, we obtain

$$\frac{d\varphi_k}{dt} = \frac{du_k}{dt} = \frac{d}{dt}A_{k\cdot}X^T = \frac{d}{dt}\sum_{l=1}^{n}a_{kl}\xi_l = \sum_{l=1}^{n}a_{kl}\frac{d\xi_l}{dt}$$

$$= \sum_{l=1}^{n}a_{kl}(\varphi_l - \Phi\xi_l) = \sum_{l=1}^{n}a_{kl}\varphi_l - \Phi\sum_{l=1}^{n}a_{kl}\xi_l = \sum_{l=1}^{n}a_{kl}\varphi_l - \Phi u_k,$$

or, finally,

$$\frac{d\varphi_k}{dt} = \sum_{l=1}^{n}a_{kl}\varphi_l - \Phi\varphi_k,$$

and (6.5) can be rewritten as

$$\frac{d\varphi_k^2}{dt} = 2\varphi_k\left(\sum_{l=1}^{n}a_{kl}\varphi_l - \Phi\varphi_k\right). \tag{6.6}$$

If, however, $u_k \leqq 0$, then φ_k is the constant zero. Consequently, the derivatives of φ_k and φ_k^2 must be zero. Therefore (6.6) remains valid in this case also.

We now sum (6.6) over all values of k:

$$\sum_{k=1}^{n} \frac{d\varphi_k^2}{dt} = \frac{d\varphi}{dt} = 2\sum_{k=1}^{n} \varphi_k \left(\sum_{l=1}^{n} a_{kl}\varphi_l - \Phi\varphi_k \right)$$

$$= 2\sum_{k=1}^{n}\sum_{l=1}^{n} a_{kl}\varphi_k\varphi_l - 2\Phi \sum_{k=1}^{n} \varphi_k^2,$$

or, using the skew symmetry of A,

$$\frac{d\Psi}{dt} = -2\Phi\Psi. \tag{6.7}$$

In addition, it is evident, since all the terms of Φ are nonnegative, that

$$\sqrt{\Psi} \leqq \Phi. \tag{6.8}$$

This implies that when Ψ is positive, Φ is also positive. Consequently, in this case the right part of (6.7) is negative. Thus, if Ψ is positive, it decreases.

Substituting the estimate for Φ that appears in (6.7) into the given inequality (6.8), we obtain

$$\frac{d\Psi}{dt} \leqq -2\Psi^{3/2},$$

or

$$\frac{d}{dt}(\Psi^{-1/2}) \geqq 1.$$

Integrating this inequality over any interval of the form $(0, t)$ where $\Psi(t)$ is positive, we find that

$$\Psi^{-1/2}(t) \geqq t + \Psi^{-1/2}(0),$$

from which it follows that

$$\Psi(t) \leqq \frac{\Psi(0)}{(1 + t\sqrt{\Psi(0)})^2}. \tag{6.9}$$

If the interval of positivity of Ψ terminates, and Ψ becomes 0, then $d\Psi/dt = 0$ by (6.7). Therefore once Ψ reduces to 0, it remains 0 thereafter. Consequently (6.9) is valid for all $t > 0$.

Since $\varphi_k^2 \leqq \Psi$, it follows from (6.9) that

$$u_k \leqq \varphi_k \leqq \frac{\sqrt{\Psi(0)}}{1 + t\sqrt{\Psi(0)}}. \tag{6.10}$$

Thus, for $k = 1, \dots, n$,

$$\lim_{t\to\infty} u_k(t) \leqq \lim_{t\to\infty} \frac{\sqrt{\Psi(0)}}{1 + t\sqrt{\Psi(0)}} = 0,$$

i.e.,

$$\lim_{t\to\infty} A_k X(t)^T \leqq 0 = v_A. \tag{6.11}$$

Consequently, *for any $\varepsilon > 0$, assigned in advance, the strategy $X(t)$ is ε-optimal for sufficiently small t.*

Let us now take an unboundedly increasing sequence of values $t_1, t_2, \ldots$ and a corresponding sequence of strategies

$$X(t_1), X(t_2), \ldots. \tag{6.12}$$

If this sequence converges to a limit X^*, it follows from (6.11) that $A_k X^{*T} \leqq 0$ for $k = 1, \ldots, n$, i.e., $X^* \in \mathcal{S}(A)$.

Besides this, by the compactness of $\boldsymbol{X}$, from any sequence of the form (6.12) we can extract a convergent subsequence.

6.6. Let us make some comments on what we have said. In the first place, our discussion is, as is easily seen, another proof of the theorem on the existence of solutions in matrix games (see § 1).

In the second place, the estimate in (6.10) for the rapidity of convergence of the strategies $X(t)$ to an optimal strategy shows that $X(t)$ is an ε-optimal strategy in Γ_A, where ε is of order $1/t$.

Finally, in the third place, although a direct solution of the system (1) is possible only in a small number of very special cases, the application of approximate methods has definite prospects.

6.7 Iterative algorithms with discrete steps. We now turn to the case when the transition from one situation to another proceeds at discrete times.

Let there be given an arbitrary (not necessarily symmetric) $m \times n$-matrix game Γ_A. Let us consider the following inductive procedure for realizing the discrete iterative process of constructing situations, as outlined in 6.1.

Let us suppose that, starting from some initial situation (X_0, Y_0) in Γ_A, at the tth step we have arrived at the situation (X_t, Y_t). Then the $(t+1)$th step of the process is as follows.

Let $\boldsymbol{X}^t$ denote the set of strategies $\tilde{X} \in \boldsymbol{X}$ for which

$$\tilde{X} A Y_t^T = \max_X X A Y_t^T, \tag{6.13}$$

and let $\boldsymbol{Y}^t$ denote the set of strategies $\tilde{Y} \in \boldsymbol{Y}_t$ for which

$$X_t A Y^T = \min_Y X_t A Y^T. \tag{6.14}$$

(Evidently there are pure strategies in both sets $\boldsymbol{X}_t$ and $\boldsymbol{Y}_t$.) If we choose $\tilde{X}_t \in \boldsymbol{X}_t$ and $\tilde{Y}_t \in \boldsymbol{Y}_t$ arbitrarily, and assign weights α_t and β_t to these strategies, we obtain the mixtures

$$X_{t+1} = (1 - \alpha_t) X_t + \alpha_t \tilde{X}_t, \tag{6.15}$$

$$Y_{t+1} = (1 - \beta_t) Y_t + \beta_t \tilde{Y}_t \tag{6.16}$$

and form the strategy (X_{t+1}, Y_{t+1}). Such processes are known as *linear iterative processes* (or *algorithms*).

It is clear that linear iterative algorithms are completely described by the following data:

a) an initial situation (X_0, Y_0);

b) a rule for selecting "temporarily improved" strategies $X_t \in \mathbf{X}_t$ and $Y_t \in \mathbf{Y}_t$;

c) sequences $\alpha_0, \alpha_1, \ldots$ and $\beta_0, \beta_1, \ldots$.

We notice particularly the special case of a *symmetric* linear iterative process, in which the sequences of weights α_t and β_t coincide.

6.8. The iterative process described by (6.13)–(6.16) can be interpreted as a construction in a supergame (see section 15 of the Introduction) over the mixed strategies of an extension of the game Γ_A to a situation $(\bar{X}, \bar{Y})$ that consists of the pair of superstrategies

$$\bar{X} = (\tilde{X}_0, \tilde{X}_1, \ldots),$$

$$\bar{Y} = (\tilde{Y}_0, \tilde{Y}_1, \ldots),$$

whose components are generated iteratively from (6.13)–(6.16).

In other words, a linear iterative process can be thought of as a sort of match, each set in which is a game Γ_A (more precisely, its mixed extension). Such a match is usually called *fictitious*, since some rule for choosing strategies in each event is imposed on the players from outside, and as a result the actions of the players are not independent, as they would be in a noncooperative game, but coordinated.

It is natural, however, to refuse to consider this match, or process, as a conflict or competition, but to think of it as an imitation of Γ_A that helps us investigate this game and expecially to find the optimal strategies in it.

6.9 Regular iterative processes. A linear iterative process is of interest for us if all convergent subsequences of the sequence of strategies

$$X_0, X_1, \ldots \text{ and } Y_0, Y_1, \ldots \tag{6.17}$$

converge to optimal strategies of the players (as usual, we consider the convergent subsequences as if they were the sequence (6.17) itself).

However, if, for some t, we have $\alpha_t \geqq \alpha > 0$ and $\beta_t \geqq \beta > 0$ for α_t in (6.15) and β_t in (6.16), then the sequences (6.17) can, for some games Γ_A, "bypass" the sets $\mathcal{S}(A)$ and $\mathcal{T}(A)$, and we do not obtain the necessary convergence. On the other hand, if

$$\lim_{t\to\infty} \alpha_t = 0, \quad \lim_{t\to\infty} \beta_t = 0, \tag{6.18}$$

and the limits are approached so rapidly that

$$\sum_{t=1}^{\infty} \alpha_t < +\infty, \quad \sum_{t=1}^{\infty} \beta_t < +\infty,$$

we may conclude that the sequences (6.17) do not even "creep" to $\mathcal{S}(A)$ or $\mathcal{T}(A)$.

Consequently, general theorems on the convergence of the type that we require are of interest only when (6.18) is satisfied and

$$\sum_{t=1}^{\infty} \alpha_t = +\infty, \quad \sum_{t=1}^{\infty} \beta_t = +\infty. \tag{6.19}$$

Linear iterative processes for which (6.18) and (6.19) are satisfied are sometimes called *regular*. It turns out that regularity of an iterative process is a sufficient condition for it to converge in the required sense.

6.10 A convergence theorem. As follows from what was said in § § 1 and 3, to find the value of a game and describe the set of optimal strategies, we need only consider the case of symmetric games. We shall therefore suppose from now on that Γ_A is symmetric, and the iterative process for solving it leads to what follows.

At the tth step, from the strategy Y_t, already found, we obtain

$$\tilde{X}_t \in \mathscr{X}_t = \{X : XAY_t^T \to \max\}, \tag{6.20}$$

after which we calculate, from the previously assigned $\alpha_t \in (0,1)$,

$$Y_{t+1} = (1-\alpha_t)Y_t + \alpha_t \tilde{X}_t. \tag{6.21}$$

Such a process has a rather reasonable intuitive representation: by the symmetry of the game, the expediency of the strategy X_t of player 1 can be adopted by player 2 to improve that player's strategy Y_t.

Theorem. *If the matrix game Γ_A is symmetric, and the iterative process described by* (6.20) *and* (6.21) *is regular, then for every $\varepsilon > 0$ there is a t_ε such that $(Y_t, Y_t) \in \mathscr{C}^\varepsilon(\Gamma_A)$ for $t > t_\varepsilon$, whatever the initial strategy Y_0 may have been.*

The proof of this theorem, apart from other considerations, justifies the fictitious game as an instructional process, since it shows that in the course of the algorithm the players actually learn to play in an optimal way.

We need a number of lemmas for the proof of the theorem.

6.11. We first give a preliminary discussion.

The set $\boldsymbol{X}$ of mixed strategies of each player in our game Γ_A can, for each $\varepsilon > 0$, be divided by hyperplanes into arbitrarily small polytopes so that, for every polytope $\boldsymbol{X}^\varepsilon$, it follows for $X', X'' \in \boldsymbol{X}^\varepsilon$ that

$$\max_{Y \in \boldsymbol{X}} |X'AY^T - X''AY^T| \leqq \varepsilon \tag{6.22}$$

(i.e., so that the diameter of $\boldsymbol{X}^\varepsilon$ in the intrinsic metric does not exceed ε). We shall call these polytopes *ε-polytopes*.

Let s and t be positive integers with $s < t$; we say that an ε-polytope $\boldsymbol{X}^\varepsilon$ is *essential* on $[s,t]$ if, in our iterative process, $\tilde{X}_u \in \boldsymbol{X}^\varepsilon$ for some u with $s \leqq u \leqq t$. In the same way, we can speak of the essentialness of an ε-polytope on $[s, \infty)$.

Let us set

$$\prod_{u=s}^{t-1}(1-\alpha_u)=\beta(s,t) \tag{6.23}$$

(note that $\beta(s,t)$ is an increasing function of s for fixed t, and a decreasing function of t for fixed s), and

$$(1-\beta(s,t))^{-1}\sum_{u=s}^{t-1}\alpha_u\tilde{X}_u\prod_{v=u+1}^{t-1}(1-\alpha_v)=Z(s,t). \tag{6.24}$$

It follows immediately from this and (6.21) that

$$Y_t=\beta(s,t)Y_s+(1-\beta(s,t))Z(s,t). \tag{6.25}$$

In addition, we set, for each u,

$$\max_{X\in\boldsymbol{X}} XAY_u^T=\tilde{X}_u AY_u^T=\lambda_u, \tag{6.26}$$

for the sake of brevity.

6.12 Lemma. *Let the set $\boldsymbol{Z}\subset\boldsymbol{X}$ of strategies intersect all the ε-polytopes $\boldsymbol{X}_\Gamma^\varepsilon$ that are essential on $[s,t]$. Then*

$$\min_{X\in\boldsymbol{X}} XAY_t^T-\beta(s,t)\max_{X\in\boldsymbol{X}} XAY_s^T\leqq\varepsilon(1-\beta(s,t)). \tag{6.27}$$

Proof. By (6.24) $Z(s,t)$ is a complex combination of some $\tilde{X}_u$ $(u=s,\dots,t-1)$. Therefore

$$0=Z(s,t)AZ(s,t)^T\geqq\min_{s\leqq u\leqq t-1}\tilde{X}_u AZ(s,t)^T,$$

i.e., for some u_0 on this interval we must have

$$\tilde{X}_{u_0}AZ(s,t)^T\leqq 0,$$

Therefore, if we select X from the same ε-polytope as $\tilde{X}_{u_0}$, we obtain $XAZ(s,t)^T\leqq\varepsilon$ by (6.22), so that

$$\min_{X\in\boldsymbol{Z}} XAZ(s,t)^T\leqq\varepsilon. \tag{6.28}$$

In addition, from (6.25) we have

$$\begin{aligned}XAY_t^T&-\beta(s,t)\max_{X\in\boldsymbol{Z}} XAY_s^T\\&=\beta(s,t)XAY_s^T+(1-\beta(s,t))XAZ(s,t)^T-\beta(s,t)\max_{X\in\boldsymbol{Z}} XAY_s^T\\&=\beta(s,t)(XAY_s^T-\max_{X\in\boldsymbol{Z}} XAY_s^T)+(1-\beta(s,t))XAZ(s,t)^T.\end{aligned}$$

Taking the minimum for $X \in Z$ and also taking account of (6.28) gives us

$$\min_{X\in Z} XAY_t^T - \beta(s,t) \max_{X\in Z} XAY_s^T$$

$$\leqq \beta(s,t)(\min_{X\in Z} XAYT_s^T - \max_{X\in Z} XAY_s^T) + (1-\beta(s,t)) \min_{X\in Z} XAZ(s,t)^T$$

$$\leqq (1-\beta(s,t))\varepsilon. \qquad \square$$

6.13 Lemma. *If* $a = \max_{i,j} |a_{ij}|$ *then for all* s *and* t $(s < t)$

$$\lambda_t < \min_{s\leqq u\leqq t} \tilde{X}_u AY_t^T \leqq 3a(1-\beta(s,t)). \tag{6.29}$$

Proof. Let the minimum in (6.29) be attained for $u = u_0$. We have

$$\lambda_t - \tilde{X}_{u_0} AY_t^T \leqq \lambda_t - \beta(u_0,t)\lambda_{u_0} + \tilde{X}_{u_0} AY_{u_0}^T - \tilde{X}_{u_0} AY_t^T$$

$$\leqq (1-\beta(u_0,t))a + \tilde{X}_{u_0} A(Y_{u_0} - Y_t)^T.$$

But, by (6.25),

$$Y_t = Y_{u_0} + (1-\beta)(u_0,t))(Z(u_0,t) - Y_{u_0}).$$

Therefore

$$\lambda_t - \tilde{X}_{u_0} AY_t^T \leqq (1-\beta(u_0,t))a + (1-\beta(u_0,t))\tilde{X}_{u_0} A(Z(u_0,t) - Y_{u_0})^T$$

$$\leqq (1-\beta(u_0,t))3a. \qquad \square$$

6.14 Lemma. *Let* $s < t, l \geqq 0$, *and suppose all* ε*-polytopes that are essential on* $[s, t+l]$ *are also essential on* $[t, t+l]$. *Then*

$$\lambda_{t+l} \leqq \beta(s,t+l)\lambda_s + 3a(1-b(t,t+l)) + \varepsilon(1-\beta(s,t+l)).$$

Proof. Let

$$Z = \{X_u : u = s, \dots, t+l\}.$$

This set satisfies the hypotheses of the lemma in 6.12. Therefore

$$\min_{X\in Z} XAY_{t+l}^T \leqq \beta(s,t+l)\lambda_s + \varepsilon(1-\beta(s,t+l)).$$

On the other hand, if we replace s by t and t by $t+l$ in the hypotheses of the lemma of 6.13, we obtain

$$\lambda_{t+l} \leqq \min_{1\leqq u\leqq t+l} X_u AY_{t+l}^T + 3a(1-\beta(t,t+l)). \tag{6.31}$$

But we have assumed that all the strategies in Z are among $\tilde{X}_t, \dots, \tilde{X}_{t+l}$. Therefore the minima in (6.30) and (6.28) coincide. Adding these two inequalities gives us what was required. $\square$

6.15 Lemma. *If X consists of a single ε-polytope, then for every s, $t > s$ and Y_s we must have*

$$\lambda_t \leqq \beta(s,t)\lambda_s + \varepsilon(1 - \beta(s,t)). \tag{6.32}$$

Proof. It is evident that under our hypotheses the ε-polytope X is essential at each step. Therefore by (6.25) we have

$$\begin{aligned} \lambda_t - \beta(s,t)\lambda_s &\leqq \lambda_t - \beta(s,t)\tilde{X}_t A Y_s^T \\ &= \tilde{X}_t A(\beta(s,t)Y_s + (1 - \beta(s,t))Z(s,t)^T - \beta(s,t)\tilde{X}_t A Y_s^T \\ &= (1 - \beta(s,t))\tilde{X}_t A Z(s,t)^T. \end{aligned} \tag{6.33}$$

But $Z(s,t) \in X$ because X is convex. Since X is an ε-polytope,

$$|\tilde{X}_t A Z(s,t)^T - Z(s,t) A Z(s,t)^T| \leqq \varepsilon,$$

i.e., since A is skew-symmetric,

$$|\tilde{X}_t A Z(s,t)^T| \leqq \varepsilon,$$

which, with (6.33), yields (6.32). □

6.16. Let us show that when X consists of more than one ε-polytope, it is sufficient to weaken the estimation given by the inequality in 6.15 by an arbitrary small amount.

Let us assume that for a given $\varepsilon > 0$ and set X there are only r essential ε-polytopes. Choose any $\varepsilon' > \varepsilon$ and form the sequence

$$\varepsilon = \varepsilon_1 < \varepsilon_2 < \ldots < \varepsilon_r < \varepsilon'.$$

Lemma. *For all k, $1 \leqq k \leqq r$, and all $\gamma \in (0,1)$, there is a positive integer $s_{k,\gamma}$ such that if $s > s_{k,\gamma}$ and $t > s$, the sequence $Y_{s+1}, \ldots, Y_t$ tends to the set X' consisting of k ε-polytopes and $\beta(s,t) \leqq \gamma$; then*

$$\lambda_t \leqq \beta(s,t)\lambda_s + \varepsilon_k(1 - \beta(s,t)). \tag{6.34}$$

Proof. The lemma is proved by induction on k.

For $k = 1$, this proposition is just the content of the preceding lemma.

Suppose that the lemma has been established for some particular k, and that X' consists of $k + 1$ polytopes. Let us take any $\gamma_0 \in (0,1)$ and an integer

$$q \geqq \frac{4a}{\gamma_0(\varepsilon_{k+1} - \varepsilon_k)}. \tag{6.35}$$

By hypothesis, for $\gamma_1 > \gamma_0^{1/q}$ there is an s_{k,γ_1} for which the conclusion of the lemma holds. But by the regularity condition (6.18) there is an s_{0,γ_1} such that $\beta(s, s+1) = 1 - \alpha_s > \gamma_1$ for $s > s_{0,\gamma_1}$.

We must find a number s_{k,γ_1} with the required properties. Let us show that $\max\{s_{k,\gamma_1}, s_{0,\gamma_1}\}$ is such a number.

For this purpose, we take an arbitrary $s > \max\{s_{k,\gamma_1}, s_{0,\gamma_1}\}$ and a $t = t_0$ such that $\beta(s,t) \leqq \gamma_0$. We partition the interval $[s,t]$ into intervals $S_1, \ldots, S_w$ and S_{w+1} by setting

$$S_1 = [t_1, t_0], \qquad \text{where} \quad t_1 = \max\{u : \beta(u, t_0) \leqq \gamma_1\},$$

$$\cdots\cdots\cdots\cdots\cdots\cdots\cdots\cdots\cdots\cdots\cdots\cdots\cdots\cdots\cdots\cdots$$

$$S_w = [t_w, t_{w-1}], \quad \text{where} \quad t_w = \max\{u : \beta(u, t_{w-1}) \leqq \gamma_1\},$$

$$S_{w+1} = [s, t_w], \qquad \text{where} \quad \beta(s, t_w) > \gamma_1.$$

Evidently the last interval is empty if $s = t_w$. For the sake of uniformity, we shall sometimes write $s = t_{w+1}$ instead of $s \neq t_w$.

Since $t_l \geqq s > s_{0,\gamma_1}$, we must have

$$\beta(t_l, t_l + 1) = 1 - \alpha_{t_l} > \gamma_1 \tag{6.36}$$

and, in addition, $\beta(s,t) \leqq \gamma_0 < \gamma_1$.

Therefore, $t_1 \geqq s$, and the sequence $t = t_0, t_1, \ldots$ decreases strictly (with intervals not less than 2). Therefore w is finite and not less than unity.

If, on some interval S_l, there turn out to be at most k ε-polytopes, we can apply the inductive hypothesis (6.34), which can be written in the form

$$\lambda_{t_l} \leqq \beta(t_l, t_{l-1})\lambda_{t_l} + \varepsilon_k(1 - \beta(t_l, t_{l+1})). \tag{6.37}$$

If, however, all the $k+1$ ε-polytopes that are essential on $[s, t_{l-1}]$ are also essential on s_l, then by the lemma of 6.14 we must have

$$\lambda_{t_{l-1}} \leqq \beta(s, t_{l-1})\lambda_s + 3a(1 - \beta(t_l, t_{l-1}) + \varepsilon(1 - \beta(s, t_{l-1})). \tag{6.38}$$

Now there are two possibilities.

1°. On each interval S_l $(l = 1, \ldots, w+1)$, at most k ε-polytopes are essential.

In this case (6.37) is satisfied for $l = 1, \ldots, w+1$. An easy induction then yields

$$\lambda_t \leqq \beta(s,t)\lambda_s + \varepsilon_k(1 - \beta(s,t)) \leqq \beta(s,t)\lambda_s + \varepsilon_{k+1}(1 - \beta(s,t)).$$

2°. There are l of the numbers $1, \ldots, w+1$ for which there are $k+1$ essential ε-polytopes on $S_l = [t_l, t_{l-1}]$. Let l_0 be the smallest of these numbers. Then (6.37) holds for each $l \leqq l_0 - 1$, and (6.38) holds for $l = l_0$; letting $\beta(t_{l_0}, t_{l_0-1}) = \beta_0$ for short, we may write the inequality in the form

$$\lambda_{t_{l_0-1}} \leqq \beta(s, t_{l_0-1})\lambda_s + 3a(1 - \beta_0) + \varepsilon_1(1 - \beta(s, t_{l_0-1})). \tag{6.39}$$

From all the inequalities of the form (6.37), we obtain by induction, as in case 1°,

$$\lambda_t \leqq \beta(t_{l_0-1}, t)\lambda_{t_{l_0-1}} + \varepsilon(1 - \beta(t_{l_0-1}, t)), \tag{6.40}$$

and after substituting the right-hand side of (6.39) for $\lambda_{t_{l_0-1}}$ and making some trivial transformations,

$$\lambda_t \leqq \beta(s,t)\lambda_s + \varepsilon_{k+1}(1 - \beta(s,t)) + 3a(1 - \beta_0) - (\varepsilon_{k+1} - \varepsilon_k)(1 - \beta(s,t)).$$

It remains to show that

$$3a(1-\beta_0)-(\varepsilon_{k+1}-\varepsilon_k)(1-\beta(s,t)) \leqq 0. \tag{6.41}$$

For this purpose we remind ourselves that s and t were originally chosen so that $\beta(s,t) \leqq \gamma$. On the other hand, we must have $\beta_0 > \gamma_1$ by the choice of l_0. In addition, because $k+1 \geqq 2$ essential ε-polytopes exist on $[t_{l_0}, t_{l-1}]$ the length of this interval cannot be less than 2, even when $l_0 = w+1$. Therefore, in all applications of (6.37) we may write

$$\gamma^2 < \beta(t_{l_0}, t_{l_0}+1)\beta(t_{l_0}+1, t_{l_0-1}) = \beta_0,$$

from which $\gamma_1 < \beta_0^{1/2} \leqq (1+\beta_0)/2$. This means that $\gamma_1 - \beta_0 = \eta < 1-\gamma_1$, where $\eta > 0$.

Therefore

$$\frac{1-\beta(s,t)}{1-\beta_0} \geqq \frac{1-\gamma_1^q}{1-\gamma_1}\frac{1-\gamma_1}{1-\gamma_1+\eta}.$$

Here the second factor on the right is not less than $1/2$. In addition, we know that when $\gamma_1 \in (0,1)$,

$$1-\gamma_1^q \geqq (1-\gamma_1)q,$$

and we have only to refer to (6.36). □

6.17. Let us now turn to the proof of the theorem formulated in 6.10.

We choose $\varepsilon > 0$ arbitrarily, and for any $\varepsilon' < \varepsilon$ (for definiteness, we may take $\varepsilon' = \varepsilon/2$) we find the number r of ε-polytopes in $\boldsymbol{X}$ that may be essential for the iterative process determined by the recurrent relations (6.20) and (6.21). Then let us further choose $\gamma < \varepsilon'$ and define, in accordance with the lemma of 6.16, $s(r,\gamma)$ so that when $s > s(r,\gamma)$ it follows from $\beta(s,t) < \gamma$ that

$$\lambda_t < \lambda_s\beta(s,t) + \varepsilon(1-\beta(s,t)).$$

By (6.19), the condition $\beta(s,t) < \gamma$ is satisfied for sufficiently large t, whatever s may be, and our choice of ε and γ yields

$$\lambda_t < \frac{\lambda_s}{a}\gamma + \varepsilon' \leqq \varepsilon. \qquad \square$$

6.18 Estimating the rapidity of convergence of the iterative process. In essence, the proof in 6.10–6.17 of the convergence of the regular iterative process is "implicitly effective" in the sense that for each $\varepsilon > 0$ we may, after overcoming purely technical difficulties, find t_ε from the statement of the convergence theorem. However, these difficulties are serious enough, and the problem of an explicit expression for the rapidity of convergence of regular iterative processes described by (6.20) and (6.21) is extremely complex. This is all the more the case for an "asymmetric" regular process described in (6.13)–(6.16). We emphasize that by "rapidity of convergence" we mean a uniform estimate, i.e. determining, for each $\varepsilon > 0$, a number t_ε of iterations that would guarantee ε-optimality of a strategy for every matrix game of a given format.

Rather few facts are known in this direction. One of these applies to a symmetric process for which

$$\alpha_t = \beta_t = \frac{1}{t+1},$$

and all the strategies $\bar{X}_t$ and $\bar{Y}_t$ are pure. A process of this kind can be interpreted as follows. Let us suppose that player 2 recognizes that player 1 has, in t plays of an $m\times n$-game Γ_A, applied the available pure strategies with cumulative frequencies $\sigma_1^{(t)}, \dots, \sigma_m^{(t)}$. This gives player 2 a plausible basis for assuming (so to speak, supposing that the law of large numbers has operated for this finite sequence of trials) that player 1 is applying a mixed strategy $X_t = (\xi_1^{(t)}, \dots, \xi_m^{(t)})$, where $\xi_i^{(t)} = \sigma_i^{(t)}/t$ $(t = 1, \dots, m)$. Consequently it is reasonable for player 2 to apply the pure strategy $\bar{j}_{i+1}$ for which $X_t\, A_{.j}$ attains its minimum.

Similarly, player 1, observing the appearance of the pure strategies of player 2 with cumulative frequencies $\tau_1^{(t)}, \dots, \tau_n^{(t)}$, supposes that player 2 applies the mixed strategy $Y_t = (\eta_1^{(t)}, \dots, \eta_n^{(t)})$, $\eta_j^{(t)} = \tau_j^{(t)}$ $(j = 1, \dots, n)$, and will play the pure strategy $\bar{i}_{t+1}$ for which the payoff $A_{\bar{i}_{t+1}.}\, Y_t^T$ attains its maximum.

These pure strategies $\bar{i}_{t+1}$ and $\bar{j}_{t+1}$ are connected for the same reasons to the strategies chosen in the preceding t steps, acquiring the same weight $1/(t+1)$.

For the sake of transparency of all the reasoning that belongs to this process, we shall discuss the case of arbitrary (i.e., not necessarily symmetric) games.

6.19. Let us choose arbitrarily $U_0 \in \mathbf{R}^m$ and $V_0 \in \mathbf{R}^n$ (we may, but do not necessarily, take $U_0 = \mathbf{0}^m$ and $V_0 = \mathbf{0}^n$) and set, for $t = 1, 2, \dots,$

$$U_t = U_0 + tAY_t^T, \tag{6.42}$$

$$V_t = V_0 + tX_t A. \tag{6.43}$$

For $t > 0$, it will be convenient to replace these inequalities by the equivalent set

$$(U_t - U_0)/t = AY_t^T, \tag{6.44}$$

$$(V_t - V_0)/t = X_t A. \tag{6.45}$$

Let us now notice (see, for example, 1.11) that

$$\min_j X_t A_{.j} \leqq v_A \leqq \max_i A_{i.}\, Y_t^T. \tag{6.46}$$

Consequently, the proximity of X_t and Y_t to optimal strategies (and therefore the error in establishing the value of the game) can be estimated by the nonnegative difference

$$\Delta(t) = \max_t A_{i.}\, Y_t^T - \min_j XtA_{.j}. \tag{6.47}$$

In order to estimate this difference, we establish some auxiliary facts.

6.20 Lemma. *If* $m + n \geqq 4$ *and* $t > 2^{(m+n-1)(m+n-2)}$ *then*

$$(2^{m+n-3} - 2^{(m+n-3)^2/(m+n-2)})t^{(m+n-3)/(m+n-2)} \geqq 1.$$

Proof. Under the hypotheses of the lemma, we have

$$\frac{m+n-3}{m+n-2} > \frac{1}{2},$$

and $(m+n)\,(m+n-3) \geqq 4$. Consequently

$$1 > \frac{1}{\sqrt{2}} + \frac{1}{16} \geqq 2^{-(m+n-3)/(m+n-2)} + 2^{-(m+n)(m+n-3)};$$

and therefore

$$\begin{aligned} 1 &< (1 - 2^{-(m+n-3)/(m+n-2)})2^{(m+n)(m+n-3)} \\ &\leqq (2^{m+n-3} - 2^{(m+n-3)^2/(m+n-2)})2^{(m+n-3)/(m+n-2)} \\ &\leqq (2^{m+n-3} - 2^{(m+n-3)^2/(m+n-2)})t^{(m+n-3)/(m+n-2)}. \quad \square \end{aligned}$$

6.21. Let us introduce

$$\Delta_{U,V}(t) = \max U_t - \min V_t,$$

$$\Delta_{V,U}(t) = \max V_t - \min U_t,$$

$$\Delta_{U,U}(t) = \max U_t - \min U_t,$$

$$\Delta_{V,V}(t) = \max V_t - \min V_t,$$

where the operators max and min, as applied to vectors, mean taking their maximum and minimum components.

Evidently

$$\Delta_{U,V}(t) + \Delta_{V,U}(t) = \Delta_{U,U}(t) + \Delta_{V,V}(t). \tag{6.48}$$

Let

$$\hat{U}_t = U_t - \max U_0 J_m^T,$$

$$\hat{V}_t = V_t - \min V_0 J_n^T.$$

Evidently

$$\max \hat{U}_0 - \min \hat{V}_0 = 0$$

and

$$\Delta_{U,V}(t) - \Delta_{\hat{U},\hat{V}}(t) = \Delta_{V,V}(0).$$

Therefore $\Delta_{U,V}(t)$ differs from $\Delta_{\hat{U},\hat{V}}(t)$ only by a single term, and we shall suppose in what follows that the transition from U and V to $\hat{U}$ and $\hat{V}$ has been made. Then we may assume that

$$\max U_0 = \min V_0 = 0, \tag{6.49}$$

which involves no loss of generality.

In accordance with to (6.44), (6.45), (6.46) and (6.49), we have

$$\frac{\min U_t}{t} \leqq \frac{\max U_0}{t} + \min AY_t^T \leqq v_A \leqq \frac{\min V_0}{t} + \max X_t A \leqq \frac{\max V_t}{t},$$

from which it follows that

$$\max V_t - \min U_t = \Delta_{V,U}(t) \geqq 0.$$

This, with (6.28), gives us

$$\Delta(t) \leqq \Delta_{U,U}(t) + \Delta_{V,V}(t). \quad \square \tag{6.50}$$

6.22. We can now turn to obtaining the estimate that is of interest.

Theorem.

$$\frac{\Delta(t)}{t} \leqq a 2^{m+n} t^{-1/(m+n-2)}, \tag{6.51}$$

where $a = \max_{i,j} |a_{ij}|$.

Proof. The proof is by induction on $m+n$.

If $m+n=2$, then A is a 1×1-matrix. If we denote its single element by α and remember that we are in the situation (6.49), we obtain

$$U_t = \alpha t, \quad V_t = \alpha t,$$

so that $\Delta(t)=0$, and (6.51) is satisfied.

Now let (6.51) hold for all $m \times n$-matrices for which $m+n<k$ $(k \geqq 3)$, and consider an $m \times n$-matrix A for which $m+n=k$.

We first establish the validity of the following alternatives:

For all integers t and $T \in (0,t)$, either

$$\Delta(t) \leqq 4aT, \tag{6.52}$$

or

$$\Delta(t) - \Delta(t-T) \leqq a2^{m+n-1}T^{1-1/(m+n-3)} \tag{6.53}$$

(where when $m+n=3$ we take $T^{1-1/(m+n-3)}=0$).

Let us suppose that (6.52) is not true, i.e. that

$$\Delta(t) > 4aT. \tag{6.54}$$

Then it follows from (6.50) that at least one of the differences on the right-hand side of that inequality is larger than $2aT$. For definiteness, let

$$\Delta_{U,U}(t) = \max U_t - \min U_t > 2aT, \tag{6.55}$$

where

$$\max U_t = U_{t,i_1}, \qquad \min U_t = U_{t,i_2}.$$

The largest change in the components of U_t in a single step does not exceed a. Hence it follows from (6.35) that, for all $w=0,1,\dots,T$, we will have

$$U_{t-w,i_1} - U_{t-w,i_2} > 0. \tag{6.56}$$

Hence it follows that, for each of the $T+1$ steps, the i_2th component cannot be maximal.

We denote by A^- the $(m-1)\times n$-matrix obtained from A by removing the row $A_{i_2\cdot}$, and by U_t^- the vector obtained from U_t by omitting its i_2th component.

Now let $\bar{U}_0 = U_{t-T}^-$ and $\bar{V}_0 = V_{t-T}^-$ and construct, according to (6.37)–(6.40), two recurrent sequences of vectors. Evidently it turns out that

$$\begin{aligned} \bar{U}_\tau &= U_{t-T+\tau}^-, \\ \bar{V}_\tau &= V_{t-T+\tau}^-, \end{aligned} \qquad \tau = 0,1,\dots,T.$$

Therefore, in accordance with what we said in 6.21, we may pass to vectors $\hat{U}_{t-T}^-$ and $\hat{V}_{t-T}^-$ and apply the inductive hypothesis, which yields

$$\max \hat{U}_t^- - \min \hat{V}_t^- \leq a2^{m+n-1}t^{-1/(m+n-3)}. \tag{6.57}$$

It remains only to observe that the left-hand side of this inequality is equal to

$$\begin{aligned} &\max U_t^- - \max U_{t-T}^- - \min V_t + \min V_{t-T} \\ &\quad = \max U_t - \min V_t - \max U_{t-T} + \min V_{t-T} = \Delta(t) - \Delta(t-T), \end{aligned}$$

and (6.53) is established.

Let us suppose that $t \leqq 2^{(m+n-1)(m+n-2)}$. Since the components of the vectors do not change by more than a at each step, it follows from (6.49) that

$$\Delta(t) \leqq 2at. \tag{6.58}$$

Therefore

$$\begin{aligned}\Delta(t) &\leqq 2at = 2at^{1/(m+n-2)} t^{1-1/(m+n-2)} \\ &\leqq 2a2^{m+n-1} t^{1-1/(m+n-2)} = a2^{m+n} t^{1-1/(m+n-2)},\end{aligned}$$

from which (6.51) follows immediately.

It remains to consider the case when

$$t > 2^{(m+n-1)(m+n-2)}. \tag{6.59}$$

Let us choose an integer T and say that t is a number of the *first kind* if (6.52) is satisfied, and of the *second kind* if (6.53) is satisfied. It follows from (6.58) that every t that does not exceed T is of the first kind.

We set $q = [t/T]$ and consider the numbers

$$t, t-T, \ldots, t-qT. \tag{6.60}$$

Since the last of these numbers is less than T, it must be a number of the first kind. Let $t - rT$ be the largest number of the first kind in the sequence (6.60). Then

$$\begin{aligned}\Delta(t) &= \sum_{s=1}^{r}(\Delta(t-(s-1)T) - \Delta(t-sT)) + \Delta(t-qT) \\ &\leqq r\left(a2^{m+n-1}T^{1-1/(m+n-3)}\right) + 4aT \leqq ta2^{m+n-1}T^{-1/(m+n-3)} + 4aT.\end{aligned}$$

It follows that when $t \geqq T$

$$\frac{\Delta t}{t} \leqq a\left(2^{m+n-1}T^{-1/(m+n-3)} + \frac{4T}{t}\right). \tag{6.61}$$

Let us now estimate the term on the right.

If $k = m + n \geqq 4$, then by the lemma in 6.20 there is an integer $T(t)$ for which

$$2^{(m+n-3)^2/(m+n-2)} t^{(m+n-3)/(m+n-2)} \leqq T(t) \leqq 2^{m+n-3} t^{(m+n-3)/(m+n-2)}. \tag{6.62}$$

Since we also have (6.59), i.e.

$$t^{1/(m+n-2)} > 2^{m+n-1},$$

we obtain

$$T(t) \leqq t^{\{(m+n-3)/(m+n-2)\}\{(1+1/(m+n-1)\}} = t^{\{(m+n-3)/(m+n)\}/\{(m+n-1)(m+n-2)\}}, \tag{6.63}$$

so that all further estimates are applicable to (6.61).

Furthermore, it follows from the left side of (6.62) that

$$2^{m+n-3}t \leqq T(t)^{(m+n-2)/(m+n-3)} = T(t)^{1+1/(m+n-3)},$$

i.e.

$$2^{m+n-1}T(t)^{-1/(m+n-3)} \leqq \frac{4T(t)}{t}, \tag{6.64}$$

and from the right-hand side of the same inequality, that

$$\frac{8T(t)}{t} \leqq 2^{m+n}t^{-1/(m+n-2)}. \tag{6.65}$$

Substituting (6.64) and (6.65) into (6.51), we obtain

$$\frac{\Delta(t)}{t} \leqq a 2^{m+n} t^{-1/(m+n-2)},$$

as required.

It remains to analyze the case $m+n=3$ under conditions (6.59), i.e. when $t > 4$. Inequality (6.62) must be satisfied for all T, $0 \leqq T \leqq t$. If we take $T = 2$ and notice that the first term on the right of (6.61) is zero, we obtain

$$\frac{\Delta(t)}{t} \leqq \frac{a-8}{t},$$

which, in the present case, is the required inequality (6.51). □

6.23 Iterative processes for bimatrix games. Attempts to extend the results on the convergence of regular iterative processes to games that are not two-person zero-sum (even for the simplest bimatrix games) cannot lead to success. The following example shows that an analog of the theorem of 6.10 is invalid even for 3 × 3-bimatrix games. This example is of a parametric character, so that it is not degenerate, and consequently the considered game is, in a sense, a typical example.

Let $\Gamma_{A,B}$ be a 3 × 3-bimatrix game with payoff matrices

$$A = \begin{bmatrix} a_1 & c_1 & b_3 \\ b_2 & a_2 & c_3 \\ c_1 & b_2 & a_3 \end{bmatrix}, \qquad B = \begin{bmatrix} \beta_1 & \gamma_1 & \alpha_1 \\ \alpha_2 & \beta_2 & \gamma_2 \\ \gamma_3 & \alpha_3 & \beta_3 \end{bmatrix},$$

where we suppose that

$$a_i > b_i > c_i \text{ and } \alpha_i > \beta_i > \gamma_i \text{ for } i = 1,2,3.$$

Elementary combinatorial analysis shows that $\Gamma_{A,B}$ has a unique equilibrium situation which is completely mixed. It is also clear that if the entries in the triples of numbers a_1, a_2, a_3; b_1, b_2, b_3; etc., are sufficiently close to each other, then the players' unique equilibrium strategies are nearly uniform (i.e., each pure strategy has probability nearly 1/3).

Let us apply to $\Gamma_{A,B}$ a process similar to that described in 6.18. Let us suppose that, for some t, the payoff $A_{i\cdot} Y_t^T$ attains a maximum for $i = 1$, i.e., $i_t = 1$ ($\tilde{X} = (1,0,0)$), and the payoff $X_t B_{\cdot j}$ attains a maximum for $j = 1$, i.e., $j_t = 1$ ($\tilde{Y} = (1,0,0)$).

This is possible only for

$$A_{1\cdot} Y_t^T \geqq A_{2\cdot} Y_t^T, \quad A_{1\cdot} Y_t^T \geqq A_{3\cdot} Y_t^T, \tag{6.66}$$

$$X_t B_{\cdot 1} \geqq X_t B_{\cdot 2}, \quad X_t B_{\cdot 1} \geqq X_t B_{\cdot 3}. \tag{6.67}$$

But then

$$A_{1\cdot} Y_{t+1}^T = A_{1\cdot} \left(\frac{1}{t+1} \tilde{Y}_t + \frac{t}{t+1} Y_t \right)^T = \frac{1}{t+1} A_{1\cdot} \tilde{Y}_t^T + \frac{t}{t+1} A_{1\cdot} Y_t^T$$

$$= \frac{1}{t+1} a_1 + \frac{t}{t+1} A_{1\cdot} Y_t^T > \frac{1}{t+1} b_1 + \frac{t}{t+1} A_{2\cdot} Y_t^T = A_{2\cdot} Y_{t+1}^T$$

and similarly $A_{1\cdot}Y_{t+1}^T > A_{3\cdot}Y_{t+1}^T$. Consequently $\tilde{i}_{t+1} = 1$ ($\tilde{X}_{t+1} = (1,0,0)$).

On the other hand, it follows from (6.67) that

$$\begin{aligned} X_{t+1}B_{\cdot 1} &= \left(\frac{1}{t+1}\tilde{X}_t + \frac{t}{t+1}X_t\right)B_{\cdot 1} \\ &= \frac{1}{t+1}\tilde{X}_tB_{\cdot 1} + \frac{t}{t+1}X_tB_{\cdot 1} = \frac{1}{t+1}\beta_1 + \frac{t}{t+1}X_tB_{\cdot 1} \\ &> \frac{1}{t+1}\gamma_1 + \frac{t}{t+1}X_tB_{\cdot 2} = X_{t+1}B_{\cdot 2}. \end{aligned}$$

Therefore $\tilde{j}_{t+1} \neq 2$ (i.e. $\tilde{Y}_{t+1} \neq (0,1,0)$). However, both variants $\tilde{j}_{t+1} = 1$ and $\tilde{j}_{t+1} = 3$ are possible (depending on which has the "overbalance": replacing β_1 by the larger number α_1 or replacing $X_tB_{\cdot 1}$ by the smaller number $X_tB_{\cdot 3}$). It follows from what we have said that after the choice of the pair $(1,1)$ of strategies it is possible either to choose the same pair again, or to choose the pair $(1,3)$. (In the present context, it is out of place to refer to such pairs of strategies as "game-situations").

Let us suppose that at some stage t_0 of the iterative process there is a transition to the pair $(1,1)$. (It is easy to see that such a transition is possible only from $(2,1)$.) Set

$$(t_0-1)A_{i\cdot}Y_{t_0-1}^T = H_i, \quad i = 1,2,3.$$

With these notations, we must have $H_1 = \max\{H_1, H_2, H_3\}$. Moreover, as we have seen, the pair $(1,1)$ must repeat some number ω_{11} of times, after which there is a transition to another pair (which will necessarily be the pair $(1,3)$). Let $(1,3)$ be chosen some number ω_{13} of times in succession. The cumulative payoff to player 1 up to the step $t_0 + \omega_{11} + \omega_{13} = t_1$ will be equal to

$$(t_1-1)A_{i\cdot}Y_{t_1-1}^T = H_i', \quad i = 1,2,3.$$

Evidently

$$\begin{aligned} H_1' &= H_1 + \omega_{11}a_1 + \omega_{13}b_3, \\ H_2' &= H_2 + \omega_{11}b_1 + \omega_{13}c_3, \qquad (6.68) \\ H_3' &= H_3 + \omega_{11}c_1 + \omega_{13}a_3. \end{aligned}$$

Since after this there occurs a transition to $(3,3)$, we must have $H_3' = \max\{H_1', H_2', H_3'\}$ and therefore $H_3' \geqq H_1'$. On the other hand, $H_3 \leqq H_1$, from which it follows that $H_3' - H_3 \geqq H_1' - H_1$, i.e., by (6.68),

$$\omega_{11}c_1 + \omega_{13}a_3 \geqq \omega_{11}a_1 + \omega_{13}b_3,$$

from which

$$\omega_{13} \geqq \frac{a_1 - c_1}{a_3 - b_3}\omega_{11}. \qquad (6.69)$$

This means that the length of stay of the process in the pair $(1,3)$ is strictly longer than its stay in $(1,1)$. In the same way, the subsequent stay in $(3,3)$ is still longer, etc.

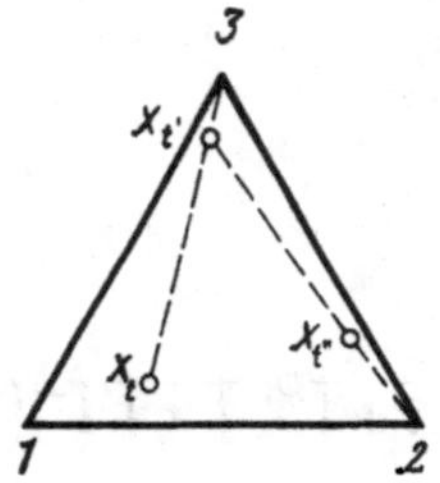

Fig. 4.18

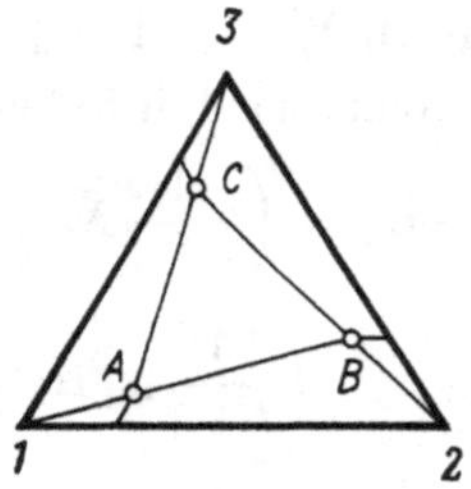

Fig. 4.19

If we continue the consideration of possible alternations of pairs, we see that they form a sequence:

$$(1,1) \to (1,3) \to (3,3) \to (3,2) \to (2,2) \to (2,1) \to (1,1). \tag{6.70}$$

We can write an inequality of the form (6.69) for each such transition. Let q be the smallest of the numbers

$$\frac{a_i - c_i}{a_i - b_i}, \quad \frac{\alpha_i - \gamma_i}{\alpha_i - \beta_i}, \qquad i = 1, 2, 3.$$

Then if we denote by ω_l the length of the process of staying with a pair of strategies that is lth from the beginning, we will have

$$\omega_{l+1} > q\omega_l. \tag{6.71}$$

We see from (6.70) that when players pass from one pair of strategies to another, they alter their strategies in turn. Therefore if we denote by ρ_r the length of the stay of a player (say, the first player), with a strategy that is rth from the beginning of the process, then it follows from (6.71) that

$$\rho_{r+1} > q^2 \rho_r.$$

Hence it follows that if $q \geqq \sqrt{3}$ (this hypothesis does not interfere with the nondegeneracy of the game; the specific number $\sqrt{3}$ was chosen to make the figures clearer), then the length of each stay exceeds at least three times the sum of the lengths of the preceding stays.

Suppose that at the tth step the process has completed a stay with the first strategy of player 1. This means that in not less than $3t$ further steps player 1 will choose the third available pure strategy. Geometrically (Figure 4.18) this means that the points corresponding to the strategies $X_{t+1}, X_{t+2}, \ldots$ will lie on a line joining X_t to vertex 3 of the simplex. In any case, let the number of the last step when player 1 selects strategy 3 be t'. Then for $X_{t'} = (\xi_{t'1}, \xi_{t'2}, \xi_{t'3})$ we will have

$$\xi_{t'1} < \frac{1}{4}\xi_{t1}, \quad \xi_{t'2} < \frac{1}{4}$$

and evidently $\xi_{t'3} < 1$.

After this, the sequence of points $X_{t'+1}, X_{t'+2}, \ldots$ turns toward vertex 2, and at some time t'' completes the stay of the process on the pure strategy 2 and for the strategy $X_{t''} = (\xi_{t''1}, \xi_{t''2}, \xi_{t''3})$ we will have

$$\xi_{t''1} < \frac{1}{4}\xi_{t'1} < \frac{1}{16}\xi_{t1} < \frac{1}{16}, \quad \xi_{t''3} < \frac{1}{4}\xi_{t'3} < \frac{1}{4}, \quad \xi_{t''2} < 1.$$

Relationships of this kind will evidently be preserved at later times when stays are completed. Therefore the points corresponding to the strategy X_t at the time of completing a stay will be found inside triangles abutting on the vertices of the simplex, as sketched in Figure 4.19. Points corresponding to all intermediate strategies X_t will be distributed on line segments joining points of these triangles. In any case, none of them will fall inside the triangle ABC that contains the unique equilibrium strategy of player 1.

It follows from what we have said that after repeated iteration of the process we cannot select from the sequence $X_1, X_2, \ldots$ of strategies any subsequence that converges to an equilibrium strategy in the game $\Gamma_{A,B}$. All the more, we cannot select from the sequence $(X_1, Y_1), (X_2, Y_2), \ldots$ of situations any subsequence that converges to an equilibrium situation.

§7 Structure of the set of solutions of a matrix game

7.1 Formulation of the problem. As we have already noticed in 1.13, the sets $\mathcal{S}(A)$ and $\mathcal{T}(A)$ of a player's optimal strategies in any matrix game Γ_A are nonempty closed convex polytopes that lie respectively in the simplexes $\boldsymbol{X}$ and $\boldsymbol{Y}$ of the mixed strategies. The examples in the course of this chapter show that the pairs of polytopes $\mathcal{S}(A)$ and $\mathcal{T}(A)$ can have very diverse forms. It follows from the same considerations that for an arbitrary pair of nonempty convex polytopes $\mathcal{U}$ and $\mathcal{V}$, lying in an $(m-1)$-dimensional simplex $\boldsymbol{X}$ and an $(n-1)$-dimensional simplex $\boldsymbol{Y}$, respectively, we cannot necessarily find an $m \times n$-game Γ_A for which $\mathcal{S}(A) = \mathcal{U}$ and $\mathcal{T}(A) = \mathcal{V}$.

The following questions naturally arise: what properties should polytopes $\mathcal{U} \subset \boldsymbol{X}$ and $\mathcal{V} \subset \boldsymbol{Y}$ have so that there will exist a game Γ_A for which $\mathcal{U} = \mathcal{S}(A)$ and $\mathcal{V} = \mathcal{T}(A)$? If such games exist, how can they be described?

There are complete answers to these questions. By what we said in 1.9, we may restrict ourselves to the case when $v_A = 0$. Consequently we shall assume this unless the contrary is explicitly stated.

7.2 Auxiliary information. Each vector $X \in \boldsymbol{X}$ and $Y \in \boldsymbol{Y}$ can be put in correspondence with a ray $(X) = \{\alpha X : \alpha \geq 0\}$ and a ray $(Y) = \{\beta Y, \beta \geqq 0\}$. The correspondence is evidently one-to-one. Then to a set of vectors (a strategy) there corresponds a cone; and to a convex set, a convex cone. Let us denote the cones corresponding to the sets $\boldsymbol{X}$ and $\boldsymbol{Y}$ by $\boldsymbol{P}$ and $\boldsymbol{Q}$, respectively.

In addition, in this section it will be convenient to consider strategies not as vectors, but as rays, and to use the terminology and results of the theory of convex cones in a Euclidean space.

Let us recall the basic definitions and results that we shall need.

We denote by $[\mathcal{R}]$ the linear hull of a subset $\mathcal{R}$ of a linear space $\mathcal{E}$, i.e., the intersection of all the linear subspaces of $\mathcal{E}$ that contain $\mathcal{R}$. If $\mathcal{R}$ is a convex cone, every vector in $[\mathcal{R}]$ can be represented as a difference $R_1 - R_2$, where R_1 and $R_2 \in \mathcal{R}$.

The *sum* $\mathcal{K}_1 + \mathcal{K}_2$ of two convex cones $\mathcal{K}_1$ and $\mathcal{K}_2$ in a linear space is the set of vectors of the form $K_1 + K_2$, where $K_1 \in \mathcal{K}_1$ and $K_2 \in \mathcal{K}_2$. A sum of convex cones is again a convex cone.

If $\mathcal{K}$ is a convex cone in the Euclidean space $\mathcal{E}$, the set of vectors $T \in \mathcal{E}$ for which $(K, T) \geqq 0$ for all $K \in \mathcal{K}$ is also a convex cone in E, called the *dual* of $\mathcal{K}$ and denoted by $\mathcal{K}^*$. Evidently the cone $(X)^*$ dual to the ray (X) is a closed half-space (for which the direction of the ray is the direction of the inner normal to the boundary).

If $\mathcal{K}$, $\mathcal{K}_1$, and $\mathcal{K}_2$ are convex cones, then $\mathcal{K}_1 \subset \mathcal{K}_2$ implies that $\mathcal{K}_2^* \subset \mathcal{K}_1^*$, $(\mathcal{K}_1 + \mathcal{K}_2)^* = \mathcal{K}_1^* \cap \mathcal{K}_2^*$, $(\mathcal{K}_1 \cap \mathcal{K}_2)^* = \mathcal{K}_1^* + \mathcal{K}_2^*$ and $\mathcal{K}^{**} = \mathcal{K}$.†)

The set $K^\perp$ of the vectors orthogonal to the vector K in the Euclidean space $\mathcal{E}$ is a hyperplane in $\mathcal{E}$. For a hyperplane in E to be a support of the convex cone $\mathcal{K}$, it is neccesary and sufficient that it has the form $T^\perp$, where $T \in \mathcal{K}^*$. The intersection of a supporting hyperplane of a cone with the cone itself is called a *face* of the cone. A face of a cone with dimension one less than the dimension of the cone is called a *bordering* face. A one-dimensional face of a convex cone is called an *extreme ray*. A convex cone is said to be *polytopic* if it is (as a cone) a union of a finite number of rays.

7.3. The following proposition on bordering faces of cones will be very important for us.

Theorem. *If a convex cone $\mathcal{K}$ is the intersection of a finite number of half-spaces:*

$$\mathcal{K} = \bigcap_j (T_j)^*,$$

then each of its bordering faces $\mathcal{F}$ must have the form

$$\mathcal{F} = \mathcal{K} \cap (T_j)^\perp.$$

Proof. From the definition we find

$$T = \mathcal{K}^* = \sum_j (T_j),$$

so that $\mathcal{F} = (T)^\perp \cap \mathcal{K}$. This T can be represented in the form

$$T = \sum_j \lambda_j T_j, \quad \lambda_j \geqq 0.$$

But $\dim \mathcal{F} = \dim \mathcal{K} - 1$. Therefore there is a $K \in \mathcal{K}$ such that

$$(\bar{K}, T) = \sum_j \lambda_j (\bar{K}, T_j) > 0,$$

i.e., for some $j = j_0$ we will have simultaneously $\lambda_{j_0} > 0$ and $(\bar{K}, T_{j_0}) > 0$. It follows that $\bar{K} \in (T_{j_0})^\perp$ and therefore $\mathcal{K} \not\subset (T_{j_0})^\perp$.

†) The last proposition is far from trivial; it is a corollary of a theorem of H. Weyl [1] on polytopes.

Now we set $\mathcal{K} \cap (T_{j_0})^{\perp} = \mathcal{G}$ and show that $\mathcal{G} = \mathcal{F}$. But it follows from $K \in \mathcal{F}$ that

$$(K,T) = \sum_j \lambda_j (K,T_j) = 0.$$

Therefore $\lambda_j (K,T) = 0$ for all j, and in particular $(K,T_{j_0}) = 0$ (since $\lambda_{j_0} > 0$). Therefore $K \in \mathcal{G}$ and consequently $\mathcal{F} \subset \mathcal{G}$.

Conversely, if we take $K \in \mathcal{G} \setminus \mathcal{F}$, then since the face $\mathcal{F}$ is bordering, it follows that

$$\dim((K) + \mathcal{F}) = \dim \mathcal{K}.$$

Hence $[\mathcal{F} + (\mathcal{K})] = [\mathcal{K}] \subset (T_{\lambda_0})^{\perp}$. Therefore $\mathcal{K} \subset (T_{\lambda_0})^{\perp}$, which contradicts what was established above. Hence $\mathcal{G} \setminus \mathcal{F} = \emptyset$, so that $\mathcal{G} = \mathcal{F}$. □

It follows from this theorem that a convex cone that is the intersection of a finite number of half-spaces has a finite number of bordering faces, and also, by induction, a finite number of faces of any dimension, including a finite number of extreme rays.

7.4 The converse problem of the theory of matrix games. Up to the end of 7.6 we shall be dealing exclusively with an $m \times n$-matrix game with matrix A. Hence the set $\mathcal{S}(A)$ of optimal strategies of player 1 in this game will simply be denoted by $\mathcal{S}$, and the set $\mathcal{T}(A)$ of optimal strategies of player 2, by $\mathcal{T}$. We recall that here we suppose that $\mathcal{S}$ and $\mathcal{T}$ are cones.

A minimal face $\mathcal{E}_1$ of a cone $\boldsymbol{P}$, containing $\mathcal{S}$, is evidently spanned by vectors corresponding to the essential (see 1.19) strategies of player 1. We shall call this an *essential face* of $\boldsymbol{P}$, and denote its dimension, i.e., the number of pure essential strategies of player 1, by e_1.*) Let the number of inessential strategies of player 1 be denoted by s_1; and the edge of $\boldsymbol{X}$ spanned by the corresponding vectors, by $\mathcal{S}_1$.

Similarly, for player 2 the essential face of $\boldsymbol{Q}$ is denoted by $\mathcal{E}_2$; its dimension, by e_2; the number of inessential strategies, by s_2; and the face generated by them, by $\mathcal{S}_2$.

The matrix A can be considered as a linear transformation from the space $\boldsymbol{P}$ to the space $\boldsymbol{Q}$. Therefore we may decompose A into 4 blocks:

$$A = \begin{bmatrix} A_e & A_2 \\ A_1 & A_s \end{bmatrix}, \tag{7.1}$$

where

$$A_e : \mathcal{E}_1 \to \mathcal{E}_2,$$

$$A_1 : \mathcal{S}_1 \to \mathcal{E}_2,$$

$$A_2 : \mathcal{E}_1 \to \mathcal{S}_2,$$

$$A_s : \mathcal{S}_1 \to \mathcal{S}_2.$$

*) We recall that the dimension of the set of optimal strategies, as originally introduced as a polytope, is less by 1 than the number of the player's essential strategies.

A bordering face of $\mathcal{S}$ (or of $\mathcal{T}$) is said to be *interior* if it contains interior points of $\mathscr{E}_1$ (or of $\mathscr{E}_2$). We denote the numbers of interior faces of $\mathcal{S}$ and $\mathcal{T}$ by f_1 and f_2.

Finally, let the dimensions of $\mathcal{S}$ and $\mathcal{T}$ be d_1 and d_2.

7.5 Theorem. *A necessary and sufficient condition for there to exist a matrix game in which the set of optimal strategies of player 1 is $\mathcal{U}$, and the set of optimal strategies of player 2 is $\mathcal{V}$, is that*

$$e_1 - d_1 = e_2 - d_2, \tag{7.2}$$

$$f_1 \leqq s_1, \; f_2 \leqq s_2. \tag{7.3}$$

Proof. *Necessity.* Consider an $e_1 \times e_2$-submatrix B of A made up of the elements in the intersections of the essential rows and essential columns of A. Evidently $\mathcal{U}$ consists of the vectors with nonnegative components for which

$$X_B B_{\cdot j} = 0 \text{ for all essential } j. \tag{7.4}$$

Since $\mathcal{U}$ contains vectors with positive components (such vectors exist, for instance the centroid of the vertices of the section of $\mathcal{U}$ by the hyperplane $XJ_{e_1} = 0$), the condition that the vectors in $\mathcal{U}$ have nonnegative components does not change the dimension of $\mathcal{U}$. Consequently, the dimension d_1 of $\mathcal{U}$ coincides with the dimension of the subspace containing $\mathcal{U}$, which by (7.4) is

$$d_1 = e_1 - \rho, \tag{7.5}$$

where ρ is the rank of B. By symmetry, we also have

$$d_2 = e_2 - \rho. \tag{7.6}$$

Substitution of (7.5) into (7.6) yields (7.2).

For the proof of (7.3), we rewrite the condition $\mathcal{U} = \mathcal{S}(A)$ in the form

$$\mathcal{U} = \bigcap_i (U_i)^* \cap \bigcap_j (AV_j)^*.$$

As a result, we have the conditions of the theorem of 7.3 on bordering faces.

By that theorem, every proper bounding face $\mathcal{F}$ of the cone $\mathcal{U}$ has one of the forms $U \cap (U_i)^\perp$ or $U \cap (AV_j)^\perp$. If $\mathcal{F}$ is interior, it cannot have the form $U \cap (U_i)^\perp$ and therefore must have the form $U \cap (AV_j)^\perp$.

Let us suppose that the strategy j is essential. Then we must have $XA_{\cdot j} = 0$ for each $X \in \mathcal{U}$, i.e., $\mathcal{U} \subset (AV_j)^\perp$, so that $\mathcal{F} = \mathcal{U} \cap (AV_j)^\perp = \mathcal{U}$, which is impossible, since $\mathcal{F}$ is a proper face of $\mathcal{U}$.

Therefore, to each bordering face $\mathcal{F}$ of $\mathcal{U}$ there corresponds a superfluous strategy of player 2. Since, in addition, different strategies correspond to different faces, we must have $f_1 \leqq s_1$.

The proof of the *sufficiency* of the theorem consists of showing that, for given convex polytopes $\mathcal{U} \subset \boldsymbol{X}$ and $\mathcal{V} \subset \boldsymbol{Y}$ that satisfy (4.2) and (4.3), we can construct an $m \times n$-matrix A for which $\mathcal{S}(A) = \mathcal{U}$ and $\mathcal{T}(A) = \mathcal{V}$.

We shall construct the matrix A in the form (7.1), block by block.

We first show that there is an $e_1 \times e_2$-matrix A_e such that the linear hulls $[\mathcal{U}]$ and $[\mathcal{V}]$ satisfy

$$[\mathcal{V}] = N(A_e), \tag{7.7}$$

$$[\mathcal{U}] = N(A_e^T). \tag{7.8}$$

In fact, the quotient space $[\mathcal{E}_2]/[\mathcal{V}]$ has dimension $e_2 - d_2$, the dimension of the intersection $\mathcal{U}^\perp \cap [\mathcal{E}_1]$ is $e_1 - d_1$, and by (7.2) these differences are equal. Consequently our spaces have the same dimension, so that there is a linear transformation

$$A_e : [\mathcal{E}_2] \to \mathcal{U}^\perp \cap [\mathcal{E}_1] \tag{7.9}$$

with kernel $[\mathcal{V}]$, and (7.7) is satisfied.

Let us select an arbitrary $Y \in \mathcal{E}_2$; since $A_e Y^T \in \mathcal{U}^\perp$ by (7.9), we have, for each $X \in \mathcal{U}$,

$$X A_e Y^T = 0. \tag{7.10}$$

Therefore each $X \in \mathcal{U}$ belongs to $N(A_e^T)$, and therefore we have

$$[\mathcal{U}] \subset N(A_e^T). \tag{7.11}$$

Now, conversely, let $X \in N(A_e^T)$. Then, for each $Y \in [\mathcal{E}_2]$, we have (7.10), i.e.,

$$X \in (A_e \mathcal{E}_2)^\perp \cap [\mathcal{E}_1] = (\mathcal{U}^\perp \cap [\mathcal{E}_1])^\perp = ((\boldsymbol{X}^\perp)^\perp \cup \mathcal{E}_1^\perp) \cap [\mathcal{E}_1] = [\mathcal{U}].$$

This means that $N(A_e^T) \subset [\mathcal{U}]$, which, together with (7.11), yields (7.8). Hence we have constructed the block A_e of the matrix A.

We now establish the existence of an $e_1 \times s_2$-matrix A_2 for which

$$A_2 \mathcal{S}_2 \subset \mathcal{U}^*, \tag{7.12}$$

$$A_2 \mathcal{S}_2 \cap \mathcal{U}^\perp = \mathbf{0}, \tag{7.13}$$

and, for every interior bordering face $\mathcal{F}$ of $\mathcal{U}$, there is a superfluous strategy $V_{\mathcal{F}}$ of player 2 for which

$$(A_2 V_{\mathcal{F}})^\perp \cap \mathcal{U} = \mathcal{F}. \tag{7.14}$$

For this purpose we find, in accordance with the theorem on bordering faces, a vector

$$Z_{\mathcal{F}} \in [\mathcal{E}_1] \cap U^*, \tag{7.15}$$

such that $(Z_{\mathcal{F}})^\perp \cap \mathcal{U} = \mathcal{F}$, and substitute it as the corresponding basis vector $V_{\mathcal{F}} \in \mathcal{S}_2$. By (7.3), here we may make different basis vectors correspond to different bordering faces.

For each such $V_{\mathcal{F}}$, we set

$$A_r V_{\mathcal{F}}^T = Z_{\mathcal{F}}^T, \tag{7.16}$$

and for basis vectors in $\mathcal{S}_2$ that do not appear in $V_{\mathcal{F}}$ we choose as the A_2-image any vector that lies inside the convex hull of the images previously defined.

The transformation A_2 that we have constructed is the one required. In fact, it follows from (7.15) and (7.16) that $A_2\mathcal{S}_2 \subset \mathcal{U}^*$, i.e., (7.12). If we had found a vector $V = \sum_{\mathcal{F}} \lambda_{\mathcal{F}} V_{\mathcal{F}}$, $\lambda_{\mathcal{F}} \geqq 0$, $\sum_{\mathcal{F}} \lambda_{\mathcal{F}} > 0$, for which $A_2 V_{\mathcal{F}}^T \in \mathcal{U}^{\perp}$, it would have followed from (7.15) and (7.16) that, for $\lambda_{\mathcal{F}} > 0$ (and such an $\mathcal{F}$ evidently exists!), we would have $A_2 V^T \in \mathcal{U}^*$. However, on the other hand, it follows from $A_2 V^T \in \mathcal{U}^*$ and $\mathcal{F} \neq \mathcal{U}$ that there is an $X \in \mathcal{U} \setminus \mathcal{F}$ such that $XAV_{\mathcal{F}}^T = 0$. Therefore $A_2 V_{\mathcal{F}}^T \in \mathcal{U}^{\perp}$. This contradiction shows that (7.13) is valid.

Finally, the (7.14) is established by the same construction.

A symmetric discussion shows that there is an $s_2 \times e_1$-matrix A_1 for which

$$\mathcal{S}_1 A_1 \subset -\mathcal{V}^*, \tag{7.17}$$

$$\mathcal{S}_1 A_1 \cap \mathcal{V}^{\perp} = \mathbf{0} \tag{7.18}$$

and for each interior bordering face $\mathcal{G}$ of the cone $\mathcal{V}$ there exists $U_{\mathcal{G}} \in \mathcal{S}_1$ such that

$$(U_{\mathcal{G}} A_1)^{\perp} \cap V = \mathcal{G}. \tag{7.19}$$

We now set

$$A = \begin{bmatrix} A_e & A_2 \\ A_1 & A_s \end{bmatrix},$$

where A_e, A_1, and A_2 were constructed above, and A_s is arbitrary, and show that

$$\mathcal{G}(A) = \mathcal{U}, \quad \mathcal{T}(A) = \mathcal{V}. \tag{7.20}$$

Let $X \in \mathcal{U}$. Then for $V \in \mathcal{E}_2$ we have, by (7.10),

$$X(A_e, A_1)V^T = X_e A_e V^T = 0,$$

and for $V \in \mathcal{S}_2$, since $AV \subset \mathcal{U}^*$,

$$X(A_2, A_s)V^T = X_e A_2 V^T \geqq 0.$$

Taking $Y \in \mathcal{V}$ arbitrarily, and carrying out a similar discussion, we find that $X \in \mathcal{S}$, $Y \in \mathcal{T}$, and $v_A = 0$.

Consequently we have

$$\mathcal{U} \subset \mathcal{S}, \quad \mathcal{V} \subset \mathcal{T}. \tag{7.21}$$

Let us establish the validity of the converse inclusions. Let us first show that $\mathcal{S} \subset [\mathcal{E}_1]$, for which we choose $U \subset \mathcal{S}_1$ arbitrarily. Then by (7.17), for some $Y \in \mathcal{V}$,

$$UAY^T = U_s A_1 Y_e^T < 0.$$

Therefore (see 1.19) the basis vector turns out to be superfluous.

Now choose $U \in \mathcal{S}(A)$. By what has already been established, $U \in [\mathcal{E}_1]$. If $V \in \mathcal{E}_2$ we must also have $UA_e V^T = 0$.

Applying (7.12), we obtain $U \in N(A_e^T) = [\mathcal{U}]$; consequently $\mathcal{S}(A) \subset [\mathcal{U}]$. Hence it follows that $[\mathcal{S}(A)] \subset [\mathcal{U}]$, and by (7.21) also

$$[\mathcal{S}(A)] = [\mathcal{U}]. \tag{7.22}$$

But $\mathcal{U}$ is the intersection of $[\mathcal{U}]$ and all its bordering half-spaces, and these half-spaces either reflect the hypotheses of nonnegativity of the components of the mixed strategies and have the form $(U_i)^*$, or, by (7.18), can be represented in the form $(A_2 V_j^T)^*$, where $V_j \in \mathcal{S}_2$. It is evident that $\mathcal{S}(A) \subset (U_i)^*$. It remains to show that $\mathcal{S}(A) \subset (A_2 V_j^T)^*$, i.e., that, for all $X \in \mathcal{S}(A)$ and $V_j \in \mathcal{S}_2$,

$$X_e A_2 V_j^T \geqq 0. \tag{7.23}$$

But it follows from (7.12) that this is the case for all $V_j \in \mathcal{S}_2$ and $X \in \mathcal{U}$; and consequently also for all $X \in [\mathcal{U}]$; it remains only to refer to (7.22). □

7.6 Description of all matrix games with given sets of solutions. In the course of proving the preceding theorem, we prepared at the same time the ground for a proof of another theorem, one that makes it possible to describe all the $m \times n$-matrices A for which $\mathcal{S}(A) = \mathcal{U}$, $\mathcal{T}(A) = \mathcal{V}$, where $\mathcal{U}$ and $\mathcal{V}$ are given convex cones that satisfy (7.2) and (7.3).

Theorem. *Let $\mathcal{U} \subset \boldsymbol{P}$, $\mathcal{V} \subset \boldsymbol{Q}$ be convex cones that satisfy* (7.6) *and* (7.7). *For there to be an $m \times n$-matrix A of the form* (7.1), *for which*

$$\mathcal{U} = \mathcal{S}(A), \quad \mathcal{V} = \mathcal{T}(A), \tag{7.24}$$

it is necessary and sufficient that the block A_e satisfies (7.7) *and* (7.8), *the block A_1 satisfies* (7.17)–(7.19), *and the block A_2 satisfies* (7.12)–(7.14) *(the block A_s can be chosen arbitrarily).*

Proof. We first observe that the preceding theorem established the existence of a matrix A with the properties just specified, as well as the sufficiency of these properties for (7.24). Consequently it remains only to establish the necessity part of the theorem.

We begin with the proof of (7.7).

An arbitrary vector $V \in [\mathcal{V}]$ can, by what was said in 7.2, be represented as a difference $Y - Y'$, where Y and $Y' \in \mathcal{V}$. Therefore, for every essential strategy i of player 1, we must have

$$A_{i\cdot} V^T = A_{i\cdot} Y^T - A_{i\cdot} Y'^T = A_{ei\cdot} Y_e^T - A_{ei\cdot} Y'^T_e = 0,$$

so that $V \in N(A_e)$, and therefore $[\mathcal{V}] \subset N(A_e)$.

To establish the converse inclusion, we select $V \in N(A_e)$, i.e., a vector such that $A_e V_e^T = 0$, and find a corresponding vector Y whose components corresponding to the essential strategies of player 2 are positive, while $A_{i}.Y^T < 0$ for all superfluous strategies i of player 1. This is possible by 1.19.

Now we may choose λ so large that

$$(V + \lambda Y)V_j^T > 0 \text{ for } V_j \in \mathscr{E}_2, \tag{7.25}$$

$$X_i A(V + \lambda Y) < 0 \text{ for } X_i \in \mathscr{S}_1. \tag{7.26}$$

In addition, since $Y \in \mathscr{E}_2$, we must have, for all $V_j \in \mathscr{S}_2$,

$$(V + \lambda Y)V_j = 0, \tag{7.27}$$

and, for all $X_i \in \mathscr{E}_1$,

$$X_i A_e (v + \lambda Y) = X_i A_e v + \lambda X_i AY^T = 0. \tag{7.28}$$

It follows from (7.25) and (7.27) that $V + \lambda Y \in \boldsymbol{Q}$, and it follows from (7.26) and (7.28) that $V + \lambda Y \in -(\boldsymbol{PA})^*$. Consequently, by what was said in 7.5, $V + \lambda Y \in \mathscr{V}$. But this means that

$$V = (V + \lambda Y) - \lambda Y \in [\mathscr{V}]$$

and (7.7) is established; (7.8) is proved similarly.

We now notice that $\mathscr{U}^* = \boldsymbol{P} + \boldsymbol{AQ}$ implies $\boldsymbol{AQ} \subset \mathscr{U}^*$, and consequently $A\mathscr{S}_2 \subset \mathscr{U}^*$. This proves (7.12).

We now turn to the proof of (7.13). Let us suppose that there is a nonzero vector $Y \in \mathscr{S}_2$ for which AY is different from zero and orthogonal to $\mathscr{U}$. Let

$$Y = \sum_{j=1}^{s_2} \lambda_j Y_j$$

be the expansion of Y in terms of basis vectors of $\mathscr{S}_2$. For all $X \in \mathscr{U}$, we have

$$XAY_j^T \geqq 0 \qquad j = 1, \ldots, s_2. \tag{7.29}$$

If AY^T is orthogonal to $\mathscr{U}$, we have, for all $X \in \mathscr{U}$,

$$XAY^T = \sum_{j=1}^{s_2} \lambda_i XAY_j^T = 0.$$

By (7.29), this is possible only when all terms of the sum are zero. But since some λ_j are positive, this means that $XAY_j^T = 0$ for some j and all $X \in \mathscr{U}$. But this is impossible when Y_j is superfluous.

The resulting contradiction establishes (7.13).

Finally, to establish (7.15) we notice that by the theorem in 7.3 each bordering face $\mathscr{F}$ of $\mathscr{U}$ is the intersection of $\mathscr{U}$ with a supporting hyperplane that has either the form $(X_i)^\perp$ or the form $(AY^T)^\perp$. Since we are concerned with interior faces of $\mathscr{U}$, there are no hyperplanes of the first kind. If

$$\mathscr{F} = \mathscr{U} \cap (AY_j^T)^\perp, \tag{7.30}$$

and the strategy Y_j is essential, then, by what was said in 1.19, $XAY_j^T = 0$ for every $X \in \mathcal{U}$, i.e., $\mathcal{U} \subset (AY_j)^{\perp}$. Then (7.30) yields $\mathcal{F} = \mathcal{U}$, which is impossible. Therefore the strategy Y_j is superfluous.

It remains only to observe that we can obtain (7.17)–(7.19) by a parallel discussion. □

7.7. As was shown in 3.5 of Chapter 2, in the "typical", i.e., nondegenerate, case of a bimatrix game there are a finite (odd) number of equilibrium situations. By the convexity of the set of the players' optimal strategies in a matrix game, and the rectangular form of the set of its saddle points, the finiteness of the number of saddle points in a matrix game implies that it has just one saddle point.

In a more complete (and more explicit) form, this property of the set $\mathcal{M}_{m\times n}$ of $m \times n$-matrix games that have a unique solution can be formulated in the following two theorems. In these formulations, the topology in the space $\mathbf{R}^{mn}$ of all $m \times n$-matrix games is evidently to be taken elementwise.

7.8 Theorem. *The set $\mathcal{M}_{m\times n}$ is open in $\mathbf{R}^{mn}$.*

Proof. Let us consider a game $\Gamma_{A^{(0)}} \in \mathcal{M}_{m\times n}$ for which $\{(X^{(0)}, Y^{(0)}\} = \mathcal{C}(A^{(0)})$. We consider a sequence $A^{(k)} \to A^{(0)}$ and set, for $k = 0, 1, \ldots$,

$$\mathfrak{x}^{(k)} = \{i : A_{i\cdot}^{(k)} Y^{(k)T} = v(A^{(k)}) \text{ for all } Y^{(k)} \in \mathcal{T}(A^{(k)})\}.$$

Since $\mathfrak{x}^{(k)} \subset \mathfrak{x}$, the number of possible sets $\mathfrak{x}^{(k)}$ is finite, and in the sequence $\mathfrak{x}^{(1)}, \mathfrak{x}^{(2)}, \ldots$ at least one set contains an infinite subsequence: $\mathfrak{x}^* = \mathfrak{x}^{(k_1)} = \mathfrak{x}^{(k-2)} = \ldots$. For any $i \in \mathfrak{x}^*$ we must have

$$A_{i\cdot}^{(k_r)} Y^{(k_r)T} = v(A^{(k_r)}), \qquad r = 1, 2, \ldots.$$

Reference to the theorem in 1.24 tells us that after passing to the limit we will have

$$A_{i\cdot}^{(0)} Y^{(0)} = v(A^{(0)}),$$

i.e., $i \in \mathfrak{x}^{(0)}$. Therefore

$$\mathfrak{x}^* \subset \mathfrak{x}^{(0)}. \tag{7.31}$$

On the other hand, let us denote by $\bar{\mathfrak{x}}^{(k)}$ the spectrum of the "most mixed" strategy of player 1 in the game $A^{(k)}$ $(k = 0, 1 \ldots)$. Let $i \in \bar{\mathfrak{x}}^{(0)}$, i.e., $\xi_i^{(0)} > 0$. We choose any $X^{(k)} \in \mathcal{S}(A^{(k)})$. By 1.24, $X^{(k)} \to X^{(0)}$, and therefore $\xi_i^{(k)} > 0$ for sufficiently large k, so that $i \in \bar{\mathfrak{x}}^{(k)}$. Therefore

$$\bar{\mathfrak{x}}^0 \subset \bar{\mathfrak{x}}^{(k)}. \tag{7.32}$$

But by 1.19 we must have $\mathfrak{x}^{(0)} = \bar{\mathfrak{x}}^{(0)}$ and $\mathfrak{x}^* = \bar{\mathfrak{x}}^{(k)}$ (for sufficiently large k). Hence the inclusions (7.31) and (7.32) yield $\mathfrak{x}^{(0)} = \mathfrak{x}^*$. This means that, from some point on, the sets of essential strategies of player 1 in the games $\Gamma_{A^{(k)}}$ and $\Gamma_{A^{(0)}}$ coincide. We can say the same for the sets of essential strategies of player 2 in these games.

According to the theorem of 4.3, the matrix of an essential subgame of $\Gamma_{A^{(0)}}$ must be square. Since $\Gamma_{A^{(k)}}$ (for large k) has the same essential subgame, the matrix of this subgame is also square and of the same order. By (7.5) its rank is equal to the difference $|x^{(0)}| - \dim \mathcal{S}(A^{(0)})$, i.e., by the uniqueness of the solution it is equal to $|x^{(0)}|$. But $A^{(k)}$ differs arbitrarily little from $A^{(0)}$. Therefore its rank is not less than the rank of $A^{(0)}$. It also evidently cannot be larger than the rank of $A^{(0)}$, i.e., than $|x^{(0)}|$ (the dimension of the submatrix does not allow it). Therefore it is equal to $|x^{(0)}|$. If we apply (7.5) to the game $A^{(k)}$ we obtain

$$\operatorname{rank} A^{(k)} = |x^{(0)}| = |x^{(k)}| - \dim \mathcal{S}(A^{(k)}),$$

and it follows from $|x^{(0)}| = |x^{(k)}|$ that $\dim \mathcal{S}(A^{(k)}) = 0$.

We obtain $\dim \mathcal{S}(A^{(k)}) = 0$ in the same way. □

7.9. We now establish a sufficient (but not necessary) purely linear-algebraic test for the uniqueness of the solution of a matrix game.

Lemma. *Let $\mathcal{M}^*_{m\times n}$ denote the set of $m \times n$-matrices A that have the following properties for every positive integer t:*

1) *The matrix obtained from any $t \times (t+1)$-submatrix of A by adjoining a unit row is nonsingular;*

2) *The matrix obtained from any $(t+1) \times t$-submatrix of A by adjoining a unit column is nonsingular.*

3) *Every $t \times t$-submatrix of A is nonsingular.*

Then $\mathcal{M}^*_{m\times n} \subset \mathcal{M}_{m\times n}$.

Proof. Let $\Gamma_A \in \mathcal{M}_{m\times n}$, $X \in \mathcal{S}(A)$, $Y \in \mathcal{T}(A)$, and let r' be the number of pure strategies of player 1 that are in the spectrum of X; r'', the number of pure strategies i of player 1 for which

$$A_{i\cdot} Y^T = v_A; \tag{7.33}$$

s' the number of pure strategies of player 2 that are in the spectrum of Y; and s'' the number of pure strategies j of player 2 for which

$$XA_{\cdot j} = v_A. \tag{7.34}$$

According to the theorem in 1.14 (complementary slackness), we must have

$$r' \leqq r'', \quad s' \leqq s''. \tag{7.35}$$

Let B be an $r' \times s''$-submatrix of A, whose rows form the spectrum of X, and let (7.34) be satisfied for every column j. We adjoin to B a last row of units, denoting the resulting $(r'+1) \times s''$-matrix by C. Evidently

$$(X_B, -v)C = X_B B - vJ_{s''} = 0^{s''},$$

If we suppose here that $r' + 1 \leqq s''$ then we may select from C an $(r' + 1) \times (r' + 1)$-submatrix D for which

$$(X_B, -v)D = 0^{r'+1},$$

and this contradicts condition 1). Therefore $r' + 1 > s''$, i.e.,

$$s'' \leqq r'. \tag{7.36}$$

Similarly, by using condition 2) and (7.33), we obtain

$$r'' \leqq s'. \tag{7.37}$$

It follows from (7.35)–(7.37) that

$$s' \leqq s'' \leqq r' \leqq r'' \leqq s';$$

hence $s' = s'' = r' = r''$. Therefore all the optimal strategies of player 1 (also of player 2) must have the same spectrum. But then for any $X', X'' \in \mathcal{S}(A)$ we must have

$$(X'_B - X''_B)B = 0^{r'},$$

and since by condition 3) the matrix B is nonsingular, $X'_B = X''_B$ and therefore $X' = X''$.

Similarly we find that $\mathcal{T}(A)$ has only one element. □

7.10. As examples of matrices in $M^*_{m\times n}$ we may take matrices of the form

$$\begin{bmatrix} a_1^{k_1}, & a_2^{k_1}, & \dots, & a_n^{k_1} \\ \dots & \dots & \dots & \dots \\ a_1^{k_m}, & a_2^{k_m}, & \dots, & a_n^{k_m} \end{bmatrix},$$

where $a_1, \dots, a_n$ are different positive numbers, not unity, and $k_1, \dots, k_m$ are different positive integers. The determinants formed either from submatrices of this matrix or from the matrices described in conditions 1) and 2) of the lemma of the preceding subsection, are Vandermonde determinants and therefore different from zero.

7.11. Lemma. *The set $\mathcal{M}^*_{m\times n}$ is everywhere dense in the space $\mathbf{R}^{mn}$ of $m\times n$-matrix games.*

Proof. Choose any $m\times n$-game Γ_A and game $\Gamma_B \in \mathcal{M}^*_{m\times n}$ (by the preceding subsection, such games exist). Let A' and B' denote square submatrices of A and B, or matrices A and B obtained from A and B by adjoining unit rows or columns, as in the hypotheses of the lemma in 7.9. The determinant

$$\det((1-\varepsilon)A' + \varepsilon B')$$

is a polynomial in ε. By hypothesis, if $\varepsilon = 1$ it is not zero. Let ε' be its smallest positive root. The number ε' depends on the choice of the submatrix of A. Since there is a finite number of variants of these submatrices, there is a smallest number ε'. Let us denote it by ε''. Then for every $\varepsilon \in (0, \varepsilon'')$ the game with payoff matrix $(1-\varepsilon)A + \varepsilon B$ belongs to $\mathcal{M}^*_{m\times n}$. □

7.12. We have arrived at our objective.

Theorem. *The set* $\mathcal{M}^*_{m\times n}$ *is everywhere dense in the set* $\mathbf{R}^{mn}$ *of* $m \times n$*-matrix games.*

The proof follows immediately from the lemmas in 7.10 and 7.11.

Notes and references for Chapter 4

§1. The lemma on two alternatives in **1.4** was stated and proved in **16.4** of J. von Neumann and O. Morgenstern's book [1]. Its application to the basic theorems of saddle points of matrix games is presented there in **17.6**. The proof of the propositions in **1.6–1.8** was given by D.J. Newman [1] and is based on a theorem of E. Stiemke [1]. The proof of the abstract minimax theorem (for vector spaces over an arbitrary ordered field) was given by H. Weyl [2]. In this connection we also recall the abstract theory of convex polytopes developed by the same author in [1]. For additional proofs of the minimax theorem for matrix games, see G. Owen [1], E. Marchi [1], I. Joó [1], and the book by G.N. Dyubin and V.G. Suzdal' [1] (see also the papers by H. König [1] and M. Neumann [1]).

The elementary properties of solutions of matrix games that are set out in **1.10–1.15** were described implicitly in § 17 of J. von Neumann and O. Morgenstern [1]. Functional properties of values of matrix games were investigated by E.Ĭ. Vilkas [2–4]. Equations in which a matrix depending on an unknown parameter appears in a functional of the value of a matrix game were considered by E.Ĭ. Vilkas and D.P. Sudzhyute [1]. J. Farkas published the lemma of **1.17** in [1]. The reduction of matrix games to their essential subgames (as well as other topics in the theory of matrix games) were described by H.F. Bohnenblust, S. Karlin and L.S. Shapley [1]. For an alternative treatment of problems of this type, see K.J. Arrow, E.J. Barankin and D. Blackwell [1]. The differentiation of the value of a game in **1.22** was suggested by H.O. Mills [1] and extended by E.G. Gol'shteĭn and S.M. Movshovich [1] to a class of infinite two-person zero-sum games. A similar construction of affine forms was carried out by G.N. Beltadze [1,2].

The method described in **1.24** for symmetrization of matrix games is implicitly mentioned at the end of subsection 17.11.3 of J. von Neumann and O. Morgenstern's monograph [1], and first fully described by D. Gale, H. Kuhn and A.W. Tucker [1]. The discussion of families of subgames of two-person zero-sum games led to characteristic functions, i.e. to the cooperative theory of these games. Questions that arise in this connection for matrix games were considered by N.N. Vorob'ev [17].

§2. The analysis in subsection 2.1 of 2×2-matrix games was presented in subsections 18.1–18.4 of J. von Neumann and O. Morgenstern [1]. The idea of the grapho-analytic approach to the solution of matrix games was first suggested by E. Borel [1]. For a detailed exposition of this type of problems, see T. Motzkin, H. Raiffa, J. Thomson and R. Thrall [1].

§3. On the evidence of G.B. Dantzig [2], the connection between linear programming and the theory of matrix games was suggested to him by J. von Neumann in the autumn of 1947. The first publication on this subject, which also contains some historical details, was given by G.B. Dantzig [1]. For the history of the question, see G.B. Dantzig [1] and his monograph [3]. Also see A.W. Tucker [1]. The theorem on the connection of the solutions of a pair of dual linear programming problems with the solution of matrix games was established by G.B. Dantzig [1]. The first transformation of a pair of dual linear programming problems into a matrix game was carried out by D. Gale, H. Kuhn and A.W. Tucker [1]. They also suggested the method of symmetrization of matrix games, described in **3.5**. Still another proof of the equivalence of the solution of a matrix game to a pair of dual linear programming programs was given H.O. Mills [2]. The equivalence of a game of hide and seek with the assignment problem of linear programming (see **3.6**) was discovered and established by J. von Neumann [4]. The linear-algebraic method of solving matrix games is described in the book by Motzkin, Raiffa, Thompson and Thrall [1].

The problem on polyhedral games in the treatment of **3.9** was proposed by H.O. Mills and solved by P. Wolfe [1]. Matrix games with restrictions were analyzed by A. Charnes [1].

§4. The theorem of **4.2** and the associated discussion in **4.3** were found by L.S. Shapley and R.N. Snow [1]. R. Restrepo [2] described a class of matrix games for which this search of matrices can be significantly shortened. The algorithm for enumerating equilibrium situations in an arbitrary bimatrix game was proposed by N.N. Vorob'ev [1]. Subsections **4.7–4.14** present an improvement of H.W. Kuhn's treatment [2].

The uniqueness condition for equilibrium situations in bimatrix games was found by V.L. Kreps [1]. The result was rediscovered by G.A. Heuer [1]. Subsequently V.L. Kreps [3] extended it to general finite noncooperative games. Similar facts for bimatrix games with a rectangular set of equilibrium situations were described by K. Isaacson and C.B. Millham [1].

§5. The propositions in **5.5** and **5.8** were obtained by H.F. Bohnenblust, S. Karlin and L.S. Shapley [1]. The case of pairwise different values of objects in the game, described in **5.10–5.15**, was developed by M. Dresher [2] (see also S. Karlin's book [3]), and in the general case presented here, by N.N. Vorob'ev [8]. The discussion of analogous games in the case of several defenders was begun by I.N. Vrublevskaya [1,2].

From later work describing the solution of matrix games on the basis of certain definitions of their structure, we call attention to the papers of I.N. Fokin [1,2], and of L.M. Brègman and I.N. Fokin [1–5]. An interesting special class of matrix games was studied by G.P. Mandrigina [1,2]. A paper by E.Ĭ. Vilkas [5] is devoted to a theoretical analysis of the domain of the solutions of a parametric matrix game.

§6. The determination of saddle points in matrix games by the method of differential equations was introduced by G.W Brown and J. von Neumann [1].

An iterative method was proposed by G.W. Brown [1], and a proof of its effectiveness was described in J. Robinson [1]. As J.M. Danskin [1] showed (also see M. Dresher's book [2]), this method can be adapted for the solution of infinite two-person zero-sum games with continuous payoff functions. Further developments in the domain of iterative methods are contained in the monograph by V.Z. Belen'kiĭ, V.A. Volkonskiĭ, S.A. Ivankov, A.B. Pomanskiĭ, and A.D. Shapiro [1], and also in the paper by S.A. Ivankov [1], from which the proof of the theorem in **6.10** was taken.

The lower bound, cited in **6.18–6.22**, for the rapidity of convergence of this method comes from H.N. Shapiro [1]. This bound guarantees only a very slow convergence of the iterative process, and there arises the natural question of improving it. It has been shown (see, for example, S. Karlin [3], p. 218) that for the tth step of the process there is an estimate of order $1/\sqrt{t}$ for the error. On the other hand, it was shown by L. G. Khachiyan [1] that for every matrix game in a rather broad class (for example, for any completely mixed game) the order of the error at the tth iteration cannot exceed $1/t$.

An essentially new iterative algorithm for solving matrix games, in which it is applied to an $m \times n$-game at each step, and instead of solving an optimization problem we solve a $k \times n$-game with $k < n$ (in practice, we may suppose that $k = 2$), has been proposed and justified by A.L. Sadovskiĭ [1]. Experiments have indicated very fast convergence of this algorithm.

R.B. Braithwaite [1] has proposed a finite iterative algorithm for solving certain classes of matrix games. Still another numerical method for finding optimal strategies in a matrix game was proposed J. von Neumann [5]. He also proposed in [6] an approach to a numerical determination of ε-optimal strategies in games of large dimension and indicated an upper bound for the number of iterations.

The example in **6.23**, showing the inapplicability of the iterative method to bimatrix games, was found by L.S. Shapley [2]. Its applicability (convergence) for a 2×2-bimatrix game was proved by K. Miyasawa [1].

§7. A description of the set of all $m \times n$-games with given sets of optimal strategies for the players was given by H.F. Bohnenblust, S. Karlin and L.S. Shapley [1]. The corresponding discussion in **7.1–7.6** is given in a clearer form obtained by D. Gale and S. Sherman [1].

The theorems of **7.7** and **7.11** come from H.F. Bohnenblust, S. Karlin and L.S. Shapley [1] (see also S. Karlin [3]).

References

Citations of the form [C,n], where n is a number, refer to the list of collections on pp. 483–484.

A. Aho, J. Hopcroft, and J. Ullman

[1] The design and analysis of computer algorithms, Addison-Wesley, Reading MA, 1976.

K.J. Arrow

[1] Social choice and individual values, Wiley, New York, 1951, 1963.

K.J. Arrow, E.W. Barankin, and D. Blackwell

[1] Admissible points of convex sets, [C3], pp. 87–92.

J.P. Aubin

[1] Mathematical methods of game and economic theory, North Holland, Amsterdam, 1979, 1982.

R.J. Auman and L.S. Shapley

[1] Values of non-atomic games, Princeton University Press, 1974.

Bachet de Meziriac

[1] Problèmes plaisants et delectables, qui se font par les nombres, Lyon, 1612.

F. Başar and G.J. Oldser

[1] Dynamic noncooperative game theory, Academic Press, London and New York, 1982.

J.-M. Bein

[1] How relevant are "irrelevant alternatives"?, Theory Decis. 7, no. 1–2, 1976, 95–105.

V.Z. Belenkiĭ, V.A. Volkonskiĭ, S.A Ivankov, A.B. Pomanskiĭ, and A.D. Shapiro

[1] Iterative methods in game theory and programming (in Russian), Nauka, Moscow, 1974.

G.N. Beltadze

[1] Sets of equilibrium situations in lexicographical noncooperative games (in Russian), Soobshch. Akad. Nauk Gruz. SSR, 98, no. 1 (1980), 41–44.

[2] Mixed extensions of finite lexicographical games (in Russian), Soobshch. Akad. Nauk Gruz. SSR, 98, no. 2 (1980), 273–276.

C. Berge

[1] Sur une convexité régulière et ses applications à la théorie des jeux, Bull. Soc. Math. France 81 (1954), 301–315.

[2] Théorie générale des jeux à n personnes, Gauthier-Villars, Paris, 1957.

D.W. Blackett

[1] Some Blotto games, Naval Res. Logist. Quart. 1 (1954), no. 1–2, 55–60.

D. Blackwell

[1] An analog of the minimax theorem for vector payoff, Pacific J. Math. 6 (1956), no. 1, 1–8.

D. Blackwell and M.A. Girshick

[1] Theory of games and statistical decisions, Wiley, New York; London, Chapman and Hall, 1954.

H.F. Bohnenblust and S. Karlin

[1] On a theorem of Ville, [C1], pp. 155–160.

H.F. Bohnenblust, S. Karlin and L.S. Shapley

[1] Solutions of discrete two-person games, [C1], pp. 51–72.

[2] Games with continuous, convex pay-off, [C1], pp. 181–192.

E. Borel

[1] La théorie du jeu et les équations intégrales à noyau symétrique, C.R. Acad. Sci. Paris 173 (1921), 1304–1308. [The theory of play and integral equations with skew symmetric kernels, Econometrica 21 (1953), no. 1. 97–100.]

[2] Sur les jeux où interviennent l'hasard et l'habilité des joueurs, Théorie des probabilités, Paris, 1924, pp. 204–224. [On games that involve chance and the skill of the players, Econometrica 21 (1953), no. 1, 101–115.]

[3] Sur les systèmes de formes linéaires à déterminant symétrique gauche et la théorie générale du jeu, C.R. Acad. Sci. Paris 184 (1927), 52–53. [On systems of linear forms of skew symmetric determinant and the general theory of play, Econometrica 21 (1953), no. 1, 116–117.]

M.M. Botvinnik

[1] An algorithm for the game of chess (in Russian), Nauka, Moscow, 1968.

[2] On the cybernetic objective of a game (in Russian), Sov. Radio, Moscow, 1975.

C.L. Bouton

[1] Nim, a game with a complete mathematical theory, Ann. of Math. 2 (1902), no. 3, 35–39.

R.B. Braithwaite

[1] A terminating iterative algorithm for solving certain games and related sets of linear equations, Naval Res. Logist. Quart. 6 (1959), no. 1, 63–74.

S.J. Brams

[1] Faith versus rationality in the Bible, in Applied game theory, ed. by Brams and others, Physica-Verlag, Wurzburg and Vienna, 1979, pp. 430–445.

[2] Biblical games: a strategic analysis of stories in the Old Testament, Cambridge, 1980.

[3] Superior beings: if they exist, how would we know?, Springer-Verlag, New York and Berlin, 1983.

L.M. Bregman and I.N. Fokin

[1] On the structure of optimal strategies in some matrix games (in Russian), Dokl. Akad. Nauk SSSR 188 (1969), no. 5, 974–977.

[2] On sums of matrix games (in Russian), Ekon. i Matem. Metody 9 (1973), no. 1, 148–154.

[3] On equivalent strategies in sums of matrix games (in Russian), [C18], pp. 15–19.

[4] A method for solving sums of matrix games (in Russian), in Study of operations and statistical modelling, no. 2, Leningrad, 1974, pp. 37–55.

[5] On equivalent strategies in sums of matrix games, (in Russian) in Study of operations and statistical modelling, no. 2, Leningrad, 1974, pp. 55–63.

G.B. Brown

[1] Iterative solution of games by fictitious play, [C1], pp. 374–376.

[2] The solution of a certain two-person zero-sum game, Operat. Res. 5, no. 1 (1957), 63–67.

G.W. Brown and J. von Neumann

[1] Solutions of games by differential equations, [C1], pp. 73–79.

V.S. Bubyalis

[1] On the problem of the structure of equilibrium situations (in Russian), Mat. metody v sotsial. nauk, no. 4 (1974), 37–42.

[2] The connection between noncooperative n-person and 3-person games (in Russian), in Current trends in game theory, pp. 18–24, Mokslas, Vilnius, 1976, pp. 18–24.

[3] A game without base equilibrium situations (in Russian), Mathematical Methods in Social Sciences, no. 7 (1976), 9–16.

[4] On equilibria in finite games, Int. Journ. Game Theory, 8, no. 2 (1979), 65–79.

R.C. Buck

[1] Preferred optimal strategies, Proc. Amer. Math. Soc. 9, no. 2 (1958), 312–314.

Bũi Công Cuoñg

[1] Some remarks on minimax theorems, Acta Math. Vietnam. 1, no. 2 (1976), 67–74.

[2] The minimax theorems and the equilibrium existence problem, 3, Toan hok 4 (1977), no. 3, 1–6.

E. Burger

[1] Einführung in die Theorie der Spiele, de Gruyter, Berlin, 1959.

A. Charnes

[1] Constrained games and linear programming, Proc. Nat. Acad. Sci. USA 38, no. 7 (1953), 639–641.

A.G. Chernyakov

[1] Stability for finite noncooperative games (in Russian), Dokl. Akad. Nauk SSSR 247, no. 4 (1979), 809–811.

[2] On the structure of the set of degenerate, finite noncooperative games, Math. Operationsforsch. Stat., Ser. Optimization 10, no. 4 (1979), 473–481.

[3] Stability for finite noncooperative games (in Russian), Math. Operationsforsch. Stat., Ser. Optimization 12, no. 1 (1981), 107–114.

H.H. Chin, T. Parthasarathy, and T.E.S. Raghavan

[1] Structure of equilibria in N-person non-cooperative games, Intern. J. Game Theory 3, no. 1 (1974), 1–19.

A.A. Cournot

[1] Recherches sur les principes mathématiques de la théorie de richesses, Paris, 1838.

J.M. Danskin

[1] Fictitious play for continuous games, Naval Res. Logist. Quart. 1, no. 4 (1954), 313–320.

G.B. Dantzig

[1] A proof of the equivalence of the programming problem and the game problem, [C2], pp. 330–335.

[2] Maximization of a linear function of variables subject to linear inequalities, [C2], pp. 339–347.

[3] Linear programming and extensions, Princeton University Press, 1963.

M.A. Davis

[1] Game theory. A nontechnical introduction, Basic Books, New York, 1970.

E.G. Davydov

[1] Methods and models in the theory of two-person zero-sum games (in Russian), Moscow State University Press, 1978.

G. Debreu

[1] A social equilibrium existence theorem, Proc. Nat. Acad. Sci. USA 38, no. 9 (1952), 886–893.

[2] Valuation equilibrium and Pareto optimum, Proc. Nat. Acad. Sci. USA 40, no. 7(1954), 588–592.

V.F. Dem'yanov and V.N. Malozemov

[1] Introduction to minimax (in Russian), Nauka, Moscow, 1972.

V.K. Domanskiĭ

[1] On an application of distribution theory to the theory of two-person zero-sum games (in Russian), Dokl. Akad. Nauk SSSR 199, no. 3 (1971), 515–518.

M. Dresher

[1] Methods of solution in game theory, Econometrica 18, no. 3 (1950), 179–181.

[2] Games of strategy, theory and applications, Prentice-Hall, Englewood Cliffs, NJ, 1961.

[3] Probability of a pure equilibrium point in n-person games, J. Combinat. Theory 8, no. 1 (1970), 134–145.

M. Dresher and S. Karlin

[1] Solution of convex games as fixed points, [C3], pp. 75–86.

M. Dresher, S. Karlin and L.S. Shapley

[1] Polynomial games, [C1], pp. 161–180.

L. Dubins and L. Savage

[1] How to gamble if you must. Inequalities for stochastic processes, McGraw-Hill, New York, 1965.

G.N. Dyubin

[1] On a set of games on the unit square with a unique solution (in Russian), Dokl. Akad. Nauk SSSR 184, no. 2 (1969), 267–269.

[2] On the structure of a set of two-person zero-sum games whose payoffs are continuous functions (in Russian), [C19], pp. 47–66.

G.N. Dyubin and V.G. Suzdal',

[1] Introduction to applied game theory (in Russian), Nauka, Moscow, 1981.

Ky Fan

[1] Minimax theorems, Proc. Nat. Acad. Sci. USA, 39, no. 1, 42–47.

J. Farkas

[1] Über die Theorie der einfachen Gleichungen, J. Reine Angew. Math. 124. no. 1 (1902), 1–24.

R. Farquharson

[1] Sur une généralisation de la notion d'equilibrium, C.R. Acad. Sci. Paris 240, no. 1, 46–48.

V.V. Fedorov

[1] On the method of penalty functions in the problem of determining a maximin (in Russian), Zh. Vychisl. Mat. i Mat. Fiz. 12, no. 2 (1972), 321–333.

[2] On the problem of seeking a sequential maximin (in Russian), Zh. Vychisl. Mat. i Mat. Fiz. 12, no. 4 (1972), 897–908.

P. Fermat

[1] Varia opera matematica D. Petri de Fermat, senatoris Tolosani, Toulouse, 1679.

P.C. Fishburn

[1] Utility theory for decision making, Wiley, New York, etc., 1970.

[2] On the foundation of game theory: The case of non-Archimedean utility, Intern. J. Game Theory, 1 no. 2 (1972), 65–71.

[3] The theory of social choice, Princeton univ. press, Princeton, 1973.

R.A. Fisher

[1] Randomisation, and an old enigma of card play, Math. Gaz. 18, no. 231 (1934), 294–298.

I.N. Fokin

[1] On the solution of some game-theory problems of the allocation of resources by means of recurrence relations (in Russian), [C18], 73–77.

[2] Some game-theory problems of the allocation of resources, [C19], 73–77.

M. Fréchet

[1] Théorie des probabilités: exposés sur les fondements et ses applications, Paris, 1952, 156–160. [Commentary on the three notes on Emile Borel, Econometrica, 21, no. 1 (1953), 118–124].

[2] Emile Borel, initiator of the theory of psychological games and its application, Econometrica 21, no. 1 (1953), 95–96.

J.W. Friedman

[1] Oligopoly and the theory of games, North Holland, Amsterdam, etc., 1977.

D. Gale and O. Gross

[1] A note on polynomial and separable games, Pacific J. Math. 8, no. 4 (1958), 735–741.

D. Gale, H. Kuhn, and A.W. Tucker

[1] Linear programming and the theory of games, [C2], pp. 317–329.

[2] On symmetric games, [C1], pp. 81–87.

[3] Reductions of game matrices, [C1], pp. 89–96.

D. Gale and S. Sherman

[1] Solutions of finite two-person games, [C1], pp. 37–49.

D. Gale and F.M. Stewart

[1] Infinite games with perfect information, [C3], pp. 245–266.

Yu.B. Germeĭer

[1] Necessary conditions for a maximin (in Russian), Zh. Prikl. Mekh. i Mat. Fiz. 9, no. 2 (1969), 432–438.

[2] Approximate reduction, with the aid of penalty functions, of the problem of determining a maximin to the problem of determining a maximum (in Russian), Zh. Prikl. Mekh. i Mat. Fiz. 9, no. 3 (1969), 730–731.

[3] On two-person games with a fixed sequence of moves (in Russian), Dokl. Akad. Nauk SSSR 198, no. 5 (1971), 1001–1004.

[4] Games with nonantagonistic interests (in Russian), Nauka, Moscow, 1976.

Yu.B. Germeĭer and N.N. Moiseev

[1] Introduction to the theory of hierarchical systems of equations (in Russian), Math. Operationsforsch. Statist. 4, no. 2 (1973), 133–154.

Yu.B. Germeĭer and I.A. Vatel'

[1] Games with hierarchical vector interests (in Russian), Izv. Akad. Nauk SSSR Tekhn. Kibern. 1974, no. 3, 54–69.

I.L. Glicksberg

[1] A further generalization of the Kakutani fixed point theorem, with application to Nash equilibrium points, Proc. Amer. Math. Soc. 3, no. 1, 170–174.

I.L. Glicksberg and O. Gross

[1] Notes on games over the square, [C3], pp. 173–182.

K. Goldberg, A.J. Goldman, and M. Neuman

[1] The probability of an equilibrium point, J. Res. Nat. Bureau Stand. Sect. B 72, no. 2 (1968), 93–101.

E.G. Gol'shteĭn and S.M. Movshovich

[1] A theorem on marginal values in two-person zero-sum games (in Russian), [C18], pp. 129–130.

V.A. Gorelik

[1] Games with similar interests (in Russian), Zh. Prikl. Mat. i Mat. Fiz. 11, no. 5 (1971), 1166–1179.

V.P. Gradusov

[1] On saddle points of functions of a certain class (in Russian), [C18], pp. 32–35.

D.Yu. Grigor'ev and N.N. Vorob'ev, Jr.

[1] Solving systems of polynomial inequalities in subexponential time, J. Symbolic Computing, 5 (1988), 37–64.

O. Gross

[1] A rational game on the square, [C6], 307–311.

W. Güth and B. Kalkofen

[1] Unique solutions for strategic games, Lecture Notes in Economics and Mathematical Systems, no. 328, Springer-Verlag, Berlin and Heidelberg, 1989.

H. Hamburger

[1] N-person prisoners' dilemma, J. Math. Sociol. 3, no. 1 (1973), 27–48.

J.C. Harsanyi

[1] A general solution for noncooperative games based on risk-dominance, [C13], 651–679.

J.C. Harsanyi and R. Selten

[1] A general theory of equilibrium selection in games, MIT Press, Cambridge MA, 1988.

E. Helly

[1] Über Systeme linearer Gleichungen mit unendlich vielen Unbekannten, Monatshefte Math. Phys. 37 (1921)

O. Helmer

[1] Open problems in game theory, Econometrica 20, no. 1 (abstract).

G.A. Heuer

[1] Uniqueness of equilibrium points in bimatrix games, Intern. J. Game Theory 8, no. 1 (1979), 13–24.

N. Howard

[1] The theory of metagames, General Systems, Yearbook of the Society for General Systems Research 11 (1966), 167–186.

[2] Paradoxes of reality: theory of metagames and political behavior, MIT Press, Cambridge, MA, 1971.

S. Huyberechts

[1] Sur le problème de l'unicité de la solution des jeux sur le carré unité, Bull. Classe Sci. Acad. Belge 44, no. 3 (1958), 200–216.

C. Huygens

[1] De ratiosiniis in ludo aleae, Oeuvres complètes, La Haye, 1925, vol. 5, pp. 35–47.

V.S. Il'ichev

[1] On polyantagonistic games (in Russian), [C17], 181–185.

K. Isaacson and C.B. Millham

[1] On a class of Nash-solvable bimatrix games and some related Nash subsets, Naval Res. Logist. Quart. 27, no. 3 (1980), 324–335.

Yu.P. Ivanilov and B.M. Mukhamediev

[1] Finding of all the equilibrium situations for a class of finite noncooperative games for several players (in Russian), Avtomatika i Telemekhanika 1978, no. 6, 94–98.

S.A. Ivankov

[1] Theorems on the convergence of iterative processes for the solution of games (in Russian), Proceedings of the third winter school on mathematical programming and related questions, 1970, Drogobych. Moscow, 1970, no. 2, 324–335.

M.J.M. Jansen and S.H. Tijs

[1] On characterizing properties of the value sets and the equilibrium-point sets of non-cooperative two-person games, Math. Operationsforsch. Statist., Ser. Optimization 12, no. 2, 263–270.

A.J. Jones

[1] Game theory: mathematical models of conflict. Mathematics and its Applications, Horwood, Chichester, 1980.

I. Joó

[1] A simple proof for von Neumann's minimax theorem, Acta Sci. Math. 42, no. 1/2 (1980), 91–94.

B. Kalkofen

[1] Gleichgewichtsauswahl in strategischen Spielen, Wirtschaftswissenschaftliche Beiträge 29, Physica-Verlag, Heidelberg, 1990.

S. Karlin

[1] Operator treatment of minimax principle, [C1], 133–154.

[2] Reduction of certain classes of games to integral equations, [C3], pp. 47–76.

[3] Mathematical Methods and Theory in Games, Programming, and Economics, Pergamon, London and Paris, 1959.

[4] On a class of games, [C3], pp. 159–171.

[5] On games described by bell-shaped kernels, [C6].

A. Kats

[1] Non-cooperative monopolistic and monopolistic market games, Intern. J. Game Theory 3, no. 3, 1974, 251–260.

J.G. Kemeny and G.L. Thompson

[1] The effect of psychological attitudes on the outcomes of games, [C6], 273–298.

L.G. Khachiyan

[1] On the rapidity of convergence of game-theoretic processes for solving matrix games (in Russian), Zh. Vychisl. Mat. i Mat. Fiz. 17, no. 6 (1977), 1421–1431.

[2] Polynomial algorithms in linear programming (in Russian), Zh. Vychisl. Mat. i Mat. Fiz. 20, no. 1 (1980), 51–68.

Khoàng Tuy

[1] On a general minimax theorem (in Russian), Dokl. Akad. Nauk SSSR 219, no. 4 (1974), 818–824.

D.M. Kilgour

[1] Equilibrium points of infinite sequential truels, Intern. J. Game Theory 6, no. 3 (1977), 167–180.

J. Kindler

[1] Unendliche Spiele mit definiten 2 × 2-Teilspielen, Arch. Math. 26. no. 6 (1975), 670–672.

[2] Minimaxtheoreme für die diskrete gemischte Erweiterung von Spielen und ein Approximationssatz, Math. Operationsforsch. Statist. 11, no. 3 (1980), 473–485.

A.Ya. Kiruta

[1] Equilibrium situations in nonatomic noncooperative games (in Russian), Matematichesky Metody v Sotsial'nykh Naukakh 1975, no. 6, 18–71.

G. Klaus

[1] Kybernetik in philosophischer Sicht, Berlin, 1965.

[2] Emanuel Lasker — ein philosophischer Vorläufer der Spieltheorie, Deutsche Z. Philos. 13, no. 8 (1965), 1976–1988.

[3] Philosophical aspects of game theory (in Russian), Voprosy Filosofii 1968, no. 8, 24–34.

[4] Spieltheorie in philosophischer Sicht, VdW, Berlin, 1968.

H. Kneser

[1] Sur un théorème fondamental de la théorie des jeux, C.R. Acad. Sci. Paris 234 (1952), 2418–2420.

H. König

[1] Über das von Neumannsche Mimimax-Theorem, Arch. Math. 19 (1968), 482–487.

N.N. Krasovskiĭ

[1] Game-theory problems on the encounter of motions (in Russian), Nauka, Moscow, 1970.

N.N. Krasovskiĭand A.I. Subbotin

[1] Positional differential games, (in Russian), Nauka, Moscow, 1974.

W. Krelle and D. Coenen

[1] Das nichtkooperative Nichtnullsummen-Zwei-Personen-Spiel, I, II, Unternehmensforschung 9 (1965), no. 2, 57–75; no. 3, 137–163.

V.L. Kreps

[1] Finite noncooperative games with dependent strategies (in Russian), [C17], pp. 211–215.

[2] Bimatrix games with unique equilibrium points, Intern. J. Game Theory 3, no. 2 (1974), 115–118.

[3] Finite N-person non-cooperative games with unique equilibrium point, Intern. J. Game Theory 10, no. 2/3 (1981), 125–129.

A.V. Krushevskiĭ

[1] Game theory, (in Russian), Vishcha Shkola, Kiev, 1977.

H.W. Kuhn

[1] Extensive games and the problem of information, [C3], 193–216.

[2] An algorithm for equilibrium points in bimatrix games, Proc. Nat. Acad. Sci. USA 47 (1961), 1657–1662.

H.W. Kuhn and A.W. Tucker

[1] Linear programming and the theory of games, [C2].

[2] John von Neumann's work in the theory of games and mathematical economics, Bull. Amer. Math. Soc. 64, no. 3 (1958), p. 2 100–122.

V.N. Lagunov

[1] Hierarchy and games (in Russian), in Existence of solutions, stability, and state of information in game theory, Kalinin State University, 1979, 65–80.

V.E. Lapitskiĭ

[1] On the uniqueness of mixed extension (in Russian), Matematichesky Metody v Sotsial'nykh Naukakh, no. 14 (1981), 9–17.

[2] On the axiomatics of equilibrium (in Russian), Matematichesky Metody v Sotsial'nikh Naukakh, no. 15 (1982), 18–26.

P.S. Laplace

[1] Essai philosophique sur les probabilités, V. Courcier, Paris, 1816.

E. Lasker

[1] Die Philosophie des Unvollendbaren, Leipzig, 1918.

C.E. Lemke and J.T. Howson

[1] Equilibrium points of bimatrix games, SIAM J. 12, no. 2 (1964), 413–423.

G. Liebscher

[1] Game theory and philosophy (in Russian), Nauchn. Dokl. Vyssh. Shkoly. Filos. Nauki 1974, no. 3, 100–109.

W.F. Lucas

[1] Some recent developments in N-person game theory, SIAM Rev. 13, no. 4 (1971), 491–523.

R.D. Luce and H. Raiffa

[1] Games and decisions. Introduction and critical survey, Wiley, New York, 1957.

M.M. Lutsenko

[1] Problem on δ-meeting on a compact set (in Russian), [C22], pp. 116–124.

[2] Problem on δ-meeting on the three-sphere (in Russian), Teor. Veroyatnost. i Mat. Statist. 23, no. 1 (1978), 198–203.

L.E. Maĭstrov

[1] The development of the concept of probability (in Russian), Nauka, Moscow, 1980.

O.A. Malafeev

[1] Stability of noncooperative games for n players (in Russian), Vestnik Leningrad. Univ. Mat. Mekh. Astron. 1978, no. 1, 50–53.

[2] Finiteness of the set of equilibria in noncooperative games (in Russian), Voprosi Mekhaniki i Protsessov Upravleniya, n. 2, Leningrad, 1978, pp. 135–142.

G.P. Mandrigina

[1] Application of game theory to the design of preventive maintenance (in Russian), Avtomat. i Vychisl. Tekhn. 1971, no. 4, 26–32.

[2] Application of game theory to the problem of the optimal allocation of resources in the design of preventive maintenance (in Russian), [C17], 300–304.

K. Manteuffel and D. Stumpe

[1] Spieltheorie, Teubner, Leipzig, 1977.

E. Marchi

[1] Otra demonstracion elemental del teorema del minimax, Rev. Un. Mat. Argentina 22, no. 1 (1978), 52–53.

[2] E-points of games, Proc. Nat. Acad. Sci. USA 57, no. 4 (1967), 878–882.

[3] Pseudo-saddle points for non-zero-sum two-person simple and generalized games, Proc. London Math. Soc. 18, no. 1 (1968), 158–168.

[4] Remarks on E-points of generalized games, An. Acad. Brasil. Ci. 41, no. 1 (1969), 21–27.

[5] Some topics on equilibria, Trans. Amer. Math. Soc. 220, 87–102.

J.C.C. McKinsey

[1] Isomorphism of games and strategic equivalence, [C1], pp. 117–130.

[2] Introduction to the theory of games, McGraw-Hill, New York, 1952.

A.S. Mikhaĭlova

[1] On some classes of games on the unit square (in Russian), [C12], pp. 426–441.

[2] Mixed duels of type 2×1 (in Russian), Math. Operationsforsch. Statist. 3, no. 2 (1972), pp. 97–102.

H.O. Mills

[1] Marginal values of matrix games and linear programs, [C5], 183–193.

[2] Equilibrium points in finite games, SIAM J. 8, no. 2 (1960), 397–402.

J. Milnor

[1] Games against Nature, [C4], pp. 49–59.

V.G. Mirkin

[1] The problem of group decisions (in Russian), Nauka, Moscow, 1974.

K. Miyasawa

[1] On the convergence of the learning process in a 2×2-non-zero-sum two-person game, Economic Research Program, Princeton University, Research Memorandum no. 33 (1961).

N.N. Moiseev

[1] Hierarchical structures and game theory (in Russian), Kibernetika, 1973, no. 6, 1–11.

R. Montmort

[1] Essai d'analyse sur le jeu de hasard, Paris, 1713.

E.H. Moore

[1] A generalization of the game called Nim, Ann. of Math. 11 (1909), 93.

O. Morgenstern

[1] The N-country problem, Fortune 63, no. 3 (1961), 136.

[2] On the application of game theory to economics, [C9], pp. 1–12.

[3] Spieltheorie und Wirtschaftswissenschaft, Oldenbourg, Munich and Vienna, 1966.

[4] The collaboration between Oskar Morgenstern and John von Neumann on the theory of games, J. Economic Literature 14 (1976), no. 3, 805–816.

T.S. Motzkin, H. Raiffa, G.L. Thompson, and R.M. Thrall

[1] The double description methods, [C3], pp. 51–73.

H. Moulin

[1] Two and three person games: a local study, Int. J. Game theory 8, no. 2 (1979), 81–107.

R. Mukundan and R. Kashyar

[1] Determination of equilibrium points in non-zero-sum noncooperative games, Intern. J. Syst. Sci. 6, no. 1 (1975), 67–80.

J. Mycielski

[1] Continuous games with perfect information, [C13], pp. 103, 111.

[2] On the axiom of indeterminateness, Fund. Math. 53, no. 2 (1964), 205–224.

J. Mycielski and H. Steinhaus

[1] A mathematical axiom contradicting the axiom of choice, [C9], 171–173; Bull. Acad. Polon. Sci. Ser. Math., Astr., Phys. 10 (1962), 1–3.

J. Mycielski and J. Swierckzowski

[1] On the Lebesgue measurability and the axiom of indeterminateness, Fund. Math. 54, no. 1 (1964), 67–71.

J.F. Nash

[1] The bargaining problem, Econometrica 18 (1950), 155–162.

[2] Non-cooperative games, Ann. of Math. 54 (1951), 286–295.

M. Neumann

[1] Bemerkungen zum von Neumannschen Minimaxtheorem, Arch. Math. 29, no. 1 (1977), 96–105.

D.J. Newman

[1] Another proof of the minimax theorem, Proc. Amer. Math. Soc. 14, no. 5, 692–693.

H. Nikaido and K. Isoda

[1] Note on non-cooperative convex games, Pacific J. Math. 5, no. 1 (1955), 807–815.

N.A. Nikitina

[1] On a class of games of statistical decision (in Russian), [C17], pp. 60–65.

D. Nowak

[1] Eine Klasse von Spielen mit verbotenen Situationen, Humboldt-Univ. Math.-Naturwiss. R. 26, no. 5 (1977), 587–594.

S.A. Orlovskiĭ

[1] Matrix games with forbidden situations (in Russian), Zh. Vychisl. Mat. i Mat. Fiz. 11, no. 3 (1971), 623–631.

[2] Games for n players with incomplete relations (in Russian), Zh. Vychisl. Mat. i Mat. Fiz. 12, no. 4 (1972), 1022–1029.

[3] On a class of games with incomplete relations (in Russian), Zh. Vychisl. Mat. i Mat. Fiz. 12, no. 6 (1972), 1420–1429.

[4] Infinite games for two players with forbidden situations (in Russian), Zh. Vychisl. Mat. i Mat. Fiz. 13 (1973), no. 3, 775–781.

[5] Equilibrium situations in noncooperative games with restrictions (in Russian), Zh. Vychisl. Mat. i Mat. Fiz. 15, no. 6 (1975), 1597–1601.

G. Owen

[1] An elementary proof of the minimax theorem, Manag. Sci. 13, no. 9 (1967), 765.

[2] Game Theory, Saunders, Philadelphia, etc., 1971.

[3] A discussion of minimax, Manag. Sci. 20, no. 9 (1974), 1316–1317.

[4] Game Theory, 2nd ed., Academic Press, New York and London, 1982.

T. Parthasarathy

[1] A note on a minimax theorem of T.T. Tie, Sankhya 27, no. 2/4, 407–408.

[2] On general minimax theorem, Math. Stud. 34, no. 3/4 (1966), 195–197.

[3] On games over the unit square, SIAM J. Appl. Math. 19, no. 2 (1970), 473–476.

[4] Selection theorems and their applications, Lecture Notes on Math., no. 263, Springer, Berlin, 1972.

T. Parthasarathy and T.E.S. Raghavan

[1] Some Topics in Two-Person Games, Amer. Elsevier, 1971.

[2] Equilibria of continuous two-person games, Pacific J. Math. 57, no. 1 (1975), 265–270.

J.E.L. Peck and A.L. Dulmage

[1] Games on a convex set, Canad. J. Math. 9, no. 3 (1957), 450–458.

V.V. Podinovskiĭ

[1] Lexicographical games (in Russian), [C18], pp. 100–103.

[2] Principle of guaranteeing a result for partial preference relations (in Russian), Zh. Vychisl. Mat. i Mat. Fiz. 19, no. 6 (1979), 1436–1450.

[3] General two-person zero-sum games (in Russian), Zh. Vychisl. Mat. i Mat. Fiz. 21, no. 5 (1981), 1140–1153.

V.V. Podinovskiĭ and V.M. Gavrilov

[1] Optimization in terms of successively applied criteria (in Russian), Sov. Radio, Moscow, 1975.

V.V. Podinovskiĭ and V.D. Nogin

[1] Pareto-optimal solutions of multicriteria problems (in Russian), Nauka, Moscow, 1982.

L.S. Pontryagin

[1] On the theory of differential games (in Russian), Uspekhi Mat. Nauk 21, no. 4 (1966), 219–274.

V.I. Ptak

[1] A combinatorial lemma on the existence of convex means and its applications to weak compactness, Proc. Symposia in Pure Math. 7 (1963), pp. 437–500.

T.E.S. Raghavan

[1] Some remarks on matrix games and nonnegative matrices, SIAM J. Appl. Math. 36, no. 1 (1979), pp. 83–85.

A. Rapoport

[1] Two-Person Game Theory. The Essential Ideas, University of Michigan Press, Ann Arbor, MI, 1966.

[2] N-Person Game Theory. Concepts and Applications, University of Michigan Press, Ann Arbor, MI, 1970.

A. Rapoport and A.M. Chammah

[1] Prisoner's Dilemma. A Study in Conflict and Cooperation, University of Michigan Press, Ann Arbor, MI, 1965.

A. Rapoport, M.J. Gruyer, and D.C. Gordon

[1] The 2×2-games, University of Michigan Press, Ann Arbor, MI, 1976.

J. Renegar

[1] A faster PSPACE algorithm for the existential theory of the reals, Proc. 29th Ann. IEEE Symp. Found. of Comput. Sci. (1988), 291–295.

R. Restrepo

[1] Tactical problems, involving several actions, [C6], 313–335.

[2] Combinatorial method for a class of matrix games, J. Appl. Probabil. 3, no. 2 (1966), 495–511.

J. Robinson

[1] An iterative method of solving a game, Ann. of Math. 54, no. 2 (1951), 296–301.

R.T. Rockafellar

[1] Saddle-points and convex analysis, In Differential Games and Related Topics, Varenna, 1970, North-Holland, Amsterdam, etc., 1971, pp. 109–119.

J.B. Rosen

[1] Existence and uniqueness of equilibrium points for concave n-person games, Econometrica 33, no. 3 (1965), 520–534.

J. Rosenmüller

[1] Kooperative Spiele und Markte, Springer, Berlin, 1971.

[2] On a generalization of the Lemke-Howson algorithm to non-cooperative N-person games, SIAM J. Appl. Math. 21,no. 1 (1971), 73–79.

[3] The Theory of Games and Markets, North-Holland, Amsterdam, 1981.

V.V. Rozen

[1] Application of the theory of binary relations to game theory (in Russian), in Mathematical Modelling of Economic Problems, Novosibirsk, 1971.

[2] Equilibrium situations in games with ordered outcomes (in Russian), [C21], pp. 114–118.

[3] Mixed extensions of games with ordered outcomes (in Russian), Zh. Vychisl. Mat. i Mat. Fiz. 16, no. 6 (1976), 1436–1450.

[4] Games with ordered outcomes (in Russian), Izv. Akad. Nauk SSSR. Tekhn. Kibernetika, 1977, no. 5, 31–37.

A.M. Rubinov

[1] A vector minimax theorem (in Russian), [C22], pp. 39–43.

W. Rupp

[1] ε-Gleichgewichtspunkte in n-Personenspiele, in R. Henn and O. Moeschlin, eds., Mathematical Economics and Game Theory, Essays in Honor of Oskar Morgenstern, Springer, Berlin, 1977, pp. 128–138.

A.L. Sadovskiĭ

[1] A monotone iterative algorithm for solving matrix games (in Russian), Dokl. Akad. Nauk SSSR 238, no. 3 (1978), 538–540.

M. Sakaguchi

[1] Values of strategic information, Rep. Stat. Appl. Res., JUSE 6, no. 1 (1959), 5–12.

H.N. Shapiro

[1] Note on a computation method in the theory of games, Comm. Pure and Appl. Math. 11, no. 4 (1958), 588–593.

L.S. Shapley

[1] Stochastic games, Proc. Nat. Acad. Sci. USA 39, no. 10 (1953), 1095–1100.

[2] Some topics in two-person games, [C13], 1–28.

[3] A note on the Lemke-Howson algorithm, Mathematical Programming Study, 1, Amsterdam, 1974, pp. 175–189.

L.S. Shapley and R.N. Snow

[1] Basic solutions of discrete two-person games, [C1], 27–36.

M. Shiffman

[1] Games of timing, [C3], pp. 97–123.

O.V. Shimel'fenig

[1] Application of the algebra of polyrelatives to game theory, (in Russian) Sibirsk. Mat. Zh. 12, no. 4 (1971), 855–879.

M. Sion

[1] On general minimax theorems, Pacific J. Math. 8, no. 1 (1958), 171–176.

M. Sion and P. Wolfe

[1] On a game without a value, [C6], pp. 299–306.

H. Skarf

[1] The approximation of fixed points of a continuous mapping, SIAM J. Appl. Math. 15, no. 5 (1967), 1328–1343.

A.I. Sobolev

[1] On a game of Borel's (in Russian), Teor. Veroyatnost. i Primenen 15, no. 3 (1970).

[2] Cooperative games (in Russian), Problemy Kibernetiki, no. 39 (1982), 201–222.

H. Steinhaus

[1] A definition for a theory of games and pursuit, Nav. Res. Logist. Quart. 7, no. 2 (1960), 105–107. (Translated from Polish, 1925.)

E. Stiemke

[1] Über positive Lösungen homogener linearen Gleichungen, Math. Ann. 76 (1915), 340–342.

F.D. Stotskiĭ

[1] On descriptive game theory (in Russian), Problemy Kibernetiki, no. 39 (1982), 201–222.

D.P. Sudzhyute

[1] The form of the spectra of equilibrium strategies of some nonantagonistic games for two players on the unit square (in Russian), Lit. Matem. Sb. 9, no. 3 (1969), pp. 687–694.

[2] Existence and form of equilibrium strategies of some nonantagonistic games of timing (in Russian), Lit. Matem. Sb. 10, no. 2 (1970), 375–389.

[3] Nonantagonistic games on the unit square with different curves of discontinuity of the payoff functions (in Russian), Lit. Matem. Sb. 12, no. 3 (1972), 165–179.

[4] Convergence of ε-equilibrium strategies to equilibrium ones in two-person zero-sum games of timing (in Russian), Lit. Matem. Sb. 14, no. 3 (1974), 195–222.

[5] On equilibria for three-person zero-sum games of timing (in Russian), Lit. Matem. Sb. 14, no. 3 (1974).

[6] On the problem of the convergence of ε-equilibria to equilibria in two-person zero-sum games of timing (in Russian), Lit. Matem. Sb. 16, no. 3, 203–215.

Mizuo Suzuki

[1] Theory of Games, Kieso Shobo, Tokyo, 1959.

E. Szép and F. Forgó

[1] Bevezedös a jätekelmälet be, Közagzd Kiadó, Budapest, 1972.

[2] Einführung in die Spieltheorie, Akad. Kiadó, Budapest 1983.

F. Szidarovszky

[1] On the oligopoly game, Budapest, 1970.

[2] On a generalization of the bimatrix game, Intern. J. Game Theory 1, no. 3 (1971–72), 205.

[3] The concave oligopoly game, K. Marx University, Budapest, 1977, 59 pp.

A. Tarski

[1] A Decision Method for Elementary Algebra and Geometry, University of California Press, Berkeley, 1951.

D.M. Thompson and G.L. Thompson

[1] A bibliography of game theory, [C7], pp. 407–453.

Teh-Tjoe Tie

[1] Minimax theorems on conditionally compact sets, Ann. Math. Stat. 34, no. 4 (1963), 1536–1540.

S.H. Tijs

[1] A characterization of the value of zero-sum two-person games, Report 1632, Department of Mathematics, Univ. of Nijmegen, The Netherlands, 1976.

A.W. Tucker

[1] Solving a matrix game by dynamical programming, IBM J. Res. Development 4, no. 5 (1960), 507–517.

E.E.C. van Damme

[1] Refinement of the Nash equilibrium concept, Lecture Notes in Economics and Mathematical Systems, no. 219, Springer-Verlag, Berlin, 1983.

[2] Stability and Perfection of Nash Equilibria, Springer-Verlag, Berlin, 1987.

E.Ĭ. Vilkas

[1] Axiomatic definition of the value of a matrix game (in Russian), Teor. Veroyatnost. i Primenen. 8, no. 3 (1963), 324–327.

[2] Solution of a functional equation with the operator of the value of a matrix game (in Russian), Lit. Matem. Sb. 3, no. 1 (1963), 61–70.

[3] Some functional properties of the value of a matrix game (in Russian), Lit. Matem. Sb. 3, no. 1 (1963), 71–76.

[4] Transformation of a matrix game and the value of the game (in Russian), Lit. Matem. Sb. 4, no. 1 (1964), 25–29.

[5] Domain of solutions of a parametric matrix game (in Russian), Lit. Matem. Sb. 4, no. 1 (1964), 31–35.

[6] Some remarks on equilibrium situations in a noncooperative n-person game (in Russian), Lit. Matem. Sb. 7, no. 4 (1967), 583–587.

[7] Equilibrium situations in noncooperative several-person games (in Russian), Lit. Mat. Sb. 7, no. 4 (1967), 589–593.

[8] Axiomatic determination of equilibrium situations and that of the value of a noncooperative n-person game (in Russian), Teor. Veroyatnost. i Primenen. 13, no. 3 (1968), 586–591.

[9] Optimality in noncooperative games. A survey of approaches (in Russian), Lit. Matem. Sb. 10, no. 3 (1970), 463–470.

[10] Utility theory and decision making (in Russian), Mat. Metody v Sotsial'nykh Naukakh, no. 1 (1971), pp. 13–60.

[11] A formalization of the problem of choosing a game-theoretic optimality criterion (in Russian), Mat. Metody v Sotsial'nykh Naukakh, no. 2 (1972), pp. 9–31.

[12] Noncooperative n-person games (in Russian), [C17], pp. 90–94.

[13] The problem of game-theoretical optimality principles, New York University, Dept. of Economics, Working paper no. 6, 1973.

[14] Optimality concepts in game theory, in Current trends in game theory (in Russian), [C21], Mokslas, Vilnius, 1976, pp. 25–43.

[15] Multicriterial optimization (in Russian), in Mat. Metody v Sotsial'nykh Naukakh, no. 7, Vilnius, 1976, pp. 181–194.

[16] Optimality and dynamics in coalitional games, [C22], pp. 181–194.

[17] Two game-theoretical theorems (in Russian), Mat. Metody v Sotsial'nykh Naukakh, no. 10, Vilnius, 1978, pp. 9–17.

[18] Optimality in games and decisions (in Russian), Nauka, Moscow, 1990.

E.Ĭ. Vilkas and D.P. Sudzhyute

[1] Classical approximation methods in game theory (in Russian), Lit. Matem. Sb. 6, no. 2 (1966), 217–225.

J. Ville

[1] Sur la théorie générale des jeux où intervient l'habilité des joueurs, in Traité du calcul des probabilités et ses applications. Applications des jeux de hasard, E. Borel et al., Gauthier-Villars, Paris, 1938, vol. IV, fasc. 2, pp. 105–113.

W. Vogel

[1] Die Annäherung guter Strategien bei einer gewissen Klasse von Spielen, Math. Z. 65, no. 3 (1956), 283–308.

J. von Neumann

[1] Zur Theorie der Gesellschaftspiele, Math. Ann. 100 (1928), 295–320. [On the theory of games of strategy, [C7], pp. 13–42.]

[2] Sur la théorie des jeux, C.R. Acad. Sci. Paris 186, no. 25 (1928), 1689–1691.

[3] Almost periodic functions in a group, I, Trans. Amer. Math. Soc. 36 (1934), 445–492.

[4] A certain zero-sum two-person game equivalent to the optimal assignment problem, [C3], pp. 5–12.

[5] Communication on the Borel notes, Econometrica 21, no. 1 (1953), 124–125.

[6] A numerical method to determine optimal strategy, Naval Res. Logist. Quart. 1 (1954), 109–115. Collected Works, v 6, Oxford e.a., 1963, pp. 82–88.

[7] A numerical method for determination of the value and the strategies of a zero-sum game with large numbers of strategies. Reviewed by H.W. Kuhn and A.W. Tucker, in Collected Works, vol. 6, pp. 96–97, Oxford, 1963.

[8] The mathematician, in Collected Works, Pergamon, New York, etc., vol. 1, 1961, pp. 1–9.

J. von Neumann and O. Morgenstern

[1] Theory of Games and Economic Behavior, Princeton University Press, 1944. 2nd ed., 1947. 3rd ed. 1953.

H. von Stackelberg

[1] Marktform und Gleichgewicht, Springer, Vienna, 1934.

N.N. Vorob'ev

[1] Equilibrium situations in bimatrix games (in Russian), Teor. Veroyatnost. i Primenen. 3, no. 3 (1958), 318–331.

[2] Finite noncooperative games (in Russian), Uspekhi Matem. Nauk 14, no. 4 (1959), 21–56.

[3] On the question of the philosophical problems of game theory (in Russian), in Cybernetics. Mentality. Life, Moscow, 1964, pp. 157–163.

[4] Games with incompletely known rules (in Russian), in International Congress of Mathematicians, 1966. Summaries of short communications. Section 13. Moscow, 1966, p. 15.

[5] Some methodological problems in game theory (in Russian), Voprosy Filosofii, 1966, no. 1, 93–103.

[6] Coalitional games (in Russian), Teor. Veroyatnost. i Primenen. 12, no. 2 (1967), 289–306.

[7] Applications of game theory in technical sciences (in Russian), in IV Internationaler Kongress über Anwendungen der Mathematik in den Ingenieurwissenschaften, Weimar, 1967, vol. 1, pp. 411–422, Berlin, 1968.

[8] The "attack and defence" game (in Russian), Lit. Matem. Sb. 8, no. 3 (1968), 436–444.

[9] Artistic modelling, conflicts, and game theory (in Russian), in Sodruzhestvo nauk i taĭny tvorchestva, Iskusstvo, Moscow, 1968, pp. 348–372.

[10] The development of game theory, in von Neumann and Morgenstern [1], Russian translation Nauka, Moscow, 1972, pp. 631–702.

[11] The present state of game theory (in Russian), Uspekhi Matem. Nauk 25, no. 2 (1970), 81–140.

[12] The development of science and game theory (in Russian), in Issledovanie operatsiĭ. Metodologicheskie aspekty, Nauka, Moscow, 1972, 9–28. Discussion: pp. 92–98, 132–135.

[13] The role of game theory in the mathematization of knowledge (in Russian), in Matematizatsiya nauchnogo znaniya, 1972, pp. 132–152.

[14] Game theory, in Scientific thought, Mouton-Unesco, Paris and The Hague, 1972, pp. 132–152.

[15] On a class of games on the unit square with discontinuous payoff function (in Russian), in [C17], pp. 95–109.

[16] Diagonal noncooperative games, in Game-theoretic problems in decision theory (in Russian), in [C19], pp. 16–23.

[17] Characteristic functions of two-person zero-sum games (in Russian), in [C18], pp. 23–31.

[18] Convergence of dynamic games (in Russian), in [C18], pp. 162–168.

[19] Applications of game theory. An essay on methodology (in Russian), as for [17], pp. 249–283.

[20] Game Theory. Lectures for Economists and Systems Scientists, Springer, New York, 1977 (Russian original, Leningrad State University Press, 1974).

[21] Entwicklung der Spieltheorie, Verlag d. Wissenschaften, Berlin, 1975.

[22] On a game-theoretic model of the optimal distribution of limited resources (in Russian), in The application of mathematics in economics, Leningrad State University Press, 1975, pp. 131–141.

[23] Mathematization of forecasting (methodological problems), in Ekonomiko-matematicheskie issledovanie zatrat i resultatov, Nauka, Moscow, 1976, pp. 43–58.

[24] Metastrategies in noncooperative games (in Russian), Math. Operationsforsch. Statist., Ser. Optimization, 9, no. 1 (1978), 43–55.

[25] Noncooperative games (in Russian), in Problems of cybernetics, no. 33, Moscow, 1978, pp. 69–90.

[26] On the question of algebraic solvability of dyadic games (in Russian), in Kombinatorno-algebraicheskie metody v prikladnoĭ matematike, Gorky, 1979, pp. 21–33.

[27] [Editor] Game theory. Annotated list of publications up to 1968 (in Russian), Nauka, Leningrad, 1976.

[28] [Editor] Game theory. Annotated list of domestic and foreign literature from 1969 to 1974 (in Russian), Nauka, Leningrad, 1980.

N.N. Vorob'ev and G.B. Epifanov

[1] Realizable payoffs in bimatrix games (in Russian), in [C17], pp. 110–113.

N.N. Vorob'ev and I.V. Romanovskiĭ

[1] Games with forbidden situations (in Russian), Vest. Leningrad. Gos. Univ. 7, no. 2 (1959), 50–54.

N.N. Vorob'ev, Jr.

[1] Two-person zero-sum games with convex payoff functions (in Russian), in Mathematical methods of optimization and structuring of systems, Kalinin, 1980, pp. 60–65.

[2] On a theorem of Bohnenblust, Karlin and Shapley (in Russian), Vest. Leningrad. Gos. Univ. 1980, no. 13, 122–124.

I.N. Vrublevskaya

[1] On a game of a single attacker against several defenders (in Russian), Lit. Matem. Sb. 8, no. 3 (1968), 445–469.

[2] Properties of the solution of a game of a single attacker against several defenders (in Russian), Lit. Matem. Sb. 10, no. 2 (1970), 235–251.

A. Wald

[1] Generalization of a theorem of J. v. Neumann concerning zero sum two-person games, Ann. of Math. 46, no. 2 (1945), 281–286.

[2] Statistical Decision Functions, Wiley, New York, 1950.

H. Weyl

[1] Elementare Theorie der konvexen Polyeder, Comment. Math. Helv. 7 (1935), 290–306. [The elementary theory of convex polyhedra [C1].]

[2] Elementary proof of a minimax theorem due to von Neumann, [C1], pp. 19–25.

R.B. Wilson

[1] Computing equilibria of n-person games, SIAM J. Appl. Math. 21, no. 1 (1971), 80–87.

P. Wolfe

[1] Determinateness of polyhedral games, [C5], pp. 195–198.

[2] The strict indeterminateness of certain infinite games, Pacific J. Math. 5 (1955), 891–897.

Wu Wen-Tsün

[1] A remark on the fundamental theorem in the theory of games, Science Record 111, no. 6 (1959), 229–233.

[2] On non-cooperative games with restricted domains of activities, Scientia Sinica 10, no. 4 (1961), 387–409.

Wu Wen-Tsün and Jiang Jia-he

[1] Essential equilibrium points of n-person noncooperative games, Scientia Sinica 11, no. 10 (1962), 1307–1322.

H. Wüthrich

[1] Ein Entscheidungsverfahren für die Theorie der reell-abgeschlossenen Körper, Lect. Notes Comp. Sci. 43 (1967), pp. 138–162.

E.B. Yanovskaya

[1] Quasiinvariant cores in two-person zero-sum games (in Russian), Dokl. Akad. Nauk SSSR 151, no. 3 (1963), 513–514.

[2] On a class of games on the unit square with unbounded payoff functions (in Russian), [C12], pp. 77–84.

[3] On the question of the uniqueness of optimal strategies in games of timing (in Russian), [C12], pp. 395–397.

[4] Minimax theorems for games on the unit square (in Russian), Teor. Veroyatnost. i Primenen. 9, no. 3 (1964), 554–555.

[5] On two-person zero-sum games played on function spaces (in Russian), Lit. Matem. Sb. 7, no. 3 (1967), 547–557.

[6] Equilibrium situations in polymatrix games (in Russian), Lit. Mat. Sb. 8, no. 2 (1968), 381–384.

[7] On games of duel type with continuous firing (in Russian), Izv. Akad. Nauk SSSR Tekhn. Kibern. 1969, no. 1, 16–19.

[8] Solution of infinite two-person games in finitely-additive strategies (in Russian), Teor. Veroyatnost. i Primenen. 15, no. 1 (1970), 162.

[9] Cores in noncooperative games, Int. J. Game Theory 1, no. 4 (1971/72) 209–215.

[10] Infinite two-person zero-sum games (in Russian), in Teor. Veroyatnost. Mat. Statist. Mat. Kibernetika, vol. 10, Moscow, 1972, pp. 75–106.

[11] On the existence of the value of a game in games on the unit square with discontinuous payoff functions (in Russian), Math. Operationsforsch. Statist. 3, no. 2 (1972), 91–96.

[12] On mixed extension of a general noncooperative game (in Russian), [C19].

[13] On the existence of a value of two-person zero-sum games with semicontinuous payoff functions (in Russian), Izv. Akad. Nauk SSSR Tekhn. Kibern. 1973, no. 6, 56–60.

[14] On the existence of equilibrium situations in non-cooperative two-person games (in Russian), [C17], pp. 354–364.

[15] Equilibrium situations in finite nonatomic games (in Russian), [C18], pp. 115–119.

[16] Equilibrium situations in games with non-Archimedean utilities (in Russian), Inst. Matem. i Kibern. Akad. Nauk Lit. SSR, Vilnius (Mathematical methods in the social sciences, 1974, no. 4, 98–118).

[17] Mixed extensions of binary relations (in Russian), Inst. Matem. i Kibern. Akad. Nauk Lit. SSR, Vilnius (Mathematical methods in the social sciences, 1975, no. 6, 152–166).

[18] Finitely additive solutions of noncooperative games (in Russian), [C21], p. 136.

[19] Equilibrium situations in general noncooperative games and in their mixed extensions (in Russian), [C22], pp. 43–65.

[20] Balanced games and $\mathfrak{K}$-equilibrium situations (in Russian), Math. Operationsforsch. Statist., Ser. Optimiz. 9, no. 1 (1978), 57–68.

[21] Two-person zero-sum games (in Russian), in Problems of Cybernetics, no. 34 (1978), 221–246.

[22] Equilibrium in meta-extensions of noncooperative games (in Russian), Math. Operationsforsch. Statist., Ser. Optimiz. 10, no. 4 (1979), 483–499.

N.J. Young

[1] On Ptak's double-limit theorems, Proc. Edinburgh Mat. Soc. 17 (1971), 193–200.

[2] Admixtures of two-person games, Proc. London Math. Soc. 25, no. 4 (1972), 736–756.

E. Zermelo

[1] Über eine Anwendung der Mengenlehre auf die Theorie des Schachspiels, Proc. Fifth Intern. Congress of Mathematicians (Cambridge,1912), Cambridge University Press, 1913, pp. 501–504.

F. Zeuthen

[1] Problems of monopoly and economic warfare, Routledge, London, 1930.

Collections

[C1] Contributions to the Theory of Games, I, Edited by H.W. Kuhn and A.W. Tucker, Annals of Math. Studies, vol. 24, Princeton University Press, 1950.

[C2] Activity Analysis of Production and Allocation, Cowles Commission Monograph no. 13, Edited by T. Koopmans et al., Wiley, New York, 1951.

[C3] Contributions to The Theory of Games, II, Edited by H.W. Kuhn and A.W. Tucker, Annals of Math. Studies, vol. 28, Princeton University Press, 1953.

[C4] Decision Processes, Edited by R.M. Thrall, C.H. Combs and R.L. Davis, Wiley, New York, 1954.

[C5] Linear Inequalities and Related Systems, Edited by H.W. Kuhn and A.W. Tucker, Annals of Math. Studies, vol. 38, Princeton University Press, 1956.

[C6] Contributions to the Theory of Games, III, Edited by M. Dresher, A.W. Tucker, and R.D. Luce, Annals of Math. Studies, vol. 39, Princeton University Press, 1957.

[C7] Contributions to the Theory of Games, IV, Edited by A.W. Tucker and R.D. Luce, Annals of Math. Studies, vol. 40, Princeton University Press, 1959.

[C8] Russian translation of [C5], with translation of S. Waida, Game Theory and Linear Programming, edited by L.V. Kantorovich and V.V. Novozhilov, Inostran. Lit., Moscow, 1959.

[C9] Recent Advances in Game Theory, Economic Research Program, Princeton, N.J., 1962.

[C10] Matrix Games (in Russian), edited by N.N. Vorob'ev, Fizmatgiz, Moscow, 1961.

[C11] Primenenie teorii igr v voennom dele, edited by V.O. Ashkenazy, Sov. Radio, Moscow, 1961.

[C12] Beskonechnye antagonisticheskie igry, edited by N.N. Vorob'ev, Fizmatgiz, Moscow, 1963.

[C13] Advances in Game Theory, edited by M. Dresher, L.S. Shapley and A.W. Tucker, Annals of Math. Studies, vol. 52, Princeton University Press, 1964.

[C14] Game Theory and Related Approaches to Social Behavior. Selections, edited by M. Shubik, Wiley, New York, 1964.

[C15] Spieltheorie, Arbeitstagung am 27. und 28. Juni 1967, Institut für höhere Studien und wissenschaftliche Forschung, Vienna.

[C16] Pozitsionnye igry, edited by N.N. Vorob'ev and N.N. Vrublevskaya, Nauka, Moscow, 1967.

[C17] Teoriya igr. Doklady na I Vsesoyuznoĭ konferentsii po teorii igr, edited by N.N. Vorob'ev, Izdatelstvo Akad. Nauk Arm. SSR Erevan, 1973.

[C18] Uspekhi teorii igr, Trudy II Vsesoyuznoĭ konferentsii po teorii igr, Erevan, 1968, edited by È. Vilkas, Mintis, Vilnius, 1973.

[C19] Teoretiko-igrovye voprosy prinyatiya resheniĭ, edited by N.N. Vorob'ev and others, Moscow, TsÈMI, 1973.

[C20] III Vsesoyuznaya konferenysiya po teorii igr, Odessa, 1974. Thesis reports.

[C21] Sovremennye napravleniya teorii igr, edited by E.Ĭ. Vilkas, and A.A. Korbut, Inst. Sots. Èk. Problem Akad. Nauk SSSR, Inst. Matem. Kibern. Akad. Nauk Lit. SSR, Mokslas, Vilnius, 1976.

[C22] Mathematical Economics and Game Theory, Essays in Honor of Oskar Morgenstern, edited by R. Henn and O. Moeschlin, Springer, Berlin, 1977.

[C23] Teoretiko-igrovye voprosy prinyatiya resheniĭ, edited by N.N. Vorob'ev, Inst. Sots.-Èk. Problem Akad. Nauk SSSR, Nauka, Leningrad, 1978.

List of Joint Authors

Index of Notations

Chapter 1

Γ (Γ', $\tilde{\Gamma}$, ...) two-person zero-sum games

$i(j, \ldots)$: players

I: set of players in a game

n: number of players in a game

x_i $(y_i, \ldots)$: strategy of player i

x_i: the set of strategies of player i in a game

$x(y, \ldots)$: situations in a game

x: the set of situations in a game

H_i: the payoff function of player i

$\Box$: end of a definition or proof

$\Gamma(A, B)$, $\Gamma_{A,B}$: bimatrix games with payoff matrices A and B

$M_{i\cdot}$: the ith row of the matrix M

$M_{\cdot j}$: the jth column of the matrix M

m, or n: the number of strategies of player m or n in a finite two-person game

x, or y: the set of all strategies of player 1 (or of player 2) in a two-person game

H: the payoff function of player 1 in a two-person zero-sum game (the payoff function of the game)

$\Gamma(A)$, Γ_A: a matrix game with payoff matrix A

φ_K, x_K, H_K, ...: K-vector, i.e., a vector the components

φ_i, x_i, H_i,... of which correspond to player i in coalition K

x_K $(y_K, \ldots)$: coalitional strategies of coalition K

x_K: the set of all coalitional strategies of coalition K

$\mathbf{R}^K$: Euclidean space whose basis vectors correspond to the players of coalition K

$\mathbf{0}^K$: the zero element of $\mathbf{R}^K$

$H_K(x)$: K-vector payoff in situation x

$W_{K,\Gamma}(y)$: the set of realizable K-vector payoffs on the set y of situations in the game Γ

$W(\Gamma)$: same as $W_{I,\Gamma}(x)$

$\mathscr{X}_i(\mathscr{Y}, \mathscr{S}_i, \ldots)$: topology (system of neighborhoods) on the set x of strategies

$\mathscr{X}(\mathscr{Y}, \mathscr{S}, \ldots)$: topology on the set x of situations

ρ_i: metric on the set x of strategies

ρ, ρ_i: metric on the set x of situations

Ξ_i: σ-algebra of measurable subsets of the set x_i of strategies
Ξ: σ-algebra of measurable subsets of the set x of situations
$\Gamma \sim \Gamma'$: strategic equivalence of games Γ and Γ'
ε_I: real I-vector
$\Gamma \approx \Gamma'$: exact homomorphism of games Γ and Γ'
$\Gamma \overset{\approx}{_A} \Gamma'$: asymptotic homomorphism of games Γ and Γ'
$\mathrm{Hom}_{\varepsilon_I}(\Gamma, \Gamma')$: class of all ε_I-homomorphisms of Γ on Γ'
$\mathcal{K}$: category of games
$\mathrm{Hom}_{\mathcal{K}}$: class of morphisms of games in the category $\mathcal{K}$
$\overset{K}{\approx}$: exact homomorphism of game in $\mathcal{K}$
$\overset{K}{\underset{A}{\approx}}$: asymptotic homomorphism of games in $\mathcal{K}$
$\rho^A_{\mathcal{K}}$: pseudometric on games that are objects in $\mathcal{K}$
Γ/x_L: factorization of Γ over coalitional strategies x_L of coalition L
φ: optimality principle
$x \underset{K}{\|} x'_K, x \| x'_K$: situations obtained from x by replacing the coalitional strategy x_K by x'_K
For K-vectors φ_K and ψ_K:
$\varphi_K < \psi_K$ means $\varphi_i < \psi_i$ for all $i \in K$
$\varphi_K \leq \psi_K$ means $\varphi_i \leqq \psi_i$ for all $i \in K$, but $\varphi_K \neq \psi_K$
$\varphi_K \leqq \psi_K$ means $\varphi_i \leqq \psi_i$ for all $i \in K$
If, in addition, $\varepsilon_K \geq 0^K$, then
$\varphi_K \overset{\varepsilon_K}{<} \psi_K$ means that $\varphi_K < \psi_K + \varepsilon_K$
$\varphi_K \overset{\varepsilon_K}{\leq} \psi_K$ means that $\varphi_K \leq \psi_K + \varepsilon_K$
$<^\circ$ denotes any of the symbols $<$, $\leq$, $\overset{\varepsilon_K}{<}$, $\overset{\varepsilon_K}{\leq}$
$\mathcal{C}_K(\Gamma)$: the set of $\mathfrak{K}$-stable situations in Γ
$\mathcal{C}_{\mathfrak{K}}(\Gamma)$: the set of K-stable situations in Γ
$\mathcal{R}_K(\Gamma)(\mathcal{R}_K(\Gamma))$: the set of K-equilibrium ($\mathfrak{K}$-equilibrium) situations in the game Γ
int: interior of a set
$\mathcal{C}_I(\Gamma)$: set of all Pareto-optimal situations in Γ
$\mathcal{C}_i(\Gamma)$: set of all i-optimal situations (admissible for player i) in Γ
$\mathcal{C}(\Gamma)$: set of all Nash-equilibrium situations (equilibrium situations) in Γ
$\mathcal{C}^{\varepsilon_K}_K(\Gamma)(\mathcal{R}^{\varepsilon_K}_K(\Gamma))$: set of ε_K-K-stable ε_K-K-equilibrium situations in Γ
$\mathcal{C}^{\varepsilon_{\mathfrak{K}}}_{\mathfrak{K}}(\Gamma)(\mathcal{R}^{\varepsilon_{\mathfrak{K}}}_{\mathfrak{K}}(\Gamma))$: set of all $\varepsilon_{\mathfrak{K}}$-$\mathfrak{K}$-stable ($\varepsilon_{\mathfrak{K}}$-$\mathfrak{K}$-equilibrium) situations in Γ

$\mathfrak{D}^0_{\mathfrak{K}}(\Gamma)$: any of the sets $\mathscr{C}_{\mathfrak{K}}(\Gamma)$, $\mathscr{R}_{\mathfrak{K}}(\Gamma)$, $\mathscr{C}^{\varepsilon\mathfrak{K}}_{\mathfrak{K}}(\Gamma)$, $(\mathscr{R}^{\varepsilon\mathfrak{K}}_{\mathfrak{K}}(\Gamma))$

ρ_H: Hausdorff metric generated by a metric ρ

$\operatorname{Gr}\varphi$: graph of the mapping φ

$C|a$: section of the set $C \subset A \times B$ by the element $a \in A$ (i.e., the set $B_a = \{b : b \in B, (a,b) \in C\}$)

$\operatorname{Pr}_A C$: projection of the set $C \subset A \times B$ on the factor A (i.e., the set $A_C = \{a : a \in A, C|a \neq \emptyset\}$)

$\operatorname{Pr}_\Gamma(f)$: projection of the metasituation f on the game Γ

M_Γ: metastrategic extension of Γ

X_i: mixed strategies of player i in a noncooperative game Γ

$\boldsymbol{X}_i$: set of mixed strategies of player i in a noncooperative game

$X(Y, \ldots)$: situations in the mixed strategies in a noncooperative game

$\boldsymbol{X}$: set of situations in the mixed strategies in a noncooperative game

$\mathscr{S}_i(\mathscr{S}_K)$: semiintrinsic topology on x_i (on x_K)

ρ_i: natural metric on x_i

$\mathscr{B}_i$: Borel σ-algebra of subsets on x_i

Chapter 2

$\operatorname{supp} X_i$: spectrum of a mixed strategy X_i

$x_i^{(j)}$: jth pure strategy of player i in a finite game

$X_i(x_i^{(j)})$: probability of the pure strategy $x_i^{(j)}$ under the conditions of the mixed strategy X_i

$X(x)$: probability of situation x under the conditions of a situation X in the mixed strategies

X (or Y): mixed strategy of player 1 (or player 2 resp.) in a two-person zero-sum game

ξ_i (or η_j): probability of the ith (or jth) pure strategy of player 1 (player 2) in a bimatrix game under the conditions of a mixed strategy X (or Y)

$\operatorname{use}_X k$: usage of strategy k in the situation X i.e., the number of players i for which $X_i(x_i^{(k)}) > 0$

$\operatorname{sym}(\Gamma)$: set of symmetric situations in the mixed strategies of the game

$0_i, 1_i$: pure strategies of player i in a dyadic game

ξ_i: probability that player i in situation X in a dyadic game chooses strategy 1_i in the situation X

$\xi^{(K)}((1-\xi)^{(K)})$ is $\prod_{i \in K} \xi_i$ (or $\prod_{i \in K}(1-\xi_i)$) for $K \subset I$ in a dyadic game

$\boldsymbol{X}_0^i(\boldsymbol{X}_1^i, \boldsymbol{X}_=^i)$: set of combinations of strategies of players in $I \setminus i$ in a dyadic game in which the strategy 0_i is preferred by player i (or strategy 1_i is preferred, or, finally, strategies 0_i and 1_i are indifferent)

J_p: the vector $(1, \ldots, 1) \in \mathbf{R}^p$

Chapter 3

Γ: two-person zero-sum game ("antagonistic" game)

$\mathfrak{x}$ (or $\mathfrak{y}$): set of strategies of player 1 (or player 2) in a two-person zero-sum game

H: payoff function of a two-person zero-sum game (of player 1)

x (or y): pure strategy of player 1 (or player 2)

H_x (or H_y): function describing the dependence of the payoff to player 1 from a strategy of player 2 (or of player 1) under a given strategy x of player 1 (strategy y of player 2)

$\mathscr{C}^{\varepsilon}(\Gamma)$: set of ε-equilibria (ε-saddle points) in Γ

$\mathscr{S}^{\varepsilon}(\Gamma)$ ($\mathscr{T}^{\varepsilon}(\Gamma)$): set of ε-equilibrium (i.e., 2ε-optimal) strategies of player 1 (of player 2) in Γ

$\bar{v}(\Gamma), \bar{v}_{\Gamma}$ ($\underline{v}(\Gamma), \underline{v}_{\Gamma}$): upper (lower) values of Γ

$\mathscr{S}(\Gamma)$ ($\mathscr{T}(\Gamma)$): set of optimal (i.e., equilibrium) strategies of player 1 (or player 2) in Γ

$\mathscr{C}(\Gamma)$: set of equilibrium situations (saddle points) in Γ

$v(\Gamma), v_{\Gamma}$: value of Γ

X (or Y): set of mixed strategies of player 1 (or player 2) in a two-person zero-sum game

R_{Γ}: (non-strict) preference relation for strategies of player 1 in the game Γ

I_{Γ} (or P_{Γ}): relation of indifference equal valuation (or of strict preference) corresponding to R_{Γ}

$R_{\mathfrak{y}}$: preference relation of strategies of player 1 in games with set $\mathfrak{y}$ of strategies of player 2

$I_{\mathfrak{y}}$ (or $P_{\mathfrak{y}}$): indifference relation (or strict preference relation) corresponding to $R_{\mathfrak{y}}$

$\mathscr{S}$ (or $\mathscr{J}$): semi-intrinsic (or intrinsic) topology on the set of a player's strategies

$\mathscr{J}(x, \varepsilon)$: ε-ball in the sense of the $\mathscr{J}$ topology with center x

ρ_1 (ρ_2) pseudometric (also metric) on the space of strategies of player 1 (of player 2), generated by the intrinsic topology

$\mathfrak{y}_{\alpha}(x)$: set of strategies of player 2, in which player 1, using strategy x, obtains a payoff not exceeding α

$v_{\mathfrak{x}'}$: $\inf_{y \in \mathfrak{y}} \sup_{x \in \mathfrak{x}'} H(x, y)$

$\mathfrak{y}_{\mathfrak{x}'}$: set of $y \in \mathfrak{y}$ on which the preceding infimum is attained

∂: operator of forming the boundary of a set

int: interior of a set

dim: dimension of a set

$\check{X}$ (or $\check{Y}$): finitely-additive strategies of player (or of player 2) supp X: spectrum of strategy X

$u(X)$ (or $v(Y)$): moment characteristic of a strategy of player 1 (or of player 2)

$\mathcal{U}_r$ (or $\mathcal{V}_r$): set of moment characteristics of strategies of player 1 (or of player 2) in a degenerate game

conv: convex hull of a set

extr: set of extreme points of a set

$\stackrel{\leq}{\circ}$ (or $\stackrel{\circ}{\leqq}$): strong (or weak) stochastic domination of distributions

Chapter 4

$\Gamma(A), \Gamma_A$: matrix game with matrix payoff A

x (or y): set $\{1, \ldots, m\}$ (or set $\{1, \ldots, n\}$) of strategies of player 1 (or of player 2) in Γ_A

i (or j): pure strategy of player 1 (or player 2) in a matrix game

X (or Y): mixed strategy of player 1 (or player 2) in a matrix game

$\boldsymbol{X}$ (or $\boldsymbol{Y}$): set of mixed strategies of player (or player 2) in a matrix game

$v(A), v_A$: value of Γ_A

$\mathcal{S}(A)$, (or $\mathcal{T}(A)$): set of optimal strategies of player (or player 2) in a mixed extension of Γ_A

$\mathcal{C}(A)$: set of saddle points in a mixed extension of Γ_A

$\Gamma(A^C)$: symmetrization of a game

$\Gamma(A/H)$: $\mathcal{S}(A) \times \mathcal{T}(A)$-subgame of Γ_H

$A^+(A, B)$ ($A^-(A, B)$): set of solutions of a system of inequalities $XA \leqq B$, $X \geqq 0$ (of a system $XA \geqq B$, $X \geqq 0$)

$P^+(A, B, C)$ (or $P^-(A, B, C)$): linear programming problem on maximization (or minimization) of a linear form XC^T with set of admissible solutions $A^+(A, B)$ (or $A^-(A, B)$)

$S^+(A,B,C)$ (or $S^-(A,B,C)$): set of optimal solutions of problem $P^+((A,B,C)$ (or of problem $P^-(A, B, C)$)

A^T: transpose of matrix A

conv: convex hull of a set

extr: set of extreme points of a set

adj M: adjoint of a square matrix M

det M: determinant of a square matrix M

$[U]$: linear span of set U

$\mathcal{K}^*$: cone conjugate to the cone $\mathcal{K}$

$k^\perp$: hyperplane composed of vectors orthogonal to k

$N(A)$: kernel of matrix A

Index